BETTER

the feel good place

Lincolnshire

COUNTY COUNCIL

Working for a better future

Lincolnshire Libraries
This book should be returned on or before the due date.

STAMFORD LIBRARY
Tel: 01522 782010

5/22

To renew or order library books please telephone 01522 782010
or visit https://prism.librarymanagementcloud.co.uk/lincolnshire/
You will require a Personal Identification Number
Ask any member of staff for this.
The above does not apply to Reader's Group Collection Stock.

EC. 199 (LIBS): RS/L5/19

D1470689

05438605

PETER FITZSIMONS

The Incredible Life of
HUBERT WILKINS
AUSTRALIA'S GREATEST EXPLORER

CONSTABLE

CONSTABLE

First published in Australia and New Zealand in 2021 by Hachette Australia,
an imprint of Hachette Australia Pty Limited

This edition published in Great Britain in 2022 by Constable

10 9 8 7 6 5 4 3 2 1

Quotes from *The Unseen Anzac* by Jeff Maynard (Scribe, 2015) and
The Last Explorer by Simon Nasht (Hachette, 2005) reproduced with permission

Photos credited (OSU) are from the Sir George Hubert Wilkins Papers, SPEC.PA.56.0006,
Byrd Polar and Climate Research Centre Archival Program, Ohio State University

A CIP catalogue record for this book
is available from the British Library

ISBN: 978-1-47213-146-1 (trade paperback)

Cover design by Luke Causby/Blue Cork
Cover images courtesy Getty Images; National Library of Australia; AAP;
US Naval History and Heritage Command
Author photo courtesy Peter Morris/Sydney Heads
Maps by Jane Macaulay

Typeset in 11.1/14.7 pt Sabon LT Pro by Bookhouse, Sydney
Printed and bound in Great Britain by Clays Ltd, Elcograf, S.p.A.

Papers used by Constable are from well-managed forests and other responsible sources

MIX
Paper from
responsible sources
FSC® C104740

Constable
An imprint of
Little, Brown Book Group
Carmelite House
50 Victoria Embankment
London EC4Y 0DZ

An Hachette UK Company
www.hachette.co.uk

www.littlebrown.co.uk

'[Wilkins is] the only man I know to whom every line of Kipling's poem "If" could be truthfully applied.'[1]

Charles Bean

To Sir Hubert Wilkins.
An incredible man, a great Australian.

CONTENTS

LIST OF MAPS

DRAMATIS PERSONAE

Caspar Middleton was one of Sir Hubert's first great friends, and a collaborator in his early films, both as actor and director.

Bernard Grant was a photographer for London's *Daily Mirror*, assigned to cover the Turkish side during the First Balkan War of 1912–13, and teamed up with George Wilkins during the campaign. (Until his knighthood, Wilkins was commonly called George.)

Vilhjálmur Stefansson, the Icelandic–American Arctic explorer was the leading exponent of the 'Friendly Arctic', the idea that if you embraced the ways of the local Indigenous people, you could live as well as them. George Wilkins joined Stefansson's scientific Canadian Arctic Expedition during the years 1913–16. It was sponsored by the (US) National Geographic Society and the American Museum of Natural History.

Natkusiak, famed hunter and guide who became an invaluable member of two of Stefansson's Arctic expeditions and a good friend of Wilkins. Much loved by the men he travelled with, not least for his offbeat and relentless sense of humour and love of practical jokes. Known as 'Billy Banksland', he was a literal lifesaver; at one point dragging an unconscious Stefansson and three others from a smoke-filled 'ice house'.

Frank Hurley, the adventurer, photographer and film-maker from Sydney was a veteran of both Sir Douglas Mawson and Sir Ernest Shackleton's Antarctic expeditions. In August 1917, he was commissioned by the War Records Section to take command of a small photographic unit assigned to cover the Western Front. Hurley worked closely with Wilkins during the following few months, establishing a significant photographic record.

Charles Bean, the pre-eminent Australian journalist of his time chronicled the Australians' experience at Gallipoli in 1915, before going with the

Diggers to the Western Front. Initially he was a correspondent only in France, but from May 1917 he also administered the Australian War Records Section, devoted to chronicling the entire Australian experience for posterity. Bean welcomed Frank Hurley and George Wilkins to France in August 1917, and worked particularly closely with the latter from that time until the close of hostilities in November 1918.

John Lachlan Cope was a medical student at Cambridge University when he joined Ernest Shackleton's 1914–17 Trans-Antarctic Expedition as biologist and surgeon. Subsequently, Cope organised and led the British Expedition to Graham Land, Antarctica, during 1920–22. George Wilkins was selected as second-in-command of this expedition.

Lucita Squier was an American screenwriter. She travelled to Russia with George Wilkins in 1922, nominally as members of a Quaker mission making a film concentrating on relief efforts in areas severely affected by famine. She would in later life appear as one of the 'witnesses' in Warren Beatty's film *Reds*.

Roald Amundsen, legendary Norwegian polar explorer; the first European to the South Pole, beating his doomed rival Captain Scott. Wilkins' hero, private adversary and admirer. Amundsen's relentless pursuit of adventure and innovation inspired Wilkins throughout his life.

Sir Ernest Shackleton, leader of three British expeditions to the Antarctic; the final one counting Wilkins as a member. Shackleton's rivalry with Roald Amundsen for the race to the South Pole ended in glorious failure and a knighthood; but his finest hour arrived in his incredible survival and leadership following his ship the *Endurance* becoming trapped and crushed in the polar ice. Wilkins dreamt of accompanying Shackleton on a polar voyage and that dream came true in a most unexpected manner.

Carl Ben Eielson was an American pioneer aviator, who conducted his first flight in Alaska in 1923, and later provided the state's first commercial flights, transporting light freight, passengers and mail. Eielson was offered the position of pilot for Wilkins' Arctic air expedition in 1925. Throughout the following few years, the two teamed up as pilot and navigator to explore the vast regions of the North.

William Randolph Hearst, a media magnate a little before anyone quite knew what that was. The model for *Citizen Kane*, he was one of Sir Hubert's key sponsors.

Lincoln Ellsworth was born into a prominent and wealthy religious family in Chicago, and both sponsored and participated in exploratory expeditions. In 1931, he committed $70,000 to Wilkins' *Nautilus* expedition.

Suzanne Bennett was an Australian-born actress who would loom large in Sir Hubert's life.

INTRODUCTION

*Australian history is almost always picturesque; indeed, it is also so
curious and strange, that it is itself the chiefest novelty the country
has to offer and so it pushes the other novelties into second and
third place. It does not read like history, but like the most beautiful
lies; and all of a fresh new sort, no mouldy old stale ones. It is full
of surprises and adventures, the incongruities, and contradictions,
and incredibilities; but they are all true, they all happened.*

Mark Twain, 1897

While writing my book on the life and times of Sir Charles Kingsford
Smith back in 2009, a particular name kept popping up, a bloke
from South Australia I'd never heard of by the name of George Wilkins.
First he appeared as a polar explorer who became – get this – a photographer in the Great War who did wonderful work with Charles Bean and
was right on the spot the moment that the Red Baron was shot down.
Then, when Kingsford Smith had to drop out of the Great Air Race of
1919, it was navigator George Wilkins who replaced him as leader on
the Blackburn Kangaroo. And when, seven years later, Kingsford Smith
was desperate to find a plane with which to fly the Pacific, who would
it be that he bought the plane from, the plane that he would rename the
mighty *Southern Cross*? Well, that would be George Wilkins.

While writing my book on Sir Douglas Mawson two years later, I came
across the story of the death of the great Sir Ernest Shackleton, on the
island of South Georgia in the South Atlantic. Who was with him on
that last expedition? George Wilkins.

When it came to my book in 2017, on the Battles of Villers-Bretonneux,
the two clutch battles the Australians fought in April 1918, with a huge
part of the war effort in the balance, the one photographer/correspondent
who was with the troops both times was . . . Wilkins again.

And you'll never guess who was the only photographer with the Australian troops as they fought the seminal battle of that war, the Battle of Le Hamel – my book of 2018 – which was General Sir John Monash's first battle in command of all Australian troops. It was, of course, George Wilkins!

He really was no less than the Forrest Gump of his era, the man in the middle of so many extraordinary historical moments, achieving far more things that play to the theme above than I knew about when I began this book.

When I decided to get down to the nitty-gritty, my first port of call was my friend Simon Nasht who wrote a wonderful book on Wilkins, *The Last Explorer*, 15 years ago. With touching and amazing generosity, he handed to me and my researchers two giant boxes of documents, with the words 'Go for your life!'

We did. Simon has provided very useful counsel since, as has the business entrepreneur Dick Smith – a Wilkins aficionado to beat them all.

My right hand on the whole project was my brilliant long-time researcher and cousin Angus FitzSimons, who loved the story as much as I did. As always, the book owes his nous, hard yakka and whimsical muse a great debt. He was a joy to work with and as ever, once I had marked a passage for Angus's attention with 'TTA', as in 'Tweak To Accuracy', I could rest assured that what came back to me tweaked, tightened and brightened could be affirmed and cross-referenced by documentation as on the money!

Over in Bunbury, Western Australia, Barb Kelly equally did wonderful research and was nearly as tireless as her indefatigable son, Lachlan, who somehow has a capacity to ferret out the finest detail gleaned from the very bottom drawer of the digital archives to bring to light long-lost aspects of the Wilkins oeuvre. Back in Sydney, I was also grateful to have the input of my own eldest son, Jake.

My warm acknowledgement also to Dr Peter Williams, the Canberra military historian who first started working with me on my book on Gallipoli, and has stayed with me thereafter. His knowledge on the military chapters was obvious but, as ever, he surprised me with how much he knew about everything else, too, from polar exploration to aviation to submarines. All up, it was a great pleasure for us all to head back to various parts of books we'd already worked on and cover again the people and places of history I already broadly understood – Mawson, the Western Front, Kingsford Smith etc. – while this time drilling deep on the Wilkins aspect of it.

I might say in passing, however, that Wilkins was a notably curious subject to work on because he had the rare concurrence of remaining remarkably humble for such a great achiever, despite having *dozens* of documented extraordinary death-defying feats on his record, while *also* leaving many scattered accounts of events that simply never occurred. How this happened is a tangled tale in itself; which began when Wilkins wrote a fictionalised version of his life for a radio serial inaccurately called 'True Adventure Thrills'. Well, the thrills part was true – as are some of the anecdotes and most of the fine detail of what various such adventures are like to experience, but the trick lay in working out what actually happened and where he had let his fancy fly.

Also confusing things was that Wilkins had the misfortune to die before completing his autobiography, leaving two rival biographers to mix Wilkins fact with his first-person fictions at will. Untangling that mess, especially with a man who frequently achieved the seemingly imposs-ible, with many witnesses attesting to it, has been a difficult task but a fascinating one. Wilkins was a remarkable man in so many ways, and so many who encountered him thought the same two things: this man is incredible and why don't more people know about him? Hopefully this book does something to bring Sir Hubert Wilkins back to where he belongs: in the pantheon of Australian heroes and adventurers as a polymath of exploration who witnessed and created history.

Beyond my usual researchers, I was blessed to be helped by many experts in the field.

Research in the USA was done by Edward Kunz, a doctoral student from Columbus, Ohio. Edward sent me many valuable documents from the Wilkins Papers in the Byrd Polar and Climate Research Center at The Ohio State University. Thanks, too, to Laura Kissel, who is the Polar Curator of that wonderful archive, and who went above and beyond to help all of my researchers.

In Adelaide, I cannot express enough gratitude for the help extended to me by the Wilkins Foundation, and specifically and particularly Dr Stephen Carthew, Philip Van Dueren and Andrew Dawe who had already done enormous work transcribing valuable documents, and not only freely handed them over, but gave valuable advice thereafter as to where other treasures might lie, and vetted the copy. I also thank the Wilkins

Foundation President Robin Turner for his general help across the board, as I do Paul Ryan, Nicole Miller, Louie-May Ryan and the 57 Films team – particularly Andrew Pulford and Olivia Butler – who created the Wilkins Foundation website which was so crucial in accessing information. Thanks also to the librarians and curators from the Queensland Museum, Jennifer High, Shannon Robinson and Donna Miller for their assistance.

Professor Paul Memmott of the University of Queensland helped add missing detail on what Wilkins saw of the Indigenous existence in the Northern Territory. Every time I do a book with accounts of historical aviation, my first port of call is the man I worked with on the Kingsford Smith biography, Peter Finlay, and he was as valuable this time, as ever.

I also thank Edward Bean LeCouteur and Anne Marie Carroll, the grandchildren of Charles Bean and owners of the copyright in his diaries and papers, for their kind permission to quote from the great man.

As ever, and as I always recount at the beginning of my historical writing, I have tried to bring the *story* part of this his*tory* alive, by putting it in the present tense, and constructing it in the manner of a novel, albeit with 2000 footnotes, give or take, as the pinpoint pillars on which the story rests. For the sake of the storytelling, I have occasionally created a direct quote from reported speech in a journal, diary or letter, and changed pronouns and tenses to put that reported speech in the present tense. I have also occasionally restored swearing words that were blanked out in the original newspaper account due to sensitivities of the time that now no longer apply. When the story required generic language – as in the words used when commanding movements in battle, I have taken the liberty of using that dialogue, to help bring the story to life. To reprise one of my favourite quotes and themes, the late, great American novelist E.L. Doctorow once noted, 'The historian will tell you what happened. The novelist will tell you what it felt like.'[1] As ever, I have tried to do both.

Always, my goal has been to determine what were the words used, based on the primary documentary evidence presented, and what the *feel* of the situation was. As much as possible I have remained faithful to the language of the day – using 'Eskimo' instead of 'Inuit', for example – staying with the imperial system of measurement and using contemporary spelling with only the odd exception where it would create too much confusion otherwise.

All books used are listed in the Bibliography, but beyond my afore-mentioned deep gratitude to Simon Nasht, allow me to pay my extra dues to the work of Lowell Thomas, who spoke to Wilkins directly back in the 1930s, and had direct access to Wilkins' own papers and many unfinished fragments of autobiography. The author whose work all of us who venture into the Wilkins world owe most, however, is Jeffrey Maynard, who is no less than the recognised expert in the field and has devoted years of his life to trawling through all things Wilkinsian, and uncovering things that would never have seen the light of day without his tireless and wonderful work. He has another book coming out on Wilkins soon, and you will find me first in the queue to buy it.

As ever, my long-time sub-editor Harriet Veitch took the fine-tooth comb to the whole thing, untangling hopelessly twisted sentences, elimi-nating many grammatical errors and giving my work a sheen which does not properly belong to it. She has strengthened my stuff for three decades now, and I warmly thank her. In all my books, I give a draft of the near finished product to my eldest brother, David, who has the best red pen in the business. When his interest flags, it is a fair bet so too will that of the reader, and I generally slash what doesn't grab him, so long as it is not key to the story. In this book, he was as astute as ever, and I record my gratitude.

My thanks also, as ever, to my highly skilled editor Deonie Fiford, who was as assiduous as ever in painstakingly polishing the whole thing from top to bottom while also honouring my request that she preserve most of the oft esoteric way I write, while providing key judgement as to just when she needed to *insist*, step in, put down the Brasso and take out the scissors.

I am also grateful to my friend and publisher, Matthew Kelly of Hachette, with whom I have worked many times over the last three decades, covering a gamut of subjects. He agrees with me, however, that I've never worked on a subject before who made a mark in so many fields as Wilkins has. Sometimes it was difficult to work out how I was going to fit it all in, but I am confident I got there.

I have loved writing this book, and hope you enjoy it.

I chiefly hope that this will help put Wilkins back where he belongs: regarded as the most extraordinary Australian who ever lived. Right now, if you can believe it, his chief public memorial in Australia is on an informational plaque in the urinals at Adelaide Airport. After reading this book, I hope you will agree that the whole airport should be named after him, as well as being next cab off the rank to be honoured on our banknotes.

For he is, the Incredible Wilkins, and here he is . . .

Peter FitzSimons
Neutral Bay, Sydney
29 June 2021

PROLOGUE

If you can dream – and not make dreams your master;
If you can think – and not make thoughts your aim . . .

Rudyard Kipling, 'If'

Theirs is a meeting between two great men who have made their mark in entirely different fields.

General Sir John Monash is the engineer and military genius who had not only been the first Australian to command the Australian Imperial Force in the Great War, but performed so brilliantly that he and his men were placed at the pointy end of the spear being ruthlessly driven straight to the heart of Germany, which helped to bring that war to an end.

And the fearless Arctic explorer Vilhjálmur Stefansson has made an enormous name for himself by living off the land – and more particularly ice – for years at a time in the northernmost climes of the planet, in the process discovering new lands and charting vast tracts of previously unexplored territory.

This evening in Adelaide in late June 1924 has been pleasant so far, a lecture from the famed 'Stef' on his Arctic adventures at the Town Hall, followed by a dinner at a nearby restaurant with these two geniuses. Stefansson – 'who looks out on the world from eyes deep-sunken, and with that faraway look characteristic of seamen and airmen, a gaze that is always bent on the horizon'[1] – mixes easily with the great and the good of the South Australian capital and he and Sir John trade tales for the after-dinner entertainment of those lucky enough to be in earshot. And yet it is an absent great Australian that most dominates the Icelandic–American's thoughts. Finally, he comes out with it.

A question that is also a rebuke to Sir John: 'Why is it that Australia does not know one of her greatest heroes?'[2]

Of course, Monash knows instantly who the visitor is referring to: Wilkins. Captain George Hubert Wilkins. Just as had happened with

Monash, Wilkins had entered Stefansson's outer orbit as a tangential meteorite of no particular consequence, only to quickly emerge as a shining star accomplishing things of such wonder and diversity that people can barely credit them.

And Stefansson wants to talk about him.

'The bravest of the brave! Australia does not yet know what a wonderful man she has produced in Captain Wilkins. The world is ringing with the fame of lesser men.'[3]

Monash could not agree more.

Why, at the conclusion of the Great War, at a time when Lawrence of Arabia was the man who had captured the world's imagination more than any other, Monash had run into Lawrence's biographer, the famed American journalist Lowell Thomas, who had a question for him. He'd heard so many stories about men of courage on the battlefields of France and Gallipoli and didn't doubt them. But when it came to dash, daring, derring-do and the luck of the Gods?

'Sir John, does Australia have any counterpart to Lawrence? Any individual who stands out from the tens of thousands who have served under you?'[4]

Sir John does not even have to think about candidates.

'Yes, there is *one*,' Monash replies. 'His war record is unique.'[5]

Monash had seen countless lives lost in battle, brave men who had met their end beneath the wheels of the war machine, hands still gripping rifles as they fell. So the notion of a man who had gone to war *without* a weapon and lived to tell the tale was impossible to ignore, an amazing anomaly.

'I understand the courage of men who charged with a rifle and bayonet in their hands,' Monash had mused to Thomas. 'But I never ceased to be amazed at the daring with which Wilkins went over the top with nothing but his camera. Time after time he kept pace with the first line of the advancing troops, taking pictures under fire as coolly as though he were at a garden party.'[6]

Despite the fact that Wilkins bore no arms, Monash had been honoured to bestow upon him a combat medal, and Wilkins had been the only photographer or correspondent of the Great War to receive one – let alone two, as he was in fact awarded a Military Cross and Bar, the equivalent of two MCs, while he also had a Mention in Dispatches, for 'meritorious action in the face of the enemy'.

So, Sir John knows precisely what Mr Stefansson means and says so. Wilkins' comparative lack of fame in his own country is extraordinary, but there it is.

Well, let Mr Stefansson tell you Australians a few things about your countryman that you clearly don't know. For the next hour, Monash sits and nods, fascinated, as Stefansson entertains the table with tales of what Wilkins had done in the Arctic, how he had fought with a polar bear and won, how he had begun the trip as a foreign photographer and ended it as a Canadian Commander, shown astonished Eskimos motion pictures on Christmas Eve, fought drunken Scotsmen to skipper a vessel to save Stefansson, found and photographed the fabled Blonde Eskimo, raced across thin ice with no sight and suffering from frostbite, survived where all others would die, where 11 of Stefansson's expedition did die, where Stefansson himself would have died had not the most junior man on the whole expedition, the humble cameraman, risen to the occasion, and led the men to his rescue!

'If I had to choose,' Mr Stefansson declares, 'I should say that Wilkins was the best man I had. He saved my life when the rest of the party gave me up for lost and wished to return.'[7]

Yes, gentlemen, that is correct. When he joined my expedition, your fellow South Australian had never set foot in the Arctic, and yet despite being in the company of grizzled veterans had finished up as the leader! Amazing!

Which brings us back to you, Sir John.

'Why haven't you told them what he is?'[8]

Again, Monash can only plead the truth. He has tried, he really has tried. Not long after the war had been over, he had been giving a public lecture at the Melbourne Town Hall and had remarked: 'If I had to select the bravest and most useful man in my army, I would select Captain Wilkins.'[9]

The next day Monash had received a phone call from Wilkins.

Yes, George?

I beg you, General, *not* to mention my name in such a manner.

After all, Wilkins says, 'You have men with VCs you might have mentioned!'[10]

Is he serious? Oh, yes.

'I beg you, do not praise me publicly.'[11]

It is extraordinary, but it is Wilkins.

It is one thing to be a hero. Monash is one, has buried many, and has known dozens personally over the years, many of whom he has been honoured to command. But a hero who not only seeks no recognition for his accomplishments, but actively seeks to hide them? Before Wilkins he had never heard of such a thing!

Humility is a virtue, but Monash declares that 'Wilkins is so aggressively modest that he seems to carry it to a fault.'[12]

As the General now defensively tells Stefansson, against military protocol, Monash personally acknowledged Wilkins' singular efforts in the preface to his own official war history.

'I would have said *much* more about him if I had known he was going to be ignored as he has been.'[13]

But Mr Stefansson won't let it go.

How can it be that they are sitting here, in Adelaide, the city in which Wilkins lived, in the state where he was born, and yet he is a stranger to the public? Is it not ludicrous?

'Captain Wilkins is a native of South Australia! Everybody who knows him is proud of him, except, apparently, *his own people*!'[14]

This conversation is had in 1924, well before Wilkins accomplished the things for which he would be most remembered.

And yet, almost a hundred years on from this conversation, Wilkins remains relatively anonymous to most of his fellow Australians despite the fact his reputation has continued to grow overseas. In Ohio and London, certain archive collections have whole sections on Wilkins, which are a cordoned-off cornucopia of faded fame. He is an explorer who vanished from the history of the nation and the world he helped create. This is all the more surprising for the fact that, beyond being an explorer, photographer and film-maker of his times, the supposedly mad theory that drove Wilkins throughout his entire life: the thought that the weather, the 'climate' some term it, was all connected, and there was a certain worry that long-term changes in that climate appeared to be thinning the polar ice . . . has rather come into its own. He was a voice in the wilderness, from the wilderness, declaring to indifferent listeners that if the weather, particularly at the polar regions, could be . . . well, tracked and monitored (the very thought laughable at the time), it might change the world and save the lives of millions of animals and people. Wilkins was a visionary who was so far ahead of his time that his time was forgotten when the world caught up to him.

Sir Hubert Wilkins was not the last of his breed, he was a breed apart; a jack of all adventure trades who mastered each in turn; a leader who never lost a man; an explorer who never admitted defeat and was only ever beaten by sabotage. Spy, Photographer, Writer, Daredevil Cameraman, Polar Explorer, Australian Explorer, War Hero, Air Ace, Mystic and Futurist; Wilkins was a man who drove a Ford through the blazing heat of unknown Australian deserts and a submarine under the unnavigated frozen Arctic. He is extraordinary and so is his story, or rather stories; for he is a man with more adventures and lives than seems possible.

He is . . . the Incredible Wilkins.

CHAPTER ONE

THE BARE BEGINNINGS

Instead of spending my days at high school or college as I had hoped, I was doomed to herd the starving animals and to rescue them from the mud of the waterholes. But they died despite all we could do. I can still shudder at the thought of these conditions. Although necessity compelled me to do many things, the one thought I have held since I was at the age of thirteen was to solve the weather problem ... while it might not be possible for man, with his limitations, to control the weather, it might be possible to learn something of its movements. With foreknowledge and fore-thought, one might then be able to save the dumb animals from suffering and the pioneering families from destruction.[1]

Sir Hubert Wilkins, *Under the North Pole*, 1931

1900, South Australia, bare essentials

In this part of the world the sun does not shine, it *beats*. No matter, these two lithe lads are upping the ante by trying to beat each other as they race along the dusty bush roads at the foot of Mount Bryan, all to a chorus of dismal sound that comes from the insistent insect life, the bountiful blowflies, and sonorous cicadas.

The boys don't hear them any more than a crab hears the ocean, a bird hears the sky. It is so hot, even the poisonous snakes that slither hither and thither in the less ferocious parts of the day have taken a siesta in whatever hollow logs they can find, while swarms of bull-ants have retreated to the shade, but still the boys keep running, their legs pounding, their lungs exploding.

Faster! *Faster*! Run until your lungs burn like the sun.

The boy who wins, Buzz Simmons, does so by a final gasp on the finish line of yonder gumtree. The boy who is nominally defeated, the one wearing the ill-fitting and supremely old-fashioned three-quarter trousers that come halfway down his calves? His name is George Hubert Wilkins.

Buzz is prepared to take a bow or at least accept a congratulatory slap on the back, but by the time he pulls up and turns round it is to see Wilkins taking off his boots. And now his shirt.

Shall we race again?

Buzz agrees, amused far more than the flock of cockatoos who rise in screeching protest as they begin. The race is run once more and, even though the logical Wilkins is now lighter, once more Buzz takes it by an agonised gasp.

Very well then.

Panting, Buzz looks up to see that Wilkins is now . . . taking off *all* of his clothes! His friend is 'left in his birthday suit'![2]

Well, that is lighter still. Again? *Very well, Wilkins.* The two boys race, faster, *faster!* Run, run, run, until your lungs give out, and this time it is a . . . *draw!*

Most of Buzz's mates would be happy with that, but not George. For this friend always sees another challenge beckoning, another chance to win. So now George has another proposal.

'Buzz, now I've got no clothes on, will you go down to the dam and I will swim against you?'[3]

This race proves to be one that George wins easily. He invariably wins in the end – because he is dogged, unconventional, determined.

Buzz admires Wilkins but is amused by him as well. He is different.

(And he always has been different. A beloved family story talks of the time when he was just five and an older sister refused to let him ride a horse by himself, insisting she hold the reins. Young George, holding the riding crop, lashed her hands forcing her to drop the reins, whereupon the horse jumped and bucked as the thrilled boy whooped in delight before slipping off the saddle, still clutching the reins with both hands. The horse kept running, swinging in so close to the barbed-wire fence that George's face was scratched, he missed death or serious injury by a matter of inches, and when they finally brought the horse to a halt and gathered George in, he hadn't stopped grinning for a week!)

Speaking of sights difficult to believe, all of the farm boys have to help their fathers plough the fields as they try to scratch a living from this scratchy ground right on the edge of the Simpson Desert, but Wilkins is the only boy Buzz has ever seen who *reads* as he walks alongside the plough horses. *Big* books! He reads and reads and reads, strange books, thick volumes, borrowed and begged, marked and mastered, all day out in the field. Why?

'I want to be an engineer,'[4] Wilkins announces. Buzz is not quite positive what an engineer does, but is fairly sure that if that is what George Wilkins wants to be, well that is what he will be.

He is different in so many ways that Buzz just about has to take his shoes off to keep count. But his will to win, his physical prowess, the *logical* way he goes about solving problems – he is always convinced there must be a way around every obstacle – his fierce determination, and his endless devouring of books, his adoration of English *and* Maths, the way he does everything with passion and purpose, are only part of it.

For there is also the fact that George's mum and dad are the age of all the other kids' grandparents. He has 12 older brothers and sisters, making him the lowest bough of a large family tree that has been growing in these parts for many, many decades.

As a matter of fact, George's father, Harry Wilkins, was the first white baby to be born in South Australia, back on 1 January 1837, just four days after the 200 first settlers of the new colony had landed at Glenelg hoping to make a go of it in the newly declared colony of the British Empire.

And so young George Wilkins continues to grow up on the family's hard scrabble farm, with the only thing separating them from debt and disaster being their long labour and lucky rain. Providing the labour is no trouble for Harry, he was born for it, and has raised his children to give plenty of the same. All of them can swing an axe and wield a shovel not long after they can walk, and if they can walk they can work. Working demonically is in the family's bones, as there is simply no other way to survive, let alone thrive, in these parts. But the rain is another matter. Drought stalks the land and makes its kills as the farm crops fail time and time again, and that previous barrier between the desert and their farmland starts to blur. Soon their land becomes little better than a dust bowl dotted with burnt wisps of wheat and the desiccated corpses of cows and sheep.

In a bad month, and there are many, stretching on into an arid eternity, the only thing that grows on the Wilkins' farm is prayer, with the devoutly Methodist mother of the family, Louisa, insisting that all her children learn the pious practices of that strict Christian path.

O Jesus, full of truth and grace
More full of grace than I of sin,
Yet once again I seek Thy face:
Open Thine arms and take me in . . .

'I doubt that it would have been possible for me to have acquired the moral and physical stamina I possess,' George Wilkins will later say, 'were it not for the splendid physique and moral training I received from my mother and father. The daily reading of the holy scripture, the family prayers and the high morals of my parents counted for much in the moulding of my career.'[5]

What goes with it is a certain dour approach, a sense that, with the Good Lord watching everything, frivolity would be conduct unbecoming.

'When I was a child,' he will recall, 'I was not allowed to play cards or play the piano . . . I always look back on that as being rather curious – living 25 miles from any semblance of a town and living under those restrictions.'[6]

What he is allowed to do, encouraged to do, is lose himself in literature; with the fantastical adventures of Jules Verne's *Twenty Thousand Leagues Under the Seas* and *Around the World in Eighty Days* becoming decidedly dog-eared favourites, as they take him into the ever expanding world of his imagination. Oh, to fly like a bird over whole continents in a matter of days! To see the life under the ocean!

In his early teens, his rebellion against the strictures of the scriptures takes a surprisingly advanced form, for one raised in a strongly Christian family in a remote regional area.

'For a year or so, attendance at a county school where morals were left to take care of themselves, and boys and girls played at intervals and walked home together,' he will record with some delicacy, 'gave [me] a knowledge of sex far greater than many men of years and fostered desires, which, for some reason or other, had reached a compelling force . . . and demanded satisfaction.'[7]

True, it is not what he was taught in Sunday School, but he will remain steadfast in his relief at the belief there is a guiding hand to the universe, a benevolent God who watches over all, inclined to help those who will help themselves – in the best sense of the phrase, and not those who take the last bikkies in the biscuit tin.

For his parents' belief in this God is a balm for adversity, and helps them sleep at night after praying that their fifth-born son, Thomas, will remain safe, while fighting for the 4th South Australian Contingent during the Boer War. To young George, and later much older George, it will be the answer to why he survived things that would have, should have, killed any other man. And yet, he puts little energy into praying for things to happen, and an enormous amount of energy into making them happen. For when it comes to physical stamina, well, young George – even by his family's high standards – is no less than a phenomenon, thriving on work and forever proving he can keep up with anyone in the family. His father may be a hard taskmaster, but George is a son who really does thrive on hard tasks. The fact that he retains a sunny disposition with each duty added is a source of wonder to his siblings. George seems to delight in drudgery – looking after a flock of 200 sheep, reaping, shearing, mending fences, riding boundaries and ridding the place of rabbits – like no other. All farms, including the Wilkins', have 'working dogs'. But George goes them one better, as a 'working boy', trained and ceaseless. He will go forever, if fed occasionally.

While a good grown man might shear from 100 to 150 sheep per day, young George can shear 75 while still little more than a boy.

A few hours walk from their farm, along dusty tracks that never quite become roads leading to it, the nearest town worthy of the name is Hallett – which is in turn no more than a stone's throw from the Black Stump, and it's only a little further from there to the Back O' Beyond. The eight Wilkins children who survive infanthood are raised in a small stone cottage. George's older siblings teach George to ride a horse almost as soon as he can walk – *squeeze your thighs tightly, George, and sway AWAY from the direction the horse wants to throw you* – and learning to ride a bucking brumby is as much a necessity as a sport; the wild horses are a free resource for those bold enough to tame them.

Some young ones his age, who not even George would take on in a competition to ride those brumbies, are the local Aboriginal lads, who were not born in the saddle, because they don't bother with saddles, but they can ride like the wind on 'scarcely broken horses through tangled scrub'.[8] George loves nothing better than riding with them.

And it doesn't stop there.

While this local tribe, the Ngadjuri people, are ignored by most of the white settlers, it is typical of George that he actively engages with them, feeling he can learn new skills, acquire knowledge from a people who have lived in these parts since forever – somehow making a go of it without prayers. And, beyond everything else, he enjoys time spent in their company.

'Many times as a small boy,' he will recall, 'I wandered through the "wurlies" as the native camps are known': George is even allowed on hunting trips with the tribe. 'At these times I would share their food, including kangaroos, wallabies, opossums, snakes and even the fat sleek yellow grubs they dug from beneath the bark of the acacia trees. Thus my appetite became accustomed to a great variety of strange things and later in life this stood me in good stead, my palate is no longer shocked by strange tastes and I'm fully aware that geographical situation accounts mostly for food preferences . . .'[9]

When the sun went down, the day did not end since, as an honoured guest, he was allowed to witness *corroborees*, 'until like them, exhausted, I would slump with them beneath the trees for a few hours rest'. Later, they would engage 'in further wild orgies of dancing, feasting and rituals which would be an adventure for most boys at the age of 10 to 13 years but to me it was the routine of my existence . . .'[10]

Fascinated, Wilkins watches the Ngadjuri people talk without speaking, using their own complex coded language with their hands, a sort of sign language for those who need to communicate at distance, often while hunting. He imitates it to the amusement and encouragement of the children, who also teach him their 'string games', devilishly clever diversions that make cat's cradles look like child's play. They can shape a kangaroo in an instant before transforming it with a flick of the wrist into a koala or kookaburra. Again and again Wilkins practises until his attempts do not bring laughter, but a nod and a smile. Not bad, not bad – for a whitefella anyway.

'No doubt,' he will later note, 'these things helped build up a character which in the time of trouble and distress has then pulled me through.'[11]

The Ngadjuri can recognise young Wilkins approaching a mile away, for he is one of the very few, and certainly the smallest, who cares about them at all to come and visit, let alone bother to try to learn from them. A rare whitefella to remember, and one who always will remember them.

'Their moral standards in general were high; they were monogamous and seldom had trouble with family affairs. Their totem system was far more intricate than any of our so-called social systems and it needed a high degree of intelligence to follow it.'[12]

While the common view of many white people at the time is that Indigenous people around the world are of inferior intelligence, Wilkins' starting point will always be the reverse, deeply aware that they know things he doesn't. He wants to learn from them.

And not just their skills.

'It seemed to me at that time that their philosophy was much more sound than that of the white man.'[13]

More formal learning is done at Mount Bryan East School, just three miles south of the Wilkins farm, and here George excels above all. Learning comes easily and he is blessed with the most crucial thing – intellectual curiosity. Never bound by the parameters of the lessons themselves, his reading moves ever wider and deeper. It is not enough for him to know things by rote, he must know the whys and wherefores, and most particularly the hows – being of a notably practical bent and very good with his hands.

Occasionally, very occasionally, the family harnesses the horses to the cart and treks to the Big Smoke of Adelaide some 150 miles to the south, but only for a truly special event. And in their whole lives to date there has never been a bigger occasion than on 29 December 1895, when a 'Commemoration Day' is held at Adelaide's Glenelg Beach, at the spot where it all began, marking the 60th celebration of the day the colony was established. All of the pioneers and settlers who first arrived that day by boat, and who are still alive, are invited. The entire beach is covered with tents, umbrellas, horses, buggies and humans – with the grey-hairs being the most venerable among them. Now, Harry Wilkins may have been inside his mother's womb, but as the firstborn son of the South Australian colony, he is, of course, front and centre on the day. Young George is proud, but the thing that entrances him, the episode he will never forget and will cherish ever afterwards, is the closing speech that day given by the great, the literally fantastic, Mark Twain.

Yes, if you can believe it, the man who is no less than the author of such classics as *The Adventures of Tom Sawyer*, *Adventures of Huckleberry Finn* and 'The Celebrated Jumping Frog of Calaveras County' is here, in person, come to visit Adelaide during his swing-through all of Australia, giving lectures as he goes. Of course, the chance to hear the most famous and beloved man in the world speak means that the entire town and all surrounds have gathered, and are even now pushing and prodding through the mass to catch his words.

Adelaide has never seen a crowd like it, as the *Adelaide Express* drolly observes: 'Everybody seemed to be there, and yet so dense was the crowd that it would be a rash matter to attempt to say whether anyone in particular was there.'[14]

The esplanades join with the seas as they 'were flowing streams of people; on the pier, one was pushed this way and that way and every way – except the right way; under the jetty families were packed like sardines, and the beach for many a hundred yards was thronged'.[15]

Observing the throng from the podium are Mark Twain and, yes, Harry Wilkins, as the Mayor attempts to belt out a greeting to the Governor.

An amused Twain watches officials trying to make the chaos orderly, with the advanced aged and huge number of the first settlers present not helping matters. Why, Twain observes, in Adelaide 'there seems to be no death-rate for the old people. There were people at the Commemoration banquet who could remember Cromwell.'[16]

Six old worthies have been selected to give short – *short* mark you, gentlemen! – speeches. The settlers have different ideas, as Twain delights to discover.

> These old settlers had all been present at the original reading of the proclamation in 1836. They showed signs of the blightings and blastings of time in their outward aspect, but they were young within; young and cheerful, and ready to talk ... and *talk* all you wanted; in their turn, and out of it. They were down for six speeches, and they made 42. The Governor and the Cabinet and the Mayor were down for 42 speeches, and they made six.[17]

The Mayor tries to get them to shut up and sit down, but with little success, for as Twain will note: 'They have splendid grit, the old settlers, splendid staying powers. But they do *not* hear well, and when they see the Mayor going through motions, which they recognise as the introducing of a speaker, they think *they* are the one, and they *all* get up together and begin to respond in the most animated way; and the more the Mayor, gesticulates, and shouts "*Sit down! Sit down!*" the more they take it for applause, and the more excited and reminiscent and enthusiastic they get.'[18]

After yet more speeches, the relieved Premier may at last give a toast: 'The Pioneers of South Australia!' (*Prolonged cheers.*)

No doubt about it, in the world of young George Wilkins, 'pioneers', those who forge a path to go where no others have gone before, are people to be honoured and he enthusiastically joins in the roars of acclaim and thunderous applause for these men who have given their lives to bring life to this colony. As the youngest progeny of the youngest pioneer, he feels particularly connected to all of them.

And oh how proud he is when his own father responds to the roars and the applause to stand, take a bow, and now shake hands with the Mayor, the Governor, the Premier and the great Mark Twain himself!

But now to what the throng have really been *aching* for, the speech by Mark Twain.

Rising to the podium, the great man pauses until the hush becomes a settled silence and begins.

'First,' he says, 'I congratulate you on your climate. Perhaps your climate has become commonplace to you, but it is not commonplace at all. I am not accustomed to climates like this, where you have beautiful spring weather in midwinter, and where snow is unknown. We who come from the overworked regions of the earth cannot describe the pleasure of finding ourselves in this restful Australia, where it is *always* holiday. (*Laughter.*) And when you have no holiday, or nothing else to do, it is *always* a horserace. (*Laughter.*) It is a peculiarly blessed land, it seems to me. And then you have an arrangement which cannot be overpraised; you place your holidays not only to dates but to what suits your own comforts. (*Laughter.*) I was through Australia when they celebrated the Prince of Wales' Birthday. They celebrated it on the 8th, the 10th, and the 11th, and skipped the 9th altogether. (*Laughter.*) I suppose there was a horserace on the 9th. (*Loud laughter.*)

'May the colony last forever and be always prosperous.' (*Cheers.*)[19]

The settlers shout, clap and rise to their feet, thrilled that Twain has paid them so touching a compliment. They see themselves as pioneering trailblazers and this legendary author has all but confirmed it.

Waiting for the crowd to settle down, Mr Twain notes that while he has been instructed to speak on behalf of 'visitors', the term doesn't fit with how he feels.

'It seems that in appointing me to respond for the visitors there is a sort of incongruity. It seems impossible for me to stand here in so chilly an attitude as that of a visitor. I have been among you in Australia for three and a half months, and from the beginning to the finish I have seemed to be regarded rather as a member of the family. (*Loud cheers.*)'[20]

It is a day to remember, a day when George Wilkins sees the magic that one man can do, a visitor from across the world who all the world knows. So impressed is he that, for the rest of his life, he will keep the newspaper clipping of the day in his wallet, the day when his father was honoured for his struggle by another man who, though a sole individual, made a *difference*.

And yet while the honour and glory accorded Wilkins Senior does make for a special day, it does nothing to alleviate the horror of the family's current struggle just to hold on to their farm as what will become known as 'the infamous Federation Drought' hits Australia in 1901. When it begins, the young nation has 106 million sheep. By the time it breaks in 1902, there are only 54 million left. Cattle losses are even worse, with 70 per cent of the animals wiped out, and the stock routes are lined like a bone highway with the carcasses of thousands of animals all being driven to find water and feed that isn't there.

George Wilkins is taken from school; he is needed to try to rescue the farm, an impossible task and inevitable defeat that his desperate father cannot bring himself to admit. As Wilkins will write in a document too painful to publish, these days and nights, these endless shocking sights, 'were the most depressing of any that I have experienced. Instead of spending my days at high school or college as I hoped, I was doomed to herd the starving animals and to rescue them from the mud of the waterholes. But they died in spite of all we could do.'[21]

It is clear to George Wilkins that he is watching 'the ruin of my family'.

Their hearty stock? Now 'starving animals gnawing bark from the trees'.[22] Their friends and fellow farmers? Now 'lean settlers with hollow cheeks and lacklustre eyes, sun bleached hair and gnarled hands, clothes saturated with the stench of the dead'.[23] It is a living nightmare.

Can no-one wake them from it?

The young Wilkins broods on it obsessively, carrying the burden of that which no man can control: the whims and whips of the weather. He has already learnt the principles of cause and effect in physics, but what about the cause and effect of the climate that controls the globe and the fate of all who live on it?

'I was so grieved about the disastrous results of the drought periods,' he will recall, 'that even as a boy I was casting about in my mind for the possibility of solving the difficulty. I wondered whether it might be possible to forecast seasonal conditions early enough to protect the animals and men from such suffering and loss . . .'[24]

The more he thinks about it, the more passionate he becomes.

'I determined to devote my life to that work.'[25]

Wilkins will do many, many things in his life, and leap from one adventure to another, but there is a sole constant: 'the one thought I have held since I was at the age of thirteen was to solve the weather problem'.[26] How that is to be done the boy does not yet know but, 'with foreknowledge and forethought, one might then be able to save the dumb animals from suffering and the pioneering families from destitution'.[27]

Leave it with him.

The idea at least provides him with enormous encouragement to concentrate on his studies to better understand science, particularly within the realms of meteorology, so as to be able to cure some of the earth's ills. In the meantime, though, the family fights on for the farm to the bitter end; in 1905 that end comes. Giving up the uneven struggle, Harry and Louisa must abandon the farm and move into a small cottage in Adelaide suburbia, with George the sole child left to fill this small home.

If they are winding down from a singularly busy life, however, George is as busy as a honeyeater on a hibiscus. With Adelaide affording educational opportunities that Hallett could not get close to – opening him to the world of literature, languages and advanced mathematics – he takes night classes in science, drawing and mathematics, while his days are spent supporting the family working as an apprentice in the electrical engineering firm of Bullock and Fulton. By virtue of his capacity for ceaseless activity, with 'each unforgiving minute' seeming to have at least 120 seconds 'worth of distance run'[28], and wrung out of it, he takes further advanced courses at the South Australian School of Mines and even enrols in the Conservatorium of Music. Perhaps he will sing opera professionally, something his parents quietly discourage him from for a reason he cannot quite fathom – they are very gentle, but insistent – and yet he won't give up. Still, he might also tour the world as a cellist or even throw himself into engineering, yes, *definitely* engineering, but maybe the other things too. He will do *everything*: 'I worked in the morning and would go to the school of mines in the afternoon and to the music school on the evening.'[29]

To go with his ever-expanding array of skills, he is nothing if not practical, and becomes the one others turn to when things go wrong. Like tonight, for example, when the band has failed to turn up at this country 'hop'.

With no fanfare, the teenage George Wilkins, as later recounted by his brother, gets 'hold of a kerosene tin, stretched a bit of kangaroo hide across it and made a pseudo-banjo on which he himself discoursed the dance music for the evening to the delectation of the company. It wasn't very melodious, but George saw to it that there was rhythm! This is a small thing, but it is typical of the man. There was no-one else in the place who would have had the initiative to carry a thing like that through.'[30]

So many things he wants to do, so little time.

As things turn out, however, his initial path into the world begins purely by chance . . .

In early 1906, when George is just 17 years old, a travelling troupe of players comes to town.

In fact, the troupe are practically troops and it is less a visit than an invasion, as there are over a hundred of them: dancers, singers, acrobats, magicians, a whole carnival complete with ropes, canvas and rousta-bouts. But the real attraction, the wonder that draws the crowd, is the new invention, 'the flicks', those amazing pictures that are in motion as you gaze. If you can believe it – *Come one, come all, step right up, step right up!* – by shining a bright light through a whole series of photos laid end on end on a long tape of film, you can actually show real-life scenes on a screen, and see people, animals, trains, everything, moving! As an invention it is just one of the extraordinary leaps forward in this gloriously modern time to be alive, right up there with mankind having learnt how to fly, how to use electricity to illuminate and power the globe, how to travel long distances in mass-produced cars or trains that span continents. You can even send Morse messages from one side of the world to the other!

Now, having heard about these 'movies', as the flicks are also known, young George can't wait to see one of them.

And yet? And yet, oddly enough, it is now when the flickering moving picture machine suddenly goes on the bloody blink, and not in a good way, leaving the screen dark, a great made-for-the-big-screen scene struck out. The problem proves to be the portable engine the impresario of this carnival is relying on to generate the power. It stone-cold refuses to work, generating no more than a few loud burps before falling into a sullen silence as the crowd's impatience grows. Again and again and again!

Is there a doctor in the house?

No, but there is a junior apprentice electrical engineer who has taken the base knowledge learnt from his father that for any engine to work there needs to be a spark and petrol, and from his studies developed a more advanced understanding of how to fix engines. For even as a small despairing crowd of carnival employees gather around the engine to argue over what to do, one of the troupers, Caspar Middleton, is watching the damn thing not moving when a wise old soul nearby gives a tip.

'There is a boy round there putting electric wires into the new theatre who could make her go.'[31]

Indeed? Once sent for, the boy duly arrives to be confronted by a highly sceptical Middleton who, as he will recount, thinks the lad looks 'like a schoolboy who had been on a dirty job'.[32]

(Funny he should say that . . .)

And you are?

George Wilkins.

Middleton looks him up and down once more.

A dirty schoolboy, yes, but a strapping and handsome one, with a full head of rather too neatly curled dark hair. He has a quick and ready smile, and if not for his piercing eyes and fixed, furrowed brow, you might think he was the type who would never take things seriously, so light is his manner.

But look, we have nothing to lose, so see what you can do.

Sure enough, as all the others wander off disconsolately, young George gets down to tin tacks, brass bolts, loose leads, leaking line and dirty spark plugs. Only Caspar Middleton stays to watch, and not because he has any hopes for Wilkins. In fact, quite the opposite.

'I was thinking to see a bit of fun, or see the thing blow up.'[33]

I mean, what chance this kid can achieve when his own mechanics have failed?

'Do you think she will go?'[34] Caspar asks Wilkins sarcastically.

'Yes, she is all right,'[35] replies Wilkins quietly, head bent and focused on tightening the last spark plug into place.

As if!

Just seconds later, however, after George gives the crank a few sharp turns, the machine gives a cough, a sputter, an utter sputter, another cough and now ... the engine ROARS into life!

Bloody hell! She *is* all right! Not only is the bloody thing working, but the roar has now settled down to a powerful purr and no longer sounds like a tantrum of tin! It is a miracle. Strangely, the boy Wilkins asks for no payment. Or at least, not in the way of cash.

'I want to look at the cinema machine,'[36] he says simply. He has heard all about this new invention and his dearest wish is not only to see some cinema itself, but to understand *how* it actually works. Very well, the novelty is shown to him and Caspar watches once more as the boy ignores him, absorbed in the wonder before him, even as he begins to appreciate the mechanics of the whirring spools of film, the flickering light, dust swirling in the beam of light projecting the picture onto the screen. Wilkins is completely silent, his concentration only broken when Caspar Middleton offers him a job.

Yes!

Wilkins makes the decision instantaneously, as the chance, 'to take the opportunity offered to move from place to place and seek distraction in the strangeness of new conditions',[37] is just too tempting to refuse.

The rest of his studies can wait. This will be learning on the spot.

It is extraordinary.

When Wilkins had woken up this morning, he had been no more than an apprentice engineer, living with his parents.

And now?

Now he is in charge of the electric lighting plant for a travelling show, with particular responsibility of keeping the cinema projector going, come what may.

All of it through a complete fluke of being in the right place at the right time with a small skill that was, nevertheless, the precise skill required.

'This chance introduction to the wonderful world of cinema,' he will later note, 'was to lead me into a series of adventures shaping the future course of my life.'[38]

To begin with, there could be no better time to be getting involved in this nascent industry, as resources flew in. For even as he is starting out, in December of this very year, 1906, the world's first feature film, *The Story of the Kelly Gang*, opens in Melbourne's Athenaeum Hall to riveted crowds and a pressing press of newspaper denizens raving at the wonder of seeing a real-life story writ large. For theatre owners, it is a revelation that leads to revolution. It is one thing to pay actors night after night in a play. But so much *more* profitable if you can pay them just once for a film – even paying handsomely – and then play the film over and over again as the crowds flock in!

'When I was with the carnival,' Wilkins will reminisce, 'we toured all the big towns – the capitals of the districts. We were carrying about 100 people.'[39]

He and Caspar Middleton soon become great friends, with such a bond that Wilkins will later characterise it as the first real deep friendship he ever had, Buzz notwithstanding.

The two become inseparable, with Caspar as much in awe of George's capacity to fix things as George admires Caspar's ability to make everyone laugh, including him. Caspar is a natural thespian, capable of mimicking anyone with an ease that belies his actual sharp powers of observation and devotion to filling out the character of any part he chooses. And the stories he can tell, of actors tatty and batty, whirls with girls and jokes with blokes from far and wide, all over Victoria and South Australia where the carnival travels. For his part, Caspar quizzes Wilkins for his tales of the bush, the things he learnt from the Ngadjuri people as a lad, the best way to ride a horse – *squeeze with the thighs, Caspar!* – and most particularly his whole bag of tricks: how does he know how to make anything work, anything go, anything happen?

Both young men, right on the cusp of adulthood, share the sense that they are not run-of-the-mill that they have bigger ambitions than most, greater abilities, and will somehow wind up on a bigger stage even if they have no clue where that stage is.

The obvious place to start though is surely in the movie world? Caspar can act, and George can work a camera, so could there be something for them there? So intoxicating is the very notion of it that it soon dominates their conversation. They could really give it a shake! For George, who has spent his life feeling different from the ordinary, the extraordinary Caspar is confirmation that there are more pathways through life than sticking to the tried and true, the straight and narrow, the expected and the commonplace. You could *do* something!

Most importantly, in terms of life experience, working with a travelling show offers experiences and puts him with people the likes of which he could only have dreamt of.

In between motion pictures, the waiting crowd is entertained by a variety of acts including 'a concert party of five singers, marionettes, a magician, [and] a pianist'.[40] The magician's name is Hatherly, and while his act is only so-so at best, when it comes to picking up young ladies he really does have the magic touch; something that young George has little success with, despite careful study from close quarters of the master at work. 'Girls of many types and degrees of culture visited the marquee,' he chronicles, 'some accompanied by their mother, some surreptitiously, and some boldly amorous . . .'[41]

When it comes to women, Wilkins is either out of luck or into trouble and – as his life experience continues to expand exponentially, both happen in one small town one unforgettable evening. Having arranged to squire a local beauty out and about on the town, the young man is doing just that only to find she has many young male admirers who are so put out at this upstart moving in on their territory, and her betrayal by agreeing to it, that they jeer and heckle them as they walk. Stay calm, George, stay *calm*. Their time will come.

And sure enough, while revenge is generally a dish best served cold, there is nothing better than serving it up electrified. For the very next evening, as chance would have it, the same youths – reminiscent of Banjo Paterson's famous verse, *Their eyes were dull, their heads were flat, they had no brains at all* – attend the picture show that very evening and lean against an iron rail next to an electricity generator. The opportunity is too good to miss. Taking a wet canvas cloth, Wilkins manages to connect the two sending a shock through the rail that throws the shrieking youths back in a heap. In the chaos and confusion, Wilkins slips away. Not long afterwards shots are heard, which sees the police rushing to the scene as the youths – who are the chief suspects – make good their escape, into the night and down to a nearby beach. (Only later will Wilkins reveal to his girl the wonderful truth: *he* was the one who fired the shots into the air, knowing he was outnumbered and this was the quickest way to bring the cavalry!)

Moving on, the carnival continues to do a roaring trade to frequently roaring crowds until one night, with a packed house, there is a sudden blackout and a cry of alarm and disappointment goes up.

'May I have your attention?' the stage director yells. 'It will only be a few minutes before gas lamps can be lit.'

Young George has been called for and indeed proves to know exactly how to fix the engine and restore the electricity, with one problem. The broken part is in the crankcase filled with burning hot oil. But he must move fast – there is no time! And so he thrusts his hands into the oil and, fighting the urge to shriek, even as the blisters instantly rise on his hands, he fixes the engine. Light is restored and the men that surround Wilkins stare at him open-mouthed: 'they saw his raw and bleeding fingers and the bloody broken blisters completely covering the back of his hands up to the wrists'.[42]

'Three cheers for the electrical bloke!'[43] yells one man. The cheers echo and Wilkins, now partly in shock and fully embarrassed, makes a small bow. He is rushed to the nearest chemist, his hands 'saturated with picric acid and wrapped in cotton wool'.[44] No less than a week of pain will pass before he can pull on his pants and boots, and do up his own buttons. As this is a carnival, however, dignity is not particularly valued, and in coming shows the ringmaster delights in telling the audience the tale of the brave camera boy who scalded himself to stop the panic of a crowd, and Wilkins is obliged to step into the limelight and show off his ghastly wounds to appropriate oohs and ahhhs from the mob.

For George Wilkins, launching himself into the world of cinema in its nascent form is more than a mere turn in the road . . . it also provides the very model on which he will base the rest of his life: take whatever interesting chance presents itself, ride your luck, show your pluck, withstand its buck and back the whole thing with your skills and elbow grease – and keep expanding those skills through your ever widening experiences. Use finesse when you can, and get a bigger hammer when you can't. And if all else fails, throw for double sixes – as you have nothing to lose!

Over the next 18 months the show tours all over the highways and byways of South Australia and Victoria, pitching the tent and – *come one, come all* – welcoming the people of the towns and the villages to come and see 'the cinema machine'.

And certainly George's main job remains to keep that machine going, come what may, primarily by keeping the gasoline engine perfectly tuned and the fuses on all the electric lighting intact, but he is also eager to expand his other skills including . . . singing?

Truly?

Truly, Caspar Middleton has heard young George sing and has noticed passing birds land, curl up their claws and *die*, rather than live in a world where such desecration is possible, but he is also amused that his camera operator has no clue of just how bad he actually is. Well, there will never be a better time for him to learn that than right now. And besides, all Caspar cares about is that his show provides entertainment, and he has learnt well over the years how entertaining complete disasters can be. So yes, by all means, George, you may sing a solo tonight, as picture slides are shown to the audience. The song shall be 'In the Valley Where the Bluebirds Sing'. Good luck, George.

'*There's a picture in my heart that lives for ever,*' he begins to warble. '*By a brook that always murmured love's sweet song . . .*'[45]

The audience look at each other. Is he serious?

George goes on . . .

> At the window I can see my sweetheart waiting
> I can see her in the valley where we parted
> In the valley where the bluebirds sing . . .

A passing tiding of magpies is struck down as if by buckshot . . .

Still the crowd is gobsmacked.

When finally George finishes, Middleton claps loudly and calls for an encore.

Wilkins calls his bluff and gives it to him.

'He was no better the second time,' Caspar will declare, but he had also been right on another count – the crowd has been entertained. Whatever else, the boss admires Wilkins' absolute nerve and complete lack of stage fright. Some people put public speaking and public performing as more frightening than death, but not this bloke. He can completely *die* on stage, but come off energised and eager to do it again, and again! Nerves of *steel*, I tell you.

And yet, though Wilkins does not add singing to the repertoire of things he does in this job, he certainly does just about everything else, and is noted as a blur of activity throughout the day and into the night, not only keeping the cinema machine working perfectly, but sorting out *all* the electric cables, fixing the trucks the carnival moves with, loading the trucks, unloading them, feeding the animals, helping everyone do anything that comes up – most particularly if he can learn a new skill from it.

In the end, however, the very nature of carnivals is that there comes a day when the carnival is over – in this case when they have gone through every population centre in two states and have no energy to go to other states where other carnivals are better established. But while the carnival declines to head to Sydney, young George Wilkins feels quite the opposite. In his time with the carnival he has gained more direction than ever in what he wants to do, which is very broadly to be more than a projectionist. He wants to make films, shoot films, be a photographer. He wants to not only master this emerging medium, he wants to be on the front lines, a pioneer – not unlike the way his dear father had been in South Australia. To do it, he knows, he needs to get to Sydney, which is the new boomtown of the new craze.

Is it a problem that he has moths in his wallet, and nary a brass razoo in either pocket?

Not for George Wilkins.

For that flitting figure in the moonlight, down by docks at Port Adelaide? It is one and the same, slipping up the gangplank of a particular steamer he is told is heading to the NSW capital, only . . .

Only to come face to face – the *moment* he sets foot on deck – with a passing ship's officer.

'Where are you off to, young fellow?'[46] asks this bemused Old Man of the Sea, in a thick Scottish accent.

'I wish to stowaway to Sydney,'[47] Wilkins replies, evenly.

Well, now the Scot has heard it all.

Still, it is so refreshing to be told something like that so bluntly that, both for his own amusement and to help the whippersnapper, the Scot decides to grant the request. Taking him to a suitable hiding spot, a canny cranny he knows well, he shoves the lad inside.

'Wait here until you are "found"!'[48] laughs the Scot.

True, the bosun is of much less kindly disposition when he finds Wilkins there the following morning when they are well out to sea, but by then Wilkins' purpose has been achieved. They are too far out to go back and the only damage done is that he is another mouth to feed, which is no real problem. The upside is they get another deckhand and one they can get to do all the worst jobs – mopping the deck, cleaning out the boilers and the heads, peeling the potatoes – and some three days later, exhausted but happy, George Wilkins, just 20 years old, sails through the heads of Sydney Harbour.

The world is his oyster even if, for the moment, he is still a bit of grit within it, not quite yet a pearl. Happily, the docks he has been deposited on are not far from Sydney's key film district along Parramatta Road. In short order he is knocking on the door of the first cinema he can find: Waddington's Pictures in Annandale, a place which has just opened, trying to ride the wave of interest in movies.

(The success of *The Story of the Kelly Gang* in Melbourne's Athenaeum Hall had indeed brought on a revolution – a turning of the world on its head turvy-topsy – that the Kellys themselves had missed out on. Yes, there is a small problem that it's not quite lifelike as the black and white figures move around in jerky motions, and there is no sound, but you can get around the latter by having title cards at the bottom of each scene to explain what is happening, together with some limited dialogue. There is enormous money to be made both by those who produce the film, and those who screen them in their theatres.)

In another hallmark of what will be an extraordinary feature of Wilkins' life, his timing could not be better. The man who answers the cinema's door is the flustered and furious owner, Frank Waddington, who cannot believe that the new projector has not only broken down, again, but that whatever ails it has just defeated the attentions of four electricians.

But what is that you say?

You think *you* can fix it?

George certainly does. His carnival days have taught him that he can fix pretty much anything, even when in far western South Australia, in the swirling lost lands north of Woop Woop and south of the Black Stump, where the only resources you have are the things you brought with you that you can fashion into a solution.

Fixing something in downtown Sydney, by comparison, will be a doddle.

'If I don't fix it, I don't want any pay,'[49] Wilkins tells Waddington.

Within the hour, Waddington is paying him a pound and gives him a job as a projectionist as well. Wilkins soon masters the job and is able to use the enormous amount of spare time he has between changing reels to work on his new passion of photography, which he engages in with the new Kodak camera he has bought with his wages.

'I liked,' Wilkins would recount, 'playing with light and shadow and colour, experimenting and trying out new ideas in taking pictures, enlarging them and colouring them.'[50]

Such passion is only outdone by the passion he frequently has for his subjects.

'They were mostly photographs of my girlfriends.'[51]

•

Wilkins! You haven't heard of him? Nice young chap, working at Waddington's these days.

News of this young fellow, so skilled in so many areas and with such a great drive to work, is not long in getting around, and they reckon he can do pretty much anything – which is useful, as the whole movie business is really taking off. Even sensible men are encouraged to be rash with their cash to try to get in on the boom.

One such man who gets interested is the well-known draper and North Sydney Alderman, Thomas Carleton, who has become fascinated by the whole industry and eager to get involved. He not only seeks George Wilkins out, practically begging an audience, but asks him for advice about making a film. How should he invest his money, who should he get involved, what roles should they have?

Well, he has come to the right place. For George's considered response is that he, the 21-year-old from South Australia, should be able to do pretty much everything required for the making of the film – writing the plot, shooting it, cutting it, 'titling' the explanatory cards, and there is really only one other person they need. Yes, a leading man.

Who better than . . . Caspar Middleton!

Caspar is duly summoned from the dusty yonder, happily accepting the offer to become this still developing new breed of person – a movie star – and he and George are soon back in business together. This time, however, instead of showing the films of others, they are making their own.

No studio? No problem.

Making it up as he goes along, Wilkins soon requisitions not only the top floor of Carleton's drapery in Walker Street, North Sydney, but also many of its sheets, which he is able to hang up to give a smooth background to the action at hand.

And . . . *action!*

Caspar wanders on, mouths a few words, to which whatever actors they might have roped in react by mouthing a few words of their own, and . . .

Cut!

Picture after picture is hastily shot and assembled. With every snip and join, every changing sequence, Wilkins begins to understand the power of film, the extraordinary capacity of moving pictures to tell a narrative so much more vividly compelling than even a hundred still shots. If, as they say, a picture is *worth* a thousand words, just ten seconds of footage can tell a million. And the most extraordinary thing of all is that once a film is captured and put together, it can be replicated and transported so that distant audiences can enjoy it all at the one time.

And yet, partly through his nature and partly through his experience of watching flickering images in a carnival tent, Wilkins knows that there are always ways to tell it better. For there's more to it than just pointing the camera and turning it on. The angle matters. The lighting matters. What you exclude in the final cut matters just as much as what you include.

But he isn't pretentious, he doesn't think of himself as an artist. He is a technician, a man determined to take the technology to the limits of possibility, capturing a scene plainly and quickly. Sometimes, however, a sense of genuine spectacle is required, and when they want a bigger canvas than the sheets behind the set . . . why not . . . Sydney Harbour?

Look we don't need to worry about the copyright on the American play *The Octoroon*, they'll never know. The harbour can pass for the Mississippi on close-up, and for a paddle-wheeler we can get that old Sydney ferry, the *Narrabeen*, to be 'set on fire'.[52]

The footage is indeed spectacular, if a little expensive for Mr Carleton who, it must be said, is more than a little underwhelmed when box-office receipts don't come close to equalling the cost of the second-hand ferry. Generally, though, the plot is secondary to the emotion, which is itself secondary to the motion, the sheer wonder of seeing pictures that move and live and breathe. Well, sometimes they breathe. As Wilkins would note with his dry wit, their motion picture company often made 'funny melodramas and sobering comedies'.[53]

Their most ambitious effort is their 1911 filmed version of a serialised story published in *The Arrow* in 1910, called *Gambler's Gold*. Two men share friendship and love for one woman. The man who wins her becomes a squatter. (Close observers might see Sydney Harbour in the background as he squats, but filming must be kept local.) The squatter then stands for long enough to accidentally kill his wife while trying to kill his friend. The End.

It is divided into more exciting chapters such as 'A Dastardly Murder', 'Foong Lee's Opium Den' and 'Great Motor Boat Chase in Sydney Harbour'.

Now that will be a *great* scene, Mr Carleton, also set on Sydney Harbour, as one motorboat with Caspar and a mob of gangsters fire their guns at the 'detectives' chasing them in the next boat. I'll have my cameras set up in the second boat, and it will be wild!

And wild it proves to be, never more than when the *real* Water Police are alerted by officers of the HMAS *Encounter*, an Australian light cruiser, that gangsters are firing on a fleeing vessel, both ships racing beneath their bow. The Water Police join the chase in their steam launch, the *Biloela* – the fastest vessel on Sydney Harbour, now punching out billowing black smoke from its enormous black funnel – providing excellent production values for no charge, although charges are contemplated when the Water Police realise just what has been going on.

It takes some time to sort out, but no matter.

A delighted Wilkins records: 'the unrehearsed scene that followed was the most realistic acting in the film'.[54] Alas, the film proves to be something less than a smash and actually records a 'pronounced loss'.[55] Finally having the major sheets with the whole thing, Thomas Carleton decides that he really is just a draper after all and, as his final act, pulls the curtains on his whole show business venture.

Certainly, it is a problem for Caspar, but George Wilkins has no time to even gnash one tooth before the Gaumont Studios mob make him an offer. You see, the thing is, George, we are in a battle with Pathe to be the dominant newsreel provider in the world. Your fictional films are fine, but what is bringing the crowds in around the world is footage of actual events, and what we need is for you to go to North Africa where there are a couple of promising wars going on to film 'striking scenes' and . . .

George . . . ? *George . . . ?*

George Wilkins is already down at Sydney docks, taking the first available ship to any port in North Africa to begin storming the world with his camera.

•

After all, it makes sense. There are so many extraordinary things in the real world they leave fictional films for dead in terms of excitement. Why, only a short time after Wilkins had arrived in Sydney, a bloke by the name of Harry Houdini had flown a plane before stunned crowds at Sydney's Rosehill racetrack, each person paying one shilling for the privilege of watching and being part of it. As reported in the *Daily Telegraph*, on the moment Houdini first became airborne, 'men tossed up their hats, women grew hysterical and wept for sheer excitement. When he landed, a hundred men rushed towards the biplane, pulled the happy aviator from the seat, and carried him, shoulder-high, mid deafening cheers and salvos.'[56]

If only Wilkins could capture wondrous stuff like that! He intends to.

And exploration is another thing. Right now, as the world waits on the results, there is a race to the South Pole between England's own Robert Falcon Scott and Norway's Roald Amundsen. Imagine the excitement of being able to capture something like that on film!

For the first time in his life Wilkins leaves Australia – a robust man ready for adventure, a wiry, strong 164 pounds of energy – on board the Barbarossa-class steamer *Friedrich Der Grosse*. It is bound for the Suez Canal, Port Said, Genoa and Algiers among other ports, and the young man can at last achieve some satisfaction for his overwhelming case of wanderlust.

It is a peculiar thing but this young man feels more at home when he's leaving it. Each new sight – whether it is the majesty of the Indian Ocean, the ancient wonder of the Red Sea, the snaking traffic of the Suez, or the blazing blue of the Mediterranean – draws him closer to his new love: 'I had courted adventure and found her.'[57]

Africa, the very name a conjuring of exotic excitement, lies before Wilkins' eyes. It is a feast, and he is a starving man whose appetite for the extraordinary never abates.

For a young fellow raised in South Australia, whose travels have taken him no further than Sydney, to find himself in Algiers is to have entered another world. It is not just the mosques that abound all around, the bizarre bazaars where you can buy everything from pearls to people, the women in long robes and masks, the camels, the swarthy characters on every corner that hover with hospitalities or murmur with menace, the sense of a searing desert trying to press into a city that is only just beating it back.

Wilkins will describe it as 'a hotbed of unrest and intrigue'.[58]

•

Returning to his ship with a spring in his step for the very possibilities of travel in exotic places, after two more weeks of steaming across the ocean blue, the *Friedrich Der Grosse* arrives in Southampton just in time for Wilkins to see another wonder of the age in action. It is the unsinkable *Titanic*, pulling out on its maiden voyage, across the Atlantic. Stunned at its size, its majesty, the young Australian feels more blessed than ever with his good fortune to be able to see it, and takes good footage of it departing.

•

As wondrous as *Titanic* is, London proves its equal, and he walks its streets with the sage wisdom of Dr Johnson ringing in his head: 'A man who is tired of London is tired of life'. Well, no matter what they may say of young Wilkins, the phrase 'tired of life' will never be uttered. He approaches every day with such purpose and passion that it's a wonder that life isn't tired of him. London is a land for a lad such as Wilkins, the mother lode of the motherland, a smorgasbord of culture high and low, a melting pot of rich and poor and black and white and everything he could possibly think of. From the flowers of Covent Garden to the lamps of the Mall, the grand walls of Buckingham Palace and the statues of Trafalgar Square, Wilkins is sure that this, *this* is where he belongs.

To a boy who thought of Adelaide as the big smoke, London is a raging fire of different sensations, most of them wonderful.

Wilkins finds his employers, the Bromhead brothers, in the Gaumont Studio at Lime Grove in Shepherd's Bush, and they are keen for him to start shooting at once. Alfred Bromhead knows talent when he sees it and though he can see little in the actors or actresses struggling with his terrible plots, he can see it in abundance in the quick eye and innovative work of his cameraman. Bromhead sets himself the target of making at least 15 films a month, so there is plenty of opportunity for that cameraman, with the strangely flattened and sunburnt vowels that mark him as an Australian, to shine.

Fictional feature films have been one thing to bring in crowds, but that art form remains in its infancy. But 'news', which covers everything from close-ups of a volcano exploding, to a lion charging, or a group of African tribesmen dancing, requires no suspension of disbelief – it is here, it is now, and you the cinema-goer are right in the middle of it!

With Alfred Bromhead's view that Wilkins might excel at gathering newsreel footage, the young Australian is given one of the more pleasant tasks of his career to date.

'Go out and show us what interests you in London.'[59]

Done. With the fresh eyes of an Antipodean looking at the greatest city on earth for the first time in detail, he spends a fortnight scouting the great locations – Westminster Abbey, the Tower, London Bridge, Big Ben – and some of the lesser known, like St Martin-in-the-Fields, Covent Garden and the Royal Observatory at Greenwich.

The next fortnight is spent revisiting them with camera in tow, taking shots at each one from just the right angle, with just the right lighting at just the right time of day. Wilkins knows photography, and his prints are so popular that the Bromhead brothers have yet another proposition: 'Would you treat thus, the principal cities of the continent?'[60]

You mean, take my cameras, at your expense, through the very cities I have always dreamed of visiting?

TAXXXXI!

Yes.

And so, 'in their service I visited the capitals of Europe; Paris, Brussels, Switzerland, St Petersburg, Scandinavia'[61] as well as every large town in Scotland and Ireland. Visiting no fewer than 27 different countries in 18 months, beyond being a great pleasure, it is a hands-on education in Western civilisation, giving him a richer understanding of European history, geography, language, cuisine and culture, not to mention contacts. But it also exposes him to one of the fresh wonders of the twentieth century, the emergence of aviation, together with the extraordinary men who fly the machines . . .

•

In June 1912 Wilkins is assigned to photograph the man who is currently the darling of the British public, a daring pilot by the name of Claude

Grahame-White. Two years earlier, when the Frenchman, Louis Blériot, had flown across the English Channel from Calais to Dover, the young Grahame-White had been so entranced he had journeyed to Paris to enrol in Blériot's flying school and returned to London with not only the ability to fly himself but with a disassembled Farman III biplane, which he promptly assembled and took to the skies as the British public roared.

Of course Wilkins has heard of him – everyone is talking about him – but he has never met him. And so when he arrives at that broad expanse of grass that is Hendon airfield, which has become the hub of British flying, he asks the first man he meets, a tall good-looking chap dressed in flying gear with a rather effete air about him, whether he knows something of this fellow, Grahame-White?

'I've been told,' Wilkins says, in a rare loquacious mood, 'that while White is one of the best pilots, he knows it too well. Several fellows have told me he's inclined to high-hat early acquaintances.'

You don't say?

I do say. It's common knowledge. In any case, I've been sent to film this conceited fellow, would you know where he is? Grahame-White?

There is an odd pause so Wilkins continues: 'I have never seen Grahame-White.'

The fellow nods.

Great.

'Would you point him out to me,' Wilkins asks hopefully, 'if he comes here today?'[62]

Again, the stranger agrees, before pointing at himself.

'That's who I am,' replies the stranger.

Luckily for Wilkins, the famous pilot is amused rather than angered and even offers to take him up for a flight.

Shortly thereafter, they are in position, Wilkins sliding his lanky frame between the flying wires to get to the seat.

It is a Farman III biplane with an ash-wood air-frame, Gnome Omega 50-horsepower rotary engine behind the occupants, and, of course, a pair of wings joined by six vertical struts on each side, braced with cross-wires. At the rear is an open rectangular box-like structure above the rear wheel.

Finally, all is in readiness. Grahame-White waves to the mechanic hovering near, who now steps forward to grip the huge, two-bladed wooden propeller.

'Switches off,' the mechanic calls.[63]

'Switches off,' the pilot affirms.

Now the mechanic pulls the propeller backwards a few turns, to suck petrol vapour into the combustion chamber.

'Contact!'

'Contact!'

With which, the mechanic gives the propeller a mighty heave in the clock-wise direction. There is a cough, a gurgle, and as the engine catches, the motor gives out the roar of a dragon, blowing blue-white castor-oil smoke out of its nostrils, the whole machine shakes and both Grahame-White and Wilkins are awash in a blast of wintry air drawn over them by the whirling propeller behind. A wave from Grahame-White and the chocks in front of the wheels are removed. Vibrating furiously, the plane rolls across the ground.

Hold on!

With every yard now, the vibration lessens as the pilot allows the plane to do what it clearly wants to do, which is to get faster, and faster still, with the air now flowing over and under the wings and wires with such speed that they begin to hum, and now *sing*, and lift and . . . and now comes the moment.

Some 150 yards down the field, first the nose starts to lift as the rattling and bumping stops completely and . . . *voila!*

With a sense of wonder like he has never experienced before, George Hubert Wilkins looks down to find the earth falling away and, as the suddenly fierce cold wind slaps his face, he realises they are . . . flying.

Flying!

Flying!

Grahame-White glances behind. From experience he knows that, at this point, many a passenger has gone green at the gills, and is prone to gurgle a half-strangled request to be taken back down . . . please . . . *now!* But there is no such sign from this Australian. He is completely entranced.

And Wilkins really is. Yes, he has a slightly uncomfortable popping sensation in his ears, but apart from that he is simply overwhelmed by the extraordinary feeling of flying like a bird and seeing the world from this high. Before his very eyes the surrounding farmlands turn into a rich mosaic of coloured squares, while looking back to the centre of London, even though he knows Big Ben and Westminster Abbey are ten miles away, he can see both clearly.

By the time they land, Wilkins, a man of many passions, now has a new one that will trump them all. Flying is the most extraordinary thing he has ever done and, he is fairly certain, ever *will* do. It is the experience of his life and he MUST do more of it.

Happily, Grahame-White proves to be a fine fellow who arranges for the young Australian to return to the Hendon flying school to get more experience in the air, with a possible view to becoming a pilot himself.

'Although I took advantage of his offer,' Wilkins will later recount, 'I never did attempt to take out a pilot's license. That would have cost more than I could afford. I was pleased to learn to fly . . .'[64]

Most importantly, such experience gives him an idea. Why not take his camera up with him, and become the first man to shoot the world from the air?

True, it will take a little time to find a pilot who agrees to his plain plane plan, for while it is one thing to take a man aloft, it is quite another to take up a man with bulky, heavy cameras as well, which, beyond the weight, affect the balance. But finally, Wilkins finds just the man: his name is J. Cyril Porte.

Porte, dressed in the mechanics' overalls that have become all the rage among these dashing knights of the sky, is precisely the kind of Devil-may-care flyer Wilkins needs, ready to take a chance, though he cheerfully tells Wilkins that the only plane he has available is a flimsy Deperdussin Monocoque racer single-seater monoplane, powered by a 160 HB cylinder engine.

Wilkins assures the pilot that this is no problem as he can simply strap himself to the front of this mechanical dragonfly, just behind the propeller, and so capture a view that no pilot has ever seen. In a clear case of 'chicken', each demonstrating their Crazy Brave courage to the other and continually upping the ante until one of them breaks, it proves to be Porte who chickens out first.

For while Wilkins indeed is able to 'strap the camera to bracing wires and ride astride the fuselage'[65] with 'my nose just a few inches from the trailing edge of the propeller'[66] to make himself and his camera effectively a part of the plane, Porte's bravado lasts no longer than having a grease-stained mechanic in tatty overalls whirl the propeller to get the engine to roar into life. As the whole machine vibrates, Porte suffers a sudden burst of sanity, and yells through the din to Wilkins: 'I don't like to chance it! She has too much power and not enough ballast. I don't know where to put the weight to keep her on an even keel.'[67]

On anyone else – as in, someone with normal courage, or even just a sense of self-preservation – this would almost definitely have worked. But Wilkins is just not made like that. He has no nerves, and does not blink. And he has one more advantage, as he will note.

'Having been brought up on horseback I thought little of the difficulty of obeying the order.'[68]

If this bastard bucks, well, he will *ride* her bare-back through the highways and byways of the cloud above – *squeeze your thighs tightly, George, and sway AWAY from the direction the horse wants to throw you* – and bring her to submission just like the Man from Snowy River. How do those words go again?

> But the man from Snowy River let the pony have his head,
> And he swung his stockwhip round and gave a cheer,
> And he raced him down the mountain like a torrent down its
>   bed,
> While the others stood and watched in very fear.

Hold on, George, hold on!

> And the man from Snowy River never shifted in his seat –
> It was grand to see that mountain horseman ride . . .

'That's all right!' Wilkins yells back, 'I . . . can move back and forth if necessary.'[69]

There is no way out for Porte other than to say to Wilkins: 'I have come to my senses, and will not risk my life, nor my plane, for your insane plan.' And he won't utter those words.

For he is J. Cyril Porte, renowned as the bravest man of the English skies. To cede would be to acknowledge Wilkins as a braver man than he.

'Are you ready?'[70]

Wilkins opens his mouth to say yes, but before he can even do so, Porte – perhaps a little vengefully – *this is what YOU asked for* – guns the four-cylinder Clerget engine, it surges and the machine slowly gathers momentum across the rough paddock, up and down, up and down, upppppppppp and awaaaaaaaay! They are flying!

. . .

'And I was blown full of wind from the propeller, almost choked in fact.'[71]

Did someone say Man from Snowy River?

Ride, you bastard, *ride!*

'The machine, unbalanced, unstable in every direction, was twisting and turning in a crazy manner. It dipped and flipped and reared, fell off on one side and then on the other.'[72]

Even as Wilkins dips and flips with the best of them, bucking and rucking the air-trough as the Deperdussin tries to throw him, twisting like a sick dragonfly trying to regain altitude, he desperately tries to focus his camera, while not admitting to himself just how mad this whole venture is, which is difficult on a plane that, yes, has wings but behaves – to add one more metaphor to the mix – 'like a barnyard fowl flying in a gale'.[73]

In this matter of life and death, the only thing Wilkins has approaching reins are the wires holding the whole apparatus together and he must grip them so tightly that they cut into his flesh and his left hand starts to bleed, while his right hand holds the camera. The flight does not last long as Porte, white with fear, brings it back to land as quickly as possible.

'We had been in the air only a few minutes,' Wilkins will recount, 'but time enough for me to become almost completely numb with fear and cold.'[74]

With the return to solid ground there is an equal return to sanity as the understanding sinks in of how close they had come to a grisly death through their mutual foolhardiness.

'We both realised it was only a miracle that brought us down alive, in spite of Porte being one of the greatest of the pioneer pilots.'[75]

The footage Wilkins has shot is useless, but a useful and most valuable lesson has been learnt by the Australian.

Say it slowly after me: *One-seater planes are for one man only. If you need to go up in a plane with a pilot, your starting point must be two-seater planes.*

By George, I think he's got it!

It is the third time in the air that proves the charm for Wilkins as he is given an assignment in France to cover a hare hunt. But it is how it is done that will make history.

'I was perched in the front of a Farman plane, on a bicycle seat with my camera strapped between my legs. I well recall the thrill of skimming along within a few feet of the ground, filming a hare as it ran before us.'

Run rabbit, run rabbit, run, run, run.

The footage is perfect and gives cinema-goers a perspective that they have never had before in their lives. As Wilkins cautiously boasts: 'These were, I believe,[76] the first motion pictures ever shot from an aeroplane.'[77]

Adventurers, by air or otherwise, fascinate Wilkins and he reads every scrap of newsprint he can find about the polar explorer Roald Amundsen, who on the 14th of December 1911 had become the first man to reach the South Pole. The English are obsessed with the doomed gallantry of Captain Scott and his men, Wilkins is obsessed with the man who succeeded, how he achieved what so many thought impossible and overcame the perils of the Pole with aplomb. He will later describe Amundsen as 'the hero of my youth'.[78]

CHAPTER TWO

AIR AND ARMS

If you can meet with Triumph and Disaster
And treat those two impostors just the same . . .

Rudyard Kipling, 'If'

As it happens, only weeks after his Porte peril in the air, Wilkins will no longer need to create danger for drama, as he has been assigned to cover an actual war where his job will be to use his camera to shoot men who are shooting each other.

Spanning South-eastern Europe, Western Asia and North Africa, the Ottoman Empire has lasted for 600 years, but now 'the sick man of Europe', as it has become tagged, appears to be facing what might be its final disintegration. For though the Sick Man limps on, its uniform still covered in the medals of gloriously victorious campaigns long gone, serious change is not only afoot, it has boots on and is marching hard straight at the old guard . . .

For now, the Muslim Turks find themselves under attack from previously vassal Christian states, fighting together for their independence: Greece, Serbia, Bulgaria and Montenegro, and these states are already well on their way to securing a third of the total European territory previously held by the Ottoman Empire.

Personally, Wilkins has no strong view either way on the whys and wherefores of the war, all he cares is that it is a genuine battle, a real war, and he will be shooting from the Turkish side of the parapet.

'I shall never forget the thrill of preparing for that assignment,' the South Australian will recount. 'To be a free-lance at a war and in company with the bold, bad, terrible Turk of the story books!'[1]

It is everything he could have ever dreamed of.

Arriving in Turkey through a combination of train, ferry and motorbike, he is thrilled to find the Constantinople of 1912 is exactly what he had imagined it to be from popular literature, and more. Built around the narrow waterway which cleaves the mountainous forms of Europe and Asia on either side, it is an axis on which whole empires have hung in the balance for three thousand years. These days it is all centuries-old mosques, mazes of narrow cobbled streets, donkeys bearing Persian carpets, teeming masses, ancient buildings, Eastern music, bazaars where everything from apple tea to exotic fabrics to herbs and spices are sold, and all of it mixed with the heady smells of hookah tobacco, exotic spices and herbs, and the heavy Eastern musk perfume favoured by the local women.

What is far less pleasing is to find himself pooled with a group of 30 international reporters 'assigned' to the Turks to cover the duration in civilisation, penned and hemmed in the civilisation of the Pera Palace Hotel and given news in only sanitised dollops approved by the Turkish government, interspersed with the odd 'outing' to the front, tightly escorted by Turkish officials and shown only that which the Turks are happy for them to see.

There *must* be a way to escape such insufferable constraint feeding them nothing more than propaganda, and Wilkins quickly falls in with two British correspondents. For if birds of a feather so flock, so too do men of action gravitate to each other, and among the cadre of correspondents and posse of photographers it is inevitable that Wilkins will find himself talking into the night, well after the safe and staid are in their beds neatly made, with men like him – men who have done many things, seen more, and only want to *do* more. They have stories to tell – *Waiter, more wine!* – and Wilkins is eager to hear them all. Two such men are from the London *Daily Mirror*, Bernard Grant and John Banister. Grant is a plucky and charming photographer with a particularly fascinating background. As Wilkins listens, rapt, the Englishman tells of how one of his former assignments was to track the famous wife-murderer Dr Hawley Harvey Crippen. You don't know the story? Well, once suspected, he fled by boat to Canada with his mistress who was disguised, get this, as a young man and dressed up in a suit! Just after he left, the wife's body, minus head, was found under the floor of his house.

Arriving on the now captured Crippen's boat, Grant learns what had first twigged the stewards that something was up. It was the fact that Dr Crippen kept opening doors for the trim lad who accompanied him to dine. Dr Crippen became one of the few murderers to be caught through exquisite manners, can you believe it? For Grant, the youngest of three brothers who work for the *Daily Mirror* as professional snappers, photography is a family affair.

Banister, a long-time correspondent in far-flung parts of the world, has picked up the ability to speak eight languages along the way – something particularly impressive to Wilkins, who loves languages and is always working to expand his own grasp – and has many a tale to tell of derring-do, done and dusted, on front lines around the world.

Speaking of which . . . the issue before us remains, gentlemen. How do we tell new stories, have new experiences, break out of Constantinople, and get to the front?

As much as possible, the three decide to combine resources.

'We could work together without rivalry,' Wilkins will recall, 'for Grant was a still photographer, Banister was a writer and I was taking movies.'[2]

(And obviously, for the snapper and the newsreel man, it is imperative they get to the front, as formal press releases from the government give them nothing.)

For the moment they must bide their time, as Wilkins, particularly, finds his feet, the speed of which deeply impresses his newfound friends. For while he remains a novice, the fact he has already had so much experience as an Australian adventurer helps a great deal in covering a cavalry war, especially given that neither Grant nor Banister has ever ridden a horse before or ever slept in a tent.

Grant is particularly impressed by the friendly young man from the fly-speck of Hallett who – at the age of just 23 – can apparently do anything with ease.

'An Australian,' he will recount, 'and one of the best companions in the world, because of his continual cheerfulness, his knowledge of handicraft, of horses, and all rough work and outdoor life. He has had many experiences in the wild places of the world, which have taught him valuable lessons, among them being the gift of leadership, instant decisions in moments of peril, and a quick way of righting something that has gone wrong.'[3]

At this point Grant has barely cleared his throat.

'Distinguished by indomitable character, always resourceful and self-reliant when an awkward job had to be tackled. If a cart broke down, it was Wilkins who set it up again. If it overturned, it was Wilkins who put it on its wheels again. If horses stampeded in a camp, it was Wilkins who would first come to the rescue. He did all these things not in any arrogant way, not in a bullying, commanding spirit, but quietly and cheerfully, as though it came natural to him and was part of his scheme of life.'[4]

Despite the esteem in which he is held, however, he cannot help but notice that he remains very low in the pecking order due to the novelty of his movie camera. To be a war correspondent is a respectable, time-honoured profession. To be a movie man is something else entirely.

'I was only an amateur newsman far beneath the "Special War Correspondent" level of dignity because I deigned to carry a camera, which no real writer, in those days would even dream of.'[5]

Ah, but Wilkins has cards to play that the others have not even dreamt of, and one such card soon emerges in rather surprising fashion. The Turks, none too pleased about these troublesome young upstart reporters who actually want to see a battle and watch a war – and who are insistent they should be free to write what they like, even when it doesn't suit their hosts – decide to make things difficult for them.

So hear this: you Gentlemen of the Press are welcome to stay within the city boundaries of Constantinople, but if you wish to venture to the battlefront you will need your own resources as we will not be providing transport – or protection. That's it. As the Turkish Army cannot look after your needs, before you are issued with a pass to travel, you must have: two horses for yourself, and two servants, with a horse for each, not to mention two months' worth of rations.

Now for many of the correspondents this presents no problem at all, as they simply have no interest in risking life and limb by actually witnessing a battle. There is a long and cosy tradition that the best place to cover a foreign war is from somewhere close to government offices, finding what information you can from your sources before going to a nice club or restaurant at night. But for the new wave, the younger ones – many of them inspired by the derring-do of Winston Churchill while reporting the Sudan and Boer Wars – it is as imperative for them to get to the front as it is for the photographers and cameramen. And this Turkish edict is a real problem, as they are in a city where every good and bad horse has already been requisitioned by the Turks.

But for George Wilkins? Beating impossible odds is a personal specialty.

He gets to work and, through some of the contacts he has already made, passes word to the police. He, *Effendi* George Wilkins, will pay no less than £50 per horse, to be paid in gold – the rough equivalent of the annual salary of any of the police who can help arrange it. True, it is a long shot, but what else can he do?

Dammit, do my eyes deceive me, or is that Greek cab driver – notable for not having a beard like nearly all the Turks – staring at me?

Wilkins is not long in finding out.

For now, no sooner has Wilkins set foot outside his hotel than the ancient Greek approaches him and, mostly through sign language but with a couple of English words to boot, gets his question out: *Are you the young man looking for horse?*

Evet benim. Yes, I am.

Please, get in my horse-drawn cab.

Wilkins is soon being drawn along, further and further into that part of the city which is even more ancient than the cabbie, the Old City, as it is known.

They go past the Obelisk of Theodosius, a reminder of an Egypt that once was but is no longer, beyond the Hagia Sophia, no longer the monument to the expansion of Christianity that it was when built, now plastered up with Islamic mosaics, a testament to the faiths fallen and risen as Constantinople, as ever, continues. On they go, through bazaars, the opium dens, and across the ground that Alexander the Great had made his own 2240 years ago when this ancient place was known as Byzantium. Finally, on the edge of the Old City, outside the massive walls Mehmet II had stormed in 1453, ending the Byzantine Empire, they get to that part where modernity has lapped at its shores, here in the form of a rather nondescript six-floor apartment block.

What is going on?

Wilkins is crushed.

All this way, and no stables, no quiet field, no exercise track, no drinking troughs.

Just this ugly building?

More out of bloody-minded curiosity than anything, and on the off-chance that maybe there has been a misunderstanding and they are going into this building to negotiate a price, the Australian follows the old man up the steep, dark flights of stairs all the way to the top-floor apartment. But it is not what he was expecting at all.

For they have no sooner got inside the apartment and into its bedroom than Wilkins sees it: a horse!

And not just any horse. This is a magnificent black stallion that looks like it could win the Melbourne Cup. But where does it perform its ablutions? It is actually more a case of where doesn't it. But no matter now.

The Australian is beyond amazed, while the old man looks at him with shrewd eyes. Just how high can he be pushed?

This horse is available to you, *Effendi*, for immediate rental for . . . shall we say £100?

Done!

Caught between pleasure at the sudden surge of wealth coming, and regret that Effendi didn't hesitate – which means he could have pushed him higher – the old man is soon counting and re-counting the golden sovereigns that Wilkins hands over. Yes, it is hell on earth to get the snorting beast down the six flights of stairs but before long he has the horse outside, is astride, and galloping in triumph back through the Old City, and soon enough back and forth before the digs of the other, now *astonished* correspondents.

This, gentlemen, this is how things can be done if one has enterprise!

'My status among the correspondents,' he will recount, 'immediately went high. They had taken me perhaps for an English suburban photographer.'[6]

So now allow me to introduce myself properly.

Wilkins. George Wilkins. Australian. World traveller. War correspondent, just like you. *Cinematographer.*

Well, we, your fellow correspondents, are suddenly very pleased to meet you. What will you drink?

It is soon clear that Wilkins' ability to rustle up Royal Ascot winners from the top floor of Constantinople apartment blocks, while impressive, is the least of it. For this fellow is nothing less than a pioneer in a whole new way of chronicling battles . . .

Stay with him.

For you see, aviation has come a long way in the last decade since Wilbur and Orville Wright had got their invention off the ground at Kitty Hawk and it was only three years ago that Louis Blériot had flown his Blériot XI across the English Channel from Calais to Dover. Already planes are being lightly used to gather intelligence about the position of the enemy, and very occasionally observers and pilots in those planes take pot shots at each other if they happen to pass.

But Wilkins has an idea for an entirely new use. Given the Turks have some British aircraft used for such scouting and for ferrying supplies, why not try a new use? Making the acquaintance of a Turkish pilot by the name of Lieutenant Fazil Bey,[7] Wilkins is able to persuade him to take him up in his two-seater Bristol Prier monoplane. After all, as Wilkins learnt to fly on English airfields, perhaps he could give Fazil, who had learned to fly in France, some tips while they take this new dual-control plane into the air?

Once they are flying, however, Wilkins expresses a keen desire to fly over enemy lines to get some really good footage. As they will risk getting shot by some really good bullets, Fazil is against the idea. Alas for him, the plane not only has dual controls but a passenger who is not afraid to use them. For Fazil is just veering the plane gracefully away from the puffs of smoke coming from the artillery up ahead when he realises the plane is heading back that way. He pushes his controls back to port, but Wilkins yanks them back to starboard.

From below, the Bulgarian soldiers see the most extraordinary thing: it is nothing less than a plane apparently attacking itself, dodging and weaving in a dogfight of one . . . as the two pilots fight to steer the same craft, one of them filming with a camera. It is a one-plane show to behold.

'With the two of us struggling,' Wilkins recounts, 'it was a miracle that the controls withstood the punishment. Between us, that plane was put through all manner of manoeuvres for which it was never intended, many of them neither of us had ever experienced before.'[8]

Once Wilkins has the footage he needs, he cedes control, and Fazil is able to get his flying steed safely back on Turkish ground, and he has no sooner brought the plane to a halt than he has leapt to the ground, the better to berate the Australian for his insane foolhardiness, but Wilkins has his measure.

Calm down, Lieutenant Bey, and surely soon, CAPTAIN Bey.

You are now, *officially*, the first man to fly your aircraft over a military firing line in a European war. And, right here, I have the footage to prove it.

'We could cook up a fine intelligence report and you will get the credit!'[9] Wilkins argues.

(And yes, we can leave out the bit about you being abducted in your own plane.)

It is a compelling argument, so compelling that Lieutenant Fazil Bey, the toast of the Turks, the pride of the side, is soon posing heroically in front of his historic plane, as the humble Wilkins is honoured – *honoured*, do you hear me? – to be able to use his camera to chronicle the great man's triumph for posterity.

The experience confirms something for Wilkins.

'It seemed that as a passenger I could do all the work I wanted and satisfy a great many more interests than if I were to handle the machine,' he will recount much later. 'That attitude has never changed and even to this day I find it much more advantageous to hire a man to whom piloting a plane is a joy, a satisfaction as well as his profession, thus leaving me free to observe and navigate, than to attempt to fly the plane which I would probably do with indifferent skill.'[10]

Not long after Wilkins' experience with Lieutenant Bey, he and his fellow correspondents are at last given permission to cover a major battle, one where the Turks expect to record a glorious victory over the Bulgarians at Kirk Kilisse, some 150 miles to the north-west of Constantinople.

It is with some excitement they all board the train with such horses as they have purloined, and servants for the five-hour journey. The tension rises among the correspondents as they get closer to the front, and they can hear the odd bursts of rolling thunder from artillery fire, and even see billowing smoke up ahead. They are on the line that the famed Orient Express travels along in peace-time – the setting for much intrigue among fiction writers, but it has never seen real drama like this.

Who is winning this battle? What will we see? Are we in danger?

As it happens, however, it is just before the correspondents get to the front that they receive the news. 'The Turks,' as Grant will chronicle, 'had been heavily defeated at Kirk Kilisse.'[11]

The results of that defeat are soon apparent – 'large numbers of soldiers in hasty retreat, and a continuous stream of refugees making for Constantinople'[12] – but, still, that is not the worst of it. For not long after their train is turned back and they stop at Chorlu, they now see their first bit of drama as a train coming from the front, filled with troops and refugees from the battleground, passes them and derails.

Four whole carriages spill from the tracks before they in turn spill many of their dazed passengers. Realising that this is where their story of the day lies, Wilkins and the British journalists John Banister and Bernard Grant get off their train and, grabbing their horses, race back. While Banister has his own stills camera and snaps off shots, Wilkins shoots the entire chaotic scene with his movie camera, spanning back and forth as he swivels the Bell and Howell 1911 model cine-camera back and forth on its tripod. For his part, Banister, whose own camera can't capture scenes that move, must look instead for moving scenes and now spies just the thing. It is a group of stoical Turkish women – all heavily veiled with their *niqabs* – tenderly tending to the wounded. Alas, as he brings his camera up to take the shot, he is suddenly struck from behind as, 'an infuriated Turk snatched his camera and threw it in a ditch half full of water'.[13]

The young Turk has precisely no feeling for the finer nuances of war correspondents, the idea that they are independent chroniclers, and sees only an enemy photographing humiliation – an enemy that must be made to pay with his life. As he draws out his dagger and prepares to strike the *ayuha alkafir* infidels, Banister cries out. Wilkins turns and acts so fast that he will later declare his action to be unconscious. Uttering his own cry, Wilkins has his gun drawn and is pointing it straight at the Turk.

Easy. Easssssy, fella.

This is *not* a Mexican stand-off.

You have a knife. I have a gun.

It is a singularly compelling argument in the way of these things.

'The man looked at the weapon, which had a wonderful effect in quieting his wrath.'[14]

The Turk's final reply is to rush 'wildly towards the other end of the train shouting something intelligible'.[15]

The Turkish words for *these filthy infidels must die!* may have been involved, Wilkins is not sure. Fortunately, one of the correspondents' highly skilled interpreters, a *dragoman* – capable of speaking and translating several European, Turkish, Arabic and Persian languages – is sure of the tenor of the man's remarks, and quietly tells them the significance.

'I advise you to join your comrades as the man has gone to get a gun.'[16]

Comrades?

What comrades?

Their own train has gone. What to do?

'We thought discretion the better part of valour.'[17]

As it happens, the Turkish authorities remain big believers in discretion and are determined that, given how this battle has turned out for the Ottoman Empire, these correspondents will have nothing to cover. Henceforth, the new edict says, they are not allowed to use their cameras. The correspondents are rounded up and carefully escorted on horseback well away from the front they want to get to.

•

Today, 31 October 1912, is George Wilkins' 24th birthday. He comes up with a plan and mutters the detail to his fellow correspondents who are equally frustrated with this turn of events and agree.

Pretending to have spilled his provisions a short way back, Wilkins asks to go with his friends to retrieve them, under guard of course. Permission is reluctantly granted for Wilkins, Bernard Grant and the Irish correspondent Francis McCullagh to head back under the watch of two Turkish soldiers, who prove to be wonderfully agreeable to the idea of taking bribes! While they look the other way, the three correspondents charge off towards the front lines of the Battle of Lule Burgas.

But will they even be able to get there?

Standing in their way now is a Turkish officer, backed by soldiers.

'Are those cameras you have?' he asks, in wavering English.

What can they do but tell the truth?

Why, yes. Yes, they are.

'You hope to get pictures,' the officer replies. 'You will more likely get Bulgarian bullets and sabre. You and your cameras are cursed and I pray to Allah you may never come back.'[18]

With which he steps out of their way.

Charmed, we are sure.

It proves to be just the beginning and, as Wilkins will describe his experiences at this time, 'I have been held up twice by deserting soldiers, who were starving, and once by a couple of Greek bushrangers, who were robbing the dead as well as the living.'[19]

These are grim and desperate times, as every hour of every day brings more life-and-death drama.

'We saw the breakdown of their organization,' Grant writes, 'both military and medical. We saw an army in the grip of cholera, when the dead lay thickly scattered over the countryside, since the bodies accumulated more quickly than they could be collected and thrown into pits.'[20]

The strangest thing?

The weather and aspect could not be more glorious. The sun shines, the birds sing, butterflies flutter by, and men die. Inevitably Wilkins and his companions gravitate to the sound of the guns.

'It came in continuous shocks of sound, the crash of great artillery bursting out repeatedly into a terrific cannonade,' Grant chronicles. 'It was obviously the noise of something greater than a skirmish of outposts or a fight between small bodies of men. That thunder of guns made my pulses beat, throbbed into my brain. I could not rest inactive and in ignorance of the awful business that was being done beyond the hills.'[21]

Following their complete rout at Kirk Kilisse, the Turks are falling back towards Constantinople, heavily pursued by the forces of the Bulgarian Third Army, soon to be joined by the Bulgarian First Army, as no fewer than 250,000 soldiers lay siege to the shattered remains of the Ottoman Army, some 100,000 men in all.

By the time Wilkins and his fellows arrive, it is the second day of the battle, and they are able to set up in relative safety on a hilltop overlooking the battleground, only to be sickened by what they see.

'Even to the commanders of the army corps engaged,' Grant records, 'it was a wild and terrible confusion of great forces hurling themselves upon other great bodies of men, sometimes pressing them back, sometimes retiring, swept by a terrific fire, losing immense numbers of men, and uncertain of the damage they were inflicting upon the opposing troops.'[22]

Wilkins' camera captures an entire Turkish artillery battery being destroyed, bloodied limbs and organs flying in every which direction, their blood turned to mist, a cloud of red and brown dust where men once stood. And now a Turkish machine gun mows down an entire line of advancing Bulgarian infantry. This is not war from storybooks, of men on horseback dashing back and forth. This is industrial-level slaughter of hundreds of men at once, soon building to thousands, as mechanised warfare truly takes hold.

On this single day, as the correspondents' cameras roll and snap, as their notebooks fill up with the shocking detail of it all, the Turks lose no fewer than 10,000 men as their fabled empire slips away on a sea of blood after four centuries of rule.

Soon enough, artillery fire is turned on the correspondents themselves, likely from the Bulgarian Army thinking the men are spotters for the Turks. For the first time, George Wilkins is surrounded by explosions as cruel shards of shrapnel whistle by his ears, the ground itself seems to be jumping all around him and the air is filled with billowing acrid smoke as the fog of war becomes all too real.

Inevitably, the Bulgarian shells will move from near misses to something worse. 'I didn't hear the one that lifted me,' Wilkins will recount. 'I had heard the one screaming as it passed scarcely over our heads, and another one that burst on the slope just below us, but as I painfully picked myself up from the dust, I saw that the third shell had buried itself directly beneath my camera.'[23]

It's a dud shell!

If you didn't know better, you would swear – I mean *swear* – that he was born under a lucky star that simply never stopped shining on him. As those shells precede the Bulgarian cavalry massing at the bottom of the hill, it is time to beat a hasty retreat down the other side, where the correspondents soon join the shocking retreat of tens of thousands of defeated Turks – ragged, bloody men, many missing limbs, others seemingly held together by tied rags used as bandages – heading back along the same road. The sickly-sweet smell of freshly spilled blood fills the air.

Wilkins cannot help himself, as his camera whirrs again, taking frame after frame of this catastrophic scene of total defeat, even as he takes mental notes for later entries in his journal to chronicle the horror of being in the middle of this raw tide of human detritus.

> Here and there one would see a man stoop and lift up his comrade and help him along a yard or so, few words were spoken, was there need of any? There were many dead to be seen with their hands still clasped or arms entwined . . . Starved and blood-stained men, supported by each other's arms, tottered by, and a man with one arm hanging by a shred of skin, and supporting a comrade who hopped along on one leg (the other was shot away at the knee), dropped almost at my feet and expired with animal-like resignation. The one-legged man fell with him . . . dragged his almost lifeless body to face the east and fervently prayed to Allah. Was this war? This the reward for men who fought? There was no thunder of guns to stir one's courage, no shout of triumph as a

gunner laid his gun anew. Would we ever get out of this dejected, downtrodden mess?[24]

As the daylight ebbs and the darkness deepens, still there is nothing for it but to push on, frequently passing sodden and sorry groups of straggling men. The odd bundles of bloody rags by the road show where – if they are together – some soldiers are taking a kip for the night. If the bundle is on its own, it is more often where a sole soldier has died and been left by his comrades. But nearly everyone keeps going if they can, stumbling and staggering in the gloom, conscious that the Bulgarian barbarians are still pressing close from behind and will surely engulf them if they don't push as hard as they can. Ideally, they will be able to make it back all the way to the safety of Constantinople, but at the very least they must make it to the Catalca Line just 30 miles out of their home city, where the word is the Ottomans are going to make their next stand.

It is left for Wilkins' companion Grant to compose the bitter epitaph for the day they have just witnessed: 'We saw regiments of young soldiers in panic being rounded up by cavalry, and we heard their cries of anguish and fear. We saw them slashed mercilessly with whips and the flat of swords, and we saw those cowed and hungry men continue their flight in spite of all ... Many of the men had had less than a week's training, and their defeat became certain.'[25]

In strict contrast however?

'Nothing could have been finer than the bearing of the trained Turkish soldier. He fought to the last, and, when the time came, retreated with quiet dignity. The army was beaten, not from lack of bravery, but by being clogged by its own masses.'[26]

Wilkins, and his two in tow, do what they can to keep it together. They walk, leading their horses by the reins to spare them the weight. They nod glumly at the men they pass and are regarded in turn, brief glances and tiny tilts of the head that speak of men who are on the verge of collapse. But many of them don't even notice Wilkins. Their heads are turned down to their feet, their hands loosely gripping the belts of the men in front of them, an impromptu chain of exhausted souls leaning on one another, their collective will all they have to just take one ... more ... step. But it is not enough for some, and every now and then someone will give up, let their hands fall limp and their legs give out, and collapse in a pile of dust and sweat. They lie there, gasping for breath until the leader of the next line picks them up and convinces them to join their own catastrophic conga line. And if they can't? If they simply don't have the energy? Then they are at least moved off the road, left to die by themselves.

Off to their right and left, occasionally, there are the lights from the lanterns of homesteads, but so many seek succour there that those lights soon go out as they either don't want to be besieged any more, or have given what they can.

As devastated as this mess of defeated men is, however, they can still be collectively dangerous to correspondents who are perceived to have the sustenance the poor brutes so desperately need. When at one point several of them together loom in the gloom, clearly about to rush but uncertain who should go first, Wilkins is quick to pull his revolver.

Pointing it at the closest man he yells: '*Yock!*'[27] meaning 'Nothing here!'

It is a stand-off.

It is obvious to all that the soldiers could kill the correspondents . . . but that the correspondents could equally kill the first soldier who tries to rush them.

For the moment both sides recede to work out the next move. For Wilkins, Grant and McCullagh the issue now is to be able to sleep, and using the full darkness as cover they are able to slip away from the mass. Pausing long enough to make sure they are not being followed, they keep going until they come to the village of Hadem Roui. Now, given that an officer backed by several soldiers prevents them entering the village proper they must make do on the outskirts, stopping, as Wilkins will recount to a friend, at 'the first clear patch of ground I found'.[28]

Given that their horses are too tired to go further, and it is starting to rain hard, they have no choice, even though they are aware of a pervasive and deeply stomach-turning smell that fills the night air, mixed with an eerie silence that is more than a little haunting. Still, so exhausted are they, it is not long before they drop to sleep and it is only when they wake at dawn they realise just where it is they have taken their rest.

All around them, the first rays of the sun reveal, are dead bodies wrapped in white shrouds! These are soldiers who have died of cholera, and have been placed here in what is effectively a quarantined graveyard, it is just no-one has had the energy to bury them yet . . .

Jesus Christ. Mother of Mary.

What now?

Well, while Wilkins has been blessed with a remarkably strong body, a spine of steel, an exacting set of eyes, and the ability to take adversity on the chin, not least of his attributes is a remarkably stiff upper lip.

It is at *least* as stiff as the stiffs that surround them. So where better to have breakfast than right here? No-one in their right mind would come within a mile of here, and they can at least eat their macabre morning meal undisturbed. True, it is more than a little odd to be wolfing down boiled eggs and bully beef amid so much deathly devastation, but needs must, pass the salt. Alas, even while they are eating and sipping their black tea, the army undertaker arrives with bullock wagons full of 'dead bodies, some naked, some half-clothed, and all jumbled up in heaps'.[29] One by one they are hauled stiffly from a truck and plopped into the shallow trenches that have been prepared as mass graves.

It is, Wilkins records, 'the most gruesome breakfast hour I have ever spent,'[30] and yet the most bizarre thing of all is that 'it did not affect us so much'.[31]

Beyond everything else it is remarkable how quickly they have become used to death.

'I saw over one hundred corpses buried in less than two hours, but that was only the start of the cholera . . . [Soon] the streets, the ground around, and even the railway-station were littered with dead and dying men. The official report of the deaths was two thousand in one day.'[32]

Sights that, just weeks ago, would have turned their stomachs to the point they really would have lost their lunch, now leave them completely unmoved.

Pass the sausages?

At least they soon find themselves back with the main body of the retreating Turkish Army and are able to navigate towards the tightly controlled 'Correspondents camp' where they meet up with their comrades, who are as ever fighting their own war to be allowed near actual news. Even now, however, their problems are not over. For one thing, with the vengeful Bulgarian Army still pushing hard at their rear, they may be overtaken at any moment. For another, it is so cold in this particular part of Turkey it surely must be true that the local wolves are killing the sheep for their wool. Though Wilkins and others still have sleeping bags, their tents are long gone, which presents a real problem – for with no protection from the bitter wind it will be impossible to sleep. Typically deciding to take matters into his own hands, Wilkins and the photographic correspondent for London's *Central News*, Herbert Baldwin, start out to look for wood in order to make a fire, only to be immediately arrested by a sentry.

The bad news is that they are frog-marched straight to the tent of his commanding officer. The good news is that, while the sentry goes in to report what has happened, they are momentarily left outside the tent where . . . they spy a pile of wood!

'*Gel bu yol!*' 'Come this way!'

The sentry is back, but the job has been done . . .

Taken into the tent, both correspondents must stand there as the officer gives them a good dressing-down, and both take it well enough, though curiously keep their hands tightly inside their greatcoats. As expected, there is no particular punishment other than to be marched back to their original camp.

'He took us to the officer in charge of the correspondents camp who reprimanded us for leaving,' Wilkins will recall, 'but who nevertheless enjoyed the warmth of the fire made with the wood and laughed with us when we told him how we obtained it . . .'[33]

Even after returning to the safety of Constantinople, Wilkins and a couple of correspondent friends simply cannot resist taking just one . . . more . . . shot. *Once more unto the breach, dear friends, once more.*

Whispers on the wind tell him that a terrible battle is soon to erupt on the front lines of Catalca, and it would be a shame for such a fight to go by without a photo taken. Wilkins, Bernard Grant and the correspondent for the *Daily Mail*, Ward Price, decide there's only one thing for it – they must head off.

But wait! Why not take with them this Maltese fellow of their acquaintance, a rather cloying, clinging fellow with the unlikely name of Godfrey – previously the first *dragoman*, interpreter, of the British Embassy, and claiming to speak seven languages. He might be just the man they need to ease their passage, both through his linguistic abilities and his connections.

'His manners were perfection,' Wilkins will recount, 'and he was well known socially in Constantinople, a member of the best clubs.'[34]

It is Wilkins who takes it upon himself to make the approach.

'This time Mr Grant, Mr Price, and I are going into the front lines,' he tells Godfrey. 'We can take only one interpreter. We can use you, if you will come along, but I warn you it's going to be dangerous.'[35]

Godfrey is delighted and not only assures the Australian of his own bravery, but 'even swore that he'd rescue us if we were captured by the Bulgars. Yes, oh, yes! There was nothing he would not do.'[36]

They race towards the story, towards trouble wherever it may be heard or seen, to the front line for the front page. How do they know when they are getting close?

'The shells were bursting in the sky,' Grant will recount, 'and the roar of artillery fire shook the ground and came in great concussions of sound . . .'[37]

For the correspondents it is manna from heaven.

'The sight of those shells made us greedy for a closer view of the battle.'[38]

Not so, Godfrey, who starts to quiver and shake, before coming up with all kinds of reasons why he cannot go on.

Finally, he says, he feels 'exactly as he did at Lule Burgas', the previous battle where the correspondents had witnessed him being somewhat 'white-livered'.

'There is no need for you to tell us that,' says Wilkins, and Grant notes the Australian 'was beginning to get very angry at this revelation of cowardice.' Wilkins continues, 'We know exactly how you feel.'[39]

Deeply annoyed, Wilkins has no sooner set up his camera to take some footage of the batteries in action – 'showing men, guns and geysers of earth where the Bulgarian shells were striking'[40] – than a Bulgarian shell lands, 'blowing a gun to smithereens, as well as several men and horses . . . The sight was ghastly enough to upset anybody.'[41]

For Godfrey – who has turned an interesting shade of green – it is the last straw, and he outright refuses to take another step towards the front lines, even though the correspondents have remounted their horses and are ready to go.

'Get up on that horse and come along!'[42] Wilkins roars.

'We'll be killed!' Godfrey whimpers back, grabbing the reins of Wilkins' horse and refusing to let go. 'By Allah, we'll all be killed!'[43]

Restrained by the heavy camera under his arm, Wilkins cannot break loose from him and becomes more than frustrated. We have come so far, and are now so close, you *must* continue with us, Godfrey.

But Godfrey will not move, and is seemingly convinced that, without an interpreter, they will have to return with him. In this he is mistaken as he is firmly told that if he leaves them, he will be leaving behind his horse, and will have to *walk* back to Constantinople. Surely he will see reason, and accompany them further?

Though Wilkins first coaxes him, then orders him and finally *threatens* him with terrible punishments, nothing will change his mind. It takes a lot for the generally mild-mannered Wilkins to lose his temper, but this is one such occasion.

'You are a coward!' he thunders, pulling his revolver out and pointing it at him. 'And you will have to pay the penalty!'[44]

Still Godfrey won't change his mind! Instead, he runs first to Ward Price, then Bernard Grant, 'but obtained no comfort in that direction either. He ran about from one to another like a frightened sheep.'[45]

Will you change your mind, do your duty, and accompany us, Godfrey?

Godfrey will not!

With shells continuing to land nearby, the air filled with smoke and screams, the atmosphere is conducive to rash action. For the first time in his life, George Wilkins points his firearm at another human and pulls the trigger!

It is Ward Price who saves Godfrey, riding up and, with his riding crop, striking Wilkins' wrist at the last moment.

'The bullet went wide,'[46] Wilkins will recount.

'We'd better go ahead without an interpreter,'[47] Price says simply.

Fair enough.

'I was too angry to care what happened, but I knew Price was right, and we were wasting time.'[48]

And yet . . .

> I have always felt to this day that I would not have been sorry if I had shot that coward that day because the fellow was after all of a type that is most contemptible. Quaking with fear he dropped to the ground and, letting him remain there, we rode on towards the front.[49]

And so a none too fond farewell is bid to Godfrey, and Wilkins enjoys the thought that he shall never see him again.

(Wilkins will later hear of Godfrey being 'discovered in one of the best British clubs in Constantinople, relating the wonderful services he had rendered me and telling of his bravery in action'[50], including saving the Australian's life, whereupon, one of Wilkins' friends, 'who happened to be there at the time, and had already heard the true story from us, took

Godfrey by the neck and the seat of the pants and threw him bodily out of the room'.[51])

Wilkins and his fellow correspondents have been out amongst the chaos of the battlefield, and bleak realities of mass retreat, for weeks on end now, and are all eager for this to end. Oh, what a relief it is to arrive back in Constantinople, to at last walk into the foyer of their hotel, to at last be safe to . . .

Oh! *Oh!* OH!

Shouts from a corner of the foyer.

It proves to be Tom Grant, the brother of Bernard Grant, who had received reports of Bernard's death which, like Mark Twain himself had once famously commented, 'had been greatly exaggerated', and had immediately come to Turkey in the hope of retrieving his brother's body. And now here he is! Along with another walking corpse, Wilkins, also duly reported dead but now watching the fraternal reunion with delight, as Bernard's brother gapes.

'For a moment,' Grant will recount, 'I believe he thought he was seeing a ghost. Then, I remember, he relieved his feelings by telling me that I needed my hair cut. He was right. It had not been done for many weeks, except for a hopelessly inefficient attempt by Wilkins with a pair of nail-scissors.'[52]

Delighted to find each other, the brothers retreat to the bar, for many, many drinks.

George Wilkins decides it is equally time to consolidate, and take stock. He has had an extraordinary time of it in these last three months and by his count has been arrested on no fewer than seven occasions, while also having had a lot of his exposed films destroyed. And yet, all up, he has now been able to get through to London no less than 900 feet of precious actual battlefield films, with more to be sent shortly to Gaumont, and static shots and moving accounts for the *Daily Chronicle*!

Time to rest, perhaps?

If only.

For soon after arriving back in Constantinople he is confronted with his next life-or-death situation. Back in his digs he is approached by a man he knows, Aziz Pasha Vrioni – an Ottoman nobleman who is secretly part of an Albanian revolutionary group – and he has a proposal. We have noted your capacity to get to and from the front with speed and efficiency. We want you to take a secret message to the Turkish general, Nazim Pasha, Ottoman Army Chief of Staff, who is leading his forces in a battle against the Bulgarians some 70 miles to the west of Constantinople.

Wilkins pauses. There is a very fine line between being a correspondent chronicling affairs, and a quasi-combatant participating in them. This clearly crosses the line.

On the other hand?

On the other hand, taking the message, with specially provided papers giving him authorisation, would guarantee that he could quickly get to the front – leaving behind the stultifying life of a correspondent waiting for government releases in the safe cities – and, in Wilkins' book, nothing else counts.

So, yes, he agrees to take the secret message forward.

Wilkins will still be recording events with his companions, but must look for the opportunity to get to Nazim Pasha. Like moths to the flame, all together, they follow the sound of the guns and at dawn of 16 November, rising before daybreak, they sniff the wind like dogs on a scent. For now it is not just the sound of the booming guns, the correspondents are close enough to smell the acrid whiff of cordite that goes with shells exploding nearby.

While the Ottoman Army makes a drive to hold the Bulgarians on the outskirts of Chorlu in two days of fierce fighting on 17 and 18 November 1912, even this eventually falls to the Bulgarian juggernaut. While some of their fellow correspondents decide to head back to Constantinople for safety's sake, Grant, Wilkins and Price agree to carefully push forward.

Alas, alas . . .

As it turns out, the secret message Wilkins has been carrying all this time proves not to be very secret at all. For no sooner has he arrived near the front line at Catalca and started to hand crank his camera to film what he will later describe as the 'first trenches used in modern warfare'[53] – while Grant and Price have moved to a nearby village – than he hears the gallop of horses coming from behind. It is two Turkish cavalry horses, bearing two Turkish cavalry officers who are looking rather grim. This is borne out when, just two tics later he is arrested as . . . a Bulgarian spy!

His camera is thrown to the ground and Wilkins is searched. No message is found, which is no surprise to Wilkins as he had disposed of the message as soon as he had seen the two men approaching with intent. That is the good news. The bad news is this means the Turks need a confession from him that he truly *is* a spy.

Wilkins declines to oblige. He insists that he is not only not a Bulgarian spy, but he has been on a mission to pass a message to Nazim Pasha.

His captors refuse to believe him. Pasha is Chief of Staff for the entire Ottoman Army and it is unbelievable that a Westerner from . . . *Australia, he says, wherever that is* . . . would be entrusted with getting a message to such an important man.

So, very well then, have it your way.

After a 'court martial' back at the Selimiye Barracks in Constantinople, which mainly consists of Wilkins being yelled at by a Turkish captain in a closed room, he is given his sentence.

Yarın şafakta vurulacaksınız. You are to be shot, tomorrow at dawn.

Under the circumstances it is little surprise that Wilkins does not sleep well. Is this really it? Has he survived so much, come through so many scrapes, only for it all to end here, in a pile of shattered bones and blood on Turkish cobblestones? Is there no chance of escape? What will his family say when they find out? Will they find out? Or is he about to disappear from the face of the earth, shot and dumped in an unmarked grave, with his family left to wonder what had happened to him?

The dull lustre in the east that is the dawn of a new day actually spells sundown on his life, and the beginning of black eternity.

For in the near distance now, there is the creak of gates, the clink and then clank of keys going into cell doors that are now opened, the tramp of hobnail boots coming down the ancient cobblestone passages, getting closer.

Kalkma zamanı. Get up.

And so it has come.

Scarcely believing that this can be happening, Wilkins is blindfolded, manacled and, with other prisoners, led down the passages out to the prison yard.

Can this really be happening?

Many of the other prisoners are weeping, wailing, and uttering prayers to Allah for their salvation. Wilkins' hands are bound behind him and then around one of the many shooting posts that he had noticed lined up against the wall on the day he had arrived.

There are shouted commands in Turkish, and the sounds of rifles being swung up, and bolts being worked. And now, a thickly accented interpreter passes on in English, complete with hisses, the barked words of the Turkish officer, a Captain, commanding the whole operation. 'Will you confess to being a spy and tell all you know? If so, your life will be spared.'[54]

Oh, really? Confess to a capital offence – spying – and they *won't* shoot him?

It sounds unlikely. And why give the bastards the satisfaction? Wilkins declines.

Very well then, you will be shot.

Orders are given. George Wilkins, beloved thirteenth and last child of Harry and Louisa Wilkins, of Oxenbould Street, Parkside, Adelaide, braces himself for eternity.

A furious fusillade of shots echoes against the walls of the prison.

To Wilkins' left and right he can hear the gurgle of blood bursting from suddenly shattered hearts. But he is untouched.

Led back to the cell, he is told he is to be given another chance and will be shot on the morrow unless he confesses between now and then. Think about it.

Wilkins does and decides that they are trying to break him. Sure enough, the following morning they come again, but this time there is no blindfold and he is marched out to the posts with two other prisoners, where the same scenario takes place, except this time he can *see* the other two as they are shot.

And on the third day, he is risen.

And taken to the courtyard, alone, with no accompanying prisoners on death row.

This is it.

The Captain who had court-martialled him now appears, intent on seeing the sentence he has pronounced actually carried out. First though, he wants his prisoner to know that this time there will be no mercy, no last chance, that he is about to be shot, and rightly so, for this is what we do to spies. Behind him, orders are being barked once more, rifle bolts being pulled back. The Captain is not whispering, he is shouting, just as he had shouted at the sentencing.

Death to the infidel!

'He was so blustering and threatening that I almost faltered,' Wilkins will recall. 'Yet something inside me made me hold out.'[55]

The Captain steps back.

Ready . . .
Must be bluffing.
Aim . . .
Good Lord.
How does it go again, oh Mother?

> *O Jesus, full of truth and grace*
> *More full of grace than I of sin,*
> *Yet once again I seek Thy face:*
> *Open Thine arms and take me in . . .*

'I very nearly did collapse, but managed to cling to the post for support and, with the last fragment of my will, kept my knees from buckling.'[56]

He cannot see the rifles raised, but he can feel each muzzle pointed precisely at his form . . .

. . .

No shots are fired.

With a final barked command, the firing squad lower their rifles.[57]

But now things get stranger still. For to Wilkins' astonishment he is now led right to the Turkish General he has been seeking all this time, Nazim Pasha, who has been waiting nearby.

All that, all that horror and pain and fear and doubt, just to get a measure of Wilkins, to see if he was worth his salt. They could have just asked.

But as it happens, there is more good news – though it's hard to top the information that you are *not* about to be killed. The war is finally over. An armistice has been declared and peace talks are in progress. The secret message that Wilkins was ready to take to his grave is now worthless.

As for why the war is suddenly over, it turns out Wilkins and Grant had unwittingly had something of an international scoop when they had accidentally slept in that makeshift graveyard of diseased corpses. For as it happens the weapon that ends the war is biological: cholera. The disease has so devastated both sides that they have lost the will to fight on, and this treaty follows in the wake of the wakes.

Despite his last-minute reprieve, Wilkins is downcast as his imprisonment has meant he has lost the chance to shoot the end of the war. He feels he has cheated Gaumont of footage they will be waiting for.

Soon after, when he arrives, at their request, in Paris and hands his footage over to their office he then disappears to his hotel to sleep the sleep of the dead, and the dead exhausted. But he has a surprise the following morning. For his bosses are delighted with all of his work; they are so enthusiastic that Wilkins can barely believe it. He feels that his film is fair at best – he did not see anywhere near as much action as he would have liked. The Parisians are determined to change his mind on this matter and remove his modesty with reality. His work is extraordinary.

'To convince me, they immediately collected all the films they had and had them shown on the screen for me. And when I had compared my work with the others, I was better pleased with myself.'[58]

Typically, Wilkins has entirely underestimated just how remarkable both he and his work have come to be regarded. It is true that his framing of shots is not particularly out of the box, and while photography has by then already come to be recognised as an art form, no-one is nominating him as Rembrandt. But, against this other work? He stands out. He shoots his camera like an expert sniper and rarely misses, taking risks that none of his competitors can get close to, getting right in the middle of actual battles, and doing such innovative things as holding his camera with one hand on his left shoulder, while sitting astride a speeding motorcycle and accelerating with his right hand, to give the viewer the feel of a battle charge. He has got so close to the soldiers, he has been able to put the viewers right in the middle of a battle, capturing the soldiers' fears before they go over the top, the tight brotherhood of blood that bonds them, their blood-lust when on the winning side, their devastation when the battle has gone against them and they have lost comrades. All up, he has produced an arresting portrait of the horrors of modern warfare that he is very nearly unaware of having painted. To Wilkins it had all just been another day with his camera. But to those used to viewing 'battle recreations' and footage shot from a boringly safe distance, it is nothing less than extraordinary.

When the lights go up, Wilkins is even told that the French founder of the company, Monsieur Leon Gaumont himself, would be *enchanté* to make his acquaintance, and the meeting occurs very shortly afterwards. In halting English, Gaumont lets Wilkins know how impressed he is with his work.

'I would talk business with you,'[59] Gaumont pronounces, but proves it by showing rather than telling. He reveals vibrantly coloured photographs the likes of which Wilkins has never seen before. Oh sure, he spent many afternoons as a young lad helping to colour in slides and give them the illusion of coloured stills, but this is different. This is real.

'You are just the one to learn all about it!'[60] says Gaumont excitedly. 'Improve on it if you can!'[61]

Wilkins accepts the challenge and another roving assignment, 'to go to different parts of the world in order to present it to the public and teach the different managers how to use the various machinery necessary'.[62]

Wilkins' first major assignment will start in a few weeks and will certainly be a pleasure, as the 'work' will be to travel to the paradise of the Caribbean and capture its delights on behalf of the Cadbury Chocolate Company. Yes, Cadbury want chocolate lovers to see the beautiful islands where their cocoa is harvested; with as many lovely locales and lovelier maidens as possible added to ensure that this film will be a 'short' in British cinemas.

Curiously, it is chocolate that is just about to lead to his next brush with fame high over London . . .

Christmas Eve 1912, London, here comes Santa Claus

True, this is not your regular kind of job for Gaumont.

But Wilkins doesn't care.

His mind is already on the Caribbean trip he is to enjoy, and he is happy to do whatever job Gaumont has in mind for now, particularly if it is an exciting one.

So here's what we want you to do, Mr Wilkins. We want you to take your camera, and film Santa Claus.

Yes, Santa Claus.

We want you to go to the gasworks at Battersea and film Santa as he takes off in a hot-air balloon, just before he sails over Hyde Park, two miles to the south, where the man from the North Pole will parachute down to deliver Sandow's Chocolates to the children waiting below. The whole thing is a joint project between Gaumont and Sandow's and for days, the London billboards have been blaring, 'Santa Claus has ballooned from the Pole to London. See him drop from the sky!'[63]

Got it?

Got it.

As it happens, 'Santa', in this case, proves to be a fellow Australian of eccentric disposition and minor London fame – born as Vincent Taylor, for such occasions as this he goes by his stage name: Captain Penfold.

Quite what he is the captain of is never made clear, but it doesn't matter. The title goes perfectly with his splendid appearance. Any man with a moustache as magnificent as his – so waxed, so perfectly trimmed and curled – *should* be a captain, or it would be a real waste.

As it happens, the other man to get in the balloon is yet another Australian. Frank Spencer is not only the owner of this 45,000 cubic foot capacity balloon, but has all of his life-savings wrapped up in it, and it will be his job to pilot Santa over the park, where the children now await.

The balloon itself is a beauty, a grand and billowing canopy that rises above them, held aloft by hydrogen and attended by workers who busy themselves with bits and bobs and ballast bags around the base of the basket.

By 12.15 pm, all is in readiness.

On Donner, on Blitzen . . .

As Santa climbs in, Wilkins lets his camera roll, careful to get good footage of the sack of Sandow's on Santa's back. And yet there is one problem. The growing gusts of wind, of varying direction, threaten to take the balloon every which way the moment they are aloft, and it will blow the basket too much from side to side.

It needs more ballast.

It is Wilkins who comes up with the answer.

(Or at least *an* answer.)

'Why don't you take me along? I can add the weight and take pictures from the air.'[64]

Spencer, trying to hold this bucking bronco of a balloon steady, takes only a split second to reply. 'Come on, Bert. Hop in!'[65]

It all happens so fast.

Wilkins put his camera over the lip and into the bottom of the basket, and now gets in himself to sit beside Santa Claus, about to be an airborne elf. Or not . . . For now, as Spencer looks up, a horribly logical thought occurs, concerning the dauntingly large lattice steel framework of the gasworks high above them. Look, if the balloon is currently beneath the tank, and the tank is directly underneath the lattice, and the balloon is slowly getting bigger . . . then . . . bloody hell!

Desperately, Spencer shouts out, 'Boys, unless you give the basket a mighty heave, she'll never clear.'[66] And now all is action and every man nearby heaves, all ropes are released and they shoot upwards . . .

'Up she goes!'[67] yell the men and, indeed, up, up, up they go . . . right into the lattice!

'The basket caught,' Wilkins will recount, 'almost capsized, and it was a miracle we were not thrown to the ground. My cameras were strapped in, but I was photographing with one when the jolt came. I was flung halfway out of the basket, but Frank, hanging on for dear life with one hand, grabbed my collar and prevented me from falling. It was a narrow escape.'[68]

After 'a few minutes of desperate struggle we managed to free the basket from the iron framework, and lurched clear'.[69]

And lurch is clearly the right word, as – *hey, ho, and up she rises!* – suddenly freed from these earthly bonds they now shoot skywards as if shot from a cannon. As the balloonists dare to look over the side, what had been houses and fields become small squares of red, and larger squares and rectangles of green, and even the mighty Thames itself is reduced to being little more than a lazy ribbon left fluttering in their trail to the . . . north-east.

But hang on, it *should* be to their north, and they should be getting closer to it.

What is going on?

Well, as any experienced balloonist would know – although sadly, no-one truly answering that description is on board – sometimes the wind changes direction as you rise, and the surest guide of where one is going is the clouds on high, not the way the ribbons on your balloon are going when you are on the ground. Well, it's too late now as the houses below rush by in a blur, until the balloon rises into such heavy clouds, 'they blotted out every sign of the great city down below'.[70]

The balloon shoots over Westminster Cathedral at a rate of 45 miles an hour, now aloft at 1200 feet.

Optimistic to a fault, sanguine to the point of stupid, Captain Penfold is a man with his head permanently in the clouds. And on that note, why don't we rise to a higher altitude above those clouds where we might have a better chance at seeing our target? Without a second thought ballast is thrown out and the balloon lurches some 2000 feet higher, up into the thin sunlight. Ah yes, they can see it rather clearly now: they are lost.

Such a life-and-death situation makes George Wilkins think of his parents and he now takes the time to dash off a quick note to them, possibly to be delivered posthumously, he is not sure, but by instinct he strikes a cheerful tone.

> *Dear Parents – You will never guess where I am writing this. It is in a balloon 3,500 ft. above London. We are still going up. It is lovely up here above the clouds which look like a great rough sea of snow – 10 minutes later – We are going down a bit to let [Captain] Penfold jump off. I must get my camera ready. Goodbye. Love, from George.*[71]

By the time he has finished, things are grim and getting grimmer. Yes, Frank Spencer is loudly despairing as it is clear to him at least that he has lost all chance of fulfilling his contract by dropping Santa Claus over Hyde Park where hundreds of children are awaiting their 'palatable, digestible, economical'[72] Sandow's chocolates.

Captain Penfold, however, is not one to despair. True, he is slightly disappointed at the way things are turning out, and even confides to Wilkins that he had been hoping to swing beneath the basket and do a trapeze act for the children. It may well have been a lovely touch, but as a now deeply annoyed Frank Spencer patiently explains – because Santa clearly doesn't get it – 'It is obvious wherever you jump, it will not be in Hyde Park.'[73]

For God's sake, we passed into Essex a while back, and we are now very likely somewhere over Chelmsford!

Ah, but Santa has a very effective reply.

Without a single word being spoken, Captain Penfold swings his legs over the side, nods to Wilkins, looks Spencer in the eye, and . . . securely holding his sack of chocolates . . . *jumps!*

For a split second, Wilkins and Spencer – somewhere in that strange triangle formed by being aghast, aggravated and amazed, all at the one time – might have been inclined to look over the side to see if his parachute opens. But for the moment there is no time.

For, the basket, released of a third of its human weight, surges upwards and 'what with the relief from his weight and the mighty shove he gave as he leaped into the air, the basket tipped and swung violently. Frank and I almost went over the side with him – and we had no parachutes.'[74]

Could there ever be a more ludicrous way to die?

Here lies George Wilkins. Born in dusty South Australia. Killed in England by Santa, on Christmas Eve.

Mercifully, they stabilise. And now, far, far below, in the tiny break in the clouds Penfold has taken advantage of – to know that he had land and not sea beneath him – they can see a parachute billowing, and a crazy quilt of fields beneath him.

Well, he should be all right.

And them?

That remains to be seen!

As Saint Nick had jumped with no warning, 'we were up to fifteen thousand feet, with the gas release cord whipping violently as we swung, and so entangled in the shrouds above us that the gas valve could not be operated'.[75]

And so they go, up, up and awaaaaay with both the up and the awaaay being equally life-threatening. There is nothing to do, Frank advises Wilkins with fear in his eyes, but wait until the gas cools.

And how long will that take?

'We might be aloft twenty-four hours, perhaps thirty,'[76] Frank estimates.

And where might they be at that point?

Who knows?

Right now they are at . . . 22,230 feet!

Call it very bloody high, for short.

'We shivered for hours and rubbed and slapped ourselves for warmth, except when we were in clear sunshine now and then as we swung from cloud top to cloud top.'[77]

By the time the clouds break, Frank Spencer and George Wilkins realise that they have been carried out over the North Sea, which they privately note is likely to become their watery grave.

Their only hope is to hang on and ride the wind.

•

It's Santa!

The kids of Little Baddow, just outside of Chelmsford, 30 miles north-east of Hyde Park, can barely believe it. One moment they had been playing hare and hound in the fields and the next thing a shout had gone up and this billowing ball of Santa had suddenly landed just near them, an instant before he had been all awash in the cloth of his parachute.

The strange thing?

Santa is sleeping!

Wake up, Santa! Wake up!

Wake up, Santa! Wake up!

As luck would have it, Captain Penfold has landed on his head, easily the thickest part of him.

Wake up, Santa!

He wakes, and groggily stands up, rubbing his head.

'Don't forget I want a popgun tomorrow!'[78] says one child earnestly, oblivious to the situation at hand.

'I want a drum!'[79] calls out another. Well, young child, right now Santa wants a brandy, but there are more pressing concerns. Part of the contract with Sandow's Chocolates was that the first man to find and shake the hand of Santa will be awarded a prize, and so it is moments later that a rather surprised Mr William Jolson is confronted by a disoriented and dishevelled jolly Saint Nick reaching out to clasp his gnarled farmer's mitts.

'I give you authority to collect the Christmas souvenir awarded by the Sandow Cocoa and Chocolate Company,'[80] Santa proudly announces.

All's well that ends well, and all that the Sandow's Chocolate and Cocoa Company cares is that the whole bizarre affair receives massive publicity, which it does; especially as two men are lost at sea . . .

•

At one point, George Wilkins is convinced that to balance his birth in one of the driest places on earth, it stands to reason his death might be somewhere right in the middle of something like the very English Channel they have been half-blown across but, mercifully, the wind had changed in the afternoon and blown them back to that green and pleasant land of England. And when the gusts die down, Frank Spencer is actually able to bring the balloon back to earth just before dusk, albeit landing in extremely rough fashion, firstly on a hard field in Essex and then being dragged along – a'bumpity-bump-bump-BUMP over rocks and 'over hedges for about 200 yards'[81] and narrowly missing a horse dragging a plough with the farmer walking behind – until they are propelled skywards once more before being dashed violently against a tree. Though the impact sees the destruction of the cameras, and Wilkins

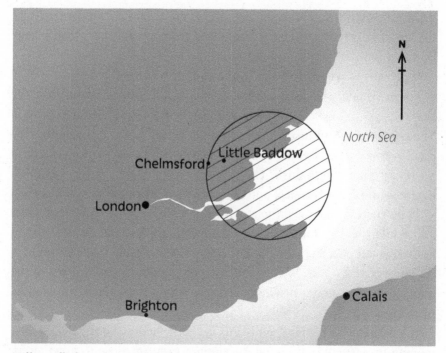

Balloon flight over south-east England, 1912

very nearly loses a finger while Spencer's hand is lacerated, little counts against the fact that they are now back in the bosom of sweet Mother Earth. Their luck has been extraordinary. They could have come down anywhere, including the North Sea, but after having been blown all over, have come down within coo-ee of London!

'At last . . . we managed to anchor the affair, deflate it and get a farmer to cart us to the nearest railroad station.'[82]

While Gaumont prove to be seriously displeased to hear that their precious movie cameras have been reduced to scrap metal, they are mortified to hear how close Wilkins had come to perishing.

'Haven't you better sense,' the London Gaumont manager asks icily, 'than to take such chances when we've got no insurance on you?'[83]

A good point, boss. Wilkins promises he will do his absolute best to stay alive in future until such times as his employers can gain insurance and be assured they will suffer no financial loss upon his death.

At least his next assignment is relatively safe and certainly earthbound by comparison as he can finally head off to cover the cocoa industry, 'from the planting of the trees in the West Indies to the dispatch of the tinsel-covered packets of chocolate from Cadbury's factory near Birmingham'.[84]

It proves, however, to be as interesting as it is confronting.

For in Trinidad, he discovers what they call 'indentured labour from the East Indies'[85] but which to his eyes looks to be not too far north of slavery. The labourers – Hindus, Tamils and others – are housed in compounds, their lives governed by extremely strict caste rules, and working six days a week, from dawn to dusk. They are paid a shilling a day for the males, nine pence for the females, and children proportionate to work done.

Only the evenings are their own, but even here the rules are strict and while there is some leisure – including games and dances – men are never supposed to see the dancing of respectable Eastern women. As innovative as ever, Wilkins borrows the makeup and dresses of the manager's sister and – disguised as a woman – films the exotic dances of the women.

It is wonderful footage, and Wilkins is thrilled as he focuses on one woman in particular, bringing the camera right up close to take her all in, in exquisite detail, until . . .

Until, he is distracted by furious shouting, getting closer.

Oh, dear.

It proves to be this very woman's husband, who had been passing, only to see his wife dancing up a storm, 'a thing naturally against the social laws',[86] and before a camera, so there will be proof of her disgrace. There will be hell to pay later that evening as he retrieves his honour by neatly slashing off her right ear with a long sharp machete.

> This was the punishment meted out by his particular caste to wives who were unfaithful. The woman screamed. Her ear has fallen to the ground and before anybody had done anything about it, one of the dogs in the compound snatched it up and ate it . . . It was almost a tragedy but as the story passed around, the woman became the heroine of the village and later proudly exhibited her earless head as evidence of distinction.[87]

Thankfully, the remaining filming is less eventful. That is until, on one sweltering hot day, so humid that not even the green and yellow parrots that abound in these parts can raise much of a cry, Wilkins has a drink with a Frenchman on the balcony of Trinidad's Queen's Park Hotel, waving away the listless curio vendors who are trying to interest them in their little carved statues, baskets and woven mats.

Wilkins and the Frenchman watch them go, while leaning back comfortably in their tall cane chairs, sipping their nearly as tall 'gin and ginger' drinks, and discussing last night's Tourist Ball, when the subject turns to a trip the Frenchman will be taking on the morrow up Venezuela's Orinoco River – just 90 miles to the south of where they now sip their gins – on a flat-bottomed stern-wheeled river boat.

'By Jove!' Wilkins says, 'I wish I were coming with you, but I hardly dare to hope for an extension of my trip and will have to content myself with a flying visit to the Northern Islands and Tobago and get back to London and submit to the vagaries of our profession.'[88]

Funny he should say that.

For even as he speaks, the latest vagary emerges in the form of Charlie the hotel clerk appearing with a cablegram from Gaumont for their most important employee.

DO ISLANDS AND TRIP SUGGESTED, TAKE INDUSTRIAL. SCENIC. [89]

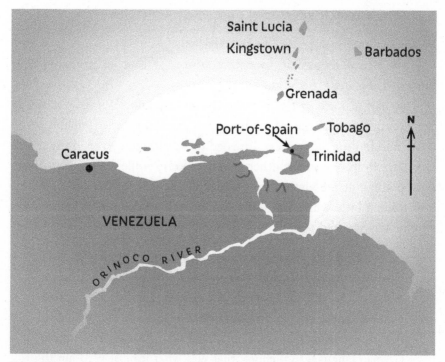

Venezuela

In layman's terms: 'Photograph by means of cinematograph camera all the most beautiful landscapes and the most interesting industries that these places afford.'[90]

Wilkins is pleased, as it means he can stay longer in this glorious part of the world. And it means he can accompany the Frenchman up the Orinoco through to Caroni, and well into the interior, offering Wilkins 'a thousand mile trip through interesting virgin country, populated with many wild tribes and affording the possibility of glorious scenic pictures not yet recorded by photograph or pen, not to speak of the lawless life surrounding the Brazilian diamond mines . . .'[91]

Both men are excited and work feverishly into the night, packing up, getting a typhoid vaccination (or at least bribing a doctor to give them a certificate to say they had), and settling accounts. And by the morrow both are ready to go, only for Wilkins to receive another shock.

For even before they have had coffee the next morning, before heading down to the wharf to board the vessel that will take them to the South American mainland, a maid enters the bedroom with yet another cablegram from Gaumont.

RETURN LONDON FIRST BOAT. IMPORTANT.[92]

Wilkins is uncharacteristically shocked.

He is a cool, calm man by nature, and it takes a lot to throw him, but this cable does, his mind whirling.

'I lay half stunned by the message,' he will recount. 'What did it mean? Something wrong at the office? Bad news from home? Or possibly some other mission for me in view. I turned these thoughts over in my mind, changing from one to the other as most likely, while feeling not a little disappointed at the alteration of my plans.'[93]

It takes a little while, but through a process of deduction he is nearly sure he has worked it out.

> It must be that I was about to be dispatched on another mission in another part of the globe. Where could it be and why? A happy thought: was not the Princess of Germany going to be married within a month or two, and was not I the only one of the English staff of operators that made any pretensions of knowledge of the German language? Of course that was what it must be, I am to

go to Berlin for the Kaiser's daughter's wedding, but this was little satisfaction as I had already seen Berlin.[94]

For yes, ever and always what Wilkins most wants is to see new things, which is why he had been so excited about visiting wild tribes in Venezuela, which would have been newsreel gold.

'Was not the trip through Venezuela that I had contemplated amidst the wild free life, my native element, and for which I long incessantly in spite of the glamour of the gay city life?'[95]

Yes, indeed.

But in the end . . .

'"*Horders is Horders*" as the sergeant said, and even if they had come some two thousand miles by cable it was my duty to obey without question.'[96]

So it is, in April 1913, he begins the journey home to London via Barbados, a long journey that almost doesn't begin, thanks to a long lunch in the hills that finishes with a mad dash by the driver weaving through a long traffic jam of stubborn mules and now – oh dear – *over* a man who was slow to get out of the way. After another mad dash to the hospital, Wilkins arrives at the dock just in time to see the steamer departing and the only way to catch up is to hire a launch.

Follow that steamer!

An amused and sardonic mate watches as Wilkins throws his bags on board and leaps off the launch to the cheers of his sloshed luncheon companions who have accompanied the Australian this far.

'You are cutting things rather fine,' says the mate, as Wilkins shows his ticket. The mate glances at the ticket again, curious, before adding, 'And, by the way the steward is looking for you with a cablegram.'[97]

What?

Hastily finding the steward, he is quick to tear open the cable.

```
WOULD YOU GO IMPORTANT [ARCTIC] EXPEDITION MEANS
TWO OR THREE YEARS ABSENCE GOOD TERMS EXCELLENT
OPPORTUNITY REPLY CABLE.
```
[98]

The Antarctic?

Something to do with the retrieval of the bodies of Scott of the Antarctic and his companions, perhaps? He is immediately interested. It would be a wonderful trip just to get there!

'London, Australia and home, New Zealand and then on into the unknown.'[99]

And yes, he had heard much about exploration in the Antarctic, the horrors experienced by the likes of Scott, the great Ernest Shackleton, the greater still Roald Amundsen, who had so comprehensively beaten Scott in the race to the South Pole, and all the men who had gone with them: 'the tales of frozen feet and fingers turning black and falling off'.[100]

But against that?

Against that, 'how could I imagine such cold when I had been spending the last two months in a temperature never less than 80 and mostly over 100. Snow all the year round! What a delightful change.'[101]

Under such circumstances, thus, it does not take long for him to send a cable back.

Taking the pen he always has on him, he writes on a piece of paper the word YES[102], and manages to get it to a sozzled friend, still bobbing in the launch below.

'Please cable this for me as soon as possible!' he calls, even as the chug-chug-chug of the steamer's engines start to dominate and they begin to make their way to the open ocean.

There is really only one problem.

Once he gets back to his cabin and excitedly pulls out the cable again to re-read it, he realises he has made a mistake. It doesn't say ANTARCTIC EXPEDITION at all. It says ARCTIC EXPEDITION!

Of course he had wanted to go on Antarctic expeditions, as he had met the great Sir Ernest Shackleton a couple of years earlier and freely offered his services for Shackleton's next trip to the South Pole. He had thought this must be it.

But now?

> My visions of the trip through the canal, home, across Australia and from New Zealand were dashed to the ground. I did not mind the trip, but I would have to start on this expedition without seeing those at home and this troubled me most.
>
> Who was it that was leading this expedition North and why? How long would I have to spend amongst the ice and snow?[103]

On the other hand?

On the other hand, while it is true that his presumed and actual destination are, well, poles apart and the polar opposite of where he had thought he was going, what can he do? He has committed to it and that is that.

And after all, the whole mix-up, he reasons, 'suits my naturally impulsive and adventure loving nature to rush headlong into the uncertain, grappling with each fresh circumstance as I met them'.[104]

It is only once back in London that he learns from the manager of Gaumont more about the trip he has signed up for – 'The Canadian Arctic Expedition' under the command of Commander Vilhjálmur Stefansson.

'What sort of fellow is he?'[105] Wilkins asks.

'I don't know much about him' the Gaumont manager replies, 'but from all appearances he is not a "Drawing Room" explorer, but a rough and ready practical man . . . I think you will get along very well with him. He is apparently a Scandinavian, judging by his name, but he speaks with a slight American accent.'[106]

But make no mistake.

'He's a man who is going to the Arctic to travel with dogs, and he'll expect you to work like a dog, or work with the dogs and eat the dogs if necessary . . .'[107]

Making further inquiries, Wilkins learns a lot more.

Vilhjálmur Stefansson is by all accounts a fellow as remarkable as he is curious, and not a little bit magnificent – a man with Icelandic parents, born in Canada, raised in North Dakota and educated at the University of Iowa, who, after rejecting his postgraduate theological studies, rose to fame for archaeological research in Iceland and polar exploration, and most particularly for being the founder of the theory called 'The Friendly Arctic'. The theory, if not necessarily the practice, is simple. Stefansson insists that all that a capable man needs to survive and thrive in the Arctic are a gun, a knife and training provided by . . . Vilhjálmur Stefansson.

A man of Iceland to the marrow of his bones, he sneers at the whole idea of explorers who rely on dragging supplies from civilisation and insists that it is quite easy, if you are expert enough, to do the Arctic equivalent of living off the land – you live off the ice and snow.

Food? That is what hunting and fishing is for.

Special clothing?

Please. Much better to live like the Eskimos – make your coats, pants and shirts from the pelts and leather of the wildlife you've slaughtered.

'We found the native method of dressing with fur clothes next to the skin to be the most suitable for extensive travel ... I also found that a woollen mask that fitted closely to the face, having two holes for the eyes and one for the mouth and nose, a great protection from the frosty wind ... Polar bear skin or domestic sheepskin mittens are the most satisfactory in comparatively warm weather, but in very cold weather ... well-fitting dog or wolf skin was about the best.'[108]

Staying warm at night?

You burn the blubber of the seal.

It is a great theory, but many of his contemporaries in the field are more than merely sceptical, they are derisive, certain that sooner or later Stefansson will get what he has coming – and it won't be pretty. One of them, Captain Fritz Wolki, is actively and publicly looking forward to it, describing Stefansson as, 'A college professor who carries books in his sledge where a sensible man carries grub.'[109]

Just a few evenings later Wilkins finds himself in a swish London restaurant with some of Gaumont's leading executives.

'This is my third farewell dinner that I have had from the firm in less than twelve months,' Wilkins writes in his diary that evening, 'but it seems as if they can't get rid of me and I have always turned up again. I hope I will do the same this time.'[110]

The more Wilkins hears about Stefansson, the more he looks forward to meeting him. As to the expedition, it is being organised by the Naval Service Department of Canada, and the plan is to start from Nome, Territory of Alaska, in mid-July 1913, and push along the northern coast of Alaska, and into Canadian territory.

They are sent to explore the borders of the Beaufort sea, north of the American continent and west of the Parry Archipelago, to look for the last unseen land on earth, to search for that dream of explorers for centuries, the thought that there might, no, *must*, be some grand mass to the north that remains undiscovered and unperturbed.

It is a place right on the edge of the known world, with a highly scattered population of native Eskimos lightly sprinkled with a hotch-potch of missionaries, hunters, rough nuts, rougher huts, trading posts and whalers – while to the north is the vast icy wonderland that defends the North Pole which, to this day, remains unconquered and unexplored.

In this exploration Stefansson is hoping Wilkins' cameras will help prove his claim that there is a race of 'Blonde Eskimos', a sight that only the Icelandic–American has seen on his last trip to the far North. (The way Stef tells the story, he had heard from a whaler and merchant named Christian Klingenberg of Eskimos with 'brown hair and rosy cheeks and light brown eyes'.[111] Fascinated, he had gone looking for them and found this elusive people near Victoria Land and was stunned to discover a seemingly different race, 'they had no trace of Icelandic or Norwegian language. They looked like Norwegians or blonde Italians'[112] and, in fact, Stefansson has a theory they are ancient descendants of the Icelandic, a long-lost European tribe, perhaps proof that the Vikings made their way across the Atlantic. His critics believe they are yet another figment of Stefansson's imagination. Well, with Wilkins' help, this time he will have proof!)

CHAPTER THREE

POLAR EXPLORATION

'If you can dream – and not make dreams your master;
If you can think – and not make thoughts your aim ...'

Rudyard Kipling, 'If'

People often ask me why I go to the polar regions. I doubt if any
one of us can say, in so many words, just why we do this or that or
the other. I might fill volumes telling why I go to the Arctic or the
Antarctic, yet the man who 'doesn't know' would not understand.[1]

Sir Hubert Wilkins, *Flying the Arctic*, 1928

The Expedition was badly organized by Stefansson; food and
equipment was loaded helter-skelter on the Karluk. Sometimes
day-long searches of the ship for needed items failed ... The frail
Karluk was not suitable for use in the Arctic.[2]

Jim Rearden, *Alaska's First Bush Pilots*

1 June 1913, Victoria, British Columbia, dandy and brandy

That tall fellow in the top hat – no, really – gazing with wonder all
around him at the Esquimalt Navy Yard on Vancouver Island? It is the
newly arrived George Wilkins, taking it all in: the whalers, the swarthy
seamen, the distant and imposing mountains peeking above the clouds.
For most arrivals it would be an arresting sight. But for a man raised
in Hallett, South Australia?

It is in the realms of wonder as he had never quite imagined – so
other-ly, it is utterly arresting.

Now, as it happens, a clearly hungover man walks past Wilkins,
wincing as the sunlight hits his face and Wilkins' words hit his ears ...

'Can you tell me where to find the *Karluk*?'[3]

The fellow looks him up and down with an odd mix of hostility and
curiosity.

'What do you want,' he says in a Scottish brogue, 'with the *Karluk*?'[4]

'I'm going on the expedition,'[5] Wilkins replies simply. He thinks it a fairly reasonable answer. But this sea dog thinks nothing of the kind. For his money it is the most unheard of thing he's ever heard of and, after looking Wilkins up and down, he sneers: 'What? In *those* clothes?'[6]

Rarely does Wilkins feel self-conscious. He has, after all, strangled music hall melodies in front of a South Australian regional audience. He has raced naked against schoolmates, been blown over London with Santa Claus, slept in an open-air cholera morgue and stared down firing squads in Constantinople. But at this moment, yes, he does feel a little ludicrous as he realises what the sailor sees: a seeming dandy about to attend a morning party, wearing spats, striped trousers and a top hat. Certainly, he had been told to report in civilian clothes, but, on reflection, he might have overdone it a little and now realises that he must look ridiculous to one seasoned by the sea. But Wilkins is a man who can gather himself quickly, and he is quick to reassure the sailor that you are not to worry, my good man, for I shall be more appropriately attired by the time we take off. Very well then, Mr Salty has a further question for him.

'Have you any money?'[7]

'Yes,' replies Wilkins before hastily adding, as the sailor smiles broadly, 'but I *never* ask for money, neither a borrower or a lender be.'[8]

That's fine. Salty has no such qualms himself and now even *smiles*, saying, 'All right, come along and let's have a drink. Then I'll take you out to the *Karluk*.'[9]

And you might be?

'I'm Mackay . . .' comes the reply, 'surgeon to the expedition.'[10]

Good God, *this* is the doctor? And the last name, *Mackay*! Wilkins realises he is talking to nothing less than exploring royalty, for this is clearly the legendary Alister Forbes Mackay, the man who had been surgeon to Shackleton on his first journey to the Antarctic, back in 1907. Among his many claims to fame, this Boer War veteran, had, on that Shackleton expedition, been the first – with two Australians, Douglas Mawson and Professor Edgeworth David – to get to the South Magnetic Pole. He's a great man to be with at the extremes and the ends of earth, but Mackay is also a Scot who likes Scotch, beer and wine – indeed anything alcoholic, as he is likely that himself.

It is, thus, some time before the two men get around to boarding the *Karluk*, both swaying lightly on the dock, well before they actually get on board . . .

Good . . . *God*.

Yes, over a drink Mackay had intimated that their vessel was not quite the *Titanic*, but nothing could have prepared Wilkins for this.

And it's not just that the stench emanating from it could peel paint, as witness its rust-bucket nature. It's *everything*.

What Wilkins sees is: 'a cramped old whaler of 250 tons, a barquentine rigged with square sails. She stank of whale oil. The quarters were like nothing I had ever seen – unpainted, crowded, smelly and swarming with cockroaches'.[11]

God help us all.

'I wondered what sort of an expedition it would be.'[12]

It is no wonder that Mackay drinks.

And Stefansson?

Where would he be?

Not yet with them.

Mackay unveils some mutterings and musings on that matter and that man. *Now listen carefully, young Wilkins.* If a house divided against itself cannot stand, then it holds to reason that an expedition divided before it even starts will never be able to stand – much less explore – in the first place. The party is already split in two, one faction belonging to Stefansson and the other belonging to Dr Rudolph Anderson, a 38-year-old zoologist employed by the Canadian Geological Survey group, with instructions, 'To ascertain if lands hitherto unknown exist, and to definitely mark any found.'[13]

The Northern Party, led by Stefansson, is set to explore said great unknown, hunting for new land in the far North under instructions from the Canadian Navy. Meanwhile the Southern Party, led by the good Dr Anderson, is to disembark and explore the fairly well known; to outline and sketch Canada's outer coast under instruction from the Canadian Geological Survey department. One might think these equal responsibilities, but Stefansson would not count himself among their number. In his eyes, this is *his* expedition, Dr Anderson is only the understudy. Dr Anderson sees this as impertinence and arrogance, but is too busy with the task at hand to really do anything about it. For the moment, they will part ways, as their focus right now is on territory, more so than authority. Still, it is not a moment too soon as the two already seem at the end of their rope and ready to string the other up at a moment's notice. And no wonder; they are fresh from three whole years in the Yukon on another expedition. So much time together would bring out a camaraderie and solidarity in other men, but for Stefansson and Anderson it has simply highlighted that the only thing they have in common is a boundless loathing for each other – which is also the one thing they are happy to explore together.

Dr Anderson has one advantage over Stef right now; the former is on board while the latter's ship has not yet come in. It means that the *Karluk* must embark on its voyage without its fearless leader, on the promise that he will join them shortly after arriving by passenger steamer. In the meantime, Wilkins gets to know some of his fellow expedition members, for the most part finding them, more than a little . . . underwhelming. He had been expecting the whole expedition to boast people of mostly scientific bent, instead what he finds is just a few men like that, together with a handful of explorers of strong repute like Mackay, while the rest are rough diamonds at best, and just rough at worst – a collection of drug addicts, alcoholics, men of mysterious past who are clearly operating under false names, a flotsam and jetsam of chancers and ne'er-do-wells.

As to Stef himself, when he arrives not long after Wilkins, he proves to be like no man Wilkins has met before, but in many ways . . . a little like Wilkins himself. Stef is educated enough to live in an ivory tower but has no desire to confine himself in one, and displays the rare combination of high intellect with enormous physical courage. Hustling and bustling about the ship, he presents as an ascetic, energetic man, whose chief regret in life is that there are only 24 hours in a day, and seven days in the week, for there is so much he wants to do.

'I found,' Wilkins will recount, 'he was not the kind of man I might have thought him to be . . . He was a college man and of too kind a nature to expect anybody to work like a dog or eat a dog.'[14]

On the other hand, Stef is nothing if not particular in his view that there are two ways of doing things: his way and the wrong way. Things, therefore, must be done *his* way and his way *alone*. It presents no problem for a neophyte like Wilkins, who is eager to learn the Stef way, but it can very quickly alienate the Arctic veterans, who are set in their ways. Stef's general manner doesn't help things. Take dinner for example, a meal where many leaders manage to bring a team together in good fellowship and high morale.

Not for Stef, though, as Wilkins observes: 'Stefansson is a moody fellow. I have never heard him tell a story in the accepted sense of the word . . . Even at his dinners, he won't have any general conversation. All through the meal somebody is talking. Somebody will miss a course and tell a story, and another man will miss a course.'[15] It is all rather odd and, as ever, it is because things are being done in the Stef way: he insists there is to be *no* general chit-chat at the table, as it is a misuse of time and a waste of attention. When one speaks, *all* must listen, don't you see?

'I don't believe there is any benefit in general spasmodic conversation,' Stef declares. 'I think it is much better to have a story where everybody can hear it.'[16]

It is as if, instead of having a convivial dinner, you are enduring a lecture. Like Stef himself, it is all very logical . . . but very unnatural and uncomfortable.

One of the things Stef needs to do right now, however, is placate the crew, for there proves to be growing disquiet on a number of things.

The hope of a new expedition is that it will be filled with familiar friends and faces, not familiar frowns and failures. And yet it proves that the latter is more common, with Mackay having upgraded from jaded disinterest to outright derision directed at each new-fangled theory that Stef feels the need to talk about. Mackay is an old-timer, stuck in his ways and yearning for the way that his Shackleton, 'The Boss', had run things. Stefansson is his boss in name only, a master to be tolerated rather than trusted. Mackay wouldn't follow him to a hearth, much less to hell and back. And the Scot is not alone in his thoughts, as even those brand new to the North are aghast that their leader has misled them before they have even departed. Take the befuddled brogue of William McKinlay, the Scottish 'meteorologist' who had in fact joined the expedition as a 'magnetician'. If this expedition can't even get job titles safe and secure, what hope have they of braving the far North?

'It was the first of many hints I was to receive as the weeks went by,' McKinlay will note, 'that not everyone shared the faith of popular Canadian newspaper writers in the suitability of Stefansson as expedition leader.'[17]

Much of the mutinous talk focuses – not completely unreasonably – on trying to take such a vessel as this on a long journey through perilous seas to a frozen wasteland known to receive far more visitors than it delivers returnees. Stefansson might talk of the 'Friendly Arctic' – that to know it is to be able to tame it and make it friendly – but they simply don't believe it. In their view the Arctic is by nature as unfriendly as a polar bear with a toothache, and rather than talk like this, they would much prefer to have a decent vessel, plentiful provisions and more water tanks filled – call them crazy – with fresh water.

Water tanks? A bemused Stefansson wants to know what on earth the sailors think they will need those for.

'You will get water from the ice,'[18] he assures them. Why, yes, you should know that ice from the Arctic Ice only keeps all its salt for a year. Ice older than that is quite tolerable as coffee; in two years good enough for tea and after three years, well then, gentlemen: 'it is fresher than any mountain brook'.[19]

In short, Stefansson does not care for their mutinous rumblings, and insists they will be going regardless.

For the Arctic will provide, gentlemen, at least for those of you who observe what I do, listen to what I say, and follow my orders. If you don't, well, yes, then you really will likely be in a lot of trouble.

I am your leader.

And yet, while Stefansson does indeed regard himself as being in total and unquestionable command of the whole expedition, Dr Anderson thinks that highly questionable indeed.

Dr Anderson is a towering pillar of establishment sensibilities, a Mason with a PhD on birds who, using that expertise, clearly has no doubt from the beginning that Stef is an odd duck likely crossbred with a strange strain of cuckoo. This will be interesting. A veteran of the Spanish–American War, Anderson has an old soldier's instinct about brewing storms and inevitable battles. Hostilities are liable to peek over present pleasantries, and Anderson knows that a fallout with Stef is more a matter of 'when' than 'if'. They are simply too different; Stef is a cock-sure adventurer with a reckless streak, Anderson is a reserved academic who would rather look before he leaps. They are a pair bound by mutual funding, not by mutual interests. Or character. Or values.

After vigorous negotiations another vessel, the *Alaska* – a rather dainty gas-fuelled schooner – is purchased as well, so that the expedition can pursue both goals at once and, ideally, support each other. Wilkins watches the machinations and manoeuvres with some interest. Of course, he has no position within the chain of command, and his only true role is observation – chronicling the expedition by taking pictures.

And yet?

And yet, he can't help himself. It is simply in his nature to seek new skills no matter what situation he finds himself in, and it is with that in mind that he approaches the captain of the *Karluk*, Bob Bartlett – an ancient mariner as tough and durable as the leather in the old boots he wears – to see if he might be given a watch of the ship, and in the process perhaps learn a little of the art and science of navigation.

Captain Bartlett would be delighted, perhaps recognising in Wilkins himself at a young age. He was a ship's master by the time he was 17 and has now been at sea for a quarter of a century. He had also been with Robert Peary's attempt at the North Pole in 1909, and knows his onions. Impressed with the young cameraman's enthusiasm, Bartlett is giving him instruction that very day and into the evening, and is amazed, over ensuing weeks of this rising summer of 1913, just how quickly Wilkins' mechanical mind masters the difficult art of navigation, one where a single degree can make the difference between life and death in this fatal realm of the world. In short order, Wilkins has learnt the workings of the quadrant, sextant, magnetic compass, charts, nautical almanac, the sounding line for depth measurement and the marine chronometer, while also displaying remarkable ease with the sometimes complex mathematics that goes with using the measurements and charts to work out exactly where in the world they are, and how much ocean they have beneath them.

All set?

Not yet.

After the *Karluk*, with the *Alaska* tight behind, moves to the port of Teller to get three dozen sledge dogs, complete with a supply of fish for dog food, Stef has another thought on how to lessen the chances still further that the 'Friendly Arctic' turns unfriendly on them. He decides it couldn't hurt to have a third whaler with them and so buys the *Mary Sachs*, a wooden schooner of 36 tons, to join their flotilla. She is a trading vessel used to transporting walrus tusks and furs, and her owner sells her for $5000 on the strict condition he be allowed to buy her back at the end of the expedition. Dubious Southern Party member Kenneth Chipman records dryly in his diary: 'Supposed to do 7 knots.'[20] We shall see, and all hands turn to hauling further supplies on board. Rolled oats are rolled on, along with sugar, pork crackling, rice, pemmican and . . . dog biscuits. (As friendly as it might be, the Arctic cannot supply everything . . .)

All three ships now arrive at Port Clarence on 19 July, to take on more supplies, while Stef also hires one 'Mr James R. Crawford', if you please – an unusually formal name for such a rough-looking fellow – to act as the mechanic cum engineer for the rather aged engines of the *Mary Sachs*.

Again, however, George Wilkins can't help but notice: despite this rather grand name complete with middle initial, the man himself looks and smells like he was born and bred in a vat of whiskey. Crawford is happy to be employed for the drinking money it will provide, but declines to get the engines on the vessel started until he is good, ready and . . . sober. This, from the look of him – his eyes are mere bloodshot slits, and his words *shlur* – will take at least a couple of days as Crawford falls into his bunk as drunk as a lord and twice or thrice as imperious.

What can the Captain of the *Mary Sachs* do?

(For there is another factor at play here, as Wilkins is coming to understand: as goes habitat, so goes the kind of men who gravitate to it. Just as refined and civilised men are drawn to refined and civilised towns where they prosper, so too do ragged rough-nuts tend to end up in the thickest pickets of the wilderness – either because they like it, or such places lack the capacity to kick them out as decent places have done. In sum, in the wild, expect a preponderance of wild men.)

So, the Captain apologises to Stef, but there is nothing he can do, as Crawford is the only man who understands how the damn engines work.

But, actually . . . ?

Actually, the man there merely to chronicle events, George Wilkins, might know of another.

'I volunteer to run the engines,'[21] he confidently tells Stef.

Stef looks him up and down, for once looking uncertain.

Are you sure?

Yes, Wilkins replies, with impressive certainty.

'Well, if you can persuade Crawford, go ahead, it will be a real service to the expedition.'[22]

It is not quite that Crawford tells Wilkins and the Captain, 'No.' It is more that through a drunken rant he gives the opinion that they both should go and perform an entirely impossible anatomical task, and he says it with such force that both Wilkins and the Captain decide he is ... a danger to shipping. So much so, that for his own good and their own, the best thing to do is to lock him in his cabin. They don't tell Stef. What he doesn't know can ... just be added to the long list of other things he doesn't know.

The main thing is that, sure enough, in his time-honoured fashion of using fencing wire, elbow grease, resourcefulness and – for tasks still resistant to his charms – big hammers, Wilkins does get the engine going and, while the *Alaska* must remain in port for further repairs, the *Mary Sachs* joins the *Karluk*.

Unfortunately, while the engine is now roaring like never before, so is Crawford. How long should they keep him locked up? The Captain thinks 48 hours should do the trick, but is mistaken, as at that time Crawford is still drunk – he must have had a supply in his cabin – and not only screaming blue murder, but threatening it.

'It was no longer safe for me to be on the same boat with him,' Wilkins chronicles, 'so I transferred back to the *Karluk*. Crawford never forgave me.'[23]

Too bad.

What now? As they are coming abreast of the settlement of Tin City, they see a rowboat approaching, about a mile and a half offshore, with a man in it making signals to attract their attention. Altering course, Bartlett brings the *Karluk* alongside and discovers the fellow has a message for Stefansson from an aviator named Fowler who was at nearby Teller with his aeroplane.

'He asked,' Bartlett will recount, 'permission to bring his machine on board the *Karluk*, accompany us for a while and later on fly from the ship to the shore. The *Karluk*'s deck was already pretty well crowded with dogs, sledges, sacks of coal and other gear, and Stefansson finally decided that it would be impossible to grant the request.'[24]

For Wilkins it is no more than a minor episode in a busy day, when everything is stunningly new, but later he will have cause to remember it as the first time he was exposed to a singularly interesting idea.

The *Karluk* chugs on through the Bering Strait towards Point Barrow on the northern tip of Alaska – lying some 320 miles north of the Arctic Circle – accompanied by the *Mary Sachs*.

At Point Hope, where they intend to trade with the Eskimos, it is noted that the *Mary Sachs* is nowhere to be seen. The *Karluk* anchors, and is quick to receive a visit from the Eskimo villagers who come out to meet them in their skin-boats to trade for whatever they can – nails, screws, axes, barrels, sea chests and cloth, for furs and sealskins, dogs, skin-boats and kayaks – while, as Bartlett notes, 'our scientists were busily engaged in writing letters, to be mailed at Point Barrow and taken back on the *Bear* which calls there once a year, usually in August'.[25]

Onwards. While progress has been satisfactory up till this point, as they near Icy Cape on 1 August 1913, they first note some ominous signs of things to come. For now, the open water becomes calmer, always an indication that its temperature is dropping towards the point of freezing, and it becomes dirtier on the surface as the freezing process begins.

The sky itself issues warning. There, straight off their weather side is the 'ice-blink', the white reflection in the sky which indicates that vast plains of ice lie before them, beyond the horizon, and soon enough 'the ice itself hove in sight about two miles away, with some larger pieces scattered here and there among the floes'.[26]

For the next few days they stick tight to the coast, while now and then Captain Bob Bartlett decides to take the *Karluk* a mile out from shore, just to keep her floating in open water. She belches steam as she paces to and fro for some time, keeping abreast of the ever shifting wind while she steers well clear of the ice pack.

By the morning of 18 August, the situation is becoming critical. For now, not only is there ice in front of them, and off their port and starboard, but it is off their stern, too, and closing in. The wind has dropped to nothing, the sea is perfectly still as the water all about them freezes.

All power, astern!

Bartlett does what he can to break through, break out, but it is useless. When you have ice in every direction, there is nowhere to break out to.

Still they try to break through, despite the fact that the *Karluk*'s chief engineer will say that the vessel had a 'coffee pot of an engine'[27] with just 150 horsepower, nowhere near powerful enough to make a sustained breakthrough and overcome the freezing forces of Mother Nature.

All on board are aware of a grinding sound as the ice they are cutting through presses ever tighter ... grinding ... grinding ... GRINDING ... until the inevitable.

They grind to a halt. Captain Bartlett again gives full thrust both forward and then in reverse, but it will move neither way. Oh, but it will soon become worse than that. For now the vessel is caught in the icy maw, that maw soon squeezes.

The *Karluk* groans in protest as rivets and joints are placed under huge pressure. This is serious, and there is a real chance that they might be crushed to a shattered shell.

Christ Almighty!

For such an eventuality, Stef's orders have been clear.

While the *Mary Sachs* and the *Alaska* go ahead to Herschel Island, 400 miles to the east – which will be their staging point for the next phase of the exploration – he and the *Karluk* will follow and join when possible. 'When', in this case, being the rather inoperative word ...

For they remain stuck there, some 15 miles offshore with not the slightest movement for the next four weeks. Winter is setting in, and so are they.

Not that the situation is without some tragic beauty.

For when it comes to the icepack that holds the ship, Wilkins becomes lyrical in his fascination with it: 'My first impression of the pack was of a fearsome, enticing, mocking monster; but soon it lost its fearsomeness and it seemed to beckon and call and reach out with pleading sadness that awakened sympathy. I am never unconscious of its pleading. It is a love to which those of us who know it must return.'[28]

On this particular evening, 19 September, everyone is engaged in their own pursuits after dinner, when Stefansson's stenographer McConnell emerges from a long consultation in Stef's cabin with news.

'George, you, Jenness and I are to go with Stef tomorrow and hunt caribou on the shore.'

'Is that official?' Jenness asks.

'Yes,' says McConnell, 'but you can go and ask yourself if you like.'[29]

Wilkins is thrilled to be part of such a trip.

'This will be an opportunity,' he exults, 'to get pictures of the ice and the Eskimos!'[30]

Ideally, he can even record vision of Eskimos seeing a white man for the first time. And images of caribou, better known as reindeer, will always be popular with Gaumont.

Without further ado he goes to get his *fusil* and *l'appareil photo* packed and ready for what will surely be an early start, before visiting Stef in turn in his cabin.

'Am I to go caribou hunting with you tomorrow?'

'Yes, you'd better come,' he says.

'How long are we to be away?'

'About a week,' he replies.

Stef has already assured Captain Bartlett of much the same time frame, 10 days at the very outside before they return, and is positive there is no real danger of them being separated. After all, the *Karluk* is going nowhere, and they will be the only ones moving, so that is a good start for the hunters being able to get back to the same spot. And yes, you, Wilkins, may as well come along for the experience.

'Can I take my camera on the sledge?'

'I don't know, but if we are successful in finding caribou we should be going backwards and forwards all winter.'

In other words, don't bother, for now. There will likely be plenty of trips back and forth, so no need to take it on this first exploratory trip.

Thinking about it though, Wilkins nevertheless gets his Moy & Bastie 35-millimetre Cinematograph camera, complete with 300-foot film magazine and spares, ready to go.

As to the Siberian huskies, their Eskimo guides are harnessing them to the sleds, in the time-honoured fashion. It is now that one of the guides, the portly Pyruak, comes into his own. A beaming rascal of a man who never saw a wall he didn't want to lean on, if not chair to sit on – most particularly when others are working hard – the whole voyage long he has delighted in teasing the white men and mocking their 'make work' ways. Still, he is so charming about it all the crew adore him, and nickname him Jim, as in the song '*Oh, Lucky Jim, how I envy him*' for his miraculous abilities to gain food and evade work at every opportunity. But now, however, they see Jim spring into action, doing what he was born to do, and has done 10,000 times in his life, though many of them are seeing it for the first time; mustering and mastering the sled dogs from alpha to omega in moments.

The 'wheel dogs', the strongest ones, go closest to the sled, 'team dogs' go next, and the 'lead dogs', which are usually female and the most intelligent, go at the front. Of course, they all bark like mad things once in harness, but that is only because they are even more eager than Stef to get going.

With all nearly set, Wilkins carries his camera down the plank, and sets it up ready to take a picture of the final preparations and departure. He passes Stef on the way, who says, 'Oh, are you going to take that?'

Taking a hint, Wilkins bows to unstated wishes and says, 'No, I don't think I will . . .'[31]

Strangely, Captain Bartlett seems in 'great spirits' to see them depart, likely at the opportunity to have unrivalled command without Stef.

'In the exuberance of his spirits,' Wilkins chronicles, '[he] seized an enamel hand bowl and kicked it as far away on the ice as his strength would allow.'[32]

Others are tense at the whole idea of splitting up into two groups as it seems against common sense, and Wilkins notes 'a prolific flow of blasphemous language'[33] as the last preparations are made. (And a sign that Stef, in practice, doesn't have quite the faith in the 'Friendly Arctic' that he has professed because he is taking a tinned supper along.)

Time to pack.

By now, of course, Wilkins' spats, striped trousers and top hat are long gone, and he has, as they say, 'gone native'.

'I started out wearing a skin shirt, fur next to my skin,' he will recall of this time, 'an outer shirt, fur outside, and a snow shirt of blue denim over these two. I had a pair of sealskin boots or "Mukluks" that reached to my knees.'[34]

In terms of warmth, the mukluks, 'made of soft sealskin, and thin pliable soles without heels'[35], are a particularly ingenious form of footwear, fashioned from the skin of the caribou or the seal – the latter being slightly more waterproof. They flex with the foot, and are remarkably light and comfortable for such a sturdy shoe, which is also lined with the fur of the fox to warm you as you walk.

And of course, like the locals, Wilkins has some dried cut grass – bartered for, from the Eskimos – lining the bottom, 'to act as a cushion as he walks over the rough sea ice and also to absorb the perspiration, for otherwise his feet would be in constant danger of freezing'.[36]

His only concession to Western ways is to take three pairs of socks, a pair of underpants and a pair of cloth pants.

Warmly attired thus, he now heads out to take 10 plates and some 100 feet or so of moving picture film to record the departure of Stef with two Eskimos, Ascetchuk and Pyruak, together with McConnell, Jenness and himself. With the chronicling complete, Wilkins just has time to hand over all the cameras, bar his still camera, to one of the crew, and then races to catch up to the just departed sledges. Hopefully, the fact he has just one small camera will help Stef to see him as one of the men, not as that damned photographer. (As Wilkins only has 18 'exposures'[37] he will have to call his shots wisely and sparsely, which should suit Stef's sensibilities.)

'It was a long run to catch up to them and I was soon warm enough to discard one of my skin shirts.'[38]

And they are off, Wilkins holding on for dear life as, in his first time on a sled, let alone pulled fast by dogs, it takes some time to work out which way to sway to keep both himself and, more importantly, the sled properly balanced.

They leave the ship at 11.30 am on 20 September and keep pushing hard across the ice and snow – pausing occasionally to cut through rough ice, or to right an overturned sled. More problematic are high ridges of ice, formed as huge slabs of ice crash together like tectonic plates and have nowhere to go but up in jagged ridges. The smaller ones the men and dogs can go over, the higher ones they must find a way round. They plan to push on for longer to try to get to solid land, but at the summit of some of the higher ridges they fail to see any sign of mountains in the distance, so they know they are not close and stop at six thirty in the evening. Wilkins' fast-tracked ways of learning to live on the ice are about to get another push, as he learns how to camp on sea ice, and who better to teach him and the other white men than Stef himself?

'We had two tents wig-wam shape,' Wilkins will recount, 'a "Burberry" supported by bamboo poles attached together at the top and a "Bell" drill one with a hole in the roof for a stove pipe.'[39]

Before settling down for the night in their sleeping bags, Stef leaves his Burberry with the Eskimos to give the white men some lessons in how to pass a comfortable, warm night, despite sleeping on a floor of snow-covered ice.

First thing: you, Jenness, are the smallest and slightest and therefore most susceptible to cold, so you must sleep in the middle, using the body warmth of your companions on either side, tightly packed, to help keep you warm. Now, to get as much insulation as possible between you and the cold, place the sheepskins we carry wool-side down first, and then place the deerskins, hair-side up on top of them, and now the sleeping bags on top of them again. Now, for a pillow. Take your boots off, and after you have carefully brushed the snow off them and put your socks inside, place them at the head of your bed, *between* the sheep and deer hides. Place your spare clothes bag on top to give you softness. Putting your boots there will also help to keep them thawed throughout the night.

But most surprising, flying in the face of any logic or reason that Wilkins can summon, is the penultimate instruction.

Take your clothes off and sleep in the nude. It is the *only* way, Stef insists. And the fact that this is the way the Eskimos do it settles any argument there might have been.

'To keep on your clothes for a night or two may be well enough but in cold weather, hoar frost will form between your clothes and the sleeping bag and soon the bag will be coated with ice on the inside.'[40]

Nightmare!

Last thing. You must, still shivering, wriggle your way to the bottom of the bag. Now fold the mouth of your sleeping bag over just so, to prevent the cold air from playing on your shoulders, and use your own hot breath to warm the whole bag.

'This is an art in itself,' Wilkins records, 'not to be learned in one night.'[41]

But they will get there, soon enough. Whatever else you might say about Stef, his knowledge really is phenomenal and, more than ever, Wilkins determines to watch him closely, to learn as much as he possibly can while he is with him.

For two days the hunting party indeed make good progress, and though there are the occasional squalls that limit visibility, as soon as it clears they are satisfied and heartened to see the *Karluk* in the distance exactly where they had left it, which is all well and good . . .

Soon they are on the next best thing to the mainland, a small solid island that looks to be about five miles off the coast, and it is here that Stef has a thought and an order over the evening meal.

'I think that McConnell and Ascetchuk had better return to the ship in the morning and bring Malloch and Mamet ashore. I want them to do some surveying along the coast.'[42]

It is a plan and the party stay up late talking.

Good God, what now?

It starts as a moderate wind from the east, before building to a growl and now a howl that waxes as the sun wanes. It shrieks, it tears at the canvas of their tent and threatens to blow them all away. Wilkins has been in some scrapes in his time, but this one is completely terrifying. If the tent does blow down they will have no protection at all, and their survival is unlikely. More terrifying still?

Shots in the distance . . .

Or at least, they *sound* like shots. The sound of ice snapping and cracking is not unlike the firing of so many .303s, and while the men know it to be a natural phenomenon, the sounds are discomforting.

Of all things they need, right at the top is the dawn, which is the only time they will be confident that they have actually made it through the night.

When they awake a little later than usual, however, and emerge from the tent, there is a real shock.

'We were astonished,' Wilkins will recount, 'to see that the ice had been moving along the shore, so it was now unsafe for McConnell and the Eskimo to venture out on the ice to reach the *Karluk*. Furthermore we did not know if the *Karluk* was still in the same place . . . we thought it possible that she would perhaps shift her position or maybe get free altogether but we expected to get back to her easily enough.'[43]

What to do?

Nothing dramatic for the moment. They must simply do what is practical until the necessary search for the *Karluk* becomes possible. Wilkins notes Stef's view, as firm as the ice is weak: 'It is useless to think of going on the rough ice in search of her.'[44] Bartlett knows the procedure; a beacon will be left, cairns constructed to show their path, ships have been lost and found regularly in the Arctic; but not if either party panics. Carry on, gentlemen, and make the best of what we have at present.

With solid land beneath them, it is time to explore the island until the ice settles down, do some hunting, and see what the morrow will bring. Wilkins shoots a couple of ducks but cannot even fetch them from the water where they fall as the ice around them is too thin to tread on.

But McConnell and Ascetchuk fare better, shooting and collecting six ducks that Wilkins will cook for supper.

All six are devoured. Stef is delighted at their hearty appetites: 'It is my policy for everyone to eat as much as he wants while there is any food at all!'[45]

This policy is popular with all, and Wilkins notes the irony of being 'lost' in the Arctic but enjoying fine dining that would be the envy of his friends at home: 'I am sure no hungry man would wish for better fare than stewed duck and rice soup.'[46]

Next morning, alas, the situation is worse. They wake to find the ice has not only drifted further but also become more fragile.

> We now begin to worry a little about getting either to the mainland or back to the ship but both of these are impossible for the ice was moving on the outside of the island and it was too thin to bear a weight on the inside between it and the shore so we were marooned on this barren piece of land not knowing how long we were to be detained there.[47]

For the moment they must sit it out for a few days, doing more hunting and waiting for a colder spell to make sure the ice is solid enough to hold them, before they try to get across the thin slush that separates them from the mainland, or make their way back to the ship. Though they treat their supplies sparingly, Wilkins does open one box of biscuits and discovers a slip of paper on which is written a request by some hopeful worker from the biscuit company who must have been told of the packet's special destination:

> *Please dear send me a picture postcard of the North Pole.*
> *Maud Rogers*
> *c/o Pophams Biscuit Company*
> *Victoria B.C.*

It is something of a problem. Stef says Wilkins must comply, the fact that he personally knows Santa quite well, changes nothing . . .

'I am afraid the shops that I will find around here do not sell picture postcards and I don't ever expect to go to the North Pole to photograph it.'[48]

Two days later, Wilkins goes out on his own to hunt for a fox, specifically a blue fox – a very rare breed, worth $120 a skin. One has been spotted by the two Eskimos earlier, using binoculars, so he knows there is one there, somewhere.

In his unsuccessful attempt to find it, Wilkins follows the fox's tracks, which lead off the island itself, before thinking better of it, and heads back to the relative safety of solid land, softly singing a favourite London music hall hit as he goes . . .

Fall in and follow me!
Fall in and follow me!
Come along and never mind the weather,
All together, stand on me, boys;
I know the way to go,
I'll take you for a spree;
You do as I do and you'll do right,
Fall in, and . . .

And . . .

'All of a sudden there was a loud hissing noise and the ice all around me began to sink under my weight . . .'[49]

Oh . . . CHRIST!

With no experience in such matters, Wilkins does the worst possible thing and immediately sets off at a dead sprint for the shores of the island just 100 yards away, meaning that every footstep cracks ice under his big smashing clod-hoppers, 'and shooting little squirts of water through it all around me . . .'[50]

He makes it, just, but it could so easily have finished in his grisly freezing, watery death.

'This is my first scare of the ice and it will tend to make me more careful in the future to look where I am going.'[51]

It is not just his body in danger of freezing as, in such climes as these, fluid in compasses have known to congeal, while it is possible for men's teeth to not only chatter, but *shatter*. Every day he survives in these climes he is learning new things, which is to the good, for any mistake could be fatal.

The news back at camp is mixed.

On the one hand there had been huge excitement with the view that they had seen the *Karluk* from a high spot on the island.

Alas, alas . . .

> After raising a lookout still higher we were not at all sure if it
> was a ship or not, for a haze was now settling down over the ice
> and the ice itself was moving in all directions. We watched until
> it was too dark to see any distance and returned to the camp still
> undecided as to whether it really was a ship or not.[52]

With all the shifting ice and changing conditions it really seems possible
they might be all alone here in the wilderness, which most of them find
unnerving. Strangely, though, not Stef. There is a calm in his approach
right now which almost makes it seem as if being lost and alone is of
no great consequence to him, as it will prove his point about surviving
in the Arctic.

Onwards.

More lessons . . .

The next day, Stef teaches Wilkins how to cook seal as, while the
Australian knows his way around a duck, something so cumbersome
as a seal is beyond his skills. Not on the griddle, and not on the boil as
some would have it done. There is just one way.

'Put the meat in the water while it is still cold and it will then be
cooked about five minutes after it's come to a boil.'[53]

And he's right. It *is* delicious, and 'ever so much better than any we
had had on board the *Karluk*. The blubber also is quite palatable and
has a taste not unlike mutton fat; and the first joint of a seal's flipper,
which has just about equal quantities of meat and blubber on it makes
for a fine dish.'[54]

Wilkins cooks what he thinks is a large meal for five but ends up
having to cook the rest of the seal as the men are so hungry. Yes, Stef had
taught them how to counter hunger by filling your stomach with 'agar
seaweed – it's harmless and distends you till you feel quite comfortable'[55]
but nothing beats actual food! As they recover from the meal, Stef tells
them some fascinating lore, of just how the names of Eskimo children
are chosen and why they are never forbidden anything or punished. For
you see what the Eskimos believe is this . . .

> When a person dies their soul remains in the house for four days,
> if a man, and five days, if a woman. It is then expelled and goes
> and remains at the graveside. When the next child is born in this

village the mother repeats a formula or chant inviting the soul to enter and take care of her child. The child is then named after the dead person, irrespective of sex or relationship. Therefore, you may find a baby being called 'Mother' by a comparatively old man even though it may be a boy and no blood relation to him. The soul of the dead person is known as their NAPATAH but after it enters the child it is known as the child's ATKA and acts as the child's protector until it reaches the age of discretion. Such being the case to reprove the child would be to reprove its protecting spirit; and the general belief is that the dead person was always a better person to anyone living; so it would really be presumptuous on the part of anyone living to reprove the spirit of the dead.[56]

Fascinating.

The following day, up at first light, Stef decides the ice is thick enough for him, Wilkins and the two Eskimos to at least head to the mainland, though not thick enough to attempt a return to the *Karluk*. Nevertheless, that boat remains on his mind as, on 29 September, he sends Wilkins back to the camp on the island to check on the two men left behind and leave a message in a beacon in a prominent position on the island, in case the *Karluk* should pursue their path there. 'I now had to face the wind,' Wilkins will recount, 'which felt awfully cold and ice soon started to form on my moustache and beard for I have not shaven since leaving home. It was a glorious sunset but I was too cold to take much notice of it and I hurried on. When I reached the tent, McConnell and Jenness were greatly amused by my appearance for I had long icicles hanging from my moustache and my beard full of ice so that I looked very much like a walrus.'[57]

The next two days are passed by going out shooting caribou, an endeavour that Wilkins finds particular for the fact that, after a herd of caribou is fired upon, the animals will usually run for five or six miles, but 'you can depend on their settling down and feeding again'[58] and a second shot and supplementary feast may be had. Other animals are harder quarry in these parts; the musk ox, for instance, herd in their hundreds but when a threat is spotted they 'huddle in a circle, the bulls on the outside with their heads down'.[59] First one bull charges, then another, then another, and when they *charge* they *keep* charging, for 'once a musk ox starts to run, he will go indefinitely without stopping'.[60]

When not hunting, the men take turns to tend to Jenness, who is ill from an attack of 'ague', a form of malaria, which he had caught on a trip to New Guinea, so weakening him that he must ride on the sledge and make no further contributions – a man freezing at the North Pole, afflicted by a disease of the tropics. And now comes a little proof that even Wilkins can have his off days, recording in his diary that evening, 'I have felt very peevish all day and afraid I have not been a very cheerful companion.'[61]

Returning to camp on 3 October, Stef makes a last attempt to look for the *Karluk*, as they all venture forth, but once more she is nowhere to be seen, hidden by an endless expanse of barren white. All they can do is head back to the mainland, and they make camp that night once more on the ice in the frozen delta of the Colville River. The tent is warm and snug, despite the fact they are on top of soft snow and ice.

> Soft snow is by far the warmest thing to camp on up here. One would think it would thaw out and wet everything but this is not the case, for after putting down the sheepskins and the deerskins on top of them, very little of your warmth gets to the snow and very little of the cold penetrates the skins, so all [that] is necessary is to make the tent as airtight as possible and the warmth of your bodies will soon heat it up even if one does not have a fire.[62]

Deciding to forge ahead, the party treks along the coast back towards Point Barrow, though the going is hard as Jenness is still ill and Wilkins finds that his feet are giving him trouble.

Despite Stef's previous assurance, there is no sign of the 'Friendly Arctic' right now and it appears quite the reverse, especially as 'we were reduced to small rations of seal oil and reindeer hair'.[63]

Reindeer hair?

Yup.

'Naturally there is no nourishment in reindeer hair, but we felt that we wanted something under our belts.'

It literally goes with the territory. Other explorers in this area – most infamously the British expedition led by Sir John Franklin in 1845 – had to eat their actual belts, and then their boots, and then each other in their attempt to survive.

Reindeer hair, on balance with human flesh, is preferable.

'The most satisfying thing we could do,' Wilkins records, 'was swallow a few tufts. This gave the sensation of something in the stomach to ease the gnawing pain.'[64]

Day after day the double-time trek continues, pushing forwards for 18 hours of every 24, every eye scanning both the shore and the sea for some sign, any sign, that they are not completely alone on the planet. Alas, all they can see is the unending endlessness of empty ice here at the ends of the earth. Getting progressively weaker from starvation, for reindeer hair can go only so far, their desperation increases. At least their Eskimo guides give them some hope when they point to fox trails congregating at one point in the snow, a clear sign the animals have been digging here, which means they must have smelt food. Led by Stef and Wilkins, the hunting party now digs too, with dreams of discovering a frozen seal or the like, only to finally discover ... the carcass of a whale, completely stripped of its nutritious blubber, with just a little rotten meat left on the bones.

The old rule applies: beggars can't be choosers, and starving beggars living on tufts of reindeer hair will devour that which in normal times they would revolt against, before cleaning their bowls, lining up for second helpings and saying thank you.

'We decided the best thing to do was to eat it raw and frozen.'[65]

Whale sushi it is; and the reluctant gourmets are thankful for small mercies: 'It was almost tasteless, and because it was frozen we got no smell.'[66]

Despite the privations, however, Wilkins cannot help but notice one thing: the stunning *beauty* of this environment!

There is a purity to the snow and ice that surrounds them. When the first rays of golden morning light reflect white from white from white, you can feel like it is the morning of creation itself. For all around is pure nature as God intended, including them, human animals making their way across a landscape entirely unmarked by man, leaving nothing behind that the next good snowfall won't cover or the next whip of wind blow away. Wilkins starts to understand why men like Stef are perpetually drawn back to this living painting, an ever changing image that is, nevertheless, just as it was a million years ago; a contradiction that can only be grasped by those who are there.

For now, however, onwards.

Mush, you huskies, *mush!*

On 5 October, however, possible salvation is at hand.

The tiny dot in the distance getting bigger in the windswept whiteness of it all soon breaks into several dots moving about each other and finally devolves into a small camp of Eskimos.

The two questions are quickly put.

Can you give us some food!

'*Ee.*' Yes.

Have they seen a ship pass?

'*Nakka.*' No.

Right now, the most important thing is the affirmative answer to the first question and these blessed people soon have a fire going for them, and are cooking frozen fish for them, and all of them are soon eating till they can eat no more in their snow house, lit by blubber lamps.

Sleep is brief as Stef rises early and rouses the others. He feels like the *Karluk* is not far away, and all he wants is a sign. They must move quickly so as not to miss it. Fare thee well, our Eskimo friends, we must away.

Which is easy for Stef – the Eskimos don't particularly care about him. Wilkins, however, must extricate himself from the womenfolk of this land who have taken a particular shine to him, and keep pressing fish upon him as gifts, which are most welcome. (The fish are frozen, which is to be expected as there are barely any other kind in these parts, where it is only life itself that can make the flesh malleable.)

Onwards they march, Stef's sudden enthusiasm infecting them all with the hope that they shall make contact with the *Karluk*, whether through message or messenger left behind for their notice. And yet, with every passing hour and still no sign, hopes fade along with the light and Cape Halkett is reached with no answer. Most in the party are joyous at reaching the mainland once more, but Stef is puzzled to see no indication at all of the *Karluk*, no beacon, no cairn, no message at all in this rendezvous.

What is even more mysterious is a report coming from an Eskimo family who affirm they have in fact seen the *Karluk*.

'We saw her drifting westward along the coast, locked in the ice.'[67]

Seeing if they could help – for they are a kind people – the Eskimos tried to approach the boat on foot, only to be stopped by open water.

But here's the thing. The Eskimos had waited there for *two days*, thinking the *Karluk* would send a boat across to them, but no boat has come, and there has been no sign of life. On the third morning they had woken up to find the *Karluk* had been carried off in the night.

It is deeply worrying, the reporting making the lost vessel sound like a frozen *Mary Celeste*.

'Stefansson was amazed,' Wilkins will recount, 'because anyone familiar with the Arctic must have known that once the ship passed Cape Halkett there would be little hope of leaving it and reaching shore.'[68]

Why on earth, on ice, did Bartlett not send men ashore? He has boats to spare, surely?

Desperately hoping there might be some misunderstanding, Stef keeps questioning the Eskimos, but their story remains the same – they could see the ship and be seen, but no man or message was sent.

'It seemed incredible that he has not at least sent a party ashore to leave a message for us.'[69]

Are they all . . . dead?

Beyond the worry of having lost contact with their vessel, Wilkins is personally devastated. He has but a single stills camera with him, the rest of the equipment is on the *Karluk*! It will take some explaining to Gaumont how their cameraman ended up on land with one stills camera, while their cameras drifted away on a lost vessel on the Arctic ice. And what he can capture on his stills camera is unlikely to bring the hordes rushing in from around the world.

So, what now?

Another trek, Stef declares. Their only option is to gird their loins, strap on their snowshoes, wrap their coats tightly around themselves and strike out for Point Barrow, 120 miles away. There lies the only trading post within 500 miles, the only place they can get supplies, more equipment and maybe even a movie camera. Their job is to survive, to trek back to civilisation as an intact gang of six. The *Karluk* is beyond their control, their reach and their comprehension for the moment.

Lead on, Macduff.

The small, exhausted group heads off the following morning with Stef indeed in the lead, while Wilkins brings up the rear. As ever, he is uncomplaining, even though by now – due to one item of Arctic apparel – he is in agony.

'I was now travelling in Eskimo *mukluks*, a soft, wrap-around footgear without support, which strains the tendons and caused the painful condition called "Achilles' heel".'[70] It is Arctic Achilles tendonitis, if you will, and every time he puts either foot down, it sends an agony of shooting pain, starting from his feet, up through his legs and making his whole body shake. To top it all off, he has 'bad blisters on the toes'.[71]

Still, no word of complaint from Wilkins, the obedient disciple.

'I thought it was an explorer's duty to suffer and be a martyr. I had yet to learn that, properly prepared, we could be as comfortable in Arctic conditions as if we were down in civilisation.'[72]

Stef is at least that comfortable right now, thriving on hardship and clearly relishing all adversity, almost as if he knows that every chapter they live through consolidates his reputation as a master of the Arctic. After all, could there be any better proof of his theories than this compulsory and comprehensive demonstration: a gun, a knife and will, that is all one needs.

Though to be fair, as Wilkins is quick to point out, their Eskimo guides help, too . . .

For their innate knowledge, borne of centuries of living and prospering in these parts, is on display, every day and every night. Like tonight, for example. When it comes to the key decision of where to pitch your tent on the ice.

'The Eskimos chose a site on a flow of "old ice" (last year's ice that had not thawed out and had been frozen in by the young ice of this year) that had about a foot of soft snow on it.'[73]

Because, of course, if it is old ice, there for at least 12 months, there is every chance it will be there for another night, too. With young ice, who knows?

They all awake, thankfully in the same co-ordinates where they slept, and trek onwards to a welcome sight. It is a small *Iglu*[74] in the Eskimo manner, with the floor lower than the surface of the snow outside and a thin trail of smoke standing out against the pristine blue skies. Stef has them steer straight towards it, certain that if, as he thinks, the *Karluk* has drifted west this group will surely have seen her. An Eskimo man comes out to meet him, soon followed by a woman and two little girls, all of them coming up to formally shake Stef's hand, just as the missionaries who have moved through these parts have taught them God's children must greet each other.

Wilkins and the others stay back, watching intently. Stef turns to them and just the small shake of his head tells it all. No ship has been seen. Wherever the *Karluk* may be, she is not west – or, if west, so far west she is likely to never be seen again.

Their new hosts are a gentle people, happy for the visitors to pass the night there if they wish within their *Iglu*, giving Wilkins a chance to observe closely just how the natives of this extraordinary land can so prosper. After the visitors haul their sledges onto the bank and unhitch the dogs, they accept the Eskimo family's invitation to enter their abode, and find it surprising.

> As we came near to it I wondered how we were going to squeeze into it; but after belabouring our boots with a stick to knock the snow off them as we saw the others doing; we went down on hands and knees and crawled inside. It was not necessary for the Eskimos to go down on their hands and knees to get inside for they after long practice could stoop down and step through the three foot high doorway. When we'd all gotten inside and distributed ourselves around on the bear and deerskins we were surprised to find how much room there was, and although it was not high enough in any part to stand upright in, quite a few people could pack themselves in while sitting down. The *Iglu* was really a tent supported by bent willow sticks and covered with snow. It was the shape of half an oval and about 15 feet long by 12 feet broad on the bottom getting smaller as it rounded off of the top.[75]

For their guests tonight, only the best, and dinner starts with a 'wooden platter full of frozen fish dumped on the floor', continues with dried meat, which Stef tells them is caribou meat, followed by 'some raw fat which we were told afterwards was quite a luxury on this part of the coast . . . the caul fat of the caribou'.[76]

Everything is eaten, including the fish scales.

'The Eskimo method . . . is to cut the fish down the back, then take the skin between their teeth and drag it off, then very soon there is nothing left of the fish but the head which is thrown into a bowl.'[77]

Not yet ready to 'go native' to that extent, Jenness, McConnell and Wilkins each try to grip the fish tightly and scrape the scales off first with their knives, 'the result being the fish was almost thawed out before we were ready to eat it. When I took a bite there was little or no taste to it but the feeling of raw meat in our mouth was unpleasant to us . . .'[78]

Oh, how the Eskimos laugh. They are always odd, these white folk – like the time, nearly a century before, when the British had first arrived in these parts and told them, the Inupiaq people, that this ancient land of *Ukpiagvik*, as it had been named by the Gods themselves, was now named 'Barrow'.

Nibbling raw fish, Wilkins and his companions are fascinated to hear the Inupiaq understanding of how the world came to be. Yes, of course they have heard the truth, about Adam and Eve and the Garden of Eden and the one and true Lord creating the world in six days and on the seventh day he rested, but still they prefer their own story – for they know it all started with Raven, that mischievous black bird who alighted on the snow to thrash and bash a snowball with his claws and wings. The snowball grew and grew and grew as snowballs do, until . . . well, you know the rest. (After all, as far as the Eskimos know, the whole world is like this – covered in snow and ice.) Raven's snowball became so large that the stars, moon and sun fell from it and what was left became the world. It is a compelling theory in this land, where the snow grows endlessly about them, unlike their appetites.

For even now, the meal has just begun and the dried meat is served with a local specialty, a bag of seal oil is brought out, which they are proudly informed is 'over 12 months old and had been hung in the sun all summer and was very nice and rancid now'.[79]

Now Wilkins is a man with a strong stomach, having, as a lad, dined on the 'fat sleek yellow grubs'[80] the Ngadjuri people in South Australia had shared with him, but this is too much even for him.

All the white men bar Stef have 'the smallest quantity possible that we took on the end of our finger and when we had tasted we sincerely wished it had been smaller . . .'[81]

None of which bothers the Eskimos remotely, as they continue to wolf it down, habitually using their hand-made and razor-sharp copper knives, rather than their teeth, to cut off each mouthful as it entered the mouth, each time so close it is sometimes a wonder they do not cut a piece of their nose off.

As to vegetables and salads, the very notion of fresh greenery is laughable.

The Eskimo diet is, needs must, based nearly exclusively on venison and fish for the reason that it is impossible to cultivate any crop in such permanent cold. The Eskimos do have small amounts of seaweed, tubers, stems and roots and berries, if they are found, but for the most part their cuisine consists of seal, grouse, caribou, walrus, polar bear, fish and eggs – with whales as a highly prized rarity when they manage to kill one, which is then preserved naturally in the ice and capable of feeding one family for an entire year with an enormous variety of dishes.

As Wilkins will note, 'Caribou . . . are scarce at all times and need a good deal of hunting for. They are in their prime from September until the end of November, but are hardly worth killing in March, April, May and June.'[82]

While 'seals are fairly numerous on the shore in all the islands . . . and can be shot while on the ice asleep'.[83]

With no choice but to stay with the hospitable Eskimos for a few days, until Jenness is strong enough to travel again, the time is passed hunting and fishing – and eating the results, with an appetite, in Wilkins' case, that quite stuns him.

> It's surprising what one can eat in the Arctic. I know I accounted
> for five fish and they weigh from one and a half to two pounds
> when taken from the water, and we had no means of telling how
> much meat there was except by the size, and the size of the piece
> we had was very much diminished when we were through eating.[84]

After two breakfasts – one Western and one Eskimo – they warmly farewell their hosts and exit into the cold before an excellent day's journey ensues, which includes Wilkins spearing a fox. Arriving by dark at a spot just eight miles west of Cape Halkett, they encounter more Eskimos and are greeted with a dinner of . . . donuts. Yes, these Eskimos are from Point Barrow, had spotted the sledge as an approaching black dot in the whiteness from miles away and have cooked up a fried treat they know the white men will enjoy. They also have: 'Mukpouras (bread fried in a pan) and some boiled polar bear.'[85] It is a varied diet indeed.

Before journeying on the next day, Wilkins attempts to make Mukpouras himself. It is not a success: 'I managed to turn out something that looked and felt like a mixture of charcoal and quartz stone and tasted about the same. I'm a pretty good fancy cook with such as oatmeal and tea, but when it comes to ordinary everyday bread, well I'm not much of a hand at that.'[86]

By this time, though the men try to appear hale and hearty there is equally no doubt that their bodies are feeling the strain at working so hard for so long, with so little rest. Three days journey after reaching Cape Halkett, Wilkins notes that to go with his own suddenly severely sore feet around the heels, the ague of Jenness has worsened further, Stef's back is out, McConnell has sprained his foot on a sharp piece of ice, and Ascetchuk's nose has started to bleed!

'. . . so that made cripples of five of us and Pyruak could not run ahead anyway, for he could not keep a straight trail for five minutes and would soon have us going in a circle. So, as I was suffering the least of all, I was ahead most the afternoon . . .'[87] as they push on towards Point Barrow.

The next day there is a breakthrough as coming to another tiny Eskimo settlement an Eskimo by the name of Anubconna tells them he had seen the *Karluk* drifting past just a week ago! Familiar with the vessel from previous visits he had tried to get out to her, but as he was the only man about at the time he could not go very far out on the ice. No matter, it at least offers the possibility that everything is all right and all the men they have left behind have indeed survived.

'This gives us some hopes that she has reached Point Barrow and broke clear from the ice.'[88]

Staying for lunch and a couple of hours rest in the relative comfort of Anubconna's *Iglu*, availing themselves of his generous larder, Wilkins proves extraordinarily popular with one member of his family.

> There was in this house an old lady who must have been quite 80 years of age. She took a great fancy to me and delighted to stroke my beard and pat me on the back and kept asking V.S. my name and other questions about me. I think he got rather tired of it. The nearest to the pronunciation of my name she could get was 'Oolikan'.[89]

Such a sweet woman, it inevitably makes the Australian think of his own aged parents, both of them now in their seventies, and no doubt fretting for his news, after not having heard from him for nearly a year. At least young George is now getting near Point Barrow, where perhaps they can all be reunited with the *Karluk* and the other men, and he can get a message out via a whaler or the like that he is still alive and going well. Desperate to get to Point Barrow, they march into the dark night . . .

So eager is Wilkins in fact to reach the Point that he momentarily gets ahead of the pack as, despite his overall inexperience, he is in the lead scouting for solid ice paths. Like this one! It is newish ice, and only just wider than a sledge, but after a few tentative steps he feels confident that it is crossable by himself and the rest of the party – and so he proceeds. Getting to the other side he waves Stef and Pyruak to follow in the lead sledge, and their sled dogs pull quickly as they race easily through but . . . Pyruak hears the tiniest *crack*, and it is enough. *It won't hold!* Stef halts their progress in an instant. The dogs howl, growl and yowl as Pyruak jumps off and runs back to yell out a warning, 'Don't stop on the young ice!' The wind whips his words and sprawls them through the air, all that the ears of the men who follow can hear is a single word. Thank God, that word is 'stop'. Desperately they pull back on the pack, and *their* sledge dogs scrape to a sudden halt. They have stopped short of the treacherous ice by centimetres. Unfortunately, poor Jenness, who has been resting his sick body as part of the sledge load, had not heard and hops off onto the edge of the ice. He sinks immediately as the others rush to grab him and the dogs before they are lost. With Jenness shaken but safe, the danger now is not from the ice but from the tongue and temper of Stef. Wilkins puts it with proper Methodist diplomacy in his diary: 'There was a little talk by V.S. (in strong language) about block-headedness in general and we proceeded on our journey.'[90]

There are many ways to die in the Arctic that cannot be helped – but when it comes to sheer blockheadedness, Stef is sure he can help, and tells the Australian what he needs to be told: Wilkins, *you fucking idiot.* Don't do that again.

At mid-afternoon the next day, 12 October – praise the Lord and pass the seal steak – they reach the trading post at Point Barrow to find 30-odd Eskimos, men and boys, playing soccer on the ice – a game that is abandoned the instant they are sighted as the players rush forward and crowd around.

This way! This way!

Mr Charlie, come quickly!

Home is the sailor, home from the sea, and the hunter home from the hill.

Just as Stef knew they would be, they are warmly greeted by the one man he knew would be there, because he is always there – Charlie Brower, a sunny trader who Stef calls 'the most northerly citizen of Uncle Sam for forty years'.[91]

A character like they just don't make anymore, Charlie is 52 years old, and stands so ramrod straight that he might as well be a far Northern oak tree – he makes the solid walls of the trading post seem like a shifty shanty in comparison. For all his strength though, his cheerful eyes – set in a face that is not just weather-beaten, but weather-*thrashed* – dart and start towards anything that interests him, which is just about everything, and most things amuse him. Oh, and while Charlie does indeed hold down an outpost of civilisation, an outpost is as close as he wants to get to the general concept. He is the post, the general store, the mayor, the alpha and omega of this cosy oasis surrounded by ice and snow. He is happy right here with his Eskimo wife, Sadie Brower Neakok, and their gaggle of children living in their grand establishment, measuring some 30 feet by 40 feet – that comes complete with a cook, Fred Hopson, who has taken his own Eskimo wife – so why would Charlie need to go anywhere else? (But yes, as Wilkins is delighted to learn, one of the few places Charlie has been that is not permanently surrounded by snow is Ballarat, where he spent a couple of years as a prospector, and he has retained a fondness for all things Australian ever since.)

Charlie's trading post serves passing whalers, missionaries and repro-bates, and he does not ask questions of the last. Sure, that means that once a year or so he has men of God supping beside men who are on first-name terms with the Devil himself, but so be it.

To these new visitors he is welcome itself, particularly delighted to see his old friend Stef, and while the Eskimos stow the sledges and take care of the dogs, he ushers them inside to give them the news.

Stef and his men are welcomed into something that is part trading post, and part refuge, first built by the US Government after a dreadful storm in the 1880s when $100,000 worth of US ships had been lost nearby, with no fewer than 30 whaling vessels wrecked, and it had been only sheer chance that two US government vessels had been nearby to rescue the whaling men. So, this place had been built as a refuge, the only Western-built structure in 400 miles, measuring 30 feet by 48 feet, featuring a large central room with bunks, a shallow-pitched gable roof and a 20-ton coal bunker on one side. Charlie had bought it from the US Government and run it ever since. To some extent, he is in the busi-ness of providing succour to lost ships.

And yes, both the *Alaska* and *Mary Sachs* had stopped by two weeks ago, and they are all fine, but since then they have not been able to get further east than Collinson Point, about 275 miles away, where he has been told by other whalers they are wintering and are in a perfectly safe place.

But no, not only has he not seen the *Karluk*, but he has heard no word of it from anyone. As they sit safely in brief respite, the thought is on each man's mind: what next? Where next? And will they see their comrades or their homes again?

CHAPTER FOUR

NORTH – AND BEYOND

If you can keep your head when all about you
Are losing theirs and blaming it on you,
If you can trust yourself when all men doubt you,
But make allowance for their doubting too . . .

Rudyard Kipling, 'If'

12 October 1913, Point Barrow, trading places with Good Time Charlie

For now, for Wilkins and the others, it is just a pleasure to be here, to be able to bunk down next to the coal stove, with a full belly and a few drams of whiskey on board, and fall asleep listening to the wind whistling by, knowing they're not still out there, not at the mercy of the elements, but here, safe and secure. Every morning, when they awake, it is to see a new cluster of Eskimos seated and staring at them, fascinated by these fresh-faced white men and their always mesmerising bright red ears and noses. (Seriously, they look to be a very strange kind of men.)

In fact, however, the Eskimos have also come to sell their own goods – furs – to the wily Charlie, and Wilkins is delighted to observe how the unchanging ritual of commerce is carefully observed in these parts.

First the fur is paraded by the successful hunter, his eyebrows upraised – *How good is THIS then, did you ever think you'd see such a large and lustrous pelt?*

Charlie's own eyebrows are also upraised, but in a different manner: *You are KIDDING, surely, and you want how much?*

Some furious negotiations now take place, all of which pale into insignificance when they get to the final fur which is the best of all. Now voices are raised, and a screaming argument takes place, until at long last agreement takes place, at which point, with funereal air Charlie ever so reluctantly hands over the cash, for the Eskimos will not barter, and insist they will only take cash.

But now to the best part of all.

The Eskimos spend every last cent buying goods from Charlie. Where else are they going to spend it? His is the only store within 400 miles.

Of course, there is a middle step that could be eliminated here, but the Eskimos delight in mimicking, or is that mocking, the white man's ways.

On the fourth morning of their stay, Wilkins goes into the kitchen of the trading post, only to hear the cook Fred Hopson swearing at his Eskimo wife, with such force it would peel paint, even as he prepares breakfast. Embarrassed at having stumbled on what must be a major domestic episode, Wilkins sits quietly in the corner waiting for the storm to pass. But it does not, it gets even stronger, and goes for a full 10 minutes, as Hopson makes comment on her parentage, her sexual disposition, her girth and her spendthrift ways, following up with suggestions for what she should do to herself if she doesn't like it.

But the lady in question does not turn a hair, nor even bat an eye, though Wilkins himself turns redder and redder, frozen, not sure what to do. Finally, the cook turns to see the horrified Wilkins staring at them and bursts out laughing before giving an explanation.

'Don't worry about her! She doesn't understand a word I say. We've been married 28 years and have 16 children, but she doesn't know a *word* of English.'[1]

Such is the secret to a long and happy marriage in the Arctic; a free and furious exchange of verbal abuse by the male letting out all his angst of the day and night, all with the blissful ignorance of his wife.

Still, after just five days at Point Barrow, Stef grows restless and wants to get going once more, to go eastwards. That is, until he doesn't, changing his mind at the last instant because he would like to 'write a few more letters'[2] and so sends the obedient, if slightly annoyed, Wilkins to fetch fish supplies instead. The peeved young Australian writes in his diary: 'I thought he might have made up his mind about it a little earlier in the day as it was now 8 p.m. and the sledges had to be unloaded and loaded in a different way'.[3] But orders is orders, and Stef orders Wilkins to take two freshly hired Eskimos and Jenness with him and go out fishing and hunting, before meeting Stef at a designated lake to the south, in a few days' time.

Yes, boss. Your instructions will be followed to the letter, as *you* write letters.

(While always a man disposed to follow orders so long as they make sense, which Stef's usually do, Wilkins is nevertheless becoming a little wary because of the leader's undoubted capacity to have things *gang aft agley*. As Oscar Wilde's Lady Bracknell might put it, to lose one party of your expedition may be regarded as a misfortune; to lose *two* looks like carelessness. Stef's faith in the friendliness of the Arctic leads to a cavalier attitude to risk, and a reluctance to acknowledge the dangers and uncertainties of any separation of men in these climes.)

Heading off, Wilkins and his men trek without incident for two days, hunting and fishing as they go, to fill their sled with fresh food, before coming to a wide windswept bay that is covered with ice. Using the few English words they possess, and many hand gestures of unmistakable meaning, his Eskimo guides strongly advise that they go around the bay. Of course they do! Wilkins is wise to them. Given they are being paid a daily rate, Wilkins figures they are simply trying to make the trip go longer by adding another 40 miles. The Australian has no interest, not least because he is not a fool and also because Stef has given him explicit and detailed instructions to walk across the bay. But all right, for the moment, and for his own amusement, Wilkins asks the Eskimos why they want to take the long way around.

'We must not cross,' they say with their hands, 'a storm is coming.'[4]

(And a big and angry storm, too, to judge by the billowing black clouds that their waving hands and ghastly expressions indicate.)

What nonsense!

He had thought they would at least come up with something a little more plausible than that. The weather is fine, with the sun shining benignly, the breeze replete with briny air, and nary a cloud in the sky. And so Wilkins makes his first command decision: *press on*. Again the guides indicate their extreme reluctance – to the point they look like they are on the edge of mutiny – but Wilkins insists.

To show them he is serious, he and Jenness and their sled dogs head out onto the lake. No more than 100 yards out though they stop and turn, to find the Eskimos still on the shore! Waving his arms, Wilkins gestures to them to come on!

Sure enough, both Eskimos now run towards him.

But it is not what he thinks.

When they get to him, they have tears running down their cheeks and exhort him: 'Come back! Come back!'[5]

What is going on?

Wilkins has his orders, and those orders trump the concerns of the ignorant locals. He pushes on, and finally one Eskimo gives up while the other gives in and they both miserably trudge on behind Wilkins and Jenness. The Australian smiles. He has successfully called their bluff, and is not even concerned when no more than 15 minutes later the wisps of wind swirling the snow around his ankle start to whirl and then get so strong they must *lean* into it.

And herein is the seed, root, branch and bloody big toppling tree of the problems they now face. For when the serious storm crashes upon them just minutes later, with a howling wind picking up the snow and hurling it into their faces, the sled dogs don't merely lean, they *charge* into it, just as the Eskimos had known they would. Perilously close to being out of control, all they can do is hold on and try to keep the sleds upright as they career out across the lake in the middle of the raging storm.

The wind and snow beat them back, the dogs drag them on. Wilkins and his men are trapped in a Sisyphean struggle with the storm and the scentless cold. It whips against their skin in a way the white men have never felt, but strangely, after a short time, Wilkins stops feeling the cold, and concludes that he must be getting used to the conditions.

It is his second potentially fatal mistake of the day . . .

When, finally, the dogs have run themselves out, by mid-afternoon on this 29th day of October 1913, the men are able to make camp on the ice, hurriedly putting up the tent.

'This was no easy matter and by the time we had it erected we were stiff with cold and were glad to get inside.'[6]

The dogs are now so exhausted they are left in their harnesses, and take shelter in the lee of the sledges.

'The wind seemed to be increasing every moment and the tent was flapping about and letting in the cold all sides.'[7]

And yet, even after they have the flaps secured and a blubber fire going and are warming inside, it is the damnedest thing.

'The Eskimos seemed to be very much concerned and soon would not answer us when we spoke to them, on looking up we saw that they were praying and the tears running down their cheeks.'[8]

What is going on?

'What is the matter,' Wilkins asks. 'What are you making such a fuss over?'[9]

'The ice upon which we are camped is not very thick,' they wail, 'and it may break and take us out to sea at any time.'[10]

Surely not!

Still, he does remember a story that Stef had told him about the death of two white men and some Eskimos, who had been carried out to sea on this very bay during a blizzard some years ago. But surely that is not going to happen this time, is it . . . ?

Hang on . . . is that it!

It suddenly feels . . . as though they are on a . . . boat . . . ?

Oh, CHRIST!

The sheer horror of their reality freezes him to the marrow in an instant. 'We felt the ice heaving up and down beneath us and knew we were afloat.'[11]

The bay ice is breaking up!

Clearly, he should have listened to the Eskimos as they knew all along what they faced. Perhaps they know what to do now then?

Wilkins looks across.

Right then.

They are crying as never before, and praying and singing snatches of every hymn the missionaries must have taught them in their youth.

The wind wails, the snow swirls, the ice cracks all around them, and the Eskimos continue to warble, '*Onwards Christian soldiers marching as to war, with the cross of Jeeessssus, marching . . . something, something, something . . . At the name of Jeeesssus, Satan's host something; On then, Christian soldiers, On to victory!*'

First, as Wilkins has already learnt on this trip, they must get through to dawn and, doing his best to ignore the sobs and the hymns, he does indeed manage to get an hour or two's sleep. Awaking at dawn, he realises that their tent is so heavily covered in snow that it comes up on one side to within a foot of the top, and on the other side to halfway up. It is still snowing so heavily they can see no more than 100 yards, so it is no small thing and very time-consuming to dig out the sleds, feed the dogs and get going once more. But at least it feels like they are no longer afloat and that their island of ice has now settled down to rejoin the rest of the frozen bay.

The order of the day remains: get off the bay! Get back onto solid land! And so they continue to push hard, straight into the cutting, freezing wind.

'It now became intensely cold and the newly fallen snow would blow against one's beard and freezing there would soon form a solid mask of ice.'

Wilkins knows how to be rid of this, thanks to Stef's instruction. The method is simple and ingenious, simply place your naked hand on the ice on your face, and melt it with the heat of your own skin and blood. Then place your hand back into your mitten, and warm it there. Rinse and repeat, as it were; and would that it were possible. As with so many of Stef's ingenious and simple theories, a small flaw of reality springs up to dash a damn fine theory. For snow does not know about Stef's theories about where it *should* go, and as Wilkins is melting his face with his hand . . . 'It blew into my mittens . . . and as soon as I removed my hand the mitten would freeze stiff, therefore I could not keep taking my hand out so often to melt the snow in my face.'[12] So Stef's method is abandoned and Wilkins finds his beard covered in a mass of ice.

> I could not open my mouth but just had a little opening between the lips through which I breathed. My forehead and nose were also covered with ice and so stiff that I could not move them by voluntary muscular movement and when I tried to rub these parts with my hand I found they were quite stiff and solid. I had not felt much pain only intense cold which gradually changed to numbness therefore I did not think that these parts could be frozen . . .[13]

Things are grim, and it is not just Wilkins who feels it.

Every now and then the dogs simply stop and yowl, curl up and refuse to go on.

Wilkins has sympathy, for he feels precisely the same. But they must move or die – and so he lashes and shouts until the animals get going once more. And yet, as the snowbanks continue to pile up in their way, moving becomes ever more difficult.

> The drift became so thick I could not see with the one eye that remained open further than the handles of the sledge from the leading dog. My other eye had become closed owing to the lashes freezing together and as soon as I would thaw the ice off them

they would freeze together again with the moisture left from my fingers.[14]

This is the ninth circle of hell.

Finally they reach solid ground, some five miles from where their guides know there is another Eskimo settlement, and they head towards it.

> When we reached the *Iglu* the men and women came out to see who were the foolish strangers travelling in this kind of weather and when one of the women came to shake hands with me she started talking excitedly and commenced to rub my forehead. I realised then for the first time what was the matter and as the men belonging to the house kindly offered to look after the dogs in the sledges I went inside to get thawed out.[15]

Oh, the *relief* of being on a solid surface once more, of being out of the wind, and in an actual shelter that comes complete with warmth from a blubber fire, and the smell of seal steaks on the griddle, together with . . .

What . . . ?

'Before I could even sit down, an Eskimo woman seized hold of me and hastily thrust me out of doors again.'[16]

What on earth is going on?

One of his guides finally – and seemingly reluctantly – explains.

Mister has terrible frostbite all over his face and hands, where the flesh has frozen solid, and with the sudden warmth is about to experience agony. As the flesh thaws, blood will be trying to circulate and it will send such bolts of pain through him that for the first 20 minutes he will be afraid he is going to die. And for the next 20 minutes he will be afraid he's not going to die.

But the only way to save his flesh and ease his suffering is for him to be brought to the warmth gradually.

The guide is right about the pain.

'My face and hands and these began to burn . . . as if somebody was holding a red hot iron to them.'[17]

Following the advice, Wilkins spends the next hour just inside the outer *Iglu*, before each hour thereafter moving a little closer down the passage until finally he is allowed the warmth of the centre. Even then there really is nothing that can be done; as the pain indeed comes with the warmth, truly biting into his hands and face as the frost fades. Tentatively he feels his face, only to hear a popping sound. It is covered in blisters.

> I had no previous experience with frostbite or thawing out, no-one had told me in particular what to do under such circumstances and I learned from bitter experience that I had done the very worst that I could have done. The thawing of deep frostbite is painful under any circumstances, but to thaw quickly against an open flame is about the worst that can be done, and soon the skin on my hands and face was rising in huge watery blisters; the pain was almost unbearable.[18]

While shaking with the agony of it, the thought occurs to Wilkins. Today, 31 October 1913, is a significant day in his life. It is his birthday. Not a Happy Birthday so far.

Today, though? Today his face burns like fire as it gradually stops freezing.

'I am twenty-five years old today, but if I live another two times twenty-five years I do not wish to pass another such day as we have had today.'[19]

Ironies abound.

'On my last birthday . . . was the first time that I had seen actual battle and was actually "under fire" but today it seemed as if I had been actually in the fire for my face presented the appearance of a gigantic blister.'[20]

For all that, right now, he is glad to be alive at all given his sheer stupidity at crossing the bay during a storm, against the clear advice of his local guides, who knew a whole lot more than he did. They remain deeply upset with him, showing their displeasure in their manner. It is not with anger but by completely withdrawing, a kind of permanent sulk that cannot be removed by apology or payment. They wish to shame him into understanding what he has done – and are successful. For even through his pain, Wilkins is deeply humiliated and has learnt a lesson he will never forget – *never* be cavalier when an Indigenous man or woman tells you anything about their own lands and ways, and if it is a matter of life and death then *lean in*.

In an attempt to inject some normality into this extraordinary day, Wilkins starts to write a letter home; although when this letter will reach any port that might be able to send it on its way, is another matter.

His mother and father will read these words, some day. Simply writing down the unreal things that have happened makes them seem more real somehow; though they still read like an adventure story you wish was happening to someone else.

> We heard on arrival that a ship, which may possibly be the *Karluk*, had been seen drifting in an ice floe towards the north-west. If that is so she may drift across to Asia or towards the North Pole, and may not get free for years. She may be crushed by ice at any moment, or continue to drift backwards and forwards past Cape Barrow all the winter and then get free next spring . . . To give you an idea of how extraordinarily difficult the ice navigation has been this year I might say that of all the ships that passed Point Barrow not one came back, and of those that reached Point Barrow only one got away and then she was wrecked. Fortunately no lives have been lost.[21]

Wilkins also decides to let his parents know a little of his Eskimo guides. 'They had a splendid treat in honour of my birthday . . . Just fancy frozen fish and whale oil for a birthday feast.'[22] He knows his family will chuckle at the thought of George, the boy who was so greedy and bold he always asked for bread and jam and *extra* cream when they went to the neighbour's farm, having a fancy feast far from his fancy. *Happy Birthday, George.*

After resting for three days, new guides – for the former ones decline to continue – take the visitors to the designated lake, where there are indeed fish aplenty, all of which are frozen within minutes of being pulled from the water.

And yet, while the physical rest is welcome, lack of activity hangs heavy in the encroaching darkness of the coming polar winter. A dark day of doing not much in particular can feel like a decade. And it is now only getting worse, as each day gets dimmer, more depressingly dismal and ever more freezing. The sun goes even lower than your spirits, the days are as short as your temper and while you understand that bleakness brings weakness, some days it really is better just to let go and feel exactly that – bleak. After all, who could not feel bleak when the only difference between you and a chunk of ice is the fact that the ice is not wearing *mukluks* – and is probably not quite as cold as you? All you can do is wait it out and . . .

And what is *that*?

Wilkins, Jenness and their Eskimo guide are in their ice house, fast asleep, near midnight, when they are awakened by something walking across the *roof*!

Convinced that one of the dogs must have got loose, Wilkins is about to tell the guide to go out and tie it up when he now hears footsteps just outside, whereupon the door is suddenly thrown open, and is followed by the head of Stef who, as recounted by Wilkins, said 'in an insinuating way: "You have all been sleeping for some hours I suppose."'[23]

Well, yes, rather.

'It was now 10 minutes to 12 by the clock,' Wilkins chronicles, 'and it gets dark about half past three in the afternoon. I don't know what else he would expect us to have been doing.'[24]

But that is just Stef.

It is as well that they are reunited in any case and the following morning, all together now, they start out on their sled to journey no less than 300 miles to Collinson Point, where they are hoping that the men of the *Alaska* and the *Mary Sachs* are waiting for them. They also hope that Stef will stick to one purpose and one point, and not divert and delay for frolics of his own interest. Their leader is a brilliant man, but perhaps not a brilliant leader. It takes the patience of Buddha, the stoicism of Job and the strength of Goliath to get through a day with Stef without rolling your eyes at his theories, sayings and dictums that flow in unending proclamation.

As ever, Wilkins learns – in this case, some of the things a leader of men should *not* do. Don't carry on. Get on with it!

21 November 1913, unsilent night

Truth be told, Stef has been getting on Wilkins' nerves more than ever lately, not least for the fact that, even when day is done and it is time for rest, his restless energy keeps going into the night. He likes to talk philosophy for hours while the other men lie in their sleeping bags, waiting for their leader to shut the hell up, even as the wind whistles outside and there is the occasional crack of ice. On this particular night Stef and Jenness have a 'theosophical' argument that goes for three hours, yes, THREE HOURS, arguing back and forth whether the Buddhist concept of reincarnation is possible. The men in general have no opinion about reincarnation, but they do know that hell exists because they are in it right now. Stuff Buddha. *Jesus Christ*, will Stef ever shut up? Thank God he does, but only to make a cup of tea then the jaw-jaw resumes, as the rest of the tent silently contemplate a couple of crucifixions they would like to see.

Despite his exhaustion, Wilkins, as ever, is up at 4.30 am and gets busy with various chores. Everyone else remains asleep, including Stef, who does not wake himself until 6.30 am.

'I thought you wanted to get an early start this morning?' Wilkins offers to the leader.

'Yes, are you nearly finished washing dishes or do you want some help?'

'I think I can do this all right,' Wilkins says pleasantly enough, before becoming more pointed, 'but if nobody else wants to go out and load the sledges I will let them do this and go out and do it myself. I don't mind getting up two hours earlier than anybody else and getting breakfast but I don't intend to wash the dishes and watch them lie around until I have finished, and then go out and load their sledges for them while they're putting on their boots.'[25]

Stef pauses.

'I don't think it is quite right either,'[26] he replies with renewed respect, before rousing the others. (It is impressive but Wilkins does note that Stef himself stays in bed. *Do as I say, not as I duvet.*)

Three weeks of trekking follows, each bitter day collapsing into the next as they collapse at night, thankfully too tired even for Stef to talk. But even the endless must have an end; and sure enough, after coming over an ice-ridge at 11 am on 15 December, there they are! Both vessels are anchored securely in a small cove, and smoke coming from the shore shows where their crews are to be found. After shouts back and forth, they receive a hearty welcome from their long-lost companions, but amidst the cheer – for they are all healthy and have lost no-one – something else is on Wilkins' mind.

> I'm extremely glad to be associated with all of these fine fellows but I can't overcome a feeling of remorse that I'm not back on board the *Karluk* and if I was given my choice I would assuredly prefer to be on the *Karluk* than here . . . I cannot reconcile myself to the fact that we shall not see her again for months if ever. There have been quite a number of boats caught in the ice under similar conditions and which have never been heard of since but I can't

believe such a fate will befall the *Karluk*; yet the fact of her in that precarious position prevents any feeling of elation that we might have had this reunion.[27]

There is one thing that cheers Wilkins, however: food. Proper food.

> The dinner tonight was a revelation to me and was far better than one would ever get in a 10/- a day hotel in London . . . We have been perfectly well satisfied with the food we have been getting at the Eskimo houses and enjoyed every meal but when seated before the table covered with so many luxuries, comparatively speaking, I wondered how I ever managed with the food cooked by myself or the Eskimos.[28]

And the man responsible for this delight?

'The kitchen is presided over by Charlie Brooks who possesses all the good qualities of a cook without the usual grouchiness that one often finds in the members of the culinary craft.'[29]

Even more satisfying for somebody who was employed by Gaumont to go to the Arctic as a cameraman: Wilkins finally gets Stef to get his hands on the most crucial element of the job – a camera. In a rather bizarre case of history repeating, the movie camera has been left on board the ship *Polar Bear* by a Mr Hudson, a reporter from the *Seattle Times*, Hudson has headed out on land in hopes of getting 'big game' pictures. And so it comes to pass that Wilkins, a cameraman who left his movie camera aboard the *Karluk* to go and hunt caribou with Stef, now gains the movie camera left behind by another snapper, who left the *Polar Bear* to go look for polar bears. Old hands think that Mr Hudson's 'hunt' is unlikely to bring success, but Wilkins' hunt is over; he can shoot and show pictures at will.

Christmas Eve 1913, Collinson Point, Blue Christmas

In most parts of the Western world, Christmas is a time of celebration, of family, giving thanks and celebrating the birth of Christ. Up here closer to where Santa Claus actually comes from, it is above all a break from the tedium and a chance to engage in good cheer, at the time when cheering up is most needed. In the middle of a polar winter, it is not easy to have a sense of occasion, but Christmas provides it and nearly everyone gets into the spirt of it. They fashion decorations out of tinfoil from food wrappings, write menus, make small gifts, place a Christmas tree in the corner, and most of all prepare a feast of rolled oats, pemmican, sugar, rice and pork crackling. Sadly, the one exception to this general rule of celebration is Andre Noram, one of the cooks, known by all as 'Norm'.

> He's been having some hallucinations lately they say and has been
> very despondent; several times threatening to do away with himself,
> and one day handed O'Neill a note and started off with his gun,
> O'Neill persuaded him to go back into the house however and he
> is become a little more rational since.[30]

As upsetting as it is for all concerned, the old hands in these parts are not surprised as some measure of 'cabin fever' – that despair of the doldrums brought on by icy isolation, the ceaseless sameness that drives a man up the very walls he stares at day after day after day – takes hold. It is so close to madness you can't really tell the difference as the whole experience razes reason with the unending season.

And yet?

And yet by cosmic coincidence Christmas near the North Pole generally comes with added oomph due to the fact that the 'Arctic midnight' comes on 21 December, a day celebrated for the fact that from now on every day will be a little longer. Only weeks from now there will be enough light at midday to match twilight in an ordinary place – like having a night-light in another room – and by early February, the sun will actually consent to appear above the horizon for the first of ever longer cameos!

Christmas approaches with the collective pulse already a little quickened and, to lighten things further, Wilkins decides to show a gathering of Eskimos some films for the first time in their lives – three reels, amounting to 4000 feet, of different scenes from around the world – and is fascinated to see their reactions. He observes that while they watch dramas and comedies and barely turn a hair, the instant he puts on a short nature film complete with lions, tigers and elephants, 'the Eskimos bolted under benches and through doors'.[31]

(It is the remarkable reverse of the famous story of Parisians watching the first commercial movies in the early 1900s. The French *bourgeoisie* barely blinked at lions or tigers, but the moment there was a vision of a train coming right at them – RIGHT AT US! – they made a mad bolt for *les* exits. Wilkins observes that people react to the threat they most recognise.)

All up, it is a great success, and not for nothing will the Eskimos come to call him '*teeum-MeeukPoeng*', meaning 'He-of-the-wonderful-eyes'[32], a reference to the camera he wields and the images he can capture.

Of course with Christmas Day being celebrated so far from home, mixed with the joy of the season there are inevitably melancholic thoughts and Wilkins personally thinks of his large family in South Australia, likely gathering around the table of his parents, with only one notable absence – himself. And so too would all his friends be with all their families.

'I wonder,' he notes in his diary, 'how many of our friends were thinking of us today and if their thoughts of what we were doing were in anyway correct.'[33]

Could they possibly imagine his current situation? Few things are so odd to an Australian as a cold Christmas, and Wilkins is in a land where it is so cold that if the temperature rises above freezing it usually means there is something wrong with your thermometer, because it is ALWAYS freezing. At home in Adelaide, of course, they will be sweltering, even as they carve up the turkey and down the plum pudding, just before retiring to the verandah to down several cold beers. Here? Well, he is about to sit down in a hut, to a Christmas lunch of, as per the written menu – OLIVES. ENDICOTT SOUP. ROAST STUFFED DUCK. GIBLET GRAVY. CRANBERRY SAUCE. MASHED POTATOES. STRING BEANS. CHRISTMAS PUDDING. MYSTICAL SAUCE. MINCE PIES. DATES. CHOCOLATES. FRUIT. COFFEE. and SMOKES – with some 20 others.[34]

Oh, and there should be one more with him, but here in the polar winter, Norm the cook is still missing, despite their searches at nearby cabins. The presumption that he has followed through on his 'threat to do away with himself' puts something of a dampener on their meal, as does their continued worries about the fate and whereabouts of those on the *Karluk*, but they do their best, aided by the careful sharing of a single bottle of whiskey that is handed around. And now for the presents!

After the third sitting is cleared, the Christmas tree is set up at the end of the table and the presents arranged in front of it.

'My Christmas presents,' Wilkins records, 'consisted of "Christmas Stories", by Dickens, a silk handkerchief from Mrs Anderson (the wife of our leader, this present was very much appreciated), a box of cigarettes and an automatic lighter, a Scotch shortbread, a mechanical puzzle and a bundle of pipe cleaners . . . It was well after midnight when we went to bed and although we had everything that was possible under the conditions to make the day enjoyable there was an underlying feeling of sadness to all our merriment.'

Christmas cheer can come at any time, and this time it comes from a female admirer in the evening.

> I was agreeably surprised by one of the Eskimo women . . . Nunia by name, came to me and made me a present of a pair of *mukluks*. I have not the least idea (where) why she did this, for I had not noticed her particularly when we stayed at their house . . . I believe I'm the only one in the outfit to receive a present from an Eskimo lady and I consider myself highly honoured.[35]

On Boxing Day a despondent search is made for the no doubt deceased Norm the cook. Three sledges full of men head out to the east, while still others walk over the tundra. Wilkins is one of these walkers, making up a two-man party with another man by the name of Cox, though neither have hopes of finding anything bar 'a huddled up corpse, for we thought that nobody could at his age, 58 years, and in the condition he was, could withstand the weather we had had, and still be walking about if he had not had any food or shelter'.[36]

But what now? Just after they climb a small bank, Wilkins lifts his field glasses to his eyes to focus on something moving in the distance and says in wonderment: 'I do believe that is Norm!'[37]

Good God! Yes, surely it is, because even though it is a far-off sight Wilkins spots the distinctive manner of Norm's walk. 'I noticed that he stepped with one foot further than the other when he was leaving the camp some days ago,'[38] Wilkins tells the still sceptical Cox. Cox's doubts and the sheer improbability of Norm being upright lead to Wilkins taking another look and changing his mind.

> I came to the conclusion that walking over the tundra would account for the halting steps and that it must be one of the Eskimos taking part in the search. For we could see that whoever it was, he had taken his arms out of his sleeves and had them under his parkie. This is what the Eskimo do in order to keep their hands warm.[39]

Cox takes a look through the field glasses now. 'It is not tall enough for Norm.' They keep walking towards the figure and . . . it is Norm!

The three men stand looking at each other in shock, before Wilkins finally speaks. 'Are you frozen?' 'In any part?'[40] adds Cox helpfully. Norm thinks for a moment. 'No, I am not,' he answers, and he might technically be right, but Wilkins and Cox can see that his face and hands have very recently been frozen. They offer to carry him but Norm says firmly that he is quite able to walk home without any assistance. And so they walk, with Cox handing the cook a bar of chocolate. 'I feel hungry,'[41] says Norm as he nibbles cautiously at the edge of the chocolate. *Well, you would, wouldn't you?* But wouldn't that be cured by eating the chocolate? Wilkins and Cox exchange a look, they are both thinking the same thing, that Norm is with them but not quite *with them*.

'I could smell the coffee from your camp this morning,' Norm informs them, although he would have been at least two miles away at the time. Is he mad? Well: 'He was as rational as any man could be after being out in such weathers we had had for the last two nights and three days when the night constitutes 22 hours a day . . .'[42]

And so the trio make their way back to camp, and as soon as he is inside a hut Norm falls in exhaustion. Two nights, three days and the incredible will to live collapse at once. The sad coda to this tale is that, having survived such an ordeal, Norm will not live long. It is not the cold that kills him, it is a bullet to the brain from his own gun. The Arctic winter can be deadly for mind as well as body, and another man is lost to its bleak weaponry.

Things settle down, as they resume their normal positions waiting for Stef to return from yet another trip and provide them with, literally, a direction. And yet as the weeks drag by with still no sign of him, Wilkins can barely bear it any longer. An excursion is called for; something to break the monotony, lest they all end up like Norm and go mad from cabin fever. Dr Anderson agrees and, after the sledges are packed, they head off on a circular recce, a polar checking of the boundary fences. They move fast until on 22 January 1914, at 2.30 pm, they reach what is known as the 'Polar Bear' Camp where they come across none other than – and Wilkins chooses his words carefully – 'the notorious Captain Mogg, the fat overbearing pigheaded ice pilot who ruled with an iron hand the ship and all the crew from sheer knowledge'.[43]

Oh, he's serious all right. And Mogg is infamous for disregarding all charts and the first principles of navigation with such things as modern sexton chronometers.

'What is the use of these things on the ship anyway?' he would say. 'For in bad weather you can't use them and in good weather you don't need them.'[44]

Wilkins is fascinated by this wildly colourful character, noting him as a 'sharp featured man . . . He's the type that one would expect to read about in one of Dickens' works . . . smoking a dirty old pipe, using it occasionally to demonstrate his arguments.'[45]

In the meantime, they have visitors.

It is a notably beautiful Eskimo woman by the name of Pungublock, who arrives with her young child. Many years ago she and her husband had accompanied Stef on some of his Arctic wanderings before the husband had died by misadventure, and she had stayed on with Stef. They had become close.

Come to think of it . . . ?

> When looking down on the child from the side there is a striking resemblance to V.S. to be seen and I believe there is some truth in the common report along the coast that it is his child. Prof seems wilfully indignant that V.S. had spoken so much about keeping the white men from intermingling with the Eskimo while he himself had a child up here . . .[46]

(Just quietly? Just quietly there is a certain irony in the fact that the man who has claimed the discovery of 'Blonde Eskimos' with an ancient European lineage has himself produced an Eskimo with very modern European ancestry. While not blond, the child is noticeably not as dark as its peers.)

Still, on this occasion Pungublock has just popped in for a visit, and will not be travelling with them. Wilkins notes that Pungublock must weigh at least 200 pounds, and quietly opines that it is unlikely she would be able to hike far in any case.

They press on.

With the end of Christmas, good cheer has faded among the waiting party, their mood spoilt by the uncertainty of the location of the rest of their brethren. The *Karluk* is still lost, perhaps forever, and there is no escaping the fact that this whole expedition could make a dog's breakfast look well organised. They have lost ships, personnel and equipment – everything so far, bar their lives.

Typically, Stef simply doesn't care what any of them thinks, and blithely announces his intention to go on with the original mission – yes, they will explore the Beaufort Sea. Those of you whingers who wish to wait for a ship that may not arrive for months may do so, but what I need are volunteers to come exploring right now.

Equally typically, George Wilkins from South Australia – though his face and hands are still covered with frostbite scabs that will stay with him for weeks, and only be completely gone a year later – is the first to put his hand in the air. Momentarily, the upright Wilkins arm looks to be a lone pine where he had been hoping to see a forest of arms, until there is at least a handful who will join.

Anything bar the sheer tedium of staying here any longer. Alas, that exact position describes them all for the slow-moving stalemate that now unfolds as Anderson argues. After all, Anderson argues – and argues some more – given that Stef has lost his men and equipment, it is Stef who must pay the price for that. For the life of him, Anderson can see no reason why his Southern Party's mission should be jeopardised.

In reply – and at some length, giving voice to a silent feud that has simmered for days, then weeks, then *months* – Stef points out that as he is the ultimate leader of the expedition, he gets to decide who goes where and with what.

Anderson replies that Stef is the leader of the Northern Party only and he is more than welcome to take what remains of it, and its equipment, to the North.

It all comes to a head on the night of 9 March 1914, when Stef gathers all his men, including Anderson, in the one tent, and takes them through it.

'Stefansson,' Wilkins records in his diary, 'explained at length that he was the LEADER of the whole expedition and Dr Anderson came to the conclusion that he wanted every order written, signed and delivered to him before he could feel justified in carrying it out.'[47]

Two days later, Stef approaches Wilkins to ask if he would be ready to leave the 'day after tomorrow together with Johansen'.[48]

No.

'I do not leave until *you* do,' Wilkins tells the leader.[49]

(Given how much better I know you now, I have no faith in any of your blandishments, your promises, that you will follow me on any given day. For we both know that, when that day comes, you will do whatever suits you on *that* day.)

Finally, on 16 March 1914 – just a week before the Equinox, so the sun now well above the horizon for over ten hours a day – Stef, Johansen, Wilkins, Chipman and Cox with their guides set out for Banks Land, some 800 miles away in the outer reaches. And yet, while it's true, as the Chinese say, that 'A journey of a thousand miles begins with a single step', so too is it true that such a journey in the land of the blizzards will be hit by one, not long after that single step is taken. In this case they are hit by a howler just a couple of spare hours after setting out, and it goes for two whole days. Then, even as they are fighting against it the best they can, their troubles soon multiply with the loss of a great deal of paraffin by accident, while three sleds break and the sled dogs exhaust themselves by pulling into a shrieking wind that simply won't stop.

Yes, despite the viciousness of the blow, they are all okay, but Stef realises that to keep going without fresh supplies is far too risky. He needs a man to go back and . . . okay, George, thank you . . . get supplies. You can take Natkusiak, my main Eskimo guide, who has been with me on all my previous expeditions. Stef will set up a camp here so they can rest themselves and the dogs and hopefully Wilkins and Natkusiak will be able to return to them within four days.

Wilkins takes an instant liking to Natkusiak or 'Billy' as Stef and the old hands call him. 'He was the most capable Eskimo I have ever met, a roly-poly little fellow, always jovial and good natured with a keen sense of humour, interested and wide awake.'[50]

Stef is equally admiring: 'In all my long travels and in everything of difficulty which I had had to undertake,' he would recount, 'Natkusiak had always been my mainstay and in many cases the only man on whom I could rely.'[51]

(As well as being an expedition bulwark, Natkusiak has a keen and elaborate sense of humour – no small thing to possess in these climes, as anybody who can consistently lift morale is worth their weight in gold. One of Natkusiak's favourite jokes is to dress up in a complete polar bear skin and 'attack' new members of the expedition from behind, starting with – what else? – a bear hug. A real one. The joke is always extremely well received by everyone, less the new recruit, who more often than not has to go and change his underwear. Another of his tricks is to show off his magic powers on first contact with other Eskimos, where he would 'light his pipe with a little stick, inhale the smoke, and after a pause blow it out through his nostrils, telling his audience that there was fire inside his body'.[52])

And yet, despite his capacity, neither he nor Wilkins can prevent a second dreadful blizzard hitting them even before they get back to the main camp, so bad that it nearly drives the dogs mad. In their frenzy the dogs break loose from Wilkins' grip and, still harnessed, rush off into the snow, dragging the broken sleds behind them. The result is that, not only must Wilkins and Natkusiak complete the journey back to Anderson on foot, but added to the list of things required now are two more:

More dogs.

New sleds.[53]

Whatever the trials and travails of the day, a glance upwards renders all problems small. For Wilkins is transfixed by the majesty of one of the most stunning marvels of the Arctic.

'On going out to home,' he chronicles in his diary, 'I saw the most beautiful aurora I have ever seen. It was moving rapidly, but formed a shape like a wheel with two broad bands stretching away from it; the streaks representing spokes were moving up and down and were most brilliantly coloured. This is the first coloration I have seen in an aurora.'[54]

This ethereal canvas of the cosmos is enough to make any man forget the mundane for a moment. But the moment must pass and Wilkins' mind goes from the glory of the sky to the challenge of keeping moving, despite the agonising pain in his feet. The key to travel in these parts, he has learnt, is working out where ice lies, where land lies, and as much as possible going from secure island to secure island, across the ice that promises safety in appearance but will deliver disaster to the unwary, for it is only with solid land beneath you that you are truly safe.

Once they've gathered the necessary items at Anderson's camp, they return with great haste to Stef's camp, anxious to impress with the speed only to find . . . open water. Getting *déjà vu* all over again, there is no trace of Stef, their camp, or any of the men, And equally there is no clue if he has moved with moving ice – as the *Karluk* had before him – or simply with impatience. Wilkins and Natkusiak spend five days trying to trace the likely path of the advance party, only to be confronted by the unwelcome sight of open sea at every turn. With no choice, all the two men can do is return with their supplies to Anderson's camp, where the raised eyebrows and muttered imprecations of Anderson himself say it all: whatever talents Stefansson has for making the Arctic friendly, his greatest talent seems to be a toss-up between losing his men and being lost by them. What can Wilkins say or do in reply? Not much at all. The point is, frankly, not unfair, and the only thing the Australian can do is wait for a message – on the presumption that Stef is, in fact, still alive to send it.

Given that this wait could be for many months, as Wilkins has 'no equipment with which to carry on my photographic work'[55] he typically sets out to make his time at Collinson Point as productive as possible by learning everything he can about the science and reality of living in the Arctic. Anderson and his men have an excellent library of science volumes, which proves to be a good place to start, and Wilkins devours them at length and at leisure. The Eskimos are more than pleased to guide this inquisitive, enthusiastic Australian whenever and wherever he likes to try to prove that he can, with practice, perform the practical matters of living in the Arctic which they do with ease every day.

The task that most fascinates and challenges Wilkins is how to hunt and kill a bear to gain the flesh and, most importantly, the pelts – something the Eskimos have done for time immemorial with a knife. Yes, these days, any fool – and in fact many fools – can shoot a bear, but this student *par excellence* of living-in-Rome-the-way-the-Romans-do has no interest in that. He wants to take down his prey in the ancient way, the way they do.

So, how is it done?

Well, it is not done by white men.

No, really, how can it be done?

By us. Not by white men.

They are happy at least to explain their technique, which turns on, once the bear is sighted, letting their sled dogs distract it, which allows the hunter to get close enough to throw himself at the bear and stab the blade of his knife beneath the bear's shoulder into its heart.

To the Eskimos' amazement, Wilkins doesn't blink. And he doesn't back off from his desire to have a go himself.

'It seemed to me that there was not much to that and an agile man with a steady hand could do it. On one day I boasted that I could do it if I had a proper knife.'[56]

Oh really?

His Eskimo guides roar with laughter, before gathering themselves to tell him that it might be better to leave honourable killing to them, while he and his fellow white men stick to their noisy guns.

'*Naga Naga*, bear too quick, white man more better stay away!'[57]

But Wilkins is so insistent that he is up to the task, the Eskimos decide to teach him a lesson. They actually sharpen his blade for him and tell him he really can have a go.

Surely *that* will shut him up?

No. Wilkins assures them that he can't wait.

And as fate has it, Wilkins' chance comes just two days later when an enormous polar bear is seen near the camp. The Eskimos unleash the dogs, which set out in hot pursuit together with Wilkins and the *whole* camp – for if the white man really is going to have a go, no-one wants to miss it – following tightly behind.

'We ran on foot to where the bear had bailed up and was dealing fairly successfully with the dogs.'[58]

Ah.

That bear.

It is not just big.

It is MASSIVE.

Standing up and swatting at the snarling dogs, it is more than 10 feet tall, bigger by half again than the biggest Eskimo.

So, what now, white man? One last chance to bail, Wilkins: 'You get killed? More better we shoot first; you stick afterwards?'[59]

'No,' says Wilkins. 'This is a good chance. I'll stick him, and if he does not drop, *then* you shoot.'[60]

Nice theory, white man. Well, the Eskimos tried. One last piece of advice, on timing: 'Now you go.'[61]

Yes, easier said than done; and the Eskimos still do not think Wilkins is man enough or mad enough to do it. They are wrong on one count as he indeed takes the knife in his right hand, gripping it tightly and steeling himself before . . .

'I dashed in and swung a vicious stab at the shaggy coat.'[62]

Alas – ALAS! – unfortunately, the knife barely cuts into the thick hide, let alone deep into the flesh to provide a kill. All that Wilkins has done is infuriate further an already enraged giant bear that is directly in front of him. Worse still? The bear is no longer distracted by yapping dogs but focused furiously on the Australian, surely sent to him by the bear Gods.

It is time to close in for the kill.

The bear knows it and sways forward, its upper right paw with the cruel claws drawn back, its snarling white teeth glistening as it unleashes a roar that would daunt the Devil.

This one swipe will take his head off, and Wilkins is powerless to do anything to stop the bear until . . .

He trips over backwards, tangled with one of the yapping sled dogs. The deadly paw misses his head by an inch, which so badly unbalances the bear as it comes back on all fours that the sled dogs take their opportunity and jump upon it, snarling, biting and drawing blood.

Mesmerised, hardly daring to believe he is still alive, Wilkins now suffers his second canine slapstick humiliation.

'I was almost winded by a dog that bounced on my stomach and then I heard a shot.'[63]

The enormous beast falls, dead before it hits the snow in a soft slump that defies its sheer size. Wilkins has bested the bear. Or at least, his companions have. Those same companions now run over, struggling to keep straight faces and not bellow with laughter. But once they see Wilkins is safe and sound, they stop trying, and soon they are doubled over, struggling for air and wiping frozen tears from their cheeks. One particularly sarcastic Eskimo makes no mystery of what they find so funny.

'That old fellow bear he plenty tough, you not strong enough to put knife through his skin. More better you try to kill baby bear. Or baby seal.'[64]

Again there are howls of laughter, so long and so loud, but also so good-hearted that even Wilkins himself must finally join in. Despite the debacle, the Eskimos really like him. He respects their ways and wants to learn from them, and not just as a way of increasing his knowledge so he can write a story about it. He puts it into action. All up, this white man has an extraordinary combination of . . . nerve *and* swerve, pluck *and* luck.

The whole episode of Wilkins attacking a bear with a knife, only to be saved by falling over a sled dog, is now turned into many songs and children's stories by the Eskimos, complete with a pantomime each time so that the English speakers know exactly which hero they are immortalising.

'I never heard the last of it,'[65] Wilkins will wryly note.

As one Eskimo declares, in a summation of pure wonder and disbelief that many in future decades will agree with: 'I don't think much of this man. He doesn't know many things he must not do. He'll do almost anything and he's always doing something.'[66]

CHAPTER FIVE

A WHALE OF A TIME

If you start a job you just have to keep on going, and I thought that one had to be hardy in those countries and put up with anything that might happen along. I thought it was an explorer's duty to put up with hardships. I thought that explorers were all heroes and martyrs and everyone expected us to be that way.[1]

Sir Hubert Wilkins

April 1914, Collinson Point, blind luck

In the spring of 1914 Wilkins' ship at last comes in, in the form of a trading schooner arriving at Collinson Point that not only has a photographer on board but, most crucially, a photographer with a spare camera for sale, for a suitably inflated price.

In the full knowledge that stories of attempting to kill a bear, no matter how entertaining, will not satisfy his employers at Gaumont, Wilkins decides to make a solo excursion to Point Barrow to photograph whales – always popular with audiences around the world. Alas, after asking the hitherto friendly Anderson for dogs, sleds and Eskimos to aid him, the wind suddenly changes and blows freezing. When it comes to lending the equipment and resources of the Southern Party, Anderson is not so smiley and can only manage to supply him with just five decrepit dogs and a half-broken sled, which Wilkins is invited to fix himself before he departs. As to Eskimos, well, Natkusiak serves at the pleasure and expense of Stef, so Wilkins may regard him as available.

Dr Anderson's disinclination to be obliging is not personal; at least not personal to him. Wilkins understands that. The problem is Dr Anderson's view of Stef, as the Northern Party's leader every action has confirmed the doctor's initial suspicions. Stef so fancies himself in a crisis he ends up creating them, and Dr Anderson will provide only what is essential for Stef, and only then if it is non-essential to his own tasks.

Very well then.

Fixing the sled is no problem, but the sled dogs prove to be in such poor condition that only a couple of days into the journey they lose the strength to pull it. So Wilkins and Natkusiak take turns pulling the load while the animals happily trot along in concert, interested observers only – unless there is a disaster in which case they can at least serve as an emergency food supply.

It is a bitter trip, just a few specks moving across endless whiteness, made even more unpleasant and worrisome one snowy day when Wilkins suddenly feels as though 'my eyeballs were coated with a dense and painful film'.[2] It starts as an odd, irritating haze, but very shortly thereafter turns so dark that he can barely see anything at all. Wilkins feels as though a desert of dust has descended upon his vision and is not only swirling all around him – but actually in his eyes, as his eyeballs feel so agonisingly gritty.

'Hi, Billy,' Wilkins yells to the foggy figure beside him, 'what's the matter with me? My eyes are full of dust but there isn't any dust here.'[3]

The Eskimo considers the question for a second before replying with a knowing nod. 'No dust. Maybe you get snow-blind now.'[4]

Maybe, indeed. For within minutes Wilkins' eyes start to burn and it feels that both eyeballs have become wrapped in sandpaper, as every movement is an *agony*. Yes, it is snow blindness, but it is not just his vision that is affected. Wilkins' limbs stiffen, his head aches, his joints are so sore that he can no longer pull the sled.

'It's no good trying, Billy!' he calls out. 'I can't lead the way. You go ahead and I will walk beside the sled.'[5]

Natkusiak does as asked, but also keeps calling to him as he walks out in front, hauling the sled onwards, knowing that the *Angayokrak*, Boss, will need constant noise to follow.

It works for a while but . . .

> Before the day was out I could not see at all and have to walk behind the sledge holding onto the handlebars. The pain was intense and the copious flow from the nose and eyes made one feel very uncomfortable . . . it was agony cooking supper tonight and I had to do most of the things by a sense of touch although inside the dark tent one's eyes do not pain quite so much . . .[6]

Still, Wilkins is acutely aware of the burden on Natkusiak, as this Eskimo hero now has to do nearly the work of two men. Natkusiak makes no comment whatsoever on how well a blind man is going to photograph some whales when they get there, but Wilkins can still feel from the Eskimo's silence his view that, *You are a very foolish man.*

In the silent watch of the night Wilkins is not sure that his friend is not absolutely correct.

Wilkins' diary tells something of the next few days.

> Friday 24th. Billy thought it better for me to stay in the tent all day and rest my eyes, for they are very much swollen, painful and red. As there is a whale carcass within a few yards of the camp for dog feed I did not object to this . . .[7]

> Sunday 26th. My eyes are still not so well as they might be and I can only keep them open for a few seconds at a time. We only reached as far as Spy Island today for it was a headwind, foggy and awfully cold. It seems good to even recognise this spot again for it is just seven months ago today since we first camped here with Stefansson after leaving the Karluk.[8]

Still they struggle on, with yet one more problem presenting itself. In the spring sunshine and Arctic winds, Wilkins now discovers a disagreeable novelty of travel in these parts at this time of year: while one side of his face becomes burnt by the sun, the other side freezes. It is a daily reminder that this is an environment like no other, with dangers in even the most mundane task. Finally arriving at Point Barrow, it is blessedly to the news that a whale has been seen that morning. The Eskimos have killed it and are cutting it up. Still, it is a good sign. Surely, if they are cutting one up now, it won't be long before they catch a live one! It would have been wonderful to film it last week but there is no use blubbering.

•

Though whales remain unhelpfully alive and at sea for the next six weeks, Wilkins is at least in his element filming the elements and the inspirational lives of the Eskimos, who have managed to prosper in an environment that seems harsh to everyone but them. It is such a magical experience, and all the more so when some actual 'magic' is captured as Wilkins films a shaman – an Eskimo holy man, an *angakok* – casting his spell across the people to bring them a bounty of fish.

Wilkins stands as part of the circle that surrounds the shaman, his camera rolling to take it all in.

'The old medicine man started to sway in time to a rhythm which he hummed, back and forth, then in a circular motion.'[9]

His body moves, his head sways, faster and faster, the rhythm rotating through the spectators. Wilkins, who had started from the point of extreme scepticism, is amazed to find that, 'the whole crowd, including myself, were hypnotised into following his movements. Our bodies swayed and our heads swung, and I soon realised that the old fellow was not only hypnotising himself, but was rapidly inducing a harmony of thought in the minds of everyone.'[10]

Oh people my people, the fish have fled but I will turn them and bring them to you as did my father and his father before him, back to the time of the Raven.

The crowd chants and chitters in their native tongue, a haunting, holy howl that cries out across the ice and snow.

'We'll get a lot of fish! We'll get a lot of fish.'

That phrase becomes a chant, becomes an anthem, becomes a roar that soars among them.

'Then suddenly,' Wilkins will recount, 'the old man started to babble in an uncanny high-pitched voice.'[11]

What is happening? Wilkins asks one villager, as Natkusiak translates the conversation back and forth.

'He is in the control of his great spirit, he is talking magic words that cannot be understood.'[12]

The shaman's words grow quicker, hurried and harried as his eyes glaze over and he seems to forget himself. He is a conduit for the divine, entranced and possessed by the spirits he speaks with, totally absorbed in when . . . CRACK!

The blunt end of a heavy, curved knife strikes him on top of the head, and his chanting stops. On the other side of the knife is a small, old woman, no malice in her eyes, only doing her duty to bring the shaman back to the corporeal plane.

Wilkins cannot decide if the shaman is the finest actor he has ever seen or if he were truly in the midst of a trance, but he doesn't have time to ponder the question as the shaman starts speaking coherently in words that all can follow.

'My magic has made many fish. I can hear them in the water.'[13]

Everyone present leans in, the shaman hard to hear, over the whistling wind that swirls about them. Suddenly, the shaman points to one man: 'You will spear many.' Then he points at another, 'You are not going to get any at all.' The shaman spins, his finger darting at a waiting woman to declare her future, 'You will find many fast between the rocks, but you must be quick to catch them.'[14] A quivering boy is next to receive his fate, 'You will catch one. Or two. Or three or four perhaps.' The man next to him is not so lucky: 'You will catch one,' and the Eskimo next to him has less luck than that: 'You will not get any, no matter how hard you try.' Finally, the shaman turns to all: 'Now, we go out and see!'[15]

Making haste to fulfil their freshly revealed destinies, the mob moves to the river. Some trudge while others trip over themselves to move faster, and Wilkins begins to understand the magic. Those told they will catch nothing are in no hurry to make it a reality, while those promised the most fish are the most insistent on getting to the river first. The women assured they would catch many by the rocks don't waste any time by the banks of the river, they rush straight to the rocks. It is a self-fulfilling prophecy if ever he has seen one. The shaman himself hangs back and graces Wilkins with a knowing nod, a wink in all but name. Wilkins speaks as Natkusiak translates for him:

'That was very clever. You made plenty of fish. You are a wonderful magician.'[16]

The shaman smiles and gives a small bow. 'It is not difficult when you know how to do it, but I *must* tell that old woman not to hit me so hard next time. Now my head aches!'[17]

Smiling at their shared joke, Wilkins has nevertheless learnt yet one more important lesson that will stand him in good stead for the rest of his life – prolonging that life many times. It is the magic of motivation: those who *believe* they will have fortune will *make* their fortune; those who are told they will fail and believe that, too, will make themselves fail. The 'magic' works for all as, one way or another, all are focused solely on fishing for the day and, magically if you will, there is a feast of fish that night.

It is time to head back to Collinson Point – no easy thing as the ice has started to melt and therefore thin in the warming spring weather. To make their way, Wilkins and Natkusiak often have to swim the dogs across pools of icy water and then strain as they each take one end of their heavy sled over their heads and march through the same pools, the water up to their chests as they wade forward.

•

Finally arriving at Collinson Point, the first thing they see is a beacon, where there is a message from Stef! He is *alive* and . . . impatient. Wilkins now reads what was written just two days after he last saw Stef. It reads:

> *Wilkins, it will be impossible for you to find me for my party have been carried fifty miles on the broken ice last night. My party is intact, and if I can find game I will push on north-east across the Beaufort Sea to Banks Land. If we cannot find game we will turn back.*[18]

And now for the second surprise.

For though Stef has not turned back, he has sent one of his best men back . . . Johansen! Johansen, accompanied by his 'ice party', greet Wilkins warmly and advise they have returned especially to pass Wilkins a message. Johansen does not know what the letter says and Wilkins is stunned when he reads it.

It is a request for Wilkins to become a commander in the expedition. Stef's manner is circumspect, but his words still shock:

> *It has occurred to me that your work can be as well done in Banks as Victoria Land for the same sort of Eskimo live from Nelson Head to Ramsay Island as those of Coronation Gulf. In case I fail to return to shore before the North Star can sail, I want therefore to ask you to take command of the North Star with Mr. Castel and some other suitable man that could be engaged.*
>
> *She should be equipped with engine room supplies and staple foods for two years at least . . . she should make her way at the first opportunity to the West coast of Banks Land as far as the Norway Island at least crossing to Prince Patrick land if possible.*[19]

Beyond everything else, Stef assures Wilkins that there will be plenty of Blonde Eskimos to film at Prince Patrick Land, and as to Dr Anderson, don't worry, Stef will personally write to him to affirm Wilkins' unexpected promotion. He finishes with something rare indeed, praise:

I have seen enough of you already to have confidence both in your judgement and good faith.
    V. Stefansson[20]

As a postscript Stef closes by advising Wilkins to get 'tracings of most if not all the Maps of the Alaska'[21], oh, and if money is of interest, 'I should be glad to put you on the payroll of the expedition.'[22]

Wilkins reels.

> I thought today would be the last day that I would have occasion to travel by sled this season and I was happy at the thought but the letter that I received at Collinson point when I arrived soon scattered these thoughts from my mind and shattered all the plans I had made for the coming winter in spring . . . This is an exceptional opportunity for me but I would rather be without it. I joined the expedition as Photographer, it is as photographer that I wished to make a reputation if any, not as leader of a relief party or navigator.[23]

As a leader, his duty will no longer be to Gaumont and the search for a snap of Blonde Eskimos, despite what Stef might promise: 'Notwithstanding the assurance that I will have equal if not better opportunities to photograph Eskimo under the new plans, I doubt if I will have the same facility or as much time as I would have if with the Southern Party and without any other responsibility than that of Photographer.'[24]

Still, he can't say no and doesn't. There is also news from Australia: 'During my absence mail had arrived and with it a letter from home, I was pleased to hear that they were all well.'[25]

Whether he will be well, and whether he is well chosen to lead, that is the test to come.

Wilkins has his orders to sail the *North Star*, do or die. Or at least, he intends to do it.

But the crew of the *Mary Sachs* steadfastly refuses to sail with any new leader on a fool's errand, and particularly not on a vessel as flimsy as the *North Star*.

Travel to Banks Land?

It is over 400 miles away, through treacherous seas.

To meet up with Stef?

Where, exactly? They are quite sure that Stef cannot and will not reach Banks Land; firstly, because there are excellent odds that he is dead and, secondly, 'If he survived at all he must have been carried far West on the ice floes. By now he is somewhere on the coast of Siberia.'[26]

Wilkins disagrees, pointing out their leader's doggedness, his ability, his track record of leaving tracks all over oceans, snowfields, glaciers, cliffs, crags and tundra, always coming out the other side, always surviving. If he says he is going to Banks Land then that is not only where he is going, it is where he will get to.

And as to where – they will meet him at the obvious place, the one Eskimo village where he will be undoubtedly living among them.

Anderson, acting as an honest broker albeit with a thumb on the tiller of scepticism, follows the debate closely and comes to admire the Australian's resolution. And it is equally clear that someone really will have to go out to try to meet Stefansson on the off-chance he actually does make it. With this in mind, Anderson takes Wilkins aside for a quiet chat where he soon enough makes a proposition: 'Well, if you're determined to go I won't stop you. Stefansson's probably dead by this time, but if you are set on going anyway why don't you take a real boat?'[27]

After all, if you take the *Mary Sachs* you are not only a much better chance of surviving, but once you get there you could actually search for Stef properly. And if he's dead or disappeared – which is probable, did he mention? – you can still do some useful exploration.

'If you succeed in getting to Banks Land, I don't believe you will find Stefansson. But you can carry on with the work if you take the *Mary Sachs* which is bigger than the *North Star*. So let me take the *North Star* for my work along the Coronation Gulf.'[28]

It is a surprisingly generous and unrefusable offer, and the two shake hands upon it, giving Wilkins his first vessel and first command.

But before he can leave, Wilkins receives letters from his employer, Gaumont, ordering him to return to Europe at once. But the die is cast; and all Wilkins can do is note that fact in his diary.

> These letters placed me in a peculiar position – I have already spent over twelve months under severe conditions without accomplishing anything for the benefit of the firm or that which I had set out to obtain. I had been separated from my outfit through unfortunate

circumstances but this had in some measure been replaced, but was not sufficiently complete to enable me to obtain satisfactory results. I can but hope for little reputation in the photographic world by continuing the Expedition, but there are the three men on the ice who, if not actually depending on some boat reaching Banks Island for their lives, would suffer great hardships and accomplish very little towards the success of the Expedition after this summer ... I could not do otherwise to continue with the plans no matter what sacrifices might mean for me photographically speaking.[29]

With Stef relying on him, Wilkins must take command of this Arctic party, no matter what his employer wants him to do. This is a matter of life and death. Gaumont will hopefully understand, should he live long enough to return to London to explain why he had taken this decision.

The news spreads that Wilkins will soon depart, which prompts a curious visit from a fellow cameraman.

'The cinematographer on board the *Hermyn*,' Wilkins notes in his diary, 'has been over talking with me most of the day, he wanted me to go over and fix his camera.'[30]

Speaking of fixing, the movie man explains his novel methods.

He seems to have taken a fairly complete set of pictures of the Alaskan Eskimo but he tells me that many of the subjects are 'faked'. They also acted the 'Rescue of the Stefansson Party from the Ice', this was in case that they met with this party and could not get the pictures, or if someone else got such pictures they could not get theirs on the market at the same time. This sort of 'Foresight' seems to them to be quite legitimate and blameless.[31]

Wilkins does not feel the same, and it is absurd to him, not to mention dishonest and tasteless, to film a 'rescue' before it has even occurred, of men who may well already be dead.

Now, having to *really* rescue someone is slightly more difficult, especially with men who refuse to follow the script. When it comes to Wilkins' own entourage, it has to be said they do not react to their latest leader and plan with eagerness. Natkusiak will do as Wilkins asks, because that is just the way he operates. (With him, most days, you could rest a full cup of tea on his emotions and, no matter what extremities he might be facing, nary spill a drop.)

Similarly, the one-eyed Swiss cook is too downtrodden to do anything other than follow orders, so he will be all right. Which leaves Stefansson's stenographer, McConnell, who, despite the fact that he has little to do without Stef actually being there to dictate anything, is extremely reluctant to travel with Wilkins. The rest of the crew of the *North Star* feel much the same. They don't care what deal he might have done with Anderson, they don't want to go to Banks Land on *any* vessel, are they clear?

So, nothing doing, Skipper, for we are doing nothing on your orders.

Well, then.

Wilkins must think of a way to force the issue in his favour.

Dum-de-dum-dum-dum.

How to get men to do what they don't want to do?

Dum-de-dum-dum-dum.

Got it! And it is not only in the realms of engineering, mathematics, medicine and physics that genius strikes.

'I procured a case of the expedition's liquor supply with which Natkusiak and the cook got the engineer and the crew of the *Mary Sachs* dead drunk. Then the three of us sailed her away from Herschel Island.'[32]

By the time the crew staggers up on deck the following morning they are well out to sea and it is too late to turn back without an armed struggle – which is too much to contemplate for men with hangovers like theirs, which feel like dwarves are sitting on their shoulders hitting them on their temples with a hammer. Most of them, thus, settle down to bitterly resentful but mostly silent co-operation. The exception is Wilkins' old friend, the Scottish engineer, Jim Crawford, who in a blue haze of abuse points out with many adjectives and not a few exclamation marks that this is the *second* time that Wilkins has taken him prisoner. Why the first time was when you locked me in my cabin, and this time you're using the whole FOOKIN' ship as a prison!

(This proves to be no more than his opening remarks as he explores language only mastered by the maritime mob after decades of practice, not to be attempted by amateurs.)

Wilkins puts it elegantly: 'I was the target of bitter words from the engineer.'[33]

Finally, however, even Crawford appears on deck as they continue on their way despite frequent explosions of thunder and wildly ferocious storms of shattering intensity – the latter from Crawford's mouth, the former from his trousers.

'None of them are congenial to me,' Wilkins notes glumly in his diary of the crew, 'and I expect to have a pretty miserable time . . .'[34]

At least there is solace in the glory of the natural world, as they continue to plough through this icy wonderland. Huge flocks of Arctic birds keep them company for some of the way, seals dart in their path before surfacing in their wake, even as bowhead whales regularly surface to survey them with a gimlet eye and their tawny head markings, only to offer a snort and cavort of complete contempt from their spouts before descending to the depths once more. And whatever else the story of the *Mary Sachs* might lack at this time – comfort, romance, fine cuisine, high fashion and intellectual discourse spring to mind – they do not lack for dark and stormy nights.

Three dreadful storms in succession hit them, starting as they pass the mouth of the Mackenzie River. They come with such wind and crashing waves, Wilkins will recount, 'we hardly knew from minute to minute whether we would be able to save the schooner'.[35]

Up one wave, down another, and look now as they are broadsided by the next, and feel much of the deck cargo starting to shift around. Good God, some of it is starting to crash into the bulwarks of the ship, and the crew must work like mad things in the teeth of this howling gale to lash extra ropes on those boxes that are moving to try and secure everything against even the . . . worst wave.

Oh.

Here it is now.

Personally, they are able to rush off the deck before it hits, but the wave is so big and so devastating that it takes out half their entire petrol supply with one crashing roar on deck. A follow-up wave narrowly misses taking away all their dogs, who are far from happy about it. Between the waves hitting, the men scramble and keep scrambling, to get the dogs secure, to lash down the remaining petrol, to man the bilge pumps, to do all of the hundreds of things that need to be done to stay afloat, stay alive, keep on course, breast the next wave, and the next one after that, and survive the next one still as it crashes over their bow once more. The one man that gives them confidence, strangely, is the brute that got them into this in the first place – George Wilkins. No matter how extreme their situation, he stays calm, giving orders, and behaving as if while this blow is a bit grim he has seen a thousand storms grimmer because he has been at sea for . . .

Well, actually, now they remember. This is his first trip as skipper. But anyway, the men are happy to follow his orders, for he seems to know what he is doing.

When at last a lull comes, with all hands exhausted, Wilkins appoints one poor sodden soul to lookout and all other men can gain some sleep.

So what is a lookout to do when everyone else is asleep and he is the only poor bastard awake when he is every bit as exhausted as everyone else. If he just closes his eyes for a moment, they won't even know . . .

Wilkins always knew Crawford would break down his door in a rage, and he was right.

For now the door burst open and it is the Scot yelling: 'Engine's awash!'[36]

The boat has drifted through the ice onto a beach and staved in her bow.

Having survived the worst of the storm they have come to grief on an icy reef in the relative calm.

> The seas were pouring in all night and all that delayed our sinking was the fact that we were now stern to on the beach. At first it seemed impossible to gain on the water but the crew worked for their lives [and] we managed to hold the water level down where we made temporary repairs, and then pump the boat almost dry.[37]

Despite that important if momentary victory, it is somewhat tempered by the fact that Wilkins now discovers that the rudder – actually, no, both the rudders – have been shaken loose from the bottom of the hull. Yes, they are still technically attached but both are practically useless for steering. Wilkins grimly records that the 'port propeller struck the ice and was broken off close to its hub'.[38]

To negotiate a ship through Arctic waters without a rudder? It is pure folly but, on the other hand . . . . right now, also pure necessity. Wilkins learns it is possible, but that luck is your main navigator, with fate and prayer both riding shotgun off each shoulder, nodding in time to your knocking knees.

> Our compass was very erratic and the steering of the boat even more so. Thompson is the only one on deck that can manage the wheel. Billy keeps the boat swinging from side to side making

almost 45 degree turns and I have seen Captain Bernard make a complete circle in one watch. In consequence of this spectacular method of proceeding we found ourselves at noon almost at the bottom of a big bay.[39]

At least, however, they finally get to a strip of clear water with no obvious dangers anywhere near, and Wilkins at last feels it is safe to get below to change his icy, soaking clothes.

It means that this time he is near naked when, rudderless, they hit a new beach. He races on deck to find the whole boat shaking vigorously as a lone engine roars in torment trying to break free. Compounding the situation is that Mr Crawford goes out in sympathy, and now becomes a lone engineer, also roaring in torment and trying to break free. For Wilkins and the rest of the crew watch aghast as the screaming Crawford, raving and rolling drunk, leaps overboard and starts to swim in the general direction of the open sea. Mr Crawford is not going to go down with the ship, or stay on the land, and his mere presence in the water is enough to make any fish within 50 feet slightly tipsy by proxy.

Under the circumstances the easiest solution would be to let the wretch go, but of course, Wilkins can't.

'I put a small boat overside and rescued him, but we had to tie him to his bunk.'[40]

Clearly, Crawford can no longer be trusted with the engines, the navigation, any bottle containing alcohol, or anything at all. It is up to Wilkins, thus, to fix the engines himself, while also supervising the rudderless navigation – no easy task. The obvious thing would be to beach the ship so as to just focus on one thing at once, but there is a clear problem with that.

'It was impossible to repair the propeller without unloading the boat; yet I dared not do that, knowing that once the cargo was unloaded, nothing would induce the crew to leave Baillie Island.'[41]

Onwards they limp, managing with but a single propeller and a single engine – happily part of the same unit – until they reach the north point of this stretch of coast. From here a path can be struck to Banks Land, but the crew of the *Mary Sachs* refuse to strike it.

Now while 'mutiny' is far and away too dramatic a word for their approach, they are certainly not backwards in coming forwards in making clear that they don't wish to go on and want Wilkins to see sweet reason.

Don't you understand, Skipper? We are on a doomed mission, risking our lives to bring supplies to a leader who is, in all probability, *already dead*. If we just make camp at this north point and anchor the boat, Anderson is sure to pass in a few months' time; and we will all get to live! Under those circumstances, we are more than happy to wait.

After all, they note: '"Stefansson is a corpse. What's the good of meeting a corpse?" And in any case the corpse won't [even] be at Banks Island. It will be somewhere out on the ocean ice . . .'[42]

Yes, but there is some misunderstanding here. For you are the crew, and I am the *Skipper*. I am certainly still alive. And I refuse to have any part of 'choosing to leave Stefansson's party to starve on Banks Land'.[43]

Still the crew won't budge, leaving Wilkins with no choice but to reach for the finest, sharpest tool in his arsenal: cold hard cash.

'The only thing I could think of was to offer them higher and higher wages.'[44]

Every man has his price, and every man gets it.

One last thing though, the crew insists.

WHAT NOW?

'Finally they agreed to go on, if I would leave a conspicuously marked record for Anderson's party, stating that, if we were lost, these payments would be given to their families.'[45]

Wilkins agrees, goes ashore, and in full view of all, places the newly written contract under a flag for Anderson to find if all hands are lost and unrewarded.

The effect is stunning. From the moment he comes back on board, and despite their previous 'mood of desperation', the crew now responds with purpose and enthusiasm. Though the die is cast, they are more than ever determined *not* to die, with Wilkins noting proudly that 'no man ever had a more anxiously willing crew. They worked without rest or food, with only the thought of keeping afloat until we could reach Banks Land.'[46]

Necessity forges unity and they forge on, but there is, of course, one exception . . . 'Crawford has been too drunk to look after the engines at all since yesterday, so it has been a long watch for me. We have searched the boat for alcohol but could find nothing except wood alcohol for lighting the primus. I do not know if he is drinking this but the can is partly empty. I have hidden it away so that he can't find that now and we will see if this has any effect.'[47]

Wilkins remains alert and on deck until, the very next day, Banks Land is finally sighted – *land ahoy!* – for'ard and northward.

And yet despite the joy evinced by the crew at reaching their principal destination, Wilkins does not join them for, despite his earnest scanning with binoculars, he can . . . spy with his little eye . . . not much at all . . .

'We can see no sign of a beacon or trace of human inhabitation.'[48]

Maybe the crew was right and this whole thing was a fool's errand after all, and they have just gone through 120 days of nautical peril for naught. Inevitably it takes no time for the crew to also realise the situation and that they were likely – sod it! – right all along! As a certain blackness settles over the vessel, Wilkins orders the ship to put in at Cape Kellett, where she can be repaired in safety.

'The men were furious with me,'[49] Wilkins notes. Such is the season, so much work does the ship require that it is obvious they will be stuck here until spring at the earliest, bearing supplies that will never be delivered – and none of them will see their families this side of eight months at least. Undeterred, unbowed and certainly unapologetic – for he remains confident that he has done what needed to be done – Wilkins draws up plans for the building of snow huts, before telling the men that, while they must wait here with the boat, he will take the dogs and one of the sleds and search the island for Stef with two of the Eskimo guides, but not his right-hand man.

'The feeling against me in camp was so strong that I thought it advisable to leave Natkusiak and the cook there.'[50] Yes, things are grim when he must leave an ally in camp to make sure the camp stays camped right there, but . . . things *are* grim. At least before leaving Wilkins spots, pursues and kills a polar bear. (Eschewing his knife, he decides to do it in the new-fashioned way, with a gun.)

'He was not so very big but will prove a welcome addition to our supply of dog food.'[51]

This is to the good as it is those dogs that are going to be relied on to cover the ground to hopefully find Stef.

And yet, after three days of hard slogs and exhausted dogs, they have covered just 30 miles, whereupon – feeling like they are three ants crawling across an eternity of whiteness, lost in the blind spot of all creation – Wilkins decides to pursue a novel search strategy. Each day he and each of the Eskimos will separate at a set point and head out in a different direction, then return to camp via a different path. This will give them six stabs at finding Stef each day, each man searching for any sign of life past or present. The first day ends in failure, with only curious white foxes seen, and the following morning they separate with a mood of doomed duty. There is little point because Stef is clearly not here, but . . .

But what is that?

Wilkins has just reached the summit of a small hill when he sees something odd and lifts his field glasses to his eyes. There, in the distance, he sees 'a little beacon built out of stones'.[52]

Panting, stumbling in his haste and excitement, Wilkins reaches the careful pile of stones and brings them tumbling down as he searches for any message left. And there it is! It is a small torn piece of paper with Stef's writing on it, attached to the back of an empty cartridge box. It reads:

MAKE CMP ¼ MI. S.E. OF HERE[53]

It is a missive left not for a rescue party but for the other men of Stef's party. True, the fact the other men clearly did not get it is troubling, but the main thing is that it is proof that Stef *did* reach his promised destination. Following the instruction Wilkins quickly finds traces of the past camp; three men and some sled dogs have been here recently.

Heading back to his day camp to await the return of the two Eskimos, he finds a message from them advising that they have actually set out on a hunt. Fine. Deciding he is too keyed up to wait, he dashes off his own message with the wonderful news, before racing back towards camp – intent on mounting a major rescue party with supplies. While Stef and his men were all alive months ago, weeks ago, days ago, there must be some chance they are on the edge now, and it is urgent to find them.

Still, there is a nagging worry as he heads back to camp.

> It would be just my luck if those people have arrived at the schooner while I'm away and [I'm] not able to take a picture of the excitement [of their arrival]. I never seem to be in the right place to get important pictures on this expedition and goodness knows these occasions are few, and to make any success in my work I should be at them all.[54]

Never one to lose an opportunity, Wilkins shoots all the game he sees on the way back – 'Five Caribou, one polar bear, one wolf . . . three foxes . . .

two bears, two seals and numerous ducks, geese and rabbits completed our supply of fresh meat'[55] – so creating, effectively, frozen food dumps in the classic manner of polar exploration, that he and the search party can use in their forthcoming venture. And now, here is the main camp.

Men, gather round!

'All hands were at first amazed,' he will chronicle, 'then some were sceptical.'[56]

After all, how very neat. In the face of a bubbling insurrection because of no sign of Stefansson, you now say you have in fact found a sign? So, where's this scrap of writing? Wilkins shows them, but only he knows Stef's handwriting and the crew of the *Mary Sachs* remains sceptical. How are they to know this is not some device of Wilkins designed to drag them out on another fool's errand to find a man who clearly is not here, that damned elusive Stef?

And yet, despite the outright disbelief of some, that Wilkins has found a trace of their missing leader, they do believe that he shot fresh meat and agree to accompany him to collect some of it. They will need it for the winter that is approaching at a rapid clip – and it will be important to gather it in to get them through comfortably.

It is a hard day indeed, retracing his steps over 30 miles and getting all the game on sleds, but by pushing on and returning through the night 'neath glittering starry skies of impossible brightness, by 4 am they know they are close to the coast by the sound of the rumbling waves crashing into the ice cliffs. And, this is looking familiar, just over yonder rise, here is the main camp!

Exhausted beyond all measure, Wilkins stumbles to his tent, only to find something extraordinary. Someone is in his bunk! More stunning still, when he brings a lantern to bear he finds it is 'Stefansson himself, sound asleep, stretched out on my sleeping bag as peacefully as though he owned it!'[57]

Far from being the worse for wear, Stef, the living exemplar of the 'Friendly Arctic', looks 'as comfortable as if nothing had happened to him in the past nine months since he had floated out to sea'.[58] And not just him, for looking around for the other two of Stef's crew to make sure they have safely arrived as well, he finds them asleep in the forecastle of the now beached *Mary Sachs*, and again bringing his lantern to bear, notes they do 'not look at all emaciated but fatter and more healthful than I had ever seen them'.[59]

Not only that, but Stef has brought all six of his sled dogs with him, and they too lie sleeping next to the sled dogs of the *Mary Sachs*, and even they are 'fatter than our dogs!'[60]

How on earth has Stef managed this? What is the story?

'I could hardly control my curiosity until morning but finally crawled into the wheelhouse of the schooner and shivered till daylight for I had left my sleeping bag at the camp.'[61]

As delighted as Wilkins is that their Commander has been found – so vindicating every decision he has made for the last four months, to risk all, stare down mutinies and give up Gaumont, to fulfil his mission to get to Banks Land on the chance they are still alive – he is shocked to find that when Stef awakes he is not of the same view!

For, far from being happy, Wilkins feels that the Arctic has become a few degrees colder just from his leader's glare.

And so it comes out. While Wilkins has been out collecting game, Crawford has been telling Stefansson how desperately the crew had been looking for him. Yes, that Wilkins had tried to thwart them at every turn but they had shouted the bumbling, procrastinating, jumped-up photographer down, and *insisted* on continuing the search. They would have been here much sooner if not for, as Wilkins would recount their thrust, 'how my stupidities and arrogance had hindered and delayed them'.[62]

And even beyond that, Stef is 'seriously annoyed'[63] to find that instead of the *North Star*, as he had instructed, Wilkins had brought the *Mary Sachs*. It is obvious to him how Anderson has clearly tricked the young man he left in charge, how Wilkins had just not been up to it after all.

Wilkins reminds Stef that Anderson made no bones about whether or not he would follow them, but he may as well be speaking with a glacier. Stef will not be thawed on this issue, and will not alter his chosen course by a jot.

Fortunately, Wilkins soon finds himself backed up by other members of the crew who throw in their lot with both Wilkins and the truth, and Stef 'after hearing the evidence of both sides agrees that I did the right thing'.[64] Wilkins is also able to make the pertinent point that, 'if I thought that somebody else would have taken charge of either boat, I would not have come myself but would have returned to London as was suggested by Gaumont Company'.[65]

But there had been no-one else, so he had come.

Too shay.

To make amends, the Icelandic giant presents Wilkins with a letter of commendation for his service, a letter that will be sent on to Canada to ask that he be put on official pay at a rate just lower – for let's not get too carried away – than Stefansson.

In the meantime, it will be as Wilkins had planned. They remain here at Cape Kellett to ride out the winter, and it is in the course of that frozen season that Stef's opinion of Wilkins moves even higher.

'I have never known anyone who worked harder than Wilkins,' he will recount flatly. 'He would be cleaning the scraps of meat off the leg bones of a wolf before breakfast and scraping the fat from a bearskin up to bedtime at night. His diaries were filled with information about the specimens he gathered, his fingers were stained with the photographic chemical used in the development of his innumerable plates and films, his mind was always alert and his response always cheerful when a new task was proposed.'[66]

He is, in short, the ideal man to make the Arctic friendlier than ever, and the best man Stef has ever encountered in such conditions. With a dozen men like Wilkins, Stef often remarks, he could achieve anything. But there is only one, incredible, Wilkins.

Wilkins, for his part, feels that his credibility and fidelity will be found wanting by the people who employed him to come to the Arctic in the first place. The facts are plain, but there is one ray of hope that the young man can see: 'I have broken the contract with the Gaumont company by coming with his party. However they may appreciate the circumstances which led to this, and continue my salary, especially if I'm able to get some pictures of the Blondes next spring.'[67]

In the meantime, oh so typically, Wilkins wants to use his time here to expand his skills, and between many hunting trips with Stef, the two compete with each other to see who can build an ice house as well as an Eskimo, with an amused Natkusiak acting as judge to these efforts. Wilkins notes, 'by the end of the winter I had learned the art of cutting the hard snow blocks with my knife to the right shape and placing them in position almost as well as an Eskimo'.[68]

All you need is a long-bladed snow-knife – something like a razor-sharp sword fashioned out of whale or reindeer bone, or even walrus tusk – and a tightly compacted snowdrift. By thrusts and sweeps, you can quickly start forming rectangular blocks, about four feet long by two feet wide and eight inches thick. Lay the first blocks in a circle on a flat spot, and now shave off at a sloping angle the top corner of each block so that the walls start to slope inward. Keep going until it meets at the top, only being careful to leave a hole in the dome for ventilation, and now fill in whatever open joints and crevasses there are with loose snow, which will quickly freeze. If it is for a long stay, you can build a semi-cylindrical passageway into it about 10 feet long, complete with some vaults to store your supplies, and a sealskin flap to prevent draughts. Do it right and you will be as snug as a bug in a rug, a baby roo in an *Iglu*.

The genius of the structure is you can even burn seal blubber in a narrow saucer to cook and generate heat, and your house won't melt, because, as Wilkins notes with wonder, the scientist in him marvelling: 'the walls being of soft snow absorb the moisture as quickly as it is formed and the temperature of the air converts the wet snow to ice on the outside of the house . . . If the temperature of the outside air is cold enough to neutralise the warmth of the interior of the wall and so keep the whole below freezing point a snow house is as comfortable and dry as any other.'[69]

Yes, it takes at least a couple of hours for a beginner to build it – while Billy can do it in an hour – but 'the snow house is decidedly more comfortable when finished than a tent'.[70]

The Australian soon gets so good at it, he builds several for specialty purposes – finding a snow house an admirable 'office' for stripping caribou.

It delights Stef, who for decades has rolled his eyes at the notion, long expounded by previous explorers, that building an *Iglu* is 'a mysterious racial quality belonging exclusively to the Eskimo'.[71] No, it is a skill that can be learnt by men who, firstly, are humble and smart enough to ask the Eskimos how to do it, and secondly, diligent enough to knuckle down and try.

Happily they settle into the new ice house, quietly musing on everything under the polar sun from how Arctic zeppelins might fare – fairly badly, says Stef – to the reasons Stef's poetry has never been published – because it is bloody awful, thinks Wilkins, very quietly.

Wilkins also talks of his notions of weather stations that might be located here and at the other end of the earth, Stef being one of the few men on earth to understand the vast difficulties and vast promise of such an endeavour.

The two are now peers and friends, as much as one can be friends with such a determined eccentric as Stef. As they continue to hunt – shooting an array of seal, Arctic fox and birds of dazzling variety – the two engage in spirited discussions about how and when the Arctic will be tamed. Having thought about it for some time, Wilkins is now convinced that a plane could fly in these lands. Yes, how to land and take off again would be problematic, but with the right kind of plane that could surely be resolved.

But the expedition leader will only go so far with his protégé.

Stef scoffs.

The aim of the game, to learn from the Arctic, can only be achieved by experiencing it, on the ground, not flying *over* it. If an Eskimo is making an *Iglu* and one man is on the ice watching it, and the other man flies over at 15,000 feet, which man learns more?

Not to put too fine a point on it . . . ?

'You cannot learn much more oceanography flying over an ocean than you learn botany flying over a botanical garden.'[72]

Stef is determined to miss the point, the one with icebergs all around and an *Iglu* upon it . . .

With the right plane, you could fly until you saw something interesting and then land on thick ice. With the right tyres, designed to get traction on that ice, you could take off again. Look around you. This whole part of the earth is, much of it, anyway, one big icy runway in every direction!

Again and again Wilkins preaches the ease of an aeroplane, consumed by the idea of returning here with one, and if not a plane then an airship, a Zeppelin! Why, one could float above where they currently freeze in pure comfort. On and on he goes until Stef has had enough of these castles of the air and makes his own case for a better machine than a flying one: 'Perhaps the submarine offers a more practical method of navigating the Arctic Ocean in comfort?'[73]

Wilkins takes pause.

Is Stef mocking him?

Actually . . . no.

He is quite serious and expands on the idea. For you see, Wilkins, the seas of the Arctic are not just blocks of ice even in the harshest winter. Beneath it all is water, and when you think about it, that fact is obvious.

'[The Arctic is] dotted every few miles with lakes of open water because of the constant shifting of ocean currents below the ice pack and of the winds above.'[74]

There is logic to this, and Wilkins immediately recognises it. But when you are under it, and need to surface, as submarines have to every 16 hours, how can one be certain of finding the open water you need? How can you be certain that you won't be trapped in an icy tomb?

Stef has thoughts on this too, returning to first principles – what is the Indigenous solution?

'Whales are known to search out thin spots in the ice above them and crash up through it for air when they failed to find a natural opening.'[75]

9 December 1914, Barrow Camp, whale of a tale

All of it is an enormous amount of food for thought, even as they continue to gather food for the men back at camp. Coming across a whale carcass half covered by snow on the beach, Wilkins strips it of snow and it gives the two intrepid explorers exactly what they need: fresh bait to catch wolves, bears and foxes. Sniffing the whale carcass in the wind, the animals keep coming, only to be shot for their trouble, whereupon the ever-industrious Wilkins prepares the skin and bones of the best specimens they have eaten, for future mounting in a museum. Not long afterwards, at a nearby site, things become even more astonishing.

A stunned Wilkins sees '63 polar bears feeding on a whale carcass; some fighting, some eating, some laying down and sleeping, some playing'.[76]

They must be shot!

With his camera!

Racing back to camp he retrieves it, but there were 'only 27 polar bears left when I got back with my moving picture camera',[77] which still should be enough to bless the merry gentlemen of Gaumont. Understandably delighted with his staggering footage, Wilkins sends it off on the first boat that leaves from Point Barrow to make its way in the US post. (Several months later, it returns, sent back by the San Francisco Post Office, because the package was 'not wrapped right'.[78] Wilkins wraps it again, and sends it again, complete with his curses whipped away on the polar winds. Alas, this time in San Francisco, the film canister is opened to check what it is, which 'fogged it'.[79] The film is sent to London, and arrives TWO YEARS after it was sent, properly wrapped but fogging useless.)

CHAPTER SIX

RUMBLE OF WAR

If you can force your heart and nerve and sinew
To serve your turn long after they are gone,
And so hold on when there is nothing in you
Except the Will which says to them: 'Hold on!'

Rudyard Kipling, 'If'

Banks Land, March 1915, *North Star*, inconstant as thou art

The time is coming soon when the expedition will need to get back to civilisation with everything it has collected. Stef is *still* insistent he wants to use the *North Star* for the job, rather than the cumbersome *Mary Sachs*, much preferring a smaller vessel to dodge the ice rather than break through it. But on the reckoning it would be too risky to send the *Mary Sachs* back to retrieve the *North Star* – as it would entail two risky journeys for two boats (and apart from that, Stef would like to keep his bird, and more particularly his *boat*, in the hand) – he has come up with another solution.

Can you, George Wilkins, go overland by sled to Coronation Gulf, to demand that Anderson release the boat to them, and then you can sail it back. You can take Natkusiak and . . . Crawford.

Really, Crawford?

Wouldn't it be easier, and simpler, to send me back with Natkusiak and a bear with a toothache?

Yes, Stef is aware that Wilkins and Crawford have a long-standing history that would make the War of the Roses blush, but there is no way around it: to come back with the *North Star*, he is going to need an engineer while he navigates.

And it is agreed.

On a freezing day in late March 1915, the three set off, with just one sled and five dogs, travelling over 'ice as thin as window glass'.[1] With spring nearing, the sledding is perilous as they are always in danger of breaking through the ice, and many a time they can hear it cracking and then see it breaking up behind them. Few things could be more calculated to keep them moving fast, and though there are many obstacles that delay them, and blizzards that stop them cold – read, freezing – they succeed in breaking the back of the journey in just 22 days. They know they are getting close, and as they build their ice house, Wilkins declares that, with any luck, they can get to Anderson's camp the following morning.

So it is that, exhausted and asleep in his *Iglu* just inside of the Arctic Circle, he is lost in one of his recurring dreams . . . engaging in an easier way to do such an exploration. By aeroplane, free as a bird, instead of endlessly trudging through it with ice-cold feet in snowshoes, and only occasionally being heavily shaken by the buffeting winds, shaken . . . shaken . . . shaken . . .

Oh.

He is actually being shaken awake. It is a trapper, someone he's never seen before, someone who has just made his way into camp. An enormous man with snow still on his beard and the squinty eyes of one who has spent too much time trying to shield himself from snow blindness.

'You one of Stefansson's party?' the fresh face asks.

'Yes,'[2] answers Wilkins still half asleep.

The stranger is pleased to hear it, and has a couple of very interesting bits of news about what has been happening in these parts.

'Wilkins and all his party went off last year and have never been heard from. Must have been lost to the last man, looking for Stefansson.'

Is that so? Oh yes.

'Stefansson is all right, so it doesn't matter a lot. Wilkins didn't amount to much.'[3]

Crawford, with malicious glee, roars with laughter as Wilkins silently fumes, giving the slightly confused recruit at least the space to go on, and keep digging.

Because everyone knows it. 'That damn fool scientist Wilkins has died after shanghaiing his crew.'[4]

Gratified, despite himself, at being called a 'scientist', even in the course of an insult, Wilkins nods. Keep going. What other news do you have?

Anything of the *Karluk*?

No, nothing known, but there is something else to tell . . .

'There's been a war in Europe!'

A *war*? Started and done?

Oh yes. Between the Brits and Germany.

'The latest news has the British advancing through Germany at a rate of twenty miles a day, so the show is over by this time, no doubt.'[5]

It is time to bring the hammer down.

'I happen to *be* Wilkins,' says Wilkins dryly. 'So how much can I believe of the rest of this "news"?'[6]

After following him to Anderson's camp, there is quick, tight, Hail fellow well met, and, far more importantly, a little more detail of this war in Europe. But for now there is a more immediate war to fight: who is to get the *North Star*?

Of course Anderson refuses to release it, whereupon Wilkins lays his trump card on the table: 'Stefansson had given me written authority making me second in command at the entire expedition, with the right, therefore, to take the boat from Anderson.'[7]

Anderson snarls, but must also bow. But make no mistake, you will get that and nothing more – no more men, and no more supplies. At least, however, Anderson provides the manpower necessary to get things started for Wilkins.

'I had her dragged out of the water and Crawford and I chipped the ice off the engine and cleaned and repaired it.'[8]

And yet, just as Wilkins' good fortune is mounting, a sole Canadian Mountie appears in this far-flung wilderness to make an arrest – they are reputed to always get their man, no matter in what remote places they might have gone into hiding. He brings with him not just a warrant but stunning news: the authorities in Ottawa have appointed Anderson as head of the expedition, on the belief that Stef is dead.

It means, suddenly, that Anderson now has written authority from their employers, Canada's Department of the Naval Service, to have control over all the men and remaining equipment which includes, obviously . . . the *North Star*.

Or does it?

In extremis, it is Wilkins who refuses to bow to Anderson's new orders – whatever they may be – at least not without a fight.

Urgently – the snowy equivalent of a bush lawyer – he puts the proposition to the men: are they to follow the written order from a distant committee who have no knowledge of what is happening here on the ground, or follow common sense? After all, the order from Ottawa is based on the premise that Stefansson is dead, which is false. Right here in my hand you can see *his* written orders.

Anderson puts to them the other view, that in the middle of this whole mess, all they can do is follow the wishes of their employers, and here there is no doubt. Yes, Stefansson is alive, but what of it? He is 500 miles away, and hardly in a position – particularly given what a debacle this whole trip has been from first to last – to try to exert the last of his authority now.

Well? When they are both done, the men talk among themselves while Anderson and Wilkins wait for the verdict.

And the winner is . . . Wilkins.

'We're not against Anderson,' is the position of the crew, 'but feel that Ottawa has acted without knowledge of the facts. Stefansson is alive and is therefore in command of his own expedition.'[9]

Anderson is far from impressed, but as a conciliatory gesture Wilkins promises to take Anderson's party to their next major camp, dictated by wherever the ice might break next, before departing with the *North Star*. As good as his word, when the ice breaks to the east a few days later, Wilkins and his men move quickly and drop Anderson and his men at the eastern shore of the Coronation Gulf, only to be hit by an extraordinary gale not long after they make their way back out to sea. At least calculating that the fierce easterly will break up the ice to the west, Wilkins decides to run before the storm, with his crew of just three – himself, Crawford and Natkusiak – to man her. It is a bold decision but pays off as the storm 'simply flung us across the gulf'.[10]

Desperate to make the most of every ounce of daylight he can squeeze from each day, Wilkins insists on pressing forward even when the last of that light is so dim that 'we could not see land at all and could barely see the white crests of waves dashed up by the storm'.[11] Their pace is nothing short of *astounding*, but as they approach the Dolphin and Union Strait – an uncharted channel that threatens jagged rocks for any sailor not wise enough to proceed with infinite caution – a choice must be made.

It seems to Wilkins that, in these conditions, with the surrounding sea no more than a mass of white foam, it matters little if they proceed slowly. It might be better to scamper through it, the way a cat scampers across a hot tin roof?

Yes, that's it.

To hell with caution, and damn the doubt. Let the devil take the hindmost, while we go full speed ahead.

And though the wind is mighty, the fact that it is behind us means that, if we keep our heads, and Crawford keeps sober, the whole journey will last no more than a few hours.

(Captain to the engine room. Captain to the engine room. Stop drinking, engine room. And full speed ahead.)

And so it proves.

'We raced through at eight or nine knots and how we ever got through alive I don't know. Luck's a strange thing.'[12]

To Wilkins' infinite relief, the icy form of Baillie Island soon appears o'er the for'ard quarter, and, upon anchoring in an icy cove and making their way ashore, Wilkins finds a message waiting for him from Stef.

There has been – oh God, how did he know? – a *change of plans*. You, Wilkins, are no longer to head for Herschel Island, but go back to Banks Land immediately and to head north from there.

Although Wilkins is generally a man of mild emotions, his thoughts are nevertheless likely a little stronger than the discreet notation in his diary.

'This surprised and puzzled me, for the object of my getting the *North Star* had been to bring supplies.'[13]

On occasion, the thought crosses his mind about that war in Europe, over now surely. Are the Balkans part of the British Empire yet? How does Constantinople fare in yet another short war? Strange to think that just a couple of years ago he was shooting the Turks with his camera . . .

•

Brooding though he may be, back to Banks Island it is, then beyond as far as he dares, following Stef's latest command. *Orders is orders.* Wilkins does as bidden, and the *North Star* makes its way to the most northern point of this forbidding Arctic coast possible, up to the tip of Cape Prince Alfred: 'Here we were farther north than any ship had ever gone before.'[14]

It is no small thing to take such a fragile vessel into such an extraordinarily beautiful and dangerous icy wonderland, and Wilkins records as much as he can in his hasty Log, including their final mooring just off the shore as they painstakingly shift the cargo onto the rocky beach before beaching the vessel itself on a high tide. All that is left to do is to set up camp and wait for Stef. Presumably, Stef is already north of

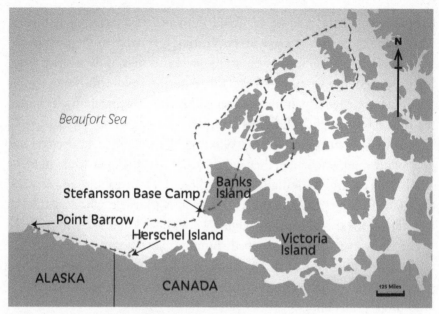

Stefansson's Arctic expedition route 1913–16

their position and will return for supplies when he needs them. They wait. And wait. And . . . wait. Months pass, the thick of winter is upon them before, out of the snowy wasteland two ice-men – all frozen furs and frozen beards – are spotted approaching through the drift being dragged by a dog team. This, surely, will prove to be Stef?

But no . . .

They are mere emissaries from the same.

'It turned out he was not north of us at all, but south-west at Cape Kellett, our camping place of the previous winter.'[15]

I see. The *North Star* is to remain where it is, as are its meagre crew. Wilkins, on the other hand, is to return with the dog team as Stef wishes to speak to him personally.

It is beyond frustrating, many more meaningless months are wasted and time does not lessen Wilkins' wonder at the uselessness of it. There is only one upside to this endless back and forth, this ceaseless criss-crossing of the land – it allows Wilkins to pursue his original purpose for coming here in the first place, trying to capture footage of the Blonde Eskimos. True, they are still proving elusive, always heard of but never seen, but Wilkins feels like he is getting closer. While he and Natkusiak search for the Blonde tribe they inevitably come across other remote tribes, some of them more troubled than others by the incursion of white people on their ancient ways. Visiting one tribe on the west coast of Victoria Land, Wilkins is blissfully unaware that, only a short time before, an Eskimo woman had sadly died during childbirth – and, so the locals believe devoutly, it had happened because a white man visiting at the time had brought evil spirits with him.

'Their law is an eye for an eye,' Wilkins will recount. 'They decided the next white man who came, they would kill. I was the next white man.'[16]

All Wilkins knows, initially, is that there is a lot of shouting between the suddenly angry tribe, and his translator, Natkusiak, who has at least negotiated one useful breakthrough: 'We will not kill him right away.'[17]

And now everyone has stopped shouting bar one man.

He is the husband of the woman who had died, who feels such a grave injustice is occurring that he grabs for a rifle, at which point Natkusiak grabs him.

Billy, what on earth is happening?

Natkusiak is too busy to reply, so Wilkins steps in to grab the rifle himself, and they are able to beat a retreat to their *Iglu* to regroup, even as, periodically, more shouting erupts outside.

'We want to kill you!'[18]

'What did they say?'[19] asks Wilkins.

Oh, nothing. They were wishing us good night.

Natkusiak doesn't have a good night at all, and lies in bed with one finger on the rifle. At least in the morning he is able to work out something of a compromise deal. They will give back the rifle, if the Eskimos agree to sell them a few sled dogs so they can quickly be on their way.

Agreed.

And Wilkins – still blithely unaware of just how extreme the situation is – cheerfully hands over the rifle to his would-be murderer. Even now, though, it is a close-run thing. For as they are leaving, some shouting behind reveals the same fellow gripping the gun, being held down by his fellows. It is only when they are a mile away, and sure they are not being pursued, that Natkusiak reveals to Wilkins the madness that has been going on for the past day. It is the first near-death experience that Wilkins has experienced without even knowing it.

Another close shave of a very different sort occurs when Wilkins decides to engage a remarkable Eskimo named Olik-tuk.

'He was the only Eskimo I have known with three wives . . . I engaged him as interpreter.'[20]

With so many wives it is no surprise that Olik-tuk speaks the language of love very well – the surprise is that he speaks the language of *free* love as, ahem, 'he was very anxious for me to share his wives'.[21] Despite the fact that, for this journey, Olik-tuk has specifically brought his two youngest and most attractive wives with him, the better to accommodate their guest, Wilkins declines . . . which is . . . delicate. While in many cultures making love to another man's wife is tantamount to an invitation to murder, in this one, as Olik-tuk sees it, *declining* to do the same when she is so graciously proffered is also highly offensive. Well, Olik-tuk will give him another chance, and takes one wife away on a brief hunting trip meaning that when Wilkins returns from his own trip hunting for reindeer it is to find that he is all alone with an admittedly very attractive, very young Eskimo woman who is looking at him expectantly.

Ah, where might your husband and your . . . ah . . . co-wife be?

'They have gone off for two or three days.'[22] Ah.

The whispering wind whistles lightly outside. It is polar cold, but the Australian is feeling more than a little hot and uncomfortable.

'I shall have to look after your clothes and boots!'[23] she declares. So you'd better take them off?

Wilkins decides to keep his clothes on, thank you, and is careful besides to do a little one-eyed snoring over the next two nights to be sure no surprise intimacy is attempted. When Olik-tuk and wife number one return, a deep freeze descends and it has nothing to do with the Arctic. 'When they came back he was very put out that I had not gone on with this woman. I had offended both him and the woman.'[24]

In fact, Olik-tuk proves so annoyed that, within a day, he has taken both wives, his hat, his coat and his umbrage and disappeared in the night. For Wilkins all that remains is a long walk to the nearest village to get another interpreter and – better not make mine a double – preferably a single man, thank you.[25]

At least lifting his spirits is that spring has sprung and, even in the Arctic, the change of seasons makes an ordinary man feel like a poet. Wilkins is no exception, recording the sights with lyrical enthusiasm:

> . . . for two months the sun has been in hiding. Hiding behind the
> bulge of the earth's equator but now at noon each day the sky

reflects a riot of evanescent colour. From grey of dawn to purple, then puce, and pink, and orange until one day a glorious band of gold insinuates itself inquiringly and glides along the ice horizon. The ice cakes sparkle and twinkling hoar-frost crystals catch up the slanting rays and hurl the myriads of spectrums to the skies. A dazzling radiance reaches from one's feet towards the sun and ends at last in his unequalled glory.[26]

In other words, the sun is shining. And Wilkins is shining too as he finally, seemingly, reaches that elusive prize, the very reason he was sent here by Gaumont in the first place . . .

For no sooner has Wilkins trudged over a ridge, he sees them in the distance. It is a group of Eskimos watching him carefully, of perhaps slightly fairer complexion. Could this be the famed Blonde Eskimos? Getting closer, he is still not sure, but gazing intently, they seem to be a bit fairer, yes?

A tiny bit. At least their facial hair is, if not quite blond, at least light brown, though the hair on their heads is black. True, 'facially Blonde Eskimos' does not have quite the ring to it, but it is certainly more accurate than the romance of their current title.

This is the much searched for tribe, it's just that they have been rather over-sold. Still, Wilkins notes that his fascination with them is equalled by theirs with him, as through him many of them are having their first encounter with another race.

It will be extraordinary footage, and he is cognisant of its significance even as he sets up his tripod and modern motion picture camera.

'For an expedition with the latest inventions of the 19th century to find itself in the midst of people who are . . . relics of the Stone Age . . . is a condition doubtless without parallel.'[27]

Unsure what this strangely white man is doing, those Eskimos who have crowded round fall back and it is only when Natkusiak arrives with the sled that things relax.

'They were reassured and quickly gathered around the sled and offer their form of greeting which is not a welcome but an assurance that they are kindly disposed.'[28]

Yes, each Blonde Eskimo gives their name and a character reference: 'I am so and so,' they say. 'I have no knife, I am a good man and do not want to fight.'[29] Pleased to meet you and likewise.

He finds himself fascinated to see that 'the blonde Eskimo are communistic in so far as food is concerned'.[30] No matter if one or other does not catch fish, they can always share in the communal catch. At least the foundation stone of their lives is familiar, and much as it is among most peoples around the world, particularly those who have not only survived but prospered in harsh environments.

'The skinning, cutting up, cleaning and cooking is always the work of the women while the men stand about and gossip when the meat is divided.'[31]

Wilkins is struck by the fiery beauty of one of these women, in particular. Her name is Kupanna and though young, as he is told the story by Natkusiak, who delightedly translates, she has already been married twice.

One day, Kupanna and her first husband had come to a new camp and another, older man, who already had two wives, had been so stunned by her loveliness he couldn't bear to be without her.

'He went to the husband and said, "Look here: I'll give you a gun for your wife" and the bargain was made.'[32]

Several days later the older man came again to the man and said, 'Well, that wife of yours is not so good as she looks, so I want my gun back.'[33] Kupanna's husband considered the offer, considered what it was like to live with Kupanna, considered the gun and then declared: 'I'll give it back all right but you have to keep the woman as well.'[34]

The Blonde Eskimos also have unique burial habits and a unique sense of humour: 'We came upon the dead body of a native stretched out upon the rocky shore, for the Blonde Eskimo never by any chance bury their dead. Beside the body were many instruments that he'd used in life, for it is the Eskimo belief that the spirit of the dead requires these things in afterlife.'[35]

How did this fellow die? Well, an old woman relates the story through Natkusiak. This dead man had been watching another Eskimo who was trying to hammer out a knife from native copper. The first man said: 'Why, you don't know how to make a knife, that thing will never be of any use.'[36]

The second man continues to make the knife. The first man keeps up his critique; the knife is too small, the edge is too dull, the copper too brittle, and then finally the second man finishes making the knife. 'I'll show you if the knife is good or not!'[37] he says and stabs the first man, who now lies dead before them. The old woman roars with laughter and is joined by all the Eskimos listening to the tale. But was the murderer not punished? Oh no, Wilkins is told: 'It served him right.'[38]

Right, *make sure to compliment their knives on the way out of town . . .*

Spending time with this tribe, who are practically untouched by contact with the white man, Wilkins is able to chronicle some more of their fascinating ways, as he will later recount in a duly reported lecture.

> In the winter time they ensconce themselves in snow huts, and when the ice goes they erect skin shelters, reminiscent of the Australian blackfellows' *wurlies* . . .
>
> Fishing, the natives make a primitive corral in a stream with stones, and the men wade in with three-pronged spears, while the women wait near the stones to catch any fish that try to make their escape through the crevices.
>
> In respect to the care of the young, it was noticeable that the babies were carried on the backs of their mothers, with a shield covering, in a manner not unlike that of our own Aborigines.[39]

Now the Blonde Eskimos have been captured by camera at last, there is nothing for it but to head off with the dog sleds to find Stef and, shattered with exhaustion, Wilkins arrives at Cape Kellett to find the leader as hale and hearty as ever, with two fascinating pieces of news.

Firstly, he has been informed by a passing Mountie, the war in Europe is not over, and far from His Majesty's forces giving the Germans a whiff of the grape before storming ashore and charging forth to haul up the Union Jack in Berlin, it seems that it has been the Germans doing most of the storming, and they are currently dug in, in France, and are resisting all attempts to push them back. It has all turned into bloody trench warfare. And the second piece of news is that, after three years of explorations and their collection canisters full, the scientific wing of the expedition – Anderson and the Southern Party – have been ordered home to Ottawa. Stef, on the other hand, has decided to continue and would like Wilkins to join him.

'You can wait until you get your mail at the *Polar Bear* [camp] before deciding whether to stay with me or go out with Anderson's party.'[40]

Given that the vessel is currently situated at Armstrong Point on Victoria Island, just 15 miles away, it sounds like a plan and Wilkins sets off by himself. When he arrives, he sees a face that looks familiar – sort of. For yes, it looks rather like John Hadley of the *Karluk* only much older, skinnier, and more haggard. The figure puts his hand out.

Hello, George.

Good God. It really is him, and what an appalling tale he has to tell – a story that Stef had known, but had declined to relate. In the wispy voice of a man reliving the horror, Hadley tells of how after Stef, Wilkins and the other four had left them, they had become aware, a couple of days later, that they were drifting across the frozen Beaufort and Chukchi seas until by the new year, the ice had crushed the *Karluk*, and it was all they could do to get their diminished supplies out before the inevitable, which had come on 11 January 1914.

What an ending for the *Karluk*, which by now had a holed hull, and was only supported topside by Mother Nature's frozen grip holding her up.[41] At 3.15 pm that day, however, she had released her frozen grip just enough for the vessel to start slipping ever lower in the water. As the decks were nearly a-wash with the sea, Captain Bartlett had ducked back on board and put on Chopin's 'Funeral March' on the Victrola record player at full blast, stepping from the vessel to the ice just as the upper railing came level with it.

At 4 pm she truly started to sink, Chopin's crescendo never more eerily perfect as, 'with the blue Canadian Government ensign at her main-topmast-head, blowing out straight and cutting the water . . . it disappeared'.[42]

As only bubbles marked her grave, Captain Bartlett had bared his head and said, 'Adios, *Karluk*!'[43]

A long trek had begun to Wrangel Island, some 80 miles away through blizzards and freezing conditions, and it was here that Bartlett insisted they should dig in and wait for a passing whaler in the coming spring.

'One man after another, [however],' Hadley recounts to the horrified Wilkins, 'took it into his head to disobey the advice of the few experienced Arctic men among them.'[44]

But how? It happened like this. A first party of four, led by *Karluk*'s first officer Sandy Anderson, was sent out by Bartlett as a scout party. They never returned. Two weeks later, the surgeon Alister Mackay has had enough waiting and enough of what he regards as Bartlett's dithering. He pens a stinging letter rejecting Bartlett's authority of them and he, along with the biologist Murray and another fellow, strike out for

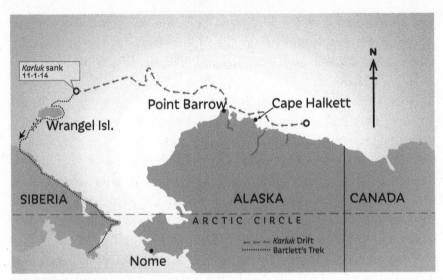

The drift of the *Karluk*

Herald Island, which Mackay thinks is just 50 miles south. They have not been seen since. It is difficult for Wilkins to think of Mackay as dead; and the vast mystery of the Arctic can always hold out hope that a man as resourceful as Mackay might return, but that is an unlikely dream. More probable is the notion that Mackay and his men are now frozen in the ocean; betrayed by fragile ice, survivors of the *Karluk* who survived for just days before mirroring its fate.

(Nobody knows it yet – or for another decade – but, against all odds, Sandy Anderson's party did make it to a hut on Herald Island, but were all asphyxiated by a faulty stove while they slept.)

Finally, in desperation, Captain Bartlett had set off with an Eskimo guide to walk all the way to the Siberian coast for help, while the 15 crew remaining had tried to survive on the island by hunting.

Captain Bartlett had made it through, and returned with the *Polar Bear*; but by the time the stranded party have been reached, both the geologists, Malloch and Mallory, have died of kidney infection, while one of the sailors, George Breddy, had shot himself after being accused of stealing food. It is a ghastly story but Wilkins does not see it as a reflection on Stef's leadership but rather on Bartlett's lack of leadership.

'Had each of them had the benefit of travelling with Stefansson and learning how to live on the resources of the Arctic as he had done for years, I feel sure that not a man would have been lost.'[45] (Those left on the *Karluk* feel very differently, with McKinlay going so far as to record his opinion that Stefansson is a 'consummate liar and cheat'.[46])

In the meantime, Wilkins is finally able to collect his mail: in fact two cablegrams and two letters. The cablegrams are from Gaumont and the *Daily Chronicle*, ordering him to return to cover the war in Europe at once. For far from being over in a matter of months, it now looks like it will be going for years.

Well, if there is a war on, then Wilkins feels strongly he must be a part of it, instead of foot-slogging across the Arctic eternity. But no, he doesn't want to take photos or capture footage for Gaumont, he wants to fight for Australia. He must get home to first see his parents and family, and then enlist.

Which brings him to the first letter. It is from his now ailing mother bearing terrible news: his beloved father has died.

Wilkins reels, stunned, grieving. The fact that his father's death was no less than 18 months earlier, somehow makes it harder to bear. He reads on.

> *It was our constant wish and our greatest desire, your dear old father's and mine, to be spared to see your safe return. I know how sad you must have been when you heard he had passed away . . . It is comforting to have so many of my children near me but I want you my son, my baby, & it is consoling to think that you will soon be on your way home. I pray that I may be spared to see you again.*[47]

The second letter follows directly from the first but should have been read before it; for it was written earlier and brings the news of the death that Mrs Wilkins presumes he must have already heard. It is a collective letter from his brothers, laying out the sad tidings and begging George to come home at once to settle their father's estate and console their mother. On the spot, Wilkins sits down to write his own formal yet moving letter to Stef, telling him the sad news, and resigning his position immediately.

> *You have a mother and the love of your parents is probably no less developed in you than most others. Imagine yourself in the position I find myself, you can if you try and I feel sure of your sympathy . . .*
>
> *In spite of the fact that it is now 18 months since the latest of these letters was written would you in my place hesitate to choose your course of action?*[48]

Of course he would not and Stef wishes him well in life and the war to come.

A long journey begins, first to get back to civilisation.

The first problem is, arriving back in Ottawa, his Canadian employer, specifically the Canadian Naval Service, require him to 'reduce my enormous mass of Arctic notes into some kind of orderly report'.[49]

Strangely, they do not remotely share his own sense of urgency about the war and insist that, as it has already been rumbling on for a couple of years, it can go at least another two weeks without him. Need they remind him, whatever arrangement he might have come to with Mr Stefansson, he has a year left on his contract with them. The compromise is that Wilkins commits to writing his report on the ship on the way home to Australia, and post it as soon as he arrives. The Canadians are duly impressed, noting that in addition to being a commander, Wilkins seems to have some artistic bent: 'George H Wilkins has made many studies with camera and cinematograph, making over 1000 film and glass plate negatives and about 9000 feet of cinematograph exposures, of Eskimo life, natural history objects and Arctic scenery and topography.'[50]

As well as that, he will be personally credited with gathering a large proportion of unique fauna and flora specimens from the Arctic regions that will soon make their way to the Canadian Museum of Nature.

Gaumont's loss is Canada's gain.

Late afternoon, 31 July 1916, Pozières on the Western Front, a man pushed to the brink pushes forward

That mysterious figure, tall and carrying himself erect, pushing forward through the deepening eventide, towards the spot where he has been told the front lines lie? In these parts a man with no rifle is an oddity indeed, and so the few figures look twice as there is no tell-tale silhouette of a muzzle poking from the top of his right shoulder. Yes, he is wearing khakis, but equally odd is that he has no cane, no holstered pistol, or any other kind of accoutrement officers so frequently bear. He bears a sense of deep responsibility . . .

Captain Charles Bean is alone, the only correspondent within coo-ee, for – a fact that annoys him – the British correspondents focus only on British exploits, such as there are, even though, as he would shortly note in his diary, 'None of them come within shellfire – much less rifle fire – and they simply don't know, I suppose.'[51]

And so it is up to him to chronicle the valour, the grandeur of the Australian experience, these extraordinary men who had done so well at Gallipoli against all odds, now doing the same on the Western Front against even greater odds. It takes a lot of careful moves, but he finally gets to the very front line to find a group of Australian survivors almost as shattered as the landscape around:

> They had been in for seven days and were nearly at the end of their powers ... They have had no fight – they have simply had to hang on in the line cut (i.e., traversed) by the German barrage ... They are quite cut off from the right and the left except for journeys overground. There they live and are slowly pounded to death – they think there are only 250 of the battalion left ... .[52]

As is ever his wont, Bean takes copious notes and asks as many questions as possible of these exhausted, shattered men – he is more than ever devoted to documenting their experience so their fellow Australians can know what they have achieved. More than ever though, he is aware that he needs help to do it. It is one thing to be the Official War Correspondent, but for photography he has had to make do with whatever is provided by the British pool photographers and they simply don't care about the Australians either. He needs photographers raised beneath the Southern Cross who actually care, who feel as proud as he does, and it is time to agitate to get some. Among other things, he is forming the idea, as he soon writes to the Australian Minister for Defence Sir George Pearce, that such photographs taken 'would eventually find a place in some national museum, when a national museum exists'.[53]

Perhaps some kind of Australian war memorial ... ?

•

For Wilkins, if only getting home was a simple matter of getting a steamer across the Pacific. Alas, no. Instead of the two-leg trip he had been hoping – from Ottawa to San Francisco to Australia – Wilkins soon finds himself on an abysmal milk-run, crawling and trawling across the waters of the globe, travelling in seas where 'favourable conditions' refer to lack of U Boats rather than the weather. The only way home proves to be via London, where, in late October 1916, right on his 28th birthday, he finds himself at a fascinating table, the background of which would leave frozen tears on your cheeks.

For when the search party had finally found the frozen remains of Scott of the Antarctic, back in November 1912 – just 11 miles from the food dump that would have saved his life – they had prised from under his frozen and emaciated right arm, his last diary, on the first page of which was written his last communication on this earth: 'Send this diary to my wife', the last word of which he had crossed out, and replaced with 'widow'.[54]

And it is that good woman, Kathleen Scott, who is hostess on this night.

A particular woman is the widow Scott. Before her husband had left for the South Pole to hopefully plant the Union Jack before Amundsen, she had written him a note on a torn piece of paper:

> Look you – when you are away South I want you to be sure that if there be a risk to take or leave, you will take it, or there is a danger for you or another man to face, it will be you who face it, just as much as before you met [our son] and me. Because man dear *we can do without you* please know for sure we can ... If there's anything you think worth doing at the cost of your life – Do it. Do you understand me? How awful if you don't.[55]

She had always been a death or glory kind of woman, not knowing Scott of the Antarctic would get both, and now it is in the basking glow of that reflected glory that her dinner table has become a veritable salon for polar explorers resident and passing through London.

Which is how Wilkins finds himself experiencing another red-letter birthday, the fates always summon contrast or consequence on that day; now it is the latter, for beside him at the dinner table is a young man roughly his age, Frank Hurley, also from Australia, with whom he has much in common. For Hurley's own professional beginnings had been to go on the famed Douglas Mawson expedition to Antarctica in 1911 as official photographer, where he had done such outstanding work, he had then been invited to join Sir Ernest Shackleton's second expedition to Antarctica in 1914. This meant he was right on the spot to take the shots for the most dramatic episode in polar exploration, the crushing of the *Endurance* and the extraordinary endurance of Shackleton and his men: who all beat incredible odds and appalling conditions and live to tell the tale, living legends all. The footage from the Mawson expedition had turned into a superb newsreel, *The Home of the Blizzard*, which had allowed the public to see for the first time just how things are in

The now restored childhood
home of George Hubert
Wilkins – a homestead in
remote Mount Bryan East,
near Hallett, South Australia.
(Regional Council of Goyder)

A studio portrait of the young
George Wilkins: on the
threshold of adventure.
(OSU)

Vilhjálmur Stefansson – Icelandic-American Arctic explorer; coiner of the term 'the Friendly Arctic'. Wilkins joined Stefansson's expedition during 1913–16.

(Wikimedia Commons)

Wilkins showing films to local Inuit people in the Arctic, Christmas 1913.

(OSU)

Captain George Hubert Wilkins, 1918. His fresh face belies the horrors that he documented in his role as an official war photographer. (OSU)

View of shell-strafed Hill 60, in Belgium, 17 August 1917, likely to have been taken by Wilkins. (AWM E02045)

Wilkins' photo of a damaged 18-pounder gun of the 25th Battery, Australian Field Artillery, Zonnebeke Valley, Belgium, 15 October 1917. (AWM E00980)

Dead and wounded Australians and Germans during the battle of Passchendaele, 12 October 1917. Wilkins and Frank Hurley both took photos of this horrific scene.
(AWM E03864)

Wilkins with Staff Sergeant William Joyce (left), standing with tripod and camera on a British Mark V tank, capturing the action. France, 4 October 1918.
(Wikimedia Commons)

A sombre photo taken by Frank Hurley, showing Wilkins looking out through the window of the shell-damaged village church at Vlamertinghe, near Ypres, upon the graves of many Australian soldiers who made the supreme sacrifice. (AWM E00849)

A photo taken by Wilkins of members of the Australian Historical Mission to Gallipoli, 1919. Charles Bean, Wilkins' boss, is second from left. The men's positions show the distance between opposing trenches in No Man's Land at Lone Pine. (AWM G01946)

The crash of the Blackburn Kangaroo ended Wilkins' chances of winning the 1919 'Great Air Race' from England to Australia. (OSU)

Photo taken by Wilkins (using the new Kodak self-timer so he could also be in the frame) of crew of the *Quest*, at the memorial cairn they built after the death of their much-loved expedition leader, Sir Ernest Shackleton, South Georgia, April 1922. Wilkins is in the back row, fourth from right. (OSU)

Members of the British Museum expedition, headed by Wilkins, to discover 'the unknown Australia', 1923. *Left to right*: Wilkins; Vladimir Kotoff, mammologist; J. Edgar Young, botanist; O.G. Cornwall, ornithologist. (OSU)

A fine display of just a few of the many hundreds of specimens collected by the team. (OSU)

Wilkins and his fellow expeditioner proudly hold up their catch: 'a flying squirrel or phalanger'. (OSU)

the Antarctic and its surrounding seas, just as Wilkins' footage of his own exploration of the Arctic had been a great success. What a rare pleasure for Wilkins to talk to a man who shares both his passions and his abilities, and the two talk until late. Hurley, in turn, is impressed with this rakish, gregarious storyteller who has such stunning tales to tell of life with the Eskimos, their customs, beliefs and staggering ability to not only survive, but prosper in such extreme climes.

'Needless to say,' Hurley will later recount, 'in the society of our charming hostess we were in thoroughly sympathetic and congenial company when it came to talk of the Polar regions . . .'[56]

For the moment it is no more than a chance encounter, and Wilkins thinks little more about it as he must soon hurry to catch his ship for home. Even now, going by the safer, longer route towards Cape Town, they are warned before they set sail to reconsider the necessity of their journey, and only the very determined remain on board. One fellow traveller tells Wilkins he is growing a beard. Why? 'I'm in such a state of nerves I don't dare shave for fear of cutting my throat.'[57]

Wilkins occupies his time on the long journey by compiling his Arctic report, careful to focus on discoveries and specimen details rather than the personalities he has encountered.

At last arriving in Adelaide in April 1917, he is tearfully reunited with his mother and the rest of the family to find that, however strong their remaining grief at the death of his father, it is countered by the joy of his safe return. No, they are not surprised that he wishes to go off again to join the war, but urge he at least take a small break?

No, George can't. Love them as he does, he is a man on a mission, and feels every bit as urgent about it as he did when he first found out about the war. History is beckoning, and Wilkins wishes to record it as it happens, to be part of it.

To a certain extent, he is already a small part of it.

For all these years, George has carried in his wallet the clipping about the day his father and his friends had been honoured by Mark Twain. Now, his mother proudly shows him, George has a couple of clippings of his own from the Adelaide *Advertiser*, ones she has cut out to give to him.

And George? How *proud* your father was when he saw them! We all were. Have a look.

AT THE WAR

AUSTRALIAN BIOGRAPH OPERATOR.

BEATS WAR CORRESPONDENTS.
UNDER SHRAPNEL FIRE.

Mr. G. H. Wilkins, a South Australian biograph photographer, who is a son of Mr. H Wilkins an old colonist, residing at Parkside, has succeeded in getting right to the front in the Balkan war and witnessing and photographing actual battle-field scenes . . .[58]

And this one . . .

A SOUTH AUSTRALIAN IN ALASKA.

THRILLING EXPERIENCES. VESSELS ADRIFT.

While people have been hearing and reading of the deeds of South Australians in the frozen south, another South Australian has been roughing it in Arctic regions. He is Mr. G. H. Wilkins, son of Mr. H. Wilkins . . .[59]

Fine. Personally, George sets absolutely no store by publicity of any kind, but if it had given his parents pleasure, that is good enough.

10 April 1917, Zurich Central train station, fellow traveller

Vladimir Lenin and his party are ready to leave Zurich, and head back to Russia, after no less than a decade in exile. Under the deal he has struck with Germany, he is to go back through Germany, under German escort, with the destination of the Russian capital of Petrograd, and if he and his Bolsheviks manage to seize power from the Tsar, then the first act of Lenin will be to get Russia out of the war with Germany, thus freeing German forces to fight on the Western Front.

On this cold morning he and his entourage of 22 fellow Bolsheviks are making their way onto the platform at Zurich station, where the special train awaits.

But what's this?

Waiting there on the platform are some other 100 Russian *emigres* living in Zurich, who know what Lenin and his group are up to, and while some have come to wish them well, most have come to register their disgust that Lenin, a Russian, has come to a deal with the very country that has brought his country to its knees.

Their catcalls begin at the mere sight of the Lenin group: 'Провокаторы!', 'Шпионы!' 'Свиньи!' 'Предатели!' – 'Provocateurs!' 'Spies!' 'Pigs!' 'Traitors!' they yell.

'The Kaiser is paying for the journey,' one Russian accuses.

'They're going to hang you . . . like German spies,'[60] shouts another.

Lenin ignores them, and ushers his group onto the designated carriage, prompting some of the more aggressive in the crowd to beat on the side of the carriage with sticks, and shout even more aggressively.

'Hiss as much as you like,' says Lenin. 'We Bolsheviks will shuffle your cards and spoil your game.'[61]

At 3.10 pm the whistle blows, and the train lurches and starts to move out of Zurich Station, to yet more jeers and cheers.

'Ilyich,' shouts one supporter to Lenin, 'take care of yourself. You're the only one we have.'[62]

Six days later, as the train at last approaches Finland Station in Petrograd just after 11 o'clock on the crisp glorious night of 16 April, a stocky figure with a luxurious black moustache is seen pacing back and forth on the platform – at least the best he can in the crush. The editor of *Pravda*, Joseph Stalin, knows Lenin well, and has been looking forward to his return. And clearly, wonderfully, Stalin is not alone. For in fact, there are perhaps as many as ten thousand of his fellow Russians come out to greet Lenin as well.

'Ура! Ура! Ура!' 'Hurrah! Hurrah! HURRAH!'

Inevitably, it is Winston Churchill who summates most memorably this moment, writing of the Germans: 'Upon the Western Front they had from the beginning used the most terrible means of offense at their disposal. They had employed poison gas on the largest scale and had invented the "*Flammenwerfer*". Nevertheless it was with a sense of awe that they turned upon Russia the most grisly of all weapons. They transported Lenin in a sealed truck like a plague bacillus from Switzerland into Russia . . .'[63]

17 April 1917, Melbourne, persist to enlist

Upon deep consideration Wilkins has decided that his best course is to apply for a commission in the Navy, 'since all that time at sea, coming on top of my experience sailing in Arctic waters, had put a touch of salt in my veins'.[64]

To a point, but only to a point, the Royal Australian Navy is pleased with his offer and offers him . . . the chance to train recruits in a motor-boat patrol. Sorry, *what?*

Training *recruits?*

Training is for men too old to fight. But he is 28, as fit as two fiddles, and having already witnessed one European war he wishes to fight in this one. The Navy, thus, is not long in receiving his terse reply: 'The war is already [more than] two years old and I want action, the sooner the better.'[65]

And yet it is the way of these things. It is not that Wilkins is famous, quite, but word of his extraordinary abilities have spread in tight circles and there are at least some people in powerful positions who are eager to help – their numbers including General Hubert John Foster, an old friend of Wilkins' late father, and young George is quickly able to arrange an appointment to see him in Melbourne.

'The only men who will get to France quickly are flyers,'[66] he tips the impatient young man. Really? Wilkins tells him all of his best flying stories, which are many and varied. Why, he has even flown in *battle* before (admittedly with the Turks). To be a pilot or to have any experience at all with planes is rare indeed at this time and General Foster puts him to the test.

'Let's drive out to the training squadron and see what you can do.'[67]

Wilkins' test flight at Point Cook is so dazzling that the training instructor declares that the young man knows almost as much as he does about planes. The report is impressive enough that General Foster personally takes it up with the Department of the Army and the Prime Minister to see if something very unconventional can be arranged. Wilkins must travel to Sydney on some personal business but by the time he arrives at his hotel a telegram awaits him from the good General:

RETURN TO MELBOURNE AT ONCE. COMMISSION GRANTED. [68]

Yes, he will be accepted into the Australian Flying Corps, with a rank of Second Lieutenant.

Wilkins takes two days to return to Melbourne to find an impatient General Foster waiting with a tailor to have him fitted for his uniform, for he has news: You, young man, are to leave for France in ten days' time.

The papers are quickly signed, as he formally joins up on 1 May 1917. For question five, he has an original answer:

5. What is your Trade or Calling? – Explorer.[69]

Pausing only to deliver a packed lecture at Adelaide's Royal Geographical Society on the 'Arctic Exploration and the Blonde Esquimaux',[70] attended by no less than the South Australian Governor, Wilkins kisses his teary mother goodbye in 1917. He sets off, first taking a ship from Sydney to Plymouth on 10 May. (Clearly, when the time between joining up and leaving is just 10 days, strings are being pulled in his favour in the highest echelons of government.)

Wilkins cannot fail to note the irony of the situation:

> I had walked 600 miles from Banks Land to reach a boat, then had journeyed from the Alaskan Arctic to the United States west coast, thence to Ottawa, to England, around the Cape of Good Hope, to Australia. Although I had my Commission in only three days, it had taken the better part of a year and a journey of some 33,000 miles before I got to my second war.[71]

Yes, Wilkins has certainly been through the wars to get to this damn war and he is still arriving three years late.

And yes, it is no small thing to be back in London, where he has so many friends and wonderful memories, but under the circumstances he is still beside himself with impatience to do his pilot training, get to France and shoot down some Huns.

But there's just one thing, George.

Military bureaucracy being what it is, before leaving for France he must undergo a compulsory medical examination.

If they insist . . .

They do.

And as it happens the military doctor . . . fails him!

The key thing is, he does not like the curious way Wilkins walks, his gait is odd. No problem, Wilkins tells him. My gait is the product of my Arctic exploration, which drove my Achilles tendon through the snow and left me with a rather distinctive stride.

Oh, an explorer? That only makes the doctor more convinced Wilkins is not fit to serve.

'Your feet are in a hopeless condition. You did not have medical atten-tion in time, and now it is too late,'[72] reports the doctor definitively. Before Wilkins can interject, he continues with the bad news.

'Nothing can be done. You will never be able to walk properly again.'[73]

Wilkins takes pause. Given that on that last 'walk', through snow, ice and wind he had done rather well, he feels free to disagree. After all, despite the fact his lope creates its own slope regardless of how level the land is, he still walks faster than anyone he has yet met.

'I averaged 15 miles a day on foot in the Arctic!'[74] Wilkins retorts.

'From a strictly medical viewpoint, such a performance is imposs-ible,'[75] the doctor declares.

And he will not budge in changing his determination that Wilkins is Medically Unfit[76] to serve in the Australian Flying Corps.

Getting desperate, Wilkins points out that shouldn't the fact he has walked across large portions of the unknown world be the point?

Still, Wilkins decides to take the high road.

'Since I already had my [Australian Flying Corps] commission, I decided not to argue the point.'[77] But, next up, a compulsory eye test, where the next doctor has more news for an increasingly annoyed Wilkins. 'You can't hold a commission as a pilot.'[78]

And why would that be?

'Because you are colour-blind.'[79]

And this colour blindness makes you, 'unfit in any capacity to be a flying officer'.[80]

As Wilkins has already not just flown, but acted as a navigation officer, he finds this puzzling. Especially so given that, in his rare times of leisure, he has proved to be a dab hand at painting watercolours and oils. Surely, being colour-blind would have become apparent before this, either in the sky or when people wondered about the unusual 'modern art' he had unwittingly produced? Well, that is a point and the military doctor agrees to consult a specialist. The specialist concludes the following:

Eyesight normal. Knowledge of colour names poor.[81]

Bottom line?

His form is marked with the words 'Unfit in any capacity as [Flying Officer] owing to defective colour vision.'[82]

Again, Wilkins does what he does best: he keeps calm, and carries on. But just quietly?

'I wonder,' he will note, 'how many good men were thus lost to their country's service because the test, rather than the tested, was faulty.'[83]

The following morning, Wilkins is ordered to report to an amused General Thomas Griffiths – the Commandant of Australian Imperial Forces HQ in London – who greets him cheerily: 'Well, I hear you're colour-blind and can't be a pilot.'[84]

Wilkins hurriedly shows him his certificates stating otherwise, only to find that Griffiths knows something he doesn't. It has already been decided from up top that Wilkins will not be a pilot in this war.

'There are other important jobs,' Griffiths says leadingly.

Such as?

'Captain Frank Hurley has been appointed official photographer for the [Commonwealth Military Forces] and Captain Bean has selected you to be Hurley's assistant and official photographer of historical events in the war.'[85]

Hur-ley?

The fellow he had met at Kathleen Scott's table just before heading home to Australia?

'That pleased me.'[86]

Also pleasing is that 'Anzac Charley' Bean has selected him. That is no small honour. Wilkins is frankly amazed that Bean has even heard of him, let alone selected him. But Wilkins certainly knows all about Bean, whose legend had spread – an honest and courageous correspondent who had been in the thick of it from the first, and is that rarest of pressmen in this war, a man who has 'the confidence and respect of both officers and men'.[87]

Some returned soldiers Wilkins had met back in Australia had already told him about the Sydney man.

'He had absolute integrity plus courage of the highest order.'[88]

Will I still be a pilot?

General Griffiths is quick to reassure him: 'You can keep your rank of Second Lieutenant in the flying corps. You can fly whenever it is desirable to get pictures from the air.'[89]

But Wilkins' assignment is broader than that. 'You will be expected to be with every section of the army in turn and photograph it in action.'[90]

Wilkins is now *very* pleased.

It is the opportunity of a lifetime, to become a chronicler for his country's participation in what has already become known as the Great

War – and for good reasons. It is a global conflagration already involving 28 countries so far and 20 million soldiers. It seems likely there will never be another conflict as grand and as terrible in the history of the world, and to be able to travel to it with his cameras is no small thing.

Given that Wilkins has a few days before catching his ship across the Channel to France, it gives him the time to find his way back to Kathleen Scott's table where he finds none other than . . .

Frank Hurley!

Delighted to encounter each other before they are due to catch the same ship to France, a jolly evening is had, with all of the night's company amused to hear of Wilkins' travails of being turned down as an officer of the Flying Corps on the grounds of colour blindness. To the amusement of all, while everyone in the room pronounces one of the roses in the room to be red, Wilkins not only says it is green, but won't back off in the face of their unanimity.

'He stuck to his guns, too,' Hurley will recount. 'Hadn't he as much right to his opinion in the matter as we had – by the subjective idealism argument? Everything depended on the individual's interpretation of what impinged on his retina – and another man would have had as much right to call the rose yellow as we had to say it was red or green. I'm not sure that George was not prepared to argue, on the same grounds, that it was his Flying Corps examiners who were colour-blind – and not he himself.'[91]

Hurley likes Wilkins' style enormously, and within two days the two have stowed their kit upon their steamship, *Princess Victoria*, on their way across the Channel. Now as two young Australians, a long way from home, they have much to discuss, and both are absorbed, leaving 'the shores of old England for the grim duties of France'.[92]

The battle beckons, and both are aware that there is a certain symmetry in the fact that, after they had done much the same job at opposite Poles of the Earth, they are now meeting in the middle, to work in concert photographing the heat of a Battle Royale . . .

CHAPTER SEVEN

MOI, AUSSIE

[Wilkins] is a completely fearless man, of sound judgement and unbeatable initiative.[1]

Frank Hurley

Wilkins has probably been in the fighting more constantly than any other officer in the Corps.[2]

Captain Charles Bean

5 pm, 21 August 1917, Boulogne-sur-Mer, France, Bean seen

Could there possibly be a more benign entry into a country at war? Crossing the Channel from England's port of Folkestone, the French port of Boulogne-sur-Mer first comes into view as a greenish blob for'ard off the bow, which soon transforms into a scene that would do Claude Monet proud with gay vessels bobbing on blue waters, people swimming and seaside cafes.

'We had a particularly fine view just before entering port, of the beach,' Hurley notes in his diary, 'where a great number of folk were bathing. Nothing seemed more distant than war, and had it not been for the great number of troops embarking and disembarking, everything was serenely peaceful.'[3]

Once Wilkins and Hurley's ship is docked, of course, things start to become a lot more serious.

For who is this man in the uniform of a captain of the Australian Imperial Force, waiting at the bottom of the gangplank and looking out for two men, one in the uniform of a captain of the AIF, the other a lieutenant of the Australian Flying Corps?

It is, of course, Captain Charles Bean, the Official War Correspondent who has recently become the Official War Historian (and seems to be both simultaneously, influencing the leaders he covers and the war he watches, through his words). Bean is technically working for the

Australian War Records Section, which he had successfully suggested be set up to chronicle the entire Australian experience in this war, and for which the two newly arrived photographers will be working. It means that Bean is now not just their colleague but also their superior officer, although remaining a civilian.[4] Wilkins, impressed, sees that he has come complete with two cars with drivers to carry their baggage, cameras and film, and in his ever no-nonsense manner now hustles them forth.

There is much to be done, and no time to be lost.

A serious man, Bean, with much to be serious about. And yet, while he had started out the war believing in British might and majesty, by now a combination of British bungling and Australian valour has seen him become a champion of the need for both of those real stories to be told. After the Battle of Pozières, in 1916, Bean had been infuriated by the lack of British acknowledgement for the Australian achievement – taking the high point of the entire Somme valley, after many failed British attempts. The Australians he had noted, 'fought the greatest battle in our history and one of the greatest in [British History] – but not a suspicion of it would you get from the English papers'.[5]

He believes the Australian Digger to have admirable fighting abilities, so great he might be able to overcome both the German steel *and* the British bungling.

Right now, the bulk of the Diggers, the men of the 1st, 2nd, 3rd, 4th and 5th Divisions are based in northern France around Ypres, where the fighting on the Western Front has been at its most fierce.

After a quick meal at a restaurant on the wharf at Boulogne-sur-Mer – Wilkins noting that the famed American General John 'Black Jack' Pershing, who will command all the American forces about to land in Europe, is with 'his staff sat at the next table'[6] – they head off.

For the next two hours, they jolt over bumpy tracks as it gets progressively darker, pushing towards the AIF HQ at Hazebrouck, some 60 miles to the east, through ever thickening traffic and soldiers.

They are travelling at such speed that they run 'into a flock of sheep'[7] in the French countryside, and even though when you have experienced what Bean has in recent years, the injured sheep and damage is minor, Bean offers the farmer recompense.

Feeling a little wild and now woolly, they are soon on their way again, and 'crossing the hill at Cassel we came within sight of the flashes of the bursting shells',[8] a kind of pulsing light on the eastern horizon, mixed in slightly delayed synchronicity with a kind of long rumbling thunder that doesn't fade.

Oh, Christ.

That must be ... ?

The Western Front, yes.

It is at that precise point, where the force of the German thrust has been equalled by the strength of the Allied parry, that both sides have dug in, while pounding each other with artillery shells and sending wave after wave of soldiers charging at each other's machine guns. It is a hell on earth such as has never existed before, and Wilkins and the others are about to move into the very heart of it.

The closer they get – no talking now, as all are left with their thoughts – the more the pulsing light scatters into stabs of flame followed by billowing smoke, from one length of the horizon to the other!

'We were,' Wilkins will recount, 'intensely impressed.'[9]

'The sky,' Hurley, for his part, chronicles, 'gradually grew brighter with the flickering quiver of hundreds of guns in action. It was the most awesome sight I have ever seen. As we neared Hazebrouck, the guns could be heard almost synchronizing with the flashing.'[10]

Good God.

Like most newcomers, Wilkins and Hurley must utter silent prayers, thinking of home and wondering: will we ever see it again? Wilkins had seen artillery in action during the Balkan War, but it had been nothing like this.

Bean?

He barely blinks.

For in terms of covering battles, Bean from the beginning had taken the same view as Wilkins at the Balkan War – the only place to report from was the front line, and he had seen a lot of it. This was in strict contrast to most of the British correspondents, for whom Bean has very little regard for not even coming within range of shell fire.[11]

Which is where you two, Captain Hurley and Lieutenant Wilkins, come in. Your job is to help me chronicle the good, the bad, the ugly of this war, and the truly great – the last of which mostly concerns the extra-ordinary effort put in by ordinary Australian men for astonishing results.

'I don't think,' Bean tells the two newly arrived men, 'the public has the least idea of the battle we are fighting.'[12]

And so we must inform them. Your pictures will capture what my words cannot, the terrible and grim realities of the fight; the conditions that the Diggers of the front endure as they wait for the fray to begin anew, and what happens when they go over the top.

Bean has deliberately selected two men with very different sensibilities. Hurley is an artist, Wilkins is a realist. Bean is frank with both men about how he judges their abilities and roles. As they continue to bump along – almost there now – he sets it out.

> Hurley, you are to photograph such scenes suitable to propaganda and press release, while you, Wilkins, must photograph actual front line scenes and incidents, showing everything the camera can represent, notwithstanding the conditions.[13]

The 'propaganda' angle, such as it is, is simply to show the Australian forces at their best and as the Australian Government would wish them to be shown. As to the gritty shots Wilkins is expected to get, there will be no time to wait for the right light.

'I would often have to photograph . . . amid the smoke and dust of battle,' Wilkins recounts. 'From a photographic point of view the film would be bad, but from a historical point of view very valuable.'[14]

Of the two photographers, Hurley stands as senior and is given a car and a driver to further his art. Wilkins will have to make do, lugging his own camera around to record battles and men as he finds them.

But, in any case, they soon must settle into their extraordinary new circumstances. After one night bunking down uncomfortably in the HQ at Hazebrouck, the next afternoon they head to the battlefield proper on the Western Front, where the Australians have been fighting around the legendary 'Hill 60', some 30 miles to the east.

Good God, such things they see, as they had never imagined existed on heaven or earth – and they were right, for this is truly hell.

For while Hill 60 sounds rustic even in the context of war, it proves to be a giant junk yard of debris and humanity; a sea of shell holes with an abandoned tide of broken rifles, gas-masks, ration boxes and used bandages to decorate it. Everywhere they look there is mud and blood, death and disaster; fresh troops marching forward to replace shattered troops coming back, not to mention Diggers with bloody stumps where their

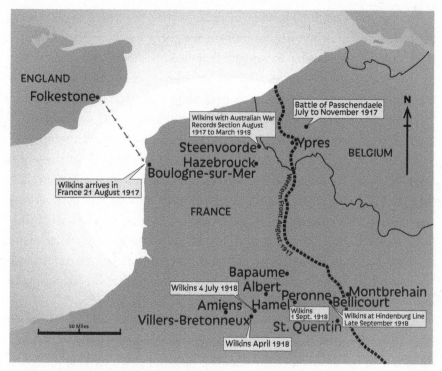

France, Belgium and the Western Front

arms and legs used to be, some of them have eyes with thousand-yard stares, others have eyes that are weeping. The moans of the wounded compete with the roar of the guns and exploding shells in the distance. Everything induces horror into Wilkins and Hurley, whose task it is to capture it. The darkest thing of all? How used to it all the other men seem to be. This horror is *normal* to them . . . which in some odd way is the most horrible thing of all.

Shaken but trying to appear unstirred, the two new recruits are about to unpack in their tent quarters, only to hear the sound of low-flying planes . . . which is quickly followed by the shattering roar of bombs exploding nearby!

Run for your lives!

The madly scrambling Hurley and Wilkins wind up in the cellar of an old French lady. 'She kept us entertained,' Wilkins says, 'telling us she often spent the whole night there. It was an exciting night for both of us, especially Hurley, because he had never experienced shell fire or bombs . . .'[15]

Neither Wilkins nor Hurley can sleep as they pop their heads out of the cellar to watch the display put on for them by the troops of both sides: 'Quite an exciting first night that was. After the air raid we sat up till morning watching the Titanic struggle against the battlefront – countless flashes from the big guns, exploding shells and signal rockets.'[16]

Those very Gorblimey things keep falling till dawn, each one shaking the ground, with none more shaken than the two Australian photographers.

The following morning, hearing of their experience, Bean does not know whether to be amused or horrified, pointing out to the men that they should locate *safety holes* at any new area they might come across. Ask around. All the old hands had just gone to their bunkers when the bombs had started falling. It is what you do in these parts. There will be more bombs and shells coming your way, and as you move around you need to reconnoitre and know where those bunkers are, so that you know where to dive when the shells start to land.

And yet it is the next night that neither of the men will forget for as long as they live, as Bean, effectively, bloods them for battle, by taking them out to the front lines through the Belgium town of Ypres . . .

Well, it used to be Ypres.

What is left now is no more than dusty roads through piles of shattered rubble, the unfortunate result of your town being situated between two armies with artillery capable of destroying buildings, and prepared to turn into pebbles that which they cannot occupy. Hurley and Wilkins don't speak, partly through shock and partly because so much army traffic is raising the dusty remains of what used to be a vibrant town.

Wilkins cannot resist filming the scene: Ypres appearing like some badly preserved Roman ruin on its last legs, the frames of grand buildings standing stark and alone surrounded by the rubble that once made up the remainder. Through the remnants march the Allied soldiers, their straight lines and order creating an irony that is inescapable. The camera turns to a fresh wreck, a chaos of timber jutting up from the ground, and then Wilkins films troops tottering through the towers of giant pick-up sticks, like a line of ants determined to make their way through any obstacle. Are these men searching for the dead or the living? Can one make such a clear distinction in Ypres? The town itself is a grotesque ghost, if you could not see it you would not believe it, and once seen you shall never forget.

On the other side, they get on the infamous Menin Road that leads straight to the front. As endless rain pours down, they travel past a succession of ominous and accurate names like 'Hell Fire Corner', 'Bloody Angle', 'Dead Man's Wood', 'Death Trap Farm', past craters the size of football fields, one of which belongs to the already infamous Hill 60. For two men who just days ago had been in London, it is a baptism of blood.

Everywhere they look there are dead bodies, pieces of dead bodies, pulverised bodies, screaming men lit by the ghastly pulsing light thrown by the flares and exploding shells of this modern and industrialised version of Dante's Hell. And every five seconds or so, another shell lands, causing more bloody carnage.

Some of the wounded lie in pools, staining the water with their blood. Dead men lie unremarked in the mud beside shattered duckboards – that wooden trail that leads men through the 'morass of slimy mud'[17] – and Wilkins and Hurley must pull their coats up to try to shut out the fumes of high explosives, and the unforgettable odour of death. Sometimes the ground beneath your feet is unaccountably spongy and it takes a while to realise you are walking over the bodies of the dead.

Both men are dazed, trying to take it all in.

From everywhere the darkness presses with the only source of light being that cast by the flares and exploding shells – just enough to allow you to navigate across the open ground and slim duckboards that separate trench from trench, even as you dodge the endless stream of grim-faced stretcher-bearers and wounded coming the other way. Once the 'front' is reached, Hurley and Wilkins discover that their headquarters is 50 feet beneath the surface, tunnelled through clay.

'There we listened to the pounding shells on top of the dugout. Every shell that struck the top drenched us with water. It shook the water down on us. We were about forty feet beneath the surface. We had a meal of bully beef, I remember.'[18]

Welcome to the Western Front.

God help us all, are we even going to make it through the night?

But to work.

They are not here just to witness such devastation, but to stand witness to it, to chronicle what is happening on the Western Front and most particularly the Australians' role in it. Everything around them is a shattered shemozzle, including both the commands that are flying about and the method of communicating them.

'All orders are given either by telephone or wireless,' Hurley notes. 'The whole place is just a network of telephone wires which are constantly being destroyed by shell-fire . . .'[19]

At Bean's insistence, both men use plate glass cameras – a process fundamentally unchanged since being, yes, developed, in 1871. The clear glass plates are coated with silver nitrate emulsion, and must be lowered into the camera, while under a black cloth you point the camera to the scene you wish to capture, before pressing the shutter. Once the image is captured the plate is removed and you write upon it the time, place and subject. Rinse and repeat, and try to ignore the bullets and shrapnel flying all around you, some of which risk being aimed right at your vertical form, a stationary, standing target.

True, a camera using film roll like a Vest Pocket Kodak would be much easier to load and a lot less fragile, but Bean does not care. His only concern is quality and he has no doubt that the plate glass cameras produce the best images, the ones which will demonstrate to the future how things really were in this hell on earth. Bean believes the right photograph can be as lasting and moving as a painting, if taken properly with the right equipment. Posterity, not convenience or ease, is their master. Hurley and Wilkins will have to carry a lot of lumbering gear with them, most particularly a tripod to ensure maximum stillness while the negative is exposed, and whole boxes of heavy and very breakable plates, but so be it. Whatever the difficulty with such photography at the best of times, both Wilkins and Hurley have done it at the worst of times, at the Poles, defeating terrible conditions to obtain excellent images. This is precisely why they have been singled out for this job.

The heroism of the soldiers they are documenting helps to inspire both Hurley and Wilkins as they set to with great enthusiasm and extraordinary courage.

Often right on the front lines, and sometimes out in front of advancing troops, Wilkins soon becomes a familiar figure to the Diggers over the coming weeks, always with a camera in hand, traipsing through trenches, not simply distinctive by his limping lope courtesy of the enduring damage done to his Achilles tendon in the polar region, but by the fact that, more often than not, he is the only bastard out there, almost always with his tripod and lugging his own big box of plates.

Even Hurley recognises that this bloke has gone from being his junior to his peer.

'I had a long talk today with Capt. Bean over affairs generally,' Hurley records in his diary, 'and the best method of running this department. Wilkins will be operating practically separately . . . It is therefore necessary he should have separate equipment.'[20]

With Wilkins proving to be capturing photographs of such quality, Bean instantly agrees and assigns both photographers 'lumpers' – soldiers to help them by doing a lot of the heavy lifting.

Wilkins visits field hospitals, slips into dugouts and ranges respectfully across freshly dug battlefield graveyards, capturing it all on film. Sometimes – in one for the old days – he hitches a ride on a plane to get aerial shots of the trenches. But mostly he is with the soldiers in the mud and blood, the death and destruction.

He and Hurley photograph the troops marching to the front, the dead brought back and the battles themselves. Life comes to have a routine, as bizarre and dangerous as it is. The photographers are given a two-room iron hut Hurley names 'The Billabong' in the rural village of Steenvoorde, about 17 miles west of Ypres and well back from front lines, and it is here that their plates are developed. At five o'clock every morning, Hurley and Wilkins depart from their base in a car, with assistants if required, and drive as close as they can to the desired front without becoming a target for enemy artillery courtesy of any observers they might have on high. Parking in a shattered farmyard, they now walk, carrying their bulky gear with them, as they scurry to the back area of the front lines, before making their way forward along communication trenches. With some delicacy Wilkins will describe the demarcation of their work: 'Hurley would take picturesque and impressive views of the troops . . . while my pictures were of actual engagements.'[21]

You handle the art, I'll handle the war.

They stay out in the field until dusk, before driving either to the Billabong or to the nearest HQ to develop the images they have captured. Then, and only then, do they have their second meal of the day. (If a random tin of bully beef can indeed be considered a 'meal'.) After dinner, they return to the Billabong, where the prints submitted to the censor the previous day are examined once more, and more particularly 'the comments on them made by the censors and headquarters staff'[22] contained in neatly written dossiers. They try to get to bed by midnight, then rise once more for the 5 am trip for the front. They are their own masters; and they are determined to be the master photographers of this war, to record images that will and should never be forgotten.

Every time the Australians launch a major attack – which is frequent in these first few harried weeks – Wilkins and Hurley are front and centre, recording it all.

During the legendary Battle of Passchendaele, which endures for three and a half months and sees five major Australian attacks to control the strategically crucial ridge south and east of Ypres – at the end of which was the devastated village of Passchendaele – Wilkins is particularly active, even while the five Australian Divisions suffer no fewer than 40,000 casualties.

On just one typical day, in mid-September, Wilkins is with Bean and another correspondent, Henry Gullett, as they approach the front. As ever, the artillery rages, the sky is dark, and the roads are choked with fresh troops going forward and the shattered remains of the relieved troops coming back.

'On a sudden,' Bean will recount, 'five yellow flares went up, one after the other as fast as they could be thrown. Within a few minutes, down came a German barrage, tremendously heavy onto the rear slope of the hill – crash, crash, crash, crash – all colours of bursts flickered with white smoke and black . . .'[23]

In peacetime, of course, one would run screaming for the hills. What do they do?

> We pressed on towards Clapham Junction but just as we crested the slope a couple of shrapnel burst straight ahead of us up the avenue up which we were walking, and then a bigger shell swished in low straight over our heads and plunged into a neighbouring shellhole and burst on that side of the route as we dived into a trench on the other. We found the trenches on the Summit . . . We pushed along the battered trench until we found an officer. The men were in little undercut pozzies in the front side of trench – we passed two, in neighbouring pozzies with a blood splashed waterproof sheet covering the entrance, & the flies buzzing round the bloodstained legs and puttees which protruded – otherwise we should not have known the men were not alive. We came at last to the officer, a poor little startled boy, horrified by the sights and the shelling . . .[24]

Throughout the day and into the evening they are close to the action. Bean takes copious notes, Wilkins sets up his camera and takes shots.

And so it goes. They are always in amongst it. (In another memorable episode that occurred almost a year later, on the Somme, some Canadian troops on the front line see that the Australians are held up by the German machine gunners and could use some help, so they send one of their tanks forth, with Canadian infantry tightly behind, all of whom wisely go to ground when the Germans open up on them from the flanks.

'Wilkins got a photograph of them . . .' Bean records, 'with the men all lying on their faces under heavy fire – he said you could see the dust spurts from the machinegun bullets and that within a quarter of an hour there were 12 men wounded there. The machinegun nests were cleared and then Wilkins walked across to where our infantry were . . . he was 800 yards in front of the infantry . . . He found a party of about 30 of the 5th Bn in a shell hole . . . They called to him to get down or he would get shot.'[25]

But Wilkins does not get down. He keeps shooting pictures even as the Canadians with the Australians are shooting back and forth with the Germans.

It is a miracle he's not shot.)

And such courage, and good fortune on his part, will be a hallmark of his time on the Western Front.

For him it is just another day chronicling the Australians in action, and is not to be remarked upon particularly.

The same goes for the day an Allied armament dump takes a direct hit. 'Wilkins and I nearly blown up by a dump which exploded only 100 yards away . . . thank God we are unscathed,'[26] Hurley notes.

Just another day . . .

And the next day, more of the same.

'Yesterday,' Hurley will chronicle, 'we damned near succeeded in having an end made to ourselves. In spite of heavy shelling by the Boche, we made an endeavour to secure a number of shell burst pictures. Many of the shells broke only a few score paces away so that we had to throw ourselves into shell holes to avoid splinters . . . I took two pictures by hiding in a dugout and then rushing out and snapping.'[27]

It is a very particular skill. The instant after the explosion, there will be shrapnel flying about. You must wait for that to pass, get out, set up, focus, open the shutter and *get* the shot before *being* shot, or cut to pieces by shrapnel when the next shell lands.

We eluded shells until just about 150 yards away when a terrific angry rocketlike shriek warned us to duck. This we did by throwing ourselves flat in a shell hole half filled with mud – a fortunate precaution, for immediately a terrific roar made us squeeze ourselves into as little bulk as possible and up went timber, stones, shells and everything else in the vicinity. A dump of 4.5 Shells had received a direct hit, the splinters rained on our helmets, and the debris and mud came down like a cloud. The frightful concussion absolutely winded us but we escaped injury and made off through the mud and water as fast as we possibly could. Egad, I've never heard such a row in my life.[28]

Another diary entry from Hurley during this period gives further furious flavour of the times.

Wilkins and I set out in the car for Hellfire Corner ... Things were reasonably quiet till we got near to Zonnebeke. But the *mud!* Trudge, trudge, sometimes to the knee in sucking tenacious slime ... shells lobbed all around and sent their splinters whizzing everywhere – God knows how anybody can escape them, and the spitting ping of machine-gun bullets that played on certain points made one wish he was a microbe; under these conditions one feels so magnified that he feels every shell the Boche fires is directed for his especial benefit. This shelled embankment of mud was a terrible sight. Every twenty paces or less lay a body. Some frightfully mutilated, without legs, arms and heads and half covered in mud and slime. I could not help thinking as Wilkins and I were trudging along this inferno and soaked to the skin, talking and living beings, might not be the next moment one of these things – it puts the wind up one at times ... Under a questionably sheltered bank lay a group of dead men. Sitting by them in little scooped out recesses sat a few living, but so emaciated by fatigue and shell shock that it was hard to differentiate.[29]

On and on, day after day, just like this.

'The work was so strenuous,' Wilkins will note, 'that Captain Bean made us take one day's leave each month in London. But for this day of rest, I doubt if we could have kept up the pace.'[30]

A selection of photographs is made once a week by Bean, Hurley and Wilkins. If the pictures 'pass' propaganda standards and are deemed fit for the press they are sent to England, if they don't, they are sent, 'to be stored in the library to be used in the history of the war that Captain Bean was compiling'.[31]

Sometimes Wilkins goes up to take shots from on high; in more ways than one.

> I would select a good pilot from an RE 8 observing squadron and fly over the lines with an escort ... But it was often necessary for me to give up my reconnaissance through being attacked by some German planes ... We had some narrow escapes in the air and had our machine riddled with bullets although we were never wounded or forced down.[32]

Bit by bit, battle by battle, Hurley and Wilkins come to know the tricks of the trade when it comes to covering a war. The first thing, early on, is *patience*.

'Often it would take me hours to get from one shell hole to another only 300 yards away. Several times I would lie out in the open and pretend I was dead to fool the Germans snipers.'[33]

The Diggers themselves give them tips.

When yers get within coo-ee of the front lines, fer *Chrissakes*, don't look up as that exposes more of your noggin to Fritz's bullets. Keeping your melon down means you gets more protection from your *helmet*.

And use your *ears*, cobber. You hear that shell whistling? That means it *isn't* coming straight at you. The truly dangerous ones are those that first *screech* and then rumble like a train, before really *roaring* – at which point you'll have just a few seconds to throw yourself into any shell hole or trench you can find or, failing that, getting lower than a snake's belly-button to escape most of the shrapnel. But, listen, if you throw yourself to the ground every time you hear a whistle, you won't do anything else, sport.

Now, the 'safest' time to be on a battlefield?

'[It is] immediately after it had been severely strafed by German artillery,' Wilkins will recount. 'There was usually a let-up for a few minutes when we could dash in, get photographs, and be out again before the attack was resumed.'[34]

Sometimes things are not so clean or easy, as happens in the Battle of Menin Road which has already begun pre-dawn with an enormous hail of artillery. No fewer than 65,000 men are waiting on the Allied side alone.

'Nothing I have heard in this world or can in the next,' Hurley notes of this artillery, 'could possibly approach its equal.'[35]

After the barrage ceases and the first wave of Allied troops goes forward, so too do Wilkins and Hurley with their cameras. Just as they leave their trench, Hurley notes 'a 5.9 inch shell landed six or seven paces from us'.[36]

He and Wilkins glance at each other helplessly, waiting for the inevitable explosion and death.

Nothing.

The shell is a dud.

Onwards.

Once, and only once, by Wilkins' own account, does the South Australian come close to losing his nerve. One horrifying night he and six soldiers are groping their way towards the front line along a duckboard above the sucking mud. Shells explode, shrapnel flies. Still they press on! But now a shell explodes just a short way in front of them, hurling shrapnel into the night. One enormous piece of shrapnel comes scything out of the darkness and hits the leading man in Wilkins' party so viciously and so *neatly* at the base of the neck that it takes his whole head off, and the man's head lands by plopping atop the post that had been right beside him when he died. That grisly head now stares back at them.

'All the rest of us roared with hysterical laughter to see his head there, stuck upon the post. At the moment it seemed hilariously funny.'[37]

Christ Almighty.

Still they push on. Bean to Brigade Headquarters at Hooge, Wilkins and Hurley to another Hellscape. Hurley's diary records another nervous night of terrifying beauty: '... we had a magnificent view of it. The whole country viewed from the 3rd story of an old ruin was so alive with gun flashes, that I can only liken them to looking over the twinkling lights of a city, and during a thunderstorm ... Naturally we got neither sleep nor rest amid this pandemonium and right glad we were to be a'move at 4 a.m.'[38]

The morning sees Wilkins and Hurley ducking and dashing from shell hole to shell hole to get to the battlefront.

They soon hear the haunting calls for help from men sinking into the mud, men who already cannot be seen and soon will not be heard. Wilkins writes: 'most of the men had been shot through the legs by machine gun fire',[39] and they simply sink into the mire and drown. Others have faster fates, because the tanks are now making their way to the front and over the wounded.

Oh God!

One tank is heading straight for three wounded men and they can't get out of the way. In an instant Wilkins is on his feet, placing himself between the lumbering monster and the men, waving his arms. Will it see him? Will it stop?

It WON'T!

At the last instant Wilkins dives out of the way, which leaves the tank to crush to death the three soldiers – whose screams of horror stop in a shattering instant – as the tank goes 'lumbering on'.[40]

Wilkins and Hurley watch in horror before another catastrophe unfolds, this time to those in that same tank: 'Before it had travelled far it was struck by an incendiary shell and caught fire. The explosion jammed the doors of the tank and we could see the tank-men slowly burn to death as we unsuccessfully tried to break open the doors.'[41]

Grim necessity is the mother of invention and the father of nothing-else-counts as Wilkins realises that the top of the now destroyed tank provides a unique viewpoint for taking photographs of the battle. And so, as Charles Bean, who is with them, later notes in his diary, Wilkins now carefully clambers onto the top of the tank, and sets up his tripod, even as shells continue to land all around and tanks continue to head their way. Wilkins indeed takes one shot but, while adjusting the tripod for his next one, a shell from hell bursts forth and explodes right under him, and he is only saved because there is a whole tank between him and the explosion.

As the plume of smoke and billowing flame calms, Bean observes what has happened.

Are you all right?

No.

No, he is NOT all right.

This time some real damage has been done. Bean waits as the groaning Wilkins regathers himself to call down his report of his grievous injuries:

'My ground-glass focusing screen is broken and a slide with two fine plates is ruined.'[42]

He is *very* annoyed.

Never mind. Wilkins repairs the camera, puts in a new slide and continues to photograph from the top of the tank.

This is a kind of bravery that Bean – himself an extremely brave man – has not witnessed before. It is the valour of the non-combatant, completely unarmed as a point of honour, doing his utmost to proceed with his task, come what may, come hell or high water, shell or shards of shrapnel, death or destruction. What they have seen that day is as awful as it is ordinary.

Lying in his bunk that night, Wilkins is simply unable to escape from the most horrific thing of many he has seen that day: the men burning in the tank in front of him. Their voices haunt him and, God help him, he can hear them now, the agonised cries of men who realise their time has come, that they are burning, will continue to burn and will die in this foreign field, never to return to their homeland, see their families, hold their wives, their girlfriends, their children, ever again. They have died before his very eyes, and but for the merest chance it would have been him not them. He tries to banish such thoughts, but does not succeed, even as the rumbling of the ongoing shell fire rumbles through the night, just as the breeze carries the fumes of the just exploded cordite.

'For a long time after,' he will recount, 'I imagined I could still hear them screaming for help in their pain, as they roasted alive.'[43]

They were brave men to even get into the tank in the first place, and deserved a better fate.

Wilkins' own bravery is noted again and again by Bean, who takes the unprecedented step of officially pushing for a non-combatant photographer to be awarded a combat decoration. Because, as Bean points out, for Wilkins, extraordinary feats of valour have become everyday:

> On September 20th early in the fight . . . a shell bursting beneath a tank from which he was operating broke some of the gear which he was carrying. In spite of this he continued to obtain records of every subject of value for Australian history . . . This work has been imposed on Lieut. Wilkins by his own sense of duty, the results being invaluable as records while he rarely obtains the credit of

publication. His demeanour has been markedly gallant, and has noticeably brought credit upon his office amongst the troops.[44]

Bean's request is refused but he keeps logging Wilkins' feats. His time will come, Bean will make sure of it.

Amidst all the horror, one thing stands out in particular for Wilkins. 'Worst of all to me was the sight of an airplane falling in flames . . .'[45]

You will hear it first, that faint droning sound, you will look up to see that, yes, it is a plane plummeting, already on fire. The pilot is at the controls, his flesh burning as the craft smashes into the ground.

'Many times I was nearby when a plane fell, and I had to photograph the charred body of the aviator.'[46]

These photos are not for press or propaganda, they are unpublishable, they are for history.

One way or another, as the weeks go by, Wilkins becomes an ever better known figure to the Diggers, the mad bastard who will 'jump the bags', on his own, ahead of the main rush and then set up his camera to photograph them charging, all the while with his back to the enemy.

What must the Germans make of it?

Some, clearly, take pot shots at him. But, surely, one reason he survives is that most of the Germans must be caught between bemusement at his foolhardiness and stunned amazement at his courage and so decline to shoot an obviously unarmed man in the back. This is certainly Wilkins' belief, declaring that the Germans can see that he is unarmed . . . and know it would not be sporting to shoot a photographer.

'The Germans frequently waved to me good-naturedly; they knew I was a photographer . . .'[47]

Other times, when things are a bit willing anyway, he uses ruses – Stef would be proud – from the natural world.

> In No Man's Land I would pretend I was dead and wait until they were not looking and get to the next shell hole. One day I spent five hours going a half mile . . . Every shell hole was full of mud and muck and rotting bodies.[48]

Occasionally the Germans take very deliberate and precise shots at his camera, which only increases Wilkins' confidence. Clearly, they are amusing themselves. If they wanted to shoot him, they could. The fact that as soon as the Allies go over the top a hail of bullets will fly and there will be no such nicety of precise aim is a fact that Wilkins refuses to focus on, preferring to focus his camera instead.

Still, it all means that Wilkins becomes perhaps the only man in the war who is actually comfortable in No Man's Land, and the soldiers will tell each other many stories of this extraordinary snapper, who has so little fear he not only takes his camera into the most dangerous situations but even walks 'in the vicinity of no-man's-land for exercise, quite oblivious to any danger'.[49]

For exercise? Yes, Wilkins will explain, he got into the habit of taking a morning walk in the Arctic and hates to break it. Word starts to get out, and one anecdote about him will later be chronicled by the famed columnist for the *Adelaide Observer*, 'Magpie'.

> It is well known by most front-line diggers that the Aussie official photographer was one of the gamest men in the war. One day he was taking the usual risks, oblivious of all considerations but that of getting a good picture. A purposeful digger was seen stalking from shell hole to shell hole. 'What in the [bloody hell] are you doing, Ginger?' yelled a cobber.
>
> 'Oh, it's all right. I'm just waiting for this photo bloke to get knocked. I want to souvenir his camera.'[50]

4 October 1917, Battle of Broodseinde Ridge, in that rich earth, a richer dust concealed

Of all the attacks launched on the Germans to date, this, the battle of Broodseinde Ridge, is one of the most mighty. No fewer than 12 Allied Divisions are involved – some 200,000 soldiers in all – including the 1st and 2nd Australian Divisions. As has become the standard, their advance at dawn will be preceded by a massive artillery bombardment on the German positions, before 'a creeping barrage' just ahead of the advancing Allies should ideally allow them to advance 1500 yards, where they can dig in once more, having pushed the Germans back.

It is something that sounds wonderful in theory but is in fact carnage in practice. Wilkins and Hurley are in position from 3 am, just in time for the Germans – who appear forewarned of the action – to send down their own barrage on Australian positions, meaning the men born beneath the Southern Cross lose a seventh of their number even before getting to the starting line.

How could the Germans have known?

It is only when the Australians jump the bags and burst forth that they realise what has happened. For there, coming towards them, are German troops charging at them, bayonets drawn!

Chaaaaarge!

In the chaotic confusion that follows the bloodshed is immense, and it is the Australians who eventually take their objectives, albeit at fearful cost of 6500 casualties. In response the Germans seem to bring to bear every gun they have on them. Diggers drop, lose limbs, die in dugouts; and Wilkins does his best to capture it all, firing off endless shots while ducking.

More disturbing still?

Above the roaring rumble of shells exploding all around them, he is suddenly aware of a strange screeching and now rumble like a train, and now a *roar* – wait, isn't that what the Diggers said . . . ?

Oh, CHRIST!

The shell explodes not 30 yards away, hurling tons of mud skywards which must come down . . .

One moment, Wilkins is standing, and the next the air is sheer knocked out of him as he is felled by a conflagration of mud, blood and bodies. It is like he is in a dumper back on one of the beaches in South Australia. He is not sure which way is up, only that he is down, and tumbling. And now bullets from the enemy are hitting them, and grenades exploding and shrapnel whistling past.

Is this it?

No. While it is a disturbing thing to be briefly buried alive, to physically claw your way out of your own grave, to emerge above ground with your last breath, to suck in those first precious gasps of oxygen to know that you're alive, alive, ALIVE . . .

A lesser man than Wilkins, or just a different man, would have taken the tip. This place is just too dangerous. Take the night off. Slip back through the lines to safety. Explain what has happened. You've just been *buried alive*.

But Wilkins is not made like that. He simply scrabbles through the mud and blood to find his camera and, most importantly, retrieve the film he has already taken, and then he keeps going.

As things turn out, precisely the same thing happens a couple of hours later. There is the whistling, the sense of the world about to stop, the explosion and then that strange sense of his whole body being effortlessly hurled about, before all goes black and somehow, somehow, once again, the bursting up from the mud. It is almost as though, *in extremis*, Mother Earth has given birth to him once more. The shells keep landing.

As before, he knows how close he has come. A few steps more, a few less, just a couple of feet deeper, and he would now be dead, his still warm body lying helpless and broken beneath tons of mud.

This time?

This time he gets the message. It is one thing to tempt fate by continuing, it is quite another to spit in her face before handing her a knife.

'By the law of averages,' he will recount, 'I should not have any more narrow escapes that day and being badly shaken I was in a hurry to get back.'[51]

The quickest way back? The duckboards, that narrow wooden path protecting the soldier from sinking into the mud permanently.

Now, it is drawing near the evening; just the time when the modus operandi of the Germans is to aim their shells a little further back to do some casual mischief to those retreating; but the risk is small and Wilkins' weariness is great. Weaving his way alone through the maze of boards he is about to reach the halfway point when – *a screech, a rumble, a roar . . . train coming!* – BAM! A shell hits and blows Wilkins into the air and into unconsciousness.

He is having a tough day.

The only men to see it happen are some Canadian artillery men and they race to what is no doubt a corpse twice over; killed by the shell and then drowned in the mud, for they can see the unmoving outline of Wilkins' still body, blown off into the water and mud, face down.

But as they approach they see a one in a million chance. Wilkins has landed in a shell hole full of water, gassed water as a matter of fact, but he is not drowned. The explosion had thrown his camera beneath his chest. Incredibly, the camera has become an instant bridge and 'in this manner my nose and mouth were kept just an inch or so above the water'.[52]

This man has the luck of the Gods.

Still blacked out – yes, unconscious, but nothing compared to the eternal blackout he had thrice escaped on the night – Wilkins is placed in an armament wagon and taken to the nearest field hospital. Waking in the middle of the night, he is told the whole story by a hovering Canadian soldier: 'Your camera saved your life!'[53]

Which is great and Wilkins is pleased to hear it, but what he most wants right now is to leave the hospital at once, as he has pictures to develop and must get back to work.

But you've been wounded! That is true, but Wilkins points out that he has been wounded many times in battle already – eight so far – this is the only time he has ever ended up carried into a field hospital on a stretcher. And what he most wants now is to get back to work. So, if you can help me with my boots, I will be on my way.

Of his own brief battlefield burials, he says little, noting only in passing he was 'twice buried by mud thrown up by shell explosions'.[54]

•

One day, on the way back to Ypres, things are hot in more ways than one. It's not just that the sun is belting down, it's that they must try and lug all their camera equipment a good two miles to the nearest transport, which is Hurley's assigned Ford car with driver – who had declined to come any closer to the front lines, what with bullets and shells whizzing all around.

True, Hurley himself has noted that, 'the Menin Road [from Ypres to the front] is like passing through the Valley of Death',[55] but he is still annoyed at the driver, Dick, and his 'holy fear of shell fire' even though he is 'absolutely regardless of reckless driving'.[56]

Yes, a strange cove, Dick.

'Rushing through narrow openings and scraping the wheels of the up and down traffic,' Hurley notes, 'makes me more windy than actually being in a barrage.'[57]

For now, they push on, Hurley getting progressively hotter in more ways than one.

Damn, Dick, and his windiness with a bit of shell fire.

'I'm going to stop any car I see and ask for a ride,' he tells Wilkins. Sure enough, as the next car approaches Hurley yells out: 'HEY! Give us a ride!' The car stops and an officer leans out: 'Jump in,'[58] he says. 'Plenty of room.'

The man's name? As they approach Wilkins can see by his uniform that this chap is a general, which is more than Hurley can be bothered to learn: 'Hurley didn't know the difference between a second-lieutenant and a general.'[59]

Ah, well, does it really matter?

As far as Hurley is concerned, he's just happy to have a ride, no matter who he is with.

And sorry, sir, your name?

Monash. General John Monash.

Neither Wilkins nor Hurley blinks, particularly.

'He was just an ordinary general and didn't amount to much, we thought,'[60] Wilkins will recount.

As for Monash, he is interested in them, for they are the famous photographers he has heard so much about and he quickly asks them about their profession and how it is done in the field. Wilkins and Hurley are impressed. 'We did notice that he seemed to show a great deal of intelligence. Most of them we talked with did not.'[61]

Yes, there really is something different about this bloke. The great British historian A.J.P. Taylor will later declare Monash to be: 'the only general of creative originality produced by the First World War'.[62] Wilkins and Hurley have had a glimpse of both those elusive qualities and are now intrigued.

'Come around and get some pictures,'[63] Monash invites them. They do go around, do get some pictures of the Australian general – an engineer from Melbourne who, they will later learn from others, including Charles Bean, despite not being regular Australian Army before the war, has been outshining all the regular generals for the stunning nature of his victories, and the low number of Australian casualties. He is a man bringing intelligence to his operations in both senses of the word. He thinks every battle plan through intricately, and devours all intelligence he can about the position and strength of the enemy, always looking for the weak points. Captain Bean has his issues, thinking Monash showy, but there is no doubt that he is effective.

'The true role of infantry,' Monash will later summate his philosophy, 'is not to expend itself upon heroic physical effort, not to wither away under merciless machine-gun fire, not to impale itself on hostile bayonets, but on the contrary, to advance under the maximum possible protection of the maximum possible array of mechanical resources, in the form of guns, machine guns, tanks, mortars and aeroplanes; to advance with as little impediment as possible; to be relieved as far as possible of the obligation to fight their way forward.'[64]

That is Monash all over, as is lunch soon enough, whereupon they go their separate ways.

•

As the weeks pass, and the battles continue, the only thing that comes close to matching the level of admiration Bean has for Wilkins is his growing level of disdain for Hurley. Hurley's approach aggravates Bean on two counts. For Captain Hurley is enamoured with the idea of composites, combining two striking images through trickery in the darkroom. For example, by overlaying the photo of the blast of a shell upon the image of troops charging bravely over the top of the trenches – does that not present a truth they have all seen, if not managed to capture with a camera?

In animated and ever louder conversations, Hurley defends his 'combination pictures', on the grounds that he is 'thoroughly convinced that it is impossible to secure effects, without resorting to composite pictures'.[65]

Perhaps, Captain Hurley, Bean insists, but it is not honest! (Wilkins, ever the diplomat, does not take sides but he is reminded of that Arctic conversation with the cinematographer of the *Hermyn,* who staged and filmed polar rescues before they happened, *just in case.* There is a danger to artists who prefer art to real life . . .)

Our job as correspondents and photographers, Bean strongly believes, is to capture reality, not construct it. The maddening thing is that Hurley's skill and eye for images is so good, his darkroom mastery so magical, that it is damn difficult to discern the real from the fake, which Hurley thinks makes his case. If Bean cannot tell which is the truth and which is the lie, then surely the lie shows a truth of battle?

There is no denying the power of images produced by Hurley such as *The Raid,* showing clearly the chaos of battle; but its falsity is known by Bean. Hurley is especially proud of one photograph – *Death the Reaper* – which he frankly states is: 'made up of two negatives. One, the foreground, shows the mud-splashed corpse of a Boche floating in a shell crater. The second is an extraordinary shell-burst, the form of which resembles Death.'[66]

In deep frustration, feeling like the integrity of his entire unit is at stake, Bean orders Hurley not to produce any more 'composites'. In his own field of work, the war correspondent is so devoted to accuracy above all else that one of his colleagues, the famed English journalist Ellis Ashmead Bartlett, had said of him, back at Gallipoli, 'Bean! O, I think Bean actually counts the bullets.'[67]

What Hurley does, making an approximation of reality and fooling the public through trickery into thinking it is real, goes against the grain and he – do you hear me? – *expressly* forbids it. Which is something of a coincidence, as Hurley *expressly* defies him.

On the current issue, inevitably the question is kicked up to High Command, who – after Hurley personally takes some photos of the English commander of the Australian forces, General William Birdwood – so strongly come down on the side of Hurley, he is rewarded with an exhibition of Australian photographs in London at the Grafton Galleries.

And here is the second thing that maddens Bean: Hurley's desire for such public recognition and acclaim, which is in such stark contrast to Wilkins. For the way Captain Bean sees it, for Hurley it is not about the work, it is about him. It does not surprise Bean that Hurley is the man who suggested the exhibition in the first place, nor that his photographs and slides make up the bulk of it. And, typically, Hurley falls out with the poor souls assigned to organise the exhibition for the fact they have not given sufficient pride of place to each and every one of his works, to the exclusion of all others. Standing on his dignity, which really does place him at least as high as his dudgeon, if not higher, Hurley works himself into such a temper that he refuses – refuses, do you hear me? – to attend the opening of the exhibition. (Even then, however, he cannot resist sneaking in later in the day to listen to the crowd's admiring comments as they pack the galleries.) Hurley recognises one face in the crowd, *by George, it is George!* Wilkins shakes Hurley's hand and shyly tells him his own piece of news. He has just been awarded a Military Cross. Hurley congratulates him but his diary shows the fury he is careful to conceal from Wilkins; a fury directed at Bean for overlooking him: 'I received nothing . . . Wilkins certainly deserved the MC but there should have been no distinction. There has been string pulling.'[68]

(Of course, because the modest Wilkins had given no details of his MC, Hurley doesn't realise that it is nothing to do with photography, and everything to do with helping the wounded on the field, dragging in the near dead from No Man's Land when no other sane man would dare. But as is Hurley's own nature, given that a close colleague has been honoured and he has not, he leaps to the angry conclusion that he is being slighted.)

Fortunately, Hurley is not in attendance when Bean also attends the exhibition a few days later. Bean is impressed but his comments regarding Hurley are muted:

> Our exhibition is easily the best I have seen although there is too much of Hurley in it. His name is on every picture, with few exceptions – including some that Wilkins took; and what should be a fine monument to the sacrifice of Australians in France is rather an advertisement for Hurley.[69]

Bean is annoyed by Hurley's selfish self-focus and is determined that when the display moves to Australia, 'I shall see that he does not have the management of this exhibition there.'[70]

The support of High Command means that Hurley is no longer controlled by Bean, but it changes naught that Bean is still in control of his work, something that brings Hurley himself to despair.

> I am having a great fight to secure the pictures for Australia. It makes me lose patience when I see all my efforts over which I spent all my time and risked my life thrown to the winds.[71]

In the meantime, Wilkins is back at the front once more, risking his life taking more photographs and could not care less who gets what credit for which photographs. All he wants to do is get on with his work, and the rest is mere background noise – something he can barely hear in any case, so overwhelming are the continuing blasts of artillery fire landing all around him.

Wilkins' extraordinary dedication is not lost on Bean. Hurley admires it too, but of Bean, by mid-November 1917 he has finally had enough. Bean's continued resistance to composite pictures, despite the fact that Hurley has the support of High Command, is wearing him down. For both Bean, the war historian requiring accurate depiction of events and Hurley, the photographer requiring artistic freedom, the matter of composites is a matter of principle.

'As this absolutely takes all possibilities of producing pictures from me,' he writes in his diary, 'I decided to tender my resignation at once. I conscientiously consider it but right to illustrate to the public the things our fellows do and how war is conducted. They can only be got by printing from a number of negatives or reenactment. This is out of reason and they prefer to let all these interesting episodes pass. This is unfair to our boys and I conscientiously could not undertake to continue the work . . . It is disheartening, after striving to secure the impossible and running all hazards to meet with little encouragement. I am unwilling and will not make a display of war pictures unless the Military people see their way clear to give me a free hand.'[72]

Bean accepts the resignation.

Hurley will leave Wilkins at least, with real regret, and after dining with him on his final night on the Western Front on 9 November 1917, records his admiration for the South Australian in his diary.

> We have both applied ourselves diligently to the work and now sense the reward of our endeavours. Wilkins is an excellent fellow, enthusiastic, conscientious and diligent. The innumerable hair breadth escapes, and our marvelous good luck I only hope will continue.[73]

Hurley will depart on 10 November for a happier realm covering the Australian forces in Egypt and the Australian Light Horse in Palestine, taking many striking and beautiful photos. (Just which ones are real as they happened, and which ones are composed from two or more shots, no one is sure.)

Bean and Wilkins continue together, chronicling this extraordinary war, just as it moves to its most astonishing, last-throw-of-the-dice phase.

With such deeds, it is no wonder that Wilkins is promoted in late 1917, but Bean is annoyed that the promotion has come so late, writing in his diary about the difficulty and politics of getting Wilkins recognised: 'Some time ago I tried very hard to get Wilkins his captaincy from Col. Dodds but Dodds wouldn't hear of it, tho' all the other colonial photographers are captains & he is the best, easily.'[74]

The thing that finally turns it? A short conversation between Wilkins and Colonel Dodds.

'Sir,' the Australian says, 'I have an offer from the Royal [Flying Corps] to manage their camera Department, with the rank of Captain and Major within a month.'[75]

Understood. The Royal Flying Corps won't steal Colonel Dodds' men, and Wilkins is consequently promoted. Such is the logic of the Army: being reluctantly promoted so that you won't get a better promotion, elsewhere.

In fact, however, it is Wilkins' entire lack of self-promotion that makes Captain Bean marvel at the South Australian. Wilkins wants simply to get on with his work, for it is the work that counts. And as a matter of fact he is not even particularly concerned if he is credited with the photos he comes up with. Time and again Bean has to gently chide Wilkins to ensure he puts his name to his death-defying photos. The standard of those photos, chronicling the devastatingly stark reality of the Diggers in modern warfare, is consistently outstanding. The risks taken to obtain them, incalculable. Yet Wilkins shrinks from having his name attached; the polar opposite of Hurley, his fellow polar explorer. There is simply nobody like Wilkins – he turns humility into a determined obscurity, when others clamour for fame or glory.

Mercifully, as Christmas of 1917 approaches, the Australians – having lost a quarter of their number killed or wounded at Passchendaele and other battles – are rested.

CHAPTER EIGHT

AUX ARMES, AUSTRALIENS, FORMEZ VOS BATAILLONS!

It was my aim to photograph every type of activity including raids, Germans surrendering . . . bodies being hurled through the air by shell explosions, men struggling from beneath debris . . . and the wrecking of most every type of equipment by shell fire.[1]

George Wilkins

Weep for the evil need to kill
and kill
Which from the golden gullies
of our land
And all the bush bred quiet of our days
Brought our young
beneath the milk-white moon
With moon-white steel to slay,
and leave them thus
Debauched of all the semblances of man

Will Dyson, official war artist, in his poem 'At Villers-Bretonneux May 1918', which is dedicated to Wilkins.

20 March 1918, Saint Quentin, Hun makes his run

That move by the Germans to get the Bolshevik leader, Vladimir Lenin, back into Russia, to foment revolution on the promise that if it is successful he will pull his country out of the war against them?

It has worked!

In the last months of 1917 the Bolsheviks had seized power, and by early March 1918 had signed an armistice with Germany, allowing the forces of Kaiser Wilhelm to bring no fewer than 50 Divisions back from the Eastern Front of Russia, and throw them into the Western Front.

Now, given that the Americans had entered the war in April 1917, and with more than 2 million of their soldiers now kitted out, trained up and at last on the high seas heading to Europe, the way forward for Germany had quickly become clear. They need to end this war *quickly*, throw all their forces into one last desperate battle to rout the French and the soldiers of the British Empire, and win the war before the weight of the Americans can be brought to bear.

The consequent battle becomes known as the *Kaiserschlacht*, the Kaiser's Battle, and it is primed to begin in the early hours of 21 March 1918.

Across a front 45 miles wide, extending from Arras to Barisis, no fewer than a total of 10,000 artillery pieces and mortars, which is half of all German guns on the Western Front, are moved into position, together with over a million shells and some half-million mortar rounds.

Just before dawn, a streak of light suddenly arcs to the heavens before . . . an enormous white flare explodes against the misty sky.

Es ist das Signal für die Artillerie-Batterien. It is the signal for the artillery batteries.

A shout goes up along the line: '*Feuer frei*, free fire.'

And again and again and again! '*Feuer!*' '*Feuer!*' '*Feuer!*'

The synchronised explosions shake the night, as hell descends on the British lines.

Captains Charles Bean and George Wilkins are in their usual head-quarters at Steenvoorde when the news comes through of the enormous German offensive 50 miles to the south, and they are immediately on the move. Bean heads to get news at the Australian Corps Headquarters, Wilkins races to the town of Bapaume, which is clearly right in the way of the coming storm. Neither, however, will be in time to chron-icle the savagery of those first hours. After five solid hours of this barrage, never before unleashed in modern warfare, no fewer than 400,000 German soldiers go over the top, and charge straight at the 100,000 British infantry of the Third and Fifth British Armies defending their positions.

Those British forces that can, fall back. Those that get caught, are mostly slaughtered.

After just a day, the Germans have advanced an extraordinary *five miles* westwards at some points, while the British have had 40,000 casualties, of whom 10,000 have been killed.

•

In the face of the British collapse, High Command turns to the one part of their forces with a record of successfully fighting the Germans – the Australians. Having done so well since their arrival on the Western Front two years earlier, and particularly brilliantly in the latter months of 1917, they have been resting up in Belgium ever since, replacing their losses with fresh troops from home, and within two days of the German breakthrough the 2nd, 3rd and 4th Australian Divisions are being rushed forward to fill the gaps left by the retreating British.

As the Australians move through the picturesque countryside – all butterflies, neat hedges, neater cottages, hares breaking cover, lowing cattle and gentle rolling hills – it is to be confronted by a rag-tag stream of villagers and remnants of the shattered British forces, moving back, and yelling warnings to the Australians – 'You can't hold them!'[2]

Well, we'll see. The Australians at least seem to think they can, and one British officer would later write of this weird mob now advancing, 'They were the first cheerful stubborn people we had met in the retreat.'[3]

For Charles Bean – who right now is a little worried to have lost contact with Wilkins – it is heartening.

As men of General John Monash's 3rd Division march into the town of Heilly, now threatened by the approaching German storm, evident by the billowing clouds of black smoke on the eastern horizon, some of the locals burst into tears at their mere sight.

'*Qui sont-ils?*'

And the answer:

'*Ce sont les Australiens . . .*'

And then the cry is taken up, first by a few, and then by many.

'*Vive l'Australie! Vive l'Australie! Vive l'Australie!*'[4]

Bean watches closely as one of the Diggers sees an old French lady gamely heading west with all her most precious possessions in a wheelbarrow.

'*Fini* retreat, *Madame*,' the Australian says in a kindly manner, trying to make her *comprenez* with what is left of his schoolboy French. '*Fini* retreat – *beaucoup Australiens ici.*'[5]

And she does understand! With a relieved smile, the old lady turns around and heads back home.

The Australians are here. Things are looking up.

Bean swells with pride, and writes it all down in his ever present notebook, hoping that, wherever Wilkins is – no doubt in the thick of the action somewhere – he is chronicling equally inspirational moments. There is no doubt the Australians are in for the fight of their lives, but the courage with which they are approaching it is really something to behold.

•

For George Wilkins on this sunny day, it is a sobering thing to be passing Pozières' legendary windmill site – the highest spot in the Somme Valley, and now marked with a monument of large white stones topped with a large dark wooden cross – for he knows that no fewer than 7000 Australian lives had been lost in taking it from the Germans.

And now he notes something intriguing.

Men in Australian uniforms!

Telling the driver to pull the Ford over to the side of the road he gets out and is thrilled to find men of – who are you? – men of the 2nd Australian Tunnelling Company, all of them flat out like lizards drinking. They are digging a trench across the line of the ridge, and a small tunnel under the road to pack with explosives.

And when Jerry comes down this road shortly, sport, they're gunna blow the bastard!

If they do it right they'll kill a few Germans, but either way the huge crater on this narrow pass of road will at least slow down their wheeled transport which is important as they are not far away! Even as they keep digging, hundreds of British infantry stream past, all but directionless, apart from the one general direction they are following: getting away from the Germans.

The British officers that Wilkins asks confess that they are without orders about where they are to go now. So they just follow the road, away, away, away . . .

As Wilkins leaves his driver momentarily to walk down the lee side of Pozières slope, he comes to more British soldiers who are bivouaced, but they seem even less disposed than their brethren to provide serious resistance.

Wilkins would recount a few of them telling him, 'He [the Germans] can have this country as far as I'm concerned.'[6] The majority wanted a rest.

After using up all the photographic plates he has, Wilkins races back to the Headquarters at Steenvoorde to get fresh supplies, where Bean is mightily relieved to see that he is alive and well, and working hard – though he knows the former always goes with the latter. After a night's sleep, Wilkins is on his way again early the next morning, heading to where the action is.

•

Two days later, the situation is even more extreme. Bean and Wilkins are together with the Australians of the 4th Brigade as they march down the cobbled road towards the town of Hébuterne in the bright afternoon sunlight, when they are met by a flood of soldiers coming the other way.

'A terrible sight met our gaze,' one Digger from the 15th Battalion would recount, 'the British Army was in full retreat, indeed everything looked desperate, they [threw] us Aussies into this huge gap . . .'[7]

Staggeringly, it is the British officers who are leading the retreating, shouting to the Australians as they pass: 'The Germans are in Souastre with armoured cars!'[8]

The exhausted British soldiers who come straggling behind also give fair if exhausted warning: 'Jerry is close behind us. And he has tanks!'[9]

The Germans continue to advance on average nine miles a day, until by 4 April 1918 they loom large on the French town of Villers-Bretonneux, a previously insignificant place, which now has enormous importance in this whole war.

It lies on a small plateau within distant sight of the railway hub town of Amiens, some six miles to the west – which is important because 60 per cent of the supplies keeping the British forces going on the front come through it. This means that if the Germans bring their artillery forward to the Villers-Bretonneux plateau, they will be able to shatter the railway hub and so throttle much of the Allied supply lines. As the British forces keep retreating, the Australians of the 9th Brigade, under the command of Brigadier Charles Rosenthal – some 3000 Diggers in all – are rushed forward with the orders from General Headquarters to hold Villers-Bretonneux at all costs.

Arriving on the afternoon of the previous day, the Australians had dug in, and started firing at the first German forces that appeared rumbling down the old Roman road from Peronne.

Aware that a battle is underway, Bean and Wilkins head in their little 10-horsepower Swift 10 car to the south-east, towards the sound of the artillery, steering so that the balls of dirty thunder continually rolling over them get progressively bigger, louder and dirtier.

Inevitably they soon find themselves in the middle of a battle as, all along the road on the outskirts of Villers-Bretonneux, Allied artillery batteries – 60-pounders and six-inch howitzers, positioned in the fields to the left – are POUNDING enemy positions. The German artillery, however, is even more willing and when Bean and Wilkins arrive in Villers-Bretonneux itself, it is to see British soldiers taking shelters in doorways, even as German shells keep shattering the buildings all around them. With the driver told to *speed up* they leave Villers-Bretonneux behind and start descending the hill towards Amiens. After pulling over and alighting, they suddenly notice many groups of men off to their left, and most of them seem to have left their weapons behind!

'. . . it was gradually borne in upon us,' Bean will note in his diary, 'that the whole countryside was retiring. Wilkins, whose eyes are better than mine, said he could see men running on the further horizon.'[10]

Bean and Wilkins watch as endless groups of retreating soldiers keep passing, one of them particularly catching Bean's eye as he appears to be little more than 'a weak looking child . . .'[11]

'Which is the road, Sir?' one of the party leaders asks.

'What road?'

'Which is the road to Amiens?' the soldier means to ask, though he actually pronounces it 'Aymeens'.

'What are you retiring for?'[12]

'There were too many Germans for us,'[13] the soldier replies simply.

Bean and Wilkins, completely dispirited, suddenly see 'bunches of men, 12 to 25 strong, walking [away from Villers-Bretonneux] . . . and a few men with bayonets still fixed to their rifles. These were Australians, the first we had seen in the crowd.'[14]

And yes, these men of the Australian 9th Brigade are obviously shattered with exhaustion but – and this is important – they still have their rifles, even if these do appear to be clogged with mud.

'We've been five days without a spell,' one of the soldiers tells them on questioning, 'and we are pretty well done up.'[15]

They need direction, and Bean is able to offer them some.

'Look, men, you Australians here,' Bean says to some, 'it's no good going on without knowing where you are going to. Hang on here a moment until an Australian officer comes along.'[16]

Fair enough.

'"Well, as we're going to stop," one of them says as he unslings his rifle and throws down his pack, "we may as well sit down."'[17]

Alas, only seconds later a 93-pound shell from a German 5.9-inch howitzer explodes just 10 yards away. The only thing that prevents a complete massacre is that the shell sinks into the heavy mud before it explodes, though one man still standing by Bean and Wilkins is wounded.

Under the circumstances, things are grim, but . . .

'The one thing that cheered us,' Bean notes, 'was the difference between our men and the British in the retreat. The British . . . were clearly panicked and quite spiritless.'[18]

But the Australians? It is obvious they'll go down, but at least they'll give it everything they've got before they do so, and they'll go down *fighting*.

But Wilkins soon sees that Bean is wrong, in one assumption.

For by dusk that evening, the Germans fight their way right to the edge of Villers-Bretonneux and the only thing standing between them and the crucial town are the 3000 Diggers of the AIF's 9th Brigade.

Go down fighting? These men clearly have no intention of going down at all!

'We saw Fritz in front of us about 500 yards away,'[19] one of the Diggers would recount.

A collective guttural roar goes up from the Australians, as a red mist of bloodlust rises.

Into 'em, boys!

After a wild charge that the Germans had simply not been expecting, a bloody bayonet battle ensues, as the two forces start clashing with flashing, slashing blades. There is no mercy as screams, groans, grunts of exertion fill the twilight, not to mention that distinctive sucking sound when blades are removed from pierced lungs followed by the gurgling geysers of lifeblood that suddenly spout high before quickly falling away, for the last time. The exhausted Germans, profoundly shocked by this sudden attack, just when it seemed Villers-Bretonneux was theirs for the taking, try to fall back. But the Australians are fresh, and suddenly liberated to finally get to close quarters with the Germans who have killed so many of their mates. They have no mercy and attack like rabid dingos going after sick rabbits.

The result is bloody carnage, with the Germans stopped – dead in their tracks, and still ebbing blood.

For the moment at least, Villers-Bretonneux is saved, and the threat to Amiens thwarted.

•

On this day, 21 April 1918, George Wilkins is heading north on the Bray–Corbie road just five miles north-east of Villers-Bretonneux when he sees a plane dog-fight for the ages going on just up ahead, with no fewer than 30 planes scrambling in the skies above – flocking, flipping, firing, diving down on each other.

One plane in particular stands out. It is a *red* Fokker triplane!

It couldn't be, could it?

Could it?

'It was so interesting we stopped our car to watch the fight,' Wilkins will recount. 'At that altitude it was almost impossible to tell the difference between the German and Allies airplanes. There must have been thirty or forty airplanes engaged in the battle . . . One plane was noticed to lag behind and then went nose down towards the fields. An enemy plane swooped down to follow it and I could tell it was Richthofen's machine.'[20]

Richthofen!

•

In his red Fokker, Baron Manfred von Richthofen – known to the British as the Red Baron, no less the deadliest pilot in the *Deutsche Luftstreitkräfte*, having accounted for no fewer than 80 Allied pilots – is closing in for the kill. His fat prize, a lone Sopwith Camel, has broken from the dog-fight to whimper back to the safety of the Allied lines.

Ruhig . . . ruhig . . . ruhig . . . Steady, steady, steady . . .

Five hundred feet away now . . . 400 . . . 300 . . . 200 . . .

As his finger tightens on the trigger his twin Spandau machine guns burst into life, his whole Fokker vibrates and – right at that instant – the Sopwith . . . jumps above the fusillade just in time.

•

The three most terrifying sights in history?

Attila the Hun galloping your way.

Captaining a fat English merchant ship, as the Spanish Armada heaves to on the eastern horizon, and . . .

And . . . while on your first patrol over enemy lines in your Sopwith Camel, seeing over your shoulder . . . the Red Baron!

In the face of it, all novice Canadian pilot Lieutenant Wilfrid May of No. 209 Squadron, Royal Air Force, can do is throw his controls wildly about.

•

Despite a series of narrow misses, the Red Baron stays tightly behind the Sopwith, and finally closes to the point where he just can't miss, and pulls the trigger.

The aircraft is suddenly hit with a deadly accurate spray of bullets, and visibly staggers in the air.

Extraordinarily though, it is *not* the Sopwith that is hit, but the red Fokker! The Red Baron has strayed over Australian lines, and from below a gunner by the name of Robert Buie – an oyster farmer before the war, from Brooklyn – just keeps pouring bullets into the red plane. Gunner Frank Wormald, standing no more than four yards from Buie, is watching with morbid fascination and will later affirm that he could see, 'plain as daylight . . . the Baron sort of shrug and sit up. I could *see* him.'[21]

The plane is going down.

•

George Wilkins watches closely as the red Fokker 'side-slips' behind a hill, momentarily obscuring his vision. Once the red Fokker hits the ground though, the Diggers rush forward, just in time to hear the German legend get out one last word before expiring: '*Kaputt*.'[22]

It is 10.50 am, on a cool, windy day.

The corpse of the Red Baron is treated with enormous respect, almost as if he was one of their own, and taken by stretcher to Hangar 3, where a guard will be placed over his corpse until his funeral on the morrow.

Arriving on the scene, Wilkins is able to catch a great deal of historic footage and shots of the crashed plane. He also captures all the Diggers and British soldiers, their respect for their fallen foe notwithstanding, as they joyously examine the wreckage of his plane; holding pieces of it high the way they might a trophy.

At least some solemnity is restored later that afternoon when official wreaths arrive from British GHQ, though it must be noted that the men carrying them cannot resist smiling broadly.

Look, we have killed the Bloody Red Baron, are we expected to be sad about it?

Only when the funeral ceremony commences the next day is the traditional decorum of death restored, with slow marches, raised rifles, prayers, *vale* to the vanquished.

Wilkins is able to film the whole sombre scene. The coffin of von Richthofen is respectfully borne by six captains of the Australian Flying Corps – the same rank as von Richthofen – and as they do a solemn slow march through the gates, the waiting honour guard of 12 Australians from No. 3 Squadron *presentttt . . . arms!*

Prayers by Chaplain George Marshall, DSO, MC, are uttered for the Baron's immortal soul, while wreaths from other Allied squadrons are laid. As von Richthofen's coffin is lowered, Wilkins' camera pans close for the final salute, with each man in the honour guard firing three shots into the air. From a different angle he records the grave being filled in, with the makeshift cross at its head being no less than the very propeller taken from the Baron's red Fokker the day before.

That night, Wilkins develops the film and copies of some of the photos are dropped by an English pilot above a German air base, together with a note.

> To the German Flying Corps,
>   *Rittmeister Baron Manfred von Richthofen was killed in aerial combat on April 21st 1918. He was buried with full military honours.*
>
> From,
> British Royal Air Force. [23]

The *Deutsche Luftstreitkräfte* is further advised that, if they wish to, they can fly unmolested over his grave on this same day between 3 pm and 6 pm, to drop their own wreaths. One of the German pilots who does precisely that has 18 kills to his credit and will shortly take over as commander of this 'Flying Circus', as von Richthofen's unit was known.

He is *Oberleutnant* Hermann Göring . . .

•

Good God.

The news could hardly be worse. Yes, three weeks earlier the charge of the Australians at the Germans approaching the outskirts of Villers-Bretonneux had saved that key town from German occupation. But then the Australians had been replaced by young British soldiers who on 24 April 1918 had . . . given way to a German advance powered by *Deutsche Panzerwagen*, German tanks, and thousands of soldiers with *flammenwerfer*, flame-throwers.

How to turn them back?

The man on horseback, prancing about, and moving back and forth among his troops shouting orders, encouragement and threats in fairly equal measure to the retreating British troops, while 'brandishing a revolver, threatening to shoot the gunners if they pulled out . . .'[24] seems to have the answer.

For Brigadier General Pompey Elliott, in command of Australia's 15th Brigade, is nothing if not encouraging, telling them, 'I will show you how Australians fight.'[25]

For there remains just one chance to prevent the disaster that would be represented by the Germans moving their artillery forward to start lobbing shells on the Amiens railway hub.

It will be for the Australians to take it back, and it is Pompey Elliott's plan that soon holds sway and Captain George Wilkins, as so often happens, the man on the spot, soon understands the plan. It will be a classic 'pincer' movement by Elliott's 15th Brigade and General Bill Glasgow's 13th Brigade.

'These two units were to advance,' Wilkins will recount, 'one from either side, and meet behind the town, cutting it off. This advance would, of course, expose them to each other's fire if they used their guns.'[26]

It will, therefore, again be mostly a job for bayonet, and the Diggers sharpen their blades accordingly.

Late that night, it is time, and the two brigades of Australians are in position in the tepid moonlight, ready to launch.

Now, if all else had been equal, Wilkins might have been able to fulfil a long-held ambition on this night, one he had unveiled to Bean a few months earlier, upon his return from London.

'He wants to get a flashlight photo of a German being ambushed in No-man's-land . . .' Bean had chronicled. 'His idea is to get out there with a flashlight, wait until they come by and then photograph them as they sometimes do with wild animals in America. There is no excitement in the war in France he says, but plenty of anxiety. He is a brave hunter of excitement; I can see that is the main motive.'[27]

But now is not the time!

For now is the time for silence and invisibility – and to remain black wraiths of vengeance in the darkness – to advance as far as they can before the Germans twig.

Just before midnight, the order hisses out, and spreads along the line: Advance!

In short order, and in good order, 3000 Australian soldiers of the 15th Brigade begin their flanking movement proper on the north side of Villers-Bretonneux, just as the 13th are doing on the south side of the town.

'The Australian troops advanced,' Wilkins will chronicle, impressed by their sheer discipline, 'practically in silence . . .'[28]

The one exception to the silence?

'It's Anzac Day,'[29] they mutter to each other, smiling. Three years earlier, eight months after the war began, Australian soldiers at Gallipoli had demonstrated to the world their valour, and they have proved it many times since.

But now, they must do it again.

'The die was now cast,' soldier Walter Downing would recount of the mood of the 15th Brigade. 'It seemed that there was nothing to do but go straight forward and die hard.'[30]

Only a short time later, however, the Germans become aware of the Australian presence – or at least a worrying noise – and so take the first defensive action.

'German flares of all kinds shot into the air,' Downing would recount, 'reds, whites, greens, bunches of golden rain.'[31]

A couple of seconds hang suspended – two ticks on the clock in real time, but an eternity of shocked silence for both the German and Australian soldiers.

The flares burst, to fully reveal 3000 Australians coming straight at them.

Gott im Himmel!

Myriad German machine guns open up, sending furious fusillade at these black phantoms of the night.

Most of the Australians are able to throw themselves to the ground in time.

Others die hard.

One Lieutenant with the 59th Battalion, goes down, shot through the leg.

'Carry on, boys,' he shouts to his men, 'I'm hit!'[32]

They carry on, roaring, and this time valour is the better part of discretion. This is no time to recede, it is time to get to grips.

As one lieutenant of the 58th Battalion would recall it, the clear voice of an officer rings out: 'Charge!'[33]

Almost as one, the Australians rise, and CHARGE, unleashing 'a savage, eager yell'.[34]

One soldier, Private Edwin Need, will recount the affair of him and his comrades as 'everyone was mad for a fight, the whole affair a bedlam, the bursting of shells, firing of flares, the town burning fiercely, and our fellows, cursing and yelling, charging on, the Fritzes flying for their lives, having very little recollection of the charge, just stabbing and thrusting with the bayonet at any thing that came in the way, and on again, taking no prisoners . . .'[35] just as the Australian Official History would chronicle, 'there was no holding the attack. The bloodthirsty cry was caught up again and again along the line, and the whole force was off at the run.'[36]

The results are as savage as have ever been recorded in the annals of Australian military history.

'A snarl came from the throat of the mob, the fierce low growl of tigers scenting blood,' Downing would memorably describe it. 'There was a howling of demons as the 57th, fighting mad, drove through the wire, through the 59th, who sprang to their sides – through their enemy.'

In an instant the sounds of this bursting bloodlust spread as the men of the 58th and the 60th out on the left join in. 'Baying like hell hounds, they also charged.'[37]

It is a roar that only gets ever bigger.

'A yell from a thousand throats split the night,'[38] one officer would recount.

On the south side of the town, some 1500 yards away, the 13th Brigade hear the yelling and know that, for them, it is now or never.

'Cheering, our men rushed straight to the muzzles of machine guns . . .'[39]

It doesn't take long.

'There was a wild rush, a short sharp clash of arms – then pandemonium.'[40]

And still they keep charging.

'The [Australians] were magnificent,' one stunned German officer would later recall. 'Nothing seemed to stop them. When our fire was heaviest, they just disappeared in shell holes and came up as soon as it slackened.'[41]

The result is inevitable.

'The Germans ran,'[42] one Digger would report with no little satisfaction.

'There they go, there they go!' some Diggers shout, as some German soldiers who have been taking shelter in a nearby trench decide to make a break for it, and run for their lives.

'The ferocious bloodlust of those boys,' Wilkins notes, 'forced the Germans to stand their ground and fight till they were killed almost to a man.'[43]

From all sides, the Australians – now joined by British troops – move in, mopping up as they go, moving from street to street, house to house, and within a couple of hours the whole town is secure for the Allies once more.

'I shall never forget the battle of Villers-Bretonneux,' Wilkins would later recount, calling it, 'perhaps the most daring manoeuvre in our experience on that front.'[44]

And as always after the battle's roar, the blood, the blast, the devastation of destruction and death, things seem very quiet on this gorgeous spring day.

'When I came back through the little village and the surrounding orchards, the perfume of apple and cherry blossoms mingled strangely with the powder smoke, and the delicate pink and white tints of the flowering trees rose like a bizarre funeral spray over the blood-soaked corpses sprawled beneath their soft branches.'[45]

•

And now, here is something new in this war. It is the Americans arriving as a force, in force, the first of the flood of 2 million that the Germans had been fearing all this time. Long wary of the 'foreign entanglements' George Washington had fervently opposed, the United States is at last joining the World at War. The 'Dough-boys' arrive fresh faced and immaculately kitted out, a strange contrast to the weary and wary Allied forces who watch them arrive so enthusiastically. Bean and Wilkins observe them carefully as they march through the town of Amiens.

One of the Australian soldiers right next to Wilkins can't help himself and calls out as they march past: 'Well, Yanks, so you've come over to win the war?'[46]

A single American soldier stops, breaks ranks and approaches the heckling Australian as all watch.

What is about to happen now? A fight?

'Aussie,' he says earnestly, 'if we do as well as you fellows have done, anyway we'll have done our share.'[47]

The mocking Australian is so moved by this sincerity that Wilkins watches him physically turn away so the Yank cannot see the emotion on his face. The Americans are here and, unaccustomed as they are – no, really – they are humble.

●

Big news. And it is has been a long time coming. Though the Australians have been under English command for all of the war so far, in the form of General Sir William Birdwood, by early 1918 an extraordinary thing was about to happen. Given that the Australian soldiers had performed so magnificently in the war to date, perhaps . . . put them under Australian command?

There had been bitter resistance but in large part through the insistence and belief of the Australian Prime Minister, the 'Little Digger' Billy Hughes, it is done and the very man that Wilkins and Hurley had had lunch with six months earlier, General John Monash, is announced in the position.

By the end of May, Monash has indeed taken command and brings with him an entirely different approach.

A man who had spent his professional life in Australia as an engineer not an Army general, he fervently wishes to find an alternative to sending waves of thousands of Australian soldiers charging across No Man's Land straight at German guns, as had been done in such notable previous battles as Fromelles and Pozières – where more than

8000 Diggers had been killed outright, nearly 2000 of them in just 14 hours at Fromelles.

Rather, he wishes to approach the German defences like the engineer he is, carefully surveilling with every tool he has to find their point of greatest weakness, before working out – in a minutely calibrated battle plan the likes of which has never been seen before in warfare – how to bring maximum forces against them, with the heavy lifting to be done by tanks, artillery and planes, rather than the Diggers.

His first battle, as it happens – at Le Hamel – just three miles northeast of Villers-Bretonneux, will be the first offensive action the Americans have participated in, with 2500 soldiers of the 33rd Division joining 7000 Australian soldiers, which will also be the first time in history that the Americans had accepted a foreign command.

And what a battle it promises to be, with the idea being to break the German line at Le Hamel.

The plan is prepared with typical Monash thoroughness and innovation, after a series of meetings where the head of all of the infantry, tank, machine-gun and aviation units work out how to smash the weak link of the German line with such successive waves of force that it will not only be broken but, once broken, the gap widened and the Allies can rush through to sow havoc behind the lines before consolidating their positions. Wilkins is impressed to see that a measure of Monash's thoroughness is that the main attack will go in at 3.10 am, the precise time when the first lustre of dawn will illuminate the silhouette of all German defenders while Monash's men will be coming from pitch blackness. Gas attacks at that precise time each day will go in for two weeks before, to train the Germans to put on their gas-masks at that time. But come the real attack, there will be no gas to contend with, just a creeping artillery barrage, followed by tanks, with infantry tight behind, and once they are through, low-flying planes dropping fresh munitions to them so they can keep going. It is something more than just the evolution of warfare and something less than revolution – but it is all Monash, whose North Star is to do heavy lead-up and intricate planning, embracing engineering principles and logic to provide maximum force to the weakest point, and in so doing protecting his men.

True, there is a problem when a very late edict by the American commander General 'Black Jack' Pershing insists that no American troops will participate after all. But the edict comes so late that Monash is able to pretend that there is no time to get it to the Americans already moving to the front lines, and while some in the back area are withdrawn, he is able to throw reserve Australian troops in to make up the difference.

As it happens, so strong is the bond formed by many of the Australians and Americans training in recent weeks that some take extreme action. George Wilkins himself comes across some of the American troops who've just got the word.

'They were most depressed,' Bean will record Wilkins' impressions. 'It is said that there are probably a fair number of Americans who will refuse to hear that order – I have no doubt of it.'[48]

But now is the hour.

After all troops are carefully moved forward under the cover of darkness, and the first wave crawl out into No Man's Land to stop at the translucent tape that has been laid for them in the moonlight, the show is ready to begin.

'Zero hour was at dawn,' Wilkins will recount of the moment the attack goes in. 'I was with the front line troops, ready to capture these Yanks in action.'[49]

Meantime, look at what the Germans have put on, just for those Yanks. Occasionally, flares rise lazily from the German lines just as they do every night so there is no cause for undue alarm, soaring skywards and throwing out an ethereal flickering light. The Australians and Americans all move into position, *'neath the rockets' red glare, the bombs bursting in air*, without incident.

And now, with Monash's carefully calibrated thoroughness, it begins.

At *precisely* 2.50 am eighteen FE2b planes of 101 Squadron RAF makes the first of what will be three sorties over the German lines. It is under the cover of the resultant roar as at 2.59 am 36 state-of-the-art Mark V tanks begin their move forward along translucent lines that have been laid out which will take them through the soldiers to deliver the first assault on the German positions.

And now at 3.02 am the artillery starts in, as has happened in the gas attacks of the last two weeks, which should see all the Germans now reach for their gas-masks.

At precisely 3.10 am that artillery fire intensifies as the first of the creeping barrage hits the German forward positions, before moving forward, and *now* the tanks hit the German lines – *more, Maestro, more!* – and *now* Monash's soldiers leave their lines and charge forward to wipe out whatever German soldiers the planes' bombs, artillery and tanks have missed. Wilkins, of course, is right there as the Australians and Americans go forward.

But now, as Wilkins observes 'midst the battle's roar – the cacophony of machine guns, tank cannons, artillery and men screaming, much of the carnage illuminated by the flares the panicked Germans have put up – something unexpected and moving happens. Even as he is duly photographing the Australians and the Yanks rushing forward; he sees one lone American remaining. The fellow is frozen in fear, 'seemingly paralysed by the shock of the barrage'.[50]

The man's face is white, entirely drained of colour. He is gasping in panic as Wilkins watches him struggle not with the enemy but his own wild emotions. But now look. For after just a few moments, the Yank 'pulled himself together and rushed out'.[51]

The man not only catches his troop but, to Wilkins' stunned amazement, overtakes them!

As ever, even as the shrapnel and bullets fly, the tanks roar and men scream, Wilkins is right in the middle of it. No matter that one piece of shrapnel lightly brushes his brow making the blood flow, he keeps going taking shots in the pulsing ethereal glow cast by the rising dawn and the endless explosions. Now and then, he puts the camera down to help a wounded Digger to a sheltering shell hole, and is about to do precisely that when, 'midst the battle's roar, he hears a louder roar still, together with shouting.

'Then, again, came that mysterious intuitive feeling of danger that I had several times before experienced and which had saved my life.'[52]

He turns to see a massive Mark V tank coming up over the rise right above him and the wounded Digger! It is *déjà vu* as the tragedy that had befallen him when he was with three Diggers at the Battle of Menin Road comes again.

> I barely had time to throw myself aside before the tank tracks came down upon the wounded man and crushed him beyond recognition while I lay within an inch of the tracks, gasping in horror.[53]

In civilian life, such a death witnessed by one who himself had a near-death experience would see tears, testimony and tribulation, going for weeks if not months. Here, in the middle of a battle, it goes for seconds.

Nothing can be done.

Wilkins retrieves his camera and moves on. Around him, the successive waves are surging forth. It looks like Monash's plan is working.

Moving forward with them he sustains another light wound in the arm.

> I set the camera on my knee and the enemy, I believe, seeing me making a second attempt [to photograph them] did not then try to shoot me. In fact they shouted and waved to me as I slithered back to my own trench.[54]

But all around him, Monash's plan proceeds like catastrophic clockwork.

In just 93 minutes the battle is won and done. General Monash will get to the bottom of what took three minutes longer than planned, later!

And yet, despite the triumph, even this most successful of battles adds to the list of the fallen. That list would have been just a little longer if not for Wilkins who, on two occasions, puts his camera down to drag two severely wounded and helpless men from the battlefield to the safety of a shell hole before moving on with his camera. For Wilkins it is not even worthy of remarking upon – he does that kind of thing all the time – but there is something different on this occasion. For although he doesn't know it, he has been carefully observed by a higher-ranking officer and will once more be recommended for the Military Cross because of it.

As ever at the conclusion of a battle, the opportunities for taking extraordinary photos – the wounded Diggers, the tears, the carnage, the relief – are myriad, and Wilkins does his best to capture them all. He crouches, frames the shot, captures it with a click and then looks up, surveying what else might be of interest.

It is while so engaged he sees him.

It is the same American soldier he had noted before, the one who had been paralysed by fear, only to have recovered just in time to not only join his comrades, but pass them. And now here he is, battle-blasted certainly, but alive, and carrying a wounded comrade on a stretcher. Fascinated, the photographer can't help himself, and becomes a journalist. Gently approaching the cove, he indicates that he had seen him earlier, and admired how he had overcome his fears. But now he wants to know: *What happened? How do you feel?* The man thinks for a moment and then raw reflection pours out.

'This is my first battle,' he tells Wilkins. 'All *night* I had been quaking with fear. When the barrage started, I just couldn't make myself go out and face the Germans. I *knew* I would be killed. It was only a greater fear that made me leave the trench, being called a "coward".'[55]

Wilkins gently tells him that he has seen many men frozen in panic on the battlefield or before it, but you, Sir, are one of those rarest of men who have been able to conquer the condition *in the moment* and be able to fight the way you have. And I congratulate you. Wilkins will never see the man again but will never forget this example.

Bravery is not proceeding without fear. It is proceeding despite fear.

•

Just four days after the success of Le Hamel and the Battle of Amiens, Monash is given a mere 24 hours notice that he is to be knighted by King George V, 'in the field', on the steps of Chateau de Bertangles, in an entire ceremony created just for him – a wartime investiture of the title he was awarded in the new year.[56] It is an extraordinary honour, the first time a commander has been so knighted on the field of battle by a British monarch since 1743. Understandably pleased, Monash makes a personal request for none other than George Wilkins to come and photograph the historical moment – which deeply annoys Charles Bean, as he has long privately thought of Monash as being 'pushy'[57] and showy and here is your proof. Is Bean seriously to lose his best photographer for even half a day all for the sake of Monash's vanity? As it happens, when arriving himself at Bertangles for the ceremony he finds that Wilkins has been employed as a quasi-stage manager, arranging some of Australia's military machinery – trucks, artillery and machine guns – around its commander.

It is quite a display, and impressive, but . . .

'. . . the front garden was simply full of guns and howitzers and M.Gs [machine-guns] and trench mortars,' Bean will note in his diary, 'facing all ways.'[58] To Bean it seems a 'damned waste of time and energy at a moment like this',[59] more particularly when, as he knows better than most, 'Wilkins was probably very tired after his work up the front.'[60]

Wilkins has a different opinion, it is an exceedingly nice change from taking photos while dodging German shells, bullets and bayonets.

For the grand occasion – a guard of honour of 600 Australian troops line the drive, greeting King George V at his arrival – General Monash is clearly quite nervous. In fact, he proves so nervous once the ceremony begins, that after King George taps one shoulder with his sword, Monash is so eager to rise once more, that the Sovereign had to quickly tap him on the other shoulder to get the job done.

Wilkins is moved by the moment. There is something inspirational in seeing an Australian recognised in this manner by an English King right by the actual battlefield where he has demonstrated his greatness. It is like a scene from one of the storybooks he read as a lad, like King Arthur with knights in shining armour, like Shakespeare with Hal ascendant, a glorious ruler bestowing an enormous honour with the gentle touch of his blade – and, like magic, the simple sword transforms the cool, detached, rational Monash into a ball of emotion.

Gathering himself, Sir John asks the Australian troops to give 'three cheers for His Majesty!', but the response is 'ragged'.[61]

Underwhelmed, Sir John tries again with a 'Come on!',[62] only to get the same unenthusiastic result.

For his part, the grumbling Charles Bean is not impressed with any of them, muttering that the whole thing is 'a lot of nonsense'.[63]

•

Where is Santa Claus when you need him?

The last time Wilkins had been in a pickle like this, Santa had at least shown the way to the exits.

But on this occasion there is no time even for that.

On this sunny August morning of 1918, Wilkins, chronicling the way battlefield intelligence is gathered, is high above the ground in an observation balloon, tethered to a field just back from the foremost trenches – the third one in a line of seven, all with observers watching as the Australians under Monash continue to push forth to Peronne – when he and all the other observers see it.

It is a German Fokker, spitting death, as an instant later one of the balloons explodes into flames and tumbles to the ground.

Frantically, the signal is given by the officer in each balloon for those below to start hauling them down to safety. Some of those in the balloons, left to their own devices, might have jumped, but military command is ahead of this, and has put out a specific order to prevent observers from parachuting at the first sign of trouble.

No observer should leave his balloon until it is actually in flames.[64]

Such an order is no problem to those in the balloon next to Wilkins as it, too, is now hit and their observers are quick to jump and pull their chutes, just as it indeed bursts into flames. The observer in the basket with Wilkins now makes an observation: 'Remember, we are *not* to jump until our balloon is on fire.'[65]

Got it.

The observer does, however, have one good idea that will still keep them on the right side of the precious regulations.

'Climb over the edge of our basket and hang on by your hands!' he tells Wilkins. 'You can be ready to let go as soon as you see our gas bag explode.'[66]

Now there is a man with a plan.

They are just in time for the Fokker to return and start firing at their balloon, and they can hear the bullets ripping into the gas bag above them, but – frustratingly – it does not burst into flames. With no such luck, they must stay there, ripe and low-hanging fruit, as the enraged German pilot turns his machine for another go.

The bullets hit, the balloon hisses and sags, but . . . no more than that.

'By some miracle, which I shall never understand, the balloon again did not catch fire.'[67]

Third time unlucky?

Yes.

The bullets hit. The balloon sways. But no explosion to give the Fokker pilot release, and themselves liberation.

'Our nerves were near the breaking-point from the suspense of waiting to drop.'[68]

And the German pilot?

He will not waste more bullets on them, preferring to destroy the next balloon along on his first go!

At this point the observer leans over the edge of the basket to have a word with Wilkins. 'Climb in again!' he urges. *Oh, for God's sake.* Yes, orders is orders, but this is ridiculous.

'It was all I could do,' Wilkins will record, 'to keep from letting go and dropping, without bothering to pull the parachute ripcord. The hardest thing I did during the entire war was to climb back into that basket . . .'[69]

Mercifully, their balloon is now lowered and Wilkins lives to shoot pictures another day . . .

'Feed your troops on victory'[70] had always been Monash's goal, and the Australian troops now feast on it as never before, coming back for second and third helpings as, under the command of Monash – an Australian leading Australians – they spearhead the whole advance of the British Expeditionary Force.

In the meantime, Bean continues to be impressed by Wilkins' extra-ordinary diligence and *elan*.

On 22 August 1918 when Monash's Australian Corps is just six miles short of Peronne, Bean happens to be walking past some Australian soldiers who are escorting wounded Germans to a battlefield cemetery, where their German countrymen will be buried. It is a touching scene, the conquerors and the conquered both paying their respects to the conquerors' conquests and the conquered's dead. It is definitely a moment worth recording, and Bean instantly looks for Wilkins to capture the reality of it, the heavy pathos that does not need to be composed. As it happens, he neither has to look far, nor wait for long, for here is Wilkins now, coming over yonder hill, with his assistant Sergeant Lowe.

And what is that you say, Captain Bean?

A burial? Well, we have already photographed it.

But how? It hasn't happened yet?

No, not that burial; another one. The Germans are burying their own at their battalion HQ, just down the hill.

Just down the hill? *You crossed enemy lines to photograph German battlefield burials?*

Yes, that is the case.

'You only had to dodge the machine guns for about fifty yards,'[71] Wilkins offers, helpfully.

•

By late August, Wilkins is able to write with some confidence to his mother in Adelaide that the end is in sight.

> *The happenings for the last few weeks make it seem that the tide has turned.*
> *And that we may hope at last that we are on the road towards making the*
> *enemy understand that we mean to fight to a finish. It may not take long*
> *now but I am sure that everyone wants to go on until the job is done.*

I do not carry arms for fighting yet I have captured two prisoners lately myself. They have been hiding when others went past and came out and surrendered to me. It is curious to feel you have the custody of another man who would have, a few minutes before, done his best to kill you but after all we are all human and have yet to find a real blood thirsty enemy. Many of them are now glad to be captured and some seem to think they have [not] much chance against our numbers.

If only the people at home could see the war as soldiers see it the war would soon be over . . .

Much love from your loving son,
George.[72]

For now, however, he must put the pen down, and busy himself getting ready for the next big push.

From 31 August to 3 September, in Monash's finest hour, it is the Australian Corps under his tight and brilliant command that all alone smashes and shatters the German line at Mont St. Quentin, quickly going on to capture even their heavily fortified positions at Peronne, which German High Command had expected to hold through the winter of 1918–19. That man, Monash. Any general with such an unprecedented record of success can expect more resources to be sent his way, and by late September and early October it is, once more, Monash's 200,000-strong Australian Corps – by now with 60,000 Americans under his command – which break through the fabled Hindenburg Line, the last prepared German defensive position – four miles of trenches and concrete fortifications fronted by deep barbed-wire entanglements and backed by heavy artillery, all of it occupied by massed German troops.

Typically, Wilkins is in the thick of it throughout, and on this day, just after the Hindenburg Line has been breached, uses the hilly topography to take some sneak pictures of the German lines. Arriving back at dusk, he is trudging through the mud, wire and ghastly remains of the battle just gone. There is no lack of subjects for still photography before him this evening, as the dead display themselves as obliging models for those who wonder what happens when fragments of a shell meet flesh. The day has been won, but their lives have been lost. Delicately, and with great reverence for the duty before him, Bean's finest records for posterity the shattered scene before him, just before his brave countrymen are to be buried in nameless graves.

They shall grow not old, as we that are left grow old,
Age shall not weary them, nor the years condemn.
At the going down of the sun, and in the morning,
We will remember them,
We will remember them.

It is while taking such photos that Wilkins comes across a disconsolate Australian captain, who is just leading his patrol on a reconnoitring party.

'I've got orders to attack tomorrow morning,'[73] he confides to this photographer who all the officers and men deeply respect. Going on, the captain bemoans the fact that heavy losses are inevitable when attacking at this particular area.

Wilkins pauses. Is he here to merely chronicle events – the traditional role for correspondents and photographers alike – or, on occasion, can he affect events, and alter outcomes? This would seem to be such an occasion.

'From my observation,' he says carefully, 'the line is very weak in this area.'[74]

Indeed?

Yes, indeed. Wilkins knows where the German defences are, because he has just woven his way through them. He knows that, right now, in this area, there are very few Germans, while the spot the captain has been told to attack is intensely defended.

The captain is now full of vigour, and angst.

'It's a stupid waste of lives and ammunition to wait for the morning barrage and attack an alerted enemy.'[75]

Wilkins agrees, and suggests an alternate strategy. 'It would be perfectly easy, to watch the German kitchens working and know when the men were eating supper and off guard.'[76]

Now, it would be against orders, obviously, to order a section to perform such an order, but a single officer, using his initiative?

'I could go out myself at suppertime,' the captain says, 'and just quietly capture them all.'[77]

Just quietly, Wilkins agrees, but perhaps more than one man is needed? Say, 40 or so volunteers?

Things start to fall into place.

At eight o'clock that evening Wilkins and the captain are in position, with their binoculars on the German kitchens, watching as dinner is served. Now is the hour. Cometh the dinner hour, cometh the man and his men.

With hand signals, the captain takes his 40 men forward, each of them with his boots wrapped in sandbags to keep quiet, while Wilkins watches them disappear in the dusk. No more than an hour later they are back ... with 300 prisoners! Quickly, the captain brings his other men forward, not to lose their lives in an attack, but simply to take occupancy of the German trenches without a shot being fired.

The following morning, the units on both sides of them duly attack the German trenches for heavy losses, but this enterprising captain has disobeyed his direct orders, and instead acted on Wilkins' information and therefore won the battle without a shot, something that will never be officially acknowledged: 'It was just a little idea of his own, and such a breach of discipline that it couldn't be made public.'[78]

•

Other Wilkins' feats are less easy to keep quiet, and will be duly reported. For it is around this same time that journalist Keith Murdoch happens to be towards the front lines as the Australian troops are mopping up the last of the German resistance operations near Chuignes. At one point he can hear a lot of German voices arguing. Getting closer he realises the subject: who should surrender first? It is the Red Cross man 'who shuffled uneasily upstairs bearing a white flag'.[79]

(As to who shall do the honours from the Australian side of things, Murdoch notes his name in his telegraphed news report.)

MR WILKINS, THE OFFICIAL PHOTOGRAPHER ... COLLECTED FORTY PRISONERS.[80]

Only a couple of days later, Murdoch sees Wilkins in far more furious action, during the Battle of Peronne. With his own head well down, he is stunned to see Wilkins 'jump the bags' just before they charge forth to take the village of St. Quentin. And now, after the first wave has passed through, Wilkins and his assistant for the day, Sergeant Jackson, join in the second wave of men attacking, stopping to photograph when Wilkins dictates, despite German machine guns firing from the shallows of the trenches ahead. Later, Bean will ask Wilkins why he and Jackson would take such a risk. The South Australian's explanation is perfectly logical, and has nothing to do with bravery: 'The second line was getting shelled and I went forward partly because I thought I would get away from the shelling. I took Sergeant Jackson with me, but we didn't avoid the shelling this way, for the Germans soon shortened onto us and made it very hot.'[81]

Peronne is taken, and Keith Murdoch is escorted through it by Wilkins: 'We pick our way gingerly about Peronne as our courageous photographer, Wilkins, says it is wise to get a photograph of a building immediately as it's impossible to say when it will go skyward.'[82]

As fast as the Australians advance onwards from Peronne, still they can never catch Wilkins, who remains in front, and after being the first into Tincourt is very lucky he isn't shot on sight as something moving in front of them. For the 3rd Pioneers have taken a battering getting into this town, the commander telling Wilkins that, on approaching, no fewer than 16 of their finest had been gunned down by the Germans before they reached the enemy trenches. His men had, in turn, gone on something of a killing rampage, wiping out every German they came across, with one notable exception; a very terrified and very young German who had thrown his hands up and run straight towards them, begging for his life.

'We hadn't the heart to shoot him,'[83] the commander tells Wilkins.

As the days go by and the forces of Monash continue to be at the pointy end of the Allies' spear, used with devastating effect, Wilkins continues to cover both the glory of the Australians, and himself in glory for his continuing splendid endeavours resulting in such graphic battlefield shots.

Bean is so impressed with Wilkins' work, his bravery and overall conduct that he writes personally to the man he had actively campaigned against, Sir John Monash, to request a decoration:

> My Dear General,
> . . . I wish to bring to your notice the exceedingly gallant work of Capt. G.H. Wilkins MC (General list). During the whole of the fighting that has occurred from the 8 August to the present, Capt. Wilkins has shown the greatest and most conspicuous gallantry. He had been in the front line in every important engagement in the long series in which Australian Corps had been engaged carrying out his duties with a coolness and efficiency which has constantly proved an example to those around him, and of great value to the force.

During the fighting of 8 August at Bayonvillers and elsewhere . . . he went out on several occasions far ahead of the front line, generally alone or with one orderly [doing] work of great value with the patrols at great risk and under close range machine gun-fire . . . This officer exhibited a devotion to duty and a disregard of personal danger worthy of the highest praise and gave an example of great value to his fellow and his country.

Captain Wilkins has probably been in the fighting more constantly than any other officer in the corps. His work has led him daily through barrage, frequently of a very severe nature. But I think that the facts above stated themselves justify a recommendation . . . for a Bar to the Military Cross.[84]

Nevertheless, whether Wilkins likes it or not, his bravery is about to become very public indeed, as his legend takes on a grandeur that cannot be ignored by the wider public.

On 27 September 1918, Wilkins is with the 27th and 29th American Divisions as they attack the famed Hindenburg Line with the Australians in support. The fact that the battlefield is covered in fog, added to the fog of war, means both attackers and defenders – not to mention correspondents and photographers – are confused about who is where, and anything could happen. At one point an entire company of Americans moves unknowingly to a point just 50 yards in front of entrenched German machine guns, which suddenly open up and cut them to pieces.

In such an extreme situation, with bullets flying and men dying, Wilkins is more than just impressed with the valour of the Yanks: 'The men were fighting with splendid courage, but they were quite lost in a country new to them. The whole thing was wild confusion.'[85]

Wild indeed. For now through the fog Wilkins sees one group of Germans running towards Americans with grenades in hand, expecting an easy slaughter, only for the Americans to rally and fight back superbly, now charging them. Shocked and disoriented themselves, some of the Germans attempt to surrender but, as Wilkins delicately puts it: 'the Americans were new to war and weren't taking any prisoners'.[86]

The bayonets keep thrusting and the bullets flying, and the Yanks keep going until a second group of Germans arrives and now captures them.

'Things were,' Wilkins will note, 'in a terrible mess. Most of the American officers were killed. They were being fired at from all quarters. In the fog they didn't know which direction they had to advance.'[87]

Wilkins has seen enough; it is time once more to be a participant, not an observer. 'I knew the country by heart and took it upon myself to get things organised as much as possible.'[88]

Things are so extreme he is, he will recount, 'tempted to pick up a gun and have a shot at the Germans myself'[89] but, for the moment at least, he decides to preserve his status as a non-combatant, and confine his efforts to getting the Americans organised.

'I discovered a lieutenant, an Australian liaison officer. Together we organised the troops and held the front line against the Germans.'[90]

That is how Wilkins briefly summarises the action in his own notes. But Charles Bean is taking notes in his daily diary, and he gives the details Wilkins modestly omits. What in fact had happened was Wilkins had raced back to Willow Trench to rally the frozen Americans, alerting them to the Germans approaching. He finds this group leaderless, their officers having been killed minutes before. One Yank is quietly cleaning his Chauchat light machine gun.

'Why aren't you shooting?'[91] asks Wilkins.

'I didn't know there was anything to shoot at,'[92] the American replies.

'Unless you use your gun now you will never be able to use it again,'[93] retorts Wilkins, stunned at their lack of engagement, even as death is hurtling towards them.

Why, even as we speak, Jerry is less than 100 yards away, preparing the bombs they will soon be hurling their way. For those explosions you heard, which you thought were the sounds of far-off shells? *They are not.* They are in fact the stick bombs being thrown over your own parapet, being directed by a German officer that Wilkins has spotted. Right *now*, you are no less than American lambs waiting for your slaughter, even while the German wolf is at the door!

'Line the parapet!'[94] Wilkins yells.

They must attack *now*, or die. And so, with Wilkins directing them, the Americans jump into action, and in just three shots manage to kill the German officer who had been so cocky as to stand up in a shell hole before directing the bombs to be thrown.

Lieutenant Herbert Boden, an Australian liaison officer from Tasmania, has just arrived to help and is amazed to see a photographer now leading a patrol of American troops. Well, he will get to that. For now, Boden does what he can to help, beginning by throwing bombs at the Germans, aiming for a quick knockout. A brief Hun halt is the result, whereupon Boden calls for his own platoon to join the defence of the parapet.

Grateful for the assistance in these extreme circumstances, Wilkins wants to show Boden where he thinks the machine guns might be located in the dense fog, so he, Boden and another man venture out to the rim of a nearby crater where Australian troops are sheltering. Before Wilkins can even point to the guns though, the Germans prove his deductions correct by firing a further hail of bullets. The three men drop to the ground, Wilkins is the only man who rises, while Boden and the other man lie there, dead.

Some Australian troops in the crater now rush out and take the machine-gun nest. As for Wilkins? Well, he assesses his scrapes in a typically detached fashion:

'Bullets had grazed the fleshy part of my chest. I had another wound in my scalp and one in my heel. My uniform was soaking with blood and I looked ghastly, but I really wasn't badly hurt.'[95]

Wilkins is at the front the next day, despite being covered in bandages. He was shot three times the day before and was 'so covered with blood that I got the sympathy, though my injuries were superficial'.[96]

This is not the view of the men he is trying to photograph, who gaze wide eyed at this bloodied and unbowed mad snapper, calmly arranging his plates to take photos of far less remarkable sights than himself. Invariably, the troops try to offer him a drink, some rum at least, but Wilkins refuses as he has a job to do, to take pictures of them. It is a bravery verging on madness, a calm that is extraordinary and a stoicism that defies belief.

A decoration follows, but Wilkins regards it as almost an embarrassment to receive it.

'A bar was added to my Military Cross because my work that day had been observed by two colonels. I was carrying my camera and that made me conspicuous.'[97]

The beginning of the citation for his bar to the Military Cross will read:

> Captain G.H. Wilkins . . . carried out his duties in a manner which brought a most marked credit to his unit amongst all the troops engaged . . . In the Battle of the Hindenburg Line of September 29th, Captain Wilkins went over shortly after the attacking Americans and found the Germans bombing the Americans back up the

trenches on the left flank. Under machine gun fire at close range he ... organised the Americans, who had lost their officers, and directed operations until the German attack was checked and the supporting troops were arriving. He then left them and continued his duties, frequently under heavy fire, during the rest of the day.[98]

Just another day.

(As Sir John Monash will lament, Wilkins carries his modesty to extraordinary lengths, seemingly determined that his valour remains obscure. A friend will later recount how, the first time he had run into Wilkins in London wearing the stars of a captain and the ribbon of the Military Cross, he had offered congratulations and asked the obvious question of how he had won it.

'Just taking a few photographs!' Wilkins replied.

Next time he had seen Wilkins there was a bar to his Military Cross. George? How, this time?

'More photographs!'[99])

•

As well as the Australians have done – having fought no fewer than 10 battles since the beginning of the *Kaiserschlacht* for 10 victories – by early October they are being withdrawn from the line, to allow them time to rebuild for the planned spring offensive of 1919.

And yet, in the second week of November, Wilkins hears an extraordinary thing. The word spreads in confined circles. There is likely to be an armistice. The Germans simply have very little fight left in them, and have been overwhelmed by the combined might of the Allies, most particularly that of the British Empire now bolstered by the American forces.

'Hearing from headquarters that the Armistice would be signed on November 11th at eleven o'clock in the morning,' Wilkins will recount, 'I decided to go to London and observe conditions ... I was one of the few who knew it was going to be signed. I went down and waited for Lloyd George to come out and read the armistice ...'[100]

So it is that Wilkins is there for the historic moment, the prelude of which is British Prime Minister Lloyd George emerging from his residence at No. 10 Downing Street just before 11 am to addresses the gathering crowd who have, like Wilkins, also heard the rumours.

'I am glad to tell you,' he says, 'that the war will be over at eleven o'clock today.' [101]

RAH! The crowd roars, at which point the Prime Minister happily heads back inside.

Sure enough, however, at 11 am . . .

The bells!

The bells of Big Ben are ringing for the first time in over four years. It is the eleventh hour of the eleventh day of the eleventh month, and Armistice is now in effect. Outside 10 Downing Street, as George Wilkins continues to drink it all in and keep his camera rolling to capture the historic moment, the crowd start to chant, 'We want Lloyd George! We want Lloyd George!'

Not long afterwards he appears at a first-floor window with two of his most senior Cabinet ministers, Bonar Law and Winston Churchill, and addresses the crowd as all fall remarkably silent, bar the sound of the bells and wild cheering in the distance.

'You are well entitled to rejoice,' the Prime Minister says. 'The people of this country and our Allies, the people of the dominions and of India, have won a great victory for humanity. The sons and daughters of the people have done it. They have won this hour of gladness and the whole country has done its duty. It has achieved a triumphant victory which the world has never seen before. Let us thank God.'[102]

And the people danced, and the people sang, and the people cried with relief. There will be very little work done on this day in the victorious nations, with one notable exception. Out on the battlefield of Fromelles in France a nattily attired 37-year-old man is to be seen poking around among the skulls and bones, rusty bayonets and broken rifles. Who else could it be other than Charles Edwin Woodrow Bean – still deeply troubled by the battle he had so narrowly missed chronicling – and now impatient to begin what will be the work of the rest of his life . . .

Fromelles is just the beginning. Bean also wants to go back to one place in particular before it is changed by time: Gallipoli, and he knows just the group of men he wants to take with him.

CHAPTER NINE

FLIGHT TO ADVENTURE

The tumult and the shouting dies;
The Captains and the Kings depart:
Still stands Thine ancient sacrifice,
An humble and a contrite heart.
Lord God of Hosts, be with us yet,
Lest we forget – lest we forget!

Kipling, *Recessional*

The only thing to do in such a case, as I had learned in early youth,
was to concentrate on some particular thing. Do one thing at a
time and make sure of it, depending on some solid fact.[1]

George Wilkins

January 1919, Versailles Peace Conference, flight of fancy

In the history of the world, there has never been such gatherings of the good and great of the day, as now assemble in the Palace of Versailles for the Peace Conference as the 27 victorious nations of the Great War decide the exact terms of Germany's surrender, and the reparations they must pay. And yet, though the defeated Germans have no seat at the conference, this does not prevent there being disputes. For there are many ...

One of the matters fiercely debated is the proper fate for the former German territory of New Guinea. President Woodrow Wilson of the United States is passionate in his advocacy for Japan, which had fought on the side of the Allies in the war, to take it over. But the irascible Australian Prime Minister Billy Hughes is equally insistent that it should be Australian territory – in part, as a buffer against what he suspects might be Japan's future territorial ambitions across South-East Asia and into the South Pacific.

Frustrated, Wilson can't help but note the relatively small number of people Hughes represents.

'After all,' he says in his New Jersey twang, 'you speak for only five million people.'

But Hughes will have none of it.

'*I* speak,' he says with enormous gravitas, 'for sixty thousand dead.'[2]

At this table, in this forum, Australia had made a far greater sacrifice than America had, and it WILL be heard. And so it is.

In the course of attending the Peace Conference, Billy Hughes has regularly flown back and forth between London and Paris. Journeying by aeroplane set him to thinking . . .

If travel like that in a plane could cut down a one-day trip to just an hour, what would it mean to Australia if an air-route could be established between, say, London and Sydney? If instead of a six-week trip by sea, the whole journey could be cut down to a matter of just a week or two? Maybe he should do something about encouraging that. After discussions with many Australian pilots, he sends a cable to the Australian Cabinet:

SEVERAL AUSTRALIAN AVIATORS ARE DESIROUS OF ATTEMPTING
FLIGHT LONDON TO AUSTRALIA [. . .] THEY ARE ALL FIRST CLASS
MEN AND VERY KEEN[3]

Late January 1919, The Nek, Diggers to dust

Traipsing over this old battleground at Gallipoli, Charles Bean tells his six companions, including George Wilkins, something of the story. They have, after all, all come here on this 'Gallipoli Mission', so soon after hostilities have ceased, in order to document whatever they can of the battlefields, and gather what 'holy relics' they can, before the passage of time erases them all.

And of all the Australian battles Bean covered, this was the most horrific, the Battle of the Nek, as it was known.

It was dawn on the morning of 7 August 1915, and the Turks were heavily entrenched at one end of this small ridge. The plan was for four waves of Australian soldiers of the Australian Light Horse to charge along this 40-yard wide ridge, the Nek, and its sloping sides, at fortifications situated above them, which Bean had described at the time, 'was like attacking an inverted frying pan from its handle'.[4]

Each wave would go two minutes apart, and they were not to shoot when charging. This was a bayonet and bombs job – they had to scramble the 27 yards and get right amongst the Turks. To force Johnny Turk to keep his head down, and smash their machine guns, a heavy naval bombardment would pound them from exactly 4 am for exactly 30 minutes at which time the first wave would go over.

But something had gone wrong!

Bean had been nearby, on his way to his dugout at Anzac Cove, and at exactly 4.23 am, the bombardment had 'cut short as if by a knife'.[5]

And then the horror of what had come next.

That seven minutes respite had allowed the Turks to not only lift their heads but get their machine guns set up, and even do a few practice bursts. They were *waiting* for the Australians!

At 4.30, the whistle had blown and, as one, the first wave of men of the 8th Light Horse, 150 horseless horsemen, charged forward, cheering and yelling as they go to ... ... 'GIVE IT TO THEM, BOYS'[6] ... ... only for those cheers to be drowned out by the instantaneous roars of 500 rifles, and five machine guns opening up on them. Bean had been just 600 yards away and would never forget the horror of the sound, 'a sudden roar of musketry and machine-gun fire, like the rush of water pouring over Niagara'.[7]

Oh, God, the *horror*.

'It was one continuous roaring tempest,' Bean would recount. 'One could not help an involuntary shiver – God help anyone that was out in that tornado. But one knew very well that men were out in it – the time put the meaning of it beyond all doubt.'[8]

The next three waves had followed in quick succession and of the 600 men who had gone over the top there had been no fewer than 234 dead and 183 wounded, for not a yard of Turkish trench gained.

And now here they are, some four years later, wandering over this same sacred ground on a wintry day, and you can still see the remains of some of the Diggers exactly where they had fallen with their face to the foe. With few further words spoken, Wilkins sets up his camera, the painter George Lambert starts to sketch, while the others spread out and start to gather what they can find: badges, bayonets, bandoliers and all the rest.

'We found the low scrub there literally strewn with their relics,' Bean will record. 'They buried more than three hundred Australians in that strip the size of three tennis courts. Their graves today mark the site of one of the bravest actions in the history of war.'[9]

But it is not just relics.

'Evidence grins coldly at us non-combatants,' Lambert will write to his wife of the sight of so many skulls. 'From the point of view of the artist-historian, The Nek is a wonderful setting to the tragedy.'[10]

It is the beginning of a highly sombre if inspirational six weeks as, together, they go all over the old battle sites.

Wilkins, having never been here before, is deeply moved by what he now realises was the extraordinary Australian achievement in landing against such a force, holding on despite everything, and time and again giving Johnny Turk one in the eye.

'Much the most impressive battlefield I've seen,'[11] says Wilkins in those first days, when they ride along the ridge to the high point, Hill 971.

After a day out on the battlefields they retreat to their abode at an old Turkish Military Hospital and talk quietly.

'Wilkins,' Bean chronicles, 'could at any time be led into deeply serious discussion of almost any subject – with rare hints of a whimsical background showing through.'[12]

Though not the loquacious kind, the circumstances of this tight little group, working so intimately together, at such a site, does lend itself to deep discussion when the day's work is done and they retreat to their makeshift camp near the sacred site of Lone Pine.

'[The men,]' Bean notes, 'could always tempt Wilkins into discussion on the aims and happiness of mankind (in particular of the Eskimos, the simplicity of whose wants, he contended, gave them the highest degree of happiness) . . . the futility of international strife and suspicion . . . on means of foretelling the weather and so perhaps enabling men to avoid the effects of drought that he had known too well as a child . . . Once, years later, I heard Wilkins criticised as an adventurer, rather than a scientist; but never was a more superficial judgement made. You had not to live with him a day to discover that the increase of knowledge for the benefit of man was the burning impulse of his life.'[13]

Bean is, by nature, a taciturn man and an extremely hard marker. With most men he is slow to offer praise. With Wilkins he cannot offer enough.

As ever, no-one works harder than the photographer. Despite suffering food-poisoning for a good deal of their time there, Bean records that

Wilkins 'was at work in his iron tank of a dark-room till far into the nights, which were often bitter. He was a born leader and . . . had largely been responsible for our success.'[14]

The result is that the Commonwealth Government now holds perhaps the most remarkable record of the war of any of the Allied forces. This is contained in 20,000 photographs and 50,000 feet of cinema film which is carefully preserved by the Australian War Museum.

Wilkins' reputation for surviving any number of scrapes against all odds precedes him, and that reputation is never more apparent than when they head through territory that has previously been mined, which is much of it.

'I used to note with amusement,' Bean will recount, 'that, as we strode and climbed about the hills, the rest of the party unconsciously followed Wilkins' lead. If he pursued a certain path, climbed a cliff in a particular way, jumped a trench or even went round left or right of a bush, the rest of us usually did the same.'[15]

•

While Wilkins' experience at Gallipoli is far removed from his time in Turkey during the Balkan War, happenstance briefly takes him back to his old stamping ground of Constantinople in late January, when he and Bean decide they need more maps of the Dardanelles in general and the Gallipoli Peninsula in particular.

On their first evening there, Wilkins takes the boss to the famed Tokatlian cabaret. They have no sooner sat down than a waiter appears with *three* bottles of champagne and a large dish of Turkish sweetmeats.

'We didn't order that,' says Wilkins, bemused and a little confused.

'It is the manager's compliments to Captain Wilkins.'

'I am surprised that the manager knows me? Who is the manager?'[16]

'He will be here in a moment, Sir.'

And indeed he is.

Who should walk in, beaming like a lighthouse on a dark night?

It is . . . Godfrey!

Yes, the very dragoman who Wilkins had narrowly missed shooting at the height of the Balkan War, seven years earlier, for his craven cowardice.

'Oh, my dear Captain Wilkins,' he exults 'how happy I am to see you again!'[17]

Again, the Australian is somewhere between bemused and confused.

'I didn't know whether to throw him over the balcony or drink his wine. But I decided to drink it and pump him for information. Neither of us mentioned the day I nearly murdered him . . .'[18]

This is possibly to Wilkins' regret, as the whole thing stirs up so many memories of Godfrey's betrayal all those years ago.

Walking back to their hotel, Wilkins, after some rumination, says to Bean, 'I knew I should have killed him.'[19]

•

Back at Gallipoli shortly afterwards, over the course of the next five weeks they see any number of sad sites, including the seaward slope of 'Baby 700' where, on 2 May 1915, the Diggers had begun their doomed quest to prise the Turks from the 3rd Ridge, a first counter-attack countered.

> In the scrub about 200 yards below the summit we found where the Australians had run into heavy fire . . . The remains of our men still lay scattered thickly under the scrub.[20]

On the left flank, towards Suvla Bay, it is even worse.

> We could see the bones of men on two hills ahead of us . . . For the first time in our mission to Gallipoli the scenes we looked on that afternoon 'got us down.' Wilkins . . . and the others felt as I did. For some reason the dissolution of the human remains . . . was not quite so complete and the number that must have been trapped [here], and the hopelessness of their situation . . . did not bear thinking of. My note says 'I have nowhere, except at The Nek seen the dead lie so thick.'[21]

The pictures carefully taken by Wilkins preserve the pathos for perpetuity. (It is a shatteringly naked, open graveyard, with the rib cages, skulls, thigh bones and the rest of once vibrant young, brave Australian men carelessly scattered about.) It is all, somehow, so *obscene*, this strewn ruin of youth lying on a foreign field. Even the naturally taciturn and self-contained Bean is affected by it, as the sheer tragedy of the 'Great War' lies exposed, as is 'the old lie' of Wilfred Owen's famous poem: *Dulce et Decorum est pro patria mori. It is a sweet and glorious thing to die for your country.*

No, it is not. For these blokes it was courageously charging forth with your mates, only to take a bullet through your neck, to gurgle and gasp your last agonised breaths and finish lying here under the Turkish sun, never to see your loved ones again, to hold your woman again, your child.

And all for what? Turkey had won this battle, but lost the war anyway. Would the dead Turks themselves say it was worth it? Will future generations?

It is a haunting place to be now, one of deep sorrow and yet still entrancing beauty, those impossible cliffs still shining in the Mediterranean sun as they have for millennia past and will forevermore, the waters of the Aegean Sea still lapping gently on the little beaches below. It all looks so serene and timeless from afar, and it is only up close that the remnants of horror reveal themselves, with Bean, Wilkins and their team doing everything they can to capture it all. For yes, it lives in the nation's mind now, but for how long? Their job is to help consolidate the memory of what happened here, and Wilkins' photos are key. And so, just a moment longer, as Wilkins checks the plates, to take one more from a different angle, and Bean checks.

Good?

Good.

Onwards.

By mid-March their Gallipoli mission is accomplished, and using Wilkins' Constantinople contacts they are able to board a train to Cairo – where the young South Australian will be able to enjoy 'a few delightful hours (as a close and happy spectator) of the Cairo riots'[22] – before a ship takes them to England. At this point Wilkins is eager to get going on the idea that has been working in him since he first talked to Stefansson about 'forming an expedition to the North Pole by aeroplane from some northern Arctic base,'[23] but, as it will be reported, 'friends of his, anxious for his life, persuaded him to accept a commission to supervise the arranging of Australian war photographs for the National Museum'.[24]

After all, how many times can a man keep risking his life before his number must inevitably come up?

No sooner have they arrived in London than both men get busy at the HQ of the Australian War Records Section in Horseferry Road in the City of Westminster, where they will spend many months categorising, detailing and compiling the immense number of photos taken of Australian troops during the war. The idea is to add archival information to each photo, when and where it was taken, ideally the names of the men in the image and the background to the shot being taken. With

Wilkins however, there is one notable omission on his photos – he rarely bothers to claim them as his own. He simply doesn't care, which is in strict contrast to Frank Hurley who sends express word: any photo taken by him *must* be labelled as such in any AIF publication.

Certainly, Frank, as you desire.

Wilkins also delights in spending time with a fine Englishwoman of artistic and theatrical bent, Miss Lorna Grace Maitland, who he had met not long after the Armistice at the table of, who else but, Lady Scott. The 31-year-old artist had been raised in India and then Bournemouth, the daughter of an English Colonel and doctor in the British Army, and has the kind of carefree spirit that pleases him while she, in turn, adores the unconventional way this Australian lives his life, the things he has done and seen, the things he still wants to do.

Another old friend Wilkins catches up with is Will Dyson, who he got to know on the Western Front through Will's service as Australia's first official war artist. No sooner has he entered the Dyson family home, however, than Will's seven-year-old daughter, Betty, announces with some gusto: 'Daddy, this man can't be an explorer, because all real explorers have beards.'[25]

Come to think of it, she might have a point?

As it happens, he has his next big trip – almost an exploration – coming up soon, and starts growing a beard immediately.

•

Not long after Prime Minister Billy Hughes had returned from Versailles, his government had announced an extraordinary prize of £10,000 for the first all-Australian crew, or solo Australian pilot, who could fly all the way from England to Australia in a plane of British make in under 30 days!

Struth! £10,000! In a world where the average worker pulls in less than £10 a week it is an astronomical sum. But Billy Hughes is serious: he wants the best and brightest to enter the race – to garner Australian attention and help open Australia to the world.

Certainly it is a risky venture, with even *The New York Times* commenting that 'Christopher Columbus did not take one-tenth the risks that these bold air pioneers will have to face ... They will be throwing dice with death.'[26]

Truth be told, however, the majority of entrants were veterans of the air war in France and had thrown dice with death most days before breakfast, up against the likes of the Red Baron, and with that background, another roll of those dice is nothing to write home about . . .

But one Australian pilot, who has had two toes shot off in the service of King and Country, Charles Kingsford Smith, does anyway.

'Should we be chosen by any chance, and pull the job through,' he writes to his parents, 'it will mean that we are made for life, because look at the big advertisement our own little venture would get out of it, to say nothing of the part of the £10,000 that would come to us.'[27]

Look, the truth is, all of Kingsford Smith's money put together plus 50 pence would barely be enough to repair the seat of his pants he's been living on for the last few years, but there are ways around that. So much press does 'The Great Race', as it is dubbed, garner that there are any number of companies eager to associate themselves with whoever they think might possibly win. One of these is the Yorkshire-based Blackburn Aircraft Company, which fancies one of their long range R.T.1 bombers might, modified, be perfect for the task at hand with the right pilots. They already have one pilot in the person of a nice young chap from Queensland by the name of Val Rendle, but after the owner, Robert Blackburn, meets Kingsford Smith he affirms that he is happy for him to join Rendle's crew, and fly their plane – the Kangaroo – to Australia.

(It is so-called because when the 75-foot wing-span, two-ton, twin Rolls-Royce Falcon II engine biplane bomber had first come off the drawing board there was a small 'pouch' below the cockpit where a machine gunner hovered in the belly of the beast.)

Making everything all the sweeter was that Robert Blackburn had told the crew that if they could get the Kangaroo to Australia, and win the race, they could keep the plane.

Alas, a problem had soon emerged.

Like many pilots who had survived the war against all odds, Kingsford Smith lived fast and loose, and Blackburn became aware that, in the lead-up to departure, he had been purchasing aircraft from government disposals and barnstorming around the country, ignoring civil air regulations and landing in fields not approved for the purpose. Reports also soon came in that he had been insuring his aircraft for more than they were worth, and there had been some lucrative crashes . . .

He was not the man they were looking for, after all, and a line was drawn through his name.

What about you, George Wilkins?

Initially, Wilkins isn't at all sure he wants to be part of such a showy spectacle as this race – he is not motivated by money, and competitions are not quite his thing – but the more he looks at it, the more he likes it. Apart from quickly getting home to see his mother and family for the first time in two years, it is the challenge he likes, and it would be no small thing to demonstrate that one really could cut down travel time from England to Australia to less than a month! Most of all though, what appeals to him is the scientific aspect.

It is like this, Blackburn: he will accept a position on the Kangaroo, on the strict proviso he can take with him all kinds of equipment to record temperatures, humidity and air pressure levels and so forth, in the course of their journey, to continue on his life's quest of pushing outwards the borders of scientific and meteorological knowledge.

Blackburn agrees, and the deal is done.

Wilkins becomes both the commander and the navigator on the flight of the Kangaroo, something which draws a little ire, and even some quick fire.

'He has had no previous experience in aerial navigation,' one member of the crew quietly objects.

'A man who can wander round the North Pole in snowstorms and blizzards for four years and not lose himself will not have much difficulty in locating Australia from the air,'[28] came the unanswerable reply, and things settle down from there.

Acquaintance with Wilkins himself immediately lifts confidence as they all get to know each other.

Lieutenant Valdemar Rendle is just 23 years old, but he has spent over half of those years being obsessed with one thing: flight. Born into a prosperous Brisbane family, the son of the prominent Doctor Richard Rendle, Val has resisted the best efforts of his father and the staff of Brisbane Boys Grammar to pursue a proper career. He had come to England to join the Royal Air Force, first as a mechanic, before working his way up through the ranks to be a combat pilot. Since the war he had been a test pilot before flying the official mail from London to Paris. A curious chance brings he and Wilkins together: another member of the Kangaroo team, Lieutenant Garney Potts, the Roo's mechanic, had been employed in aeroplane construction with Claude Grahame-White, that famed and vain pilot that Wilkins first flew with many years ago. Speaking of pilots, Lieutenant Reg Williams will act as Rendle's reserve, 'the second pilot', something he is more than qualified to do having been an instructor at the civilian Richmond Flying School. And so the Australians are connected, and Val is a torrent of enthusiasm as he outlines the scheme that has been obsessing him for over a year: the flight from the motherland to their homeland.

Their Kangaroo is going to hop all the way home from their starting point of London's Hendon, to Leeds, then Lyon, and then whatever flat spaces – more often than not, the middle of racecourses – they can find in the likes of Brindisi, Salonika, Nicosia, Aleppo, Baghdad, Bushehr, Chabbar, Karachi, Baroda, Benares, Calcutta, Rangoon, Penang, Singapore, Batavia, Kupang and finally . . . Darwin!

But, some modifications first.

That the Kangaroo was made for war is not in doubt, but, given imagination, its weapons can be put to novel and scientific use. Take the machine-gun turret, for example. Wilkins has turned it into something ingeniously effective: 'Captain Wilkins is introducing a novel position finder, made up of the circular mounting of an aerial machine gun, which can be used in the clouds so rapidly that it is adjustable and permits the definition of the position from a momentary glance at the land'.[29]

As for where the plane's bomb load used to be, there is now 100 gallons of petrol and space for another man, a man who will soon get used to the smell of petrol.

Inevitably the fact that a man who had been a polar explorer, then a highly regarded war photographer, is now commanding such a venture as this brings with it more public notoriety back in Australia. Fred Cutlack, one well-known journalist who had got to know Wilkins during the war, now tries to give the readership of the *Sydney Morning Herald* and syndicated papers, a flavour of the man.

'Captain G. H. Wilkins, commanding the Blackburn Kangaroo machine,' he writes, 'is one of the most romantic characters of the war – and the war brought out many of them. There are some men who thrive on the excitement of danger, who seem not to know what fear is, or to have a nerve in their system Wilkins is one of them. He does not mind whether it is polar exploration or searching the jungle with a cinematograph, or photographing in No Man's Land, or aeroplaning anywhere on earth. His life has held enough adventure to satisfy any ordinary man: but Wilkins will pursue adventure to his dying day. If he had been a combatant during the last war, he would have gone on winning the VC till he lost his life.'[30]

Such esteem is not confined to his homeland however, and it is a measure of the respect in which Wilkins is held within the London establishment that, at the crew's departure from London's Hendon aerodrome at 10.37 on the morning of 21 November 1919, he is carrying the cabled best wishes of no less than the future King of England, Prince Albert, and the former First Lord of the Admiralty, the Right Honourable Winston Churchill. Which is to the good. All their 'good lucks', 'bon voyages' and 'tally-hos' are very much appreciated. But can the crew actually do the job they have set out to do? Val Rendle's sister Dorothy and his wife, Alice, certainly think so, both of them there to kiss him good luck and say bon voyage, and his new bride is delighted to share her thoughts with the men of the press: 'I am as keen as Val on this flight, and would go myself without further preparation if there was room!'[31]

Yes, but will they make it, Mrs Rendle?

'Sure, they'll land in Australia,' says Alice as Val grins at her enthusiasm. 'He's full of infectious confidence.'[32]

Done with their spruiking, both young women now climb into the plane as Val taxis around the runway for their pleasure and for the amusement of the crowd and Wilkins. *Is this a stunt, Captain Wilkins?*

'I wouldn't do this as a stunt,'[33] he assures the dubious pen wielders. 'I've most definite and scientific reasons for going.' Such as? 'I think we will add materially to air research and lighten future difficulties of flights to Australia.'[34] The cash prize, is also to the good: science doesn't fund itself. These last few weeks, they have all been working hard to be at full readiness.

'She's not pretty,' Captain Wilkins says, 'but is all British, and thoroughly sound.'[35]

At least there had been one attempt to pretty her up a little with 'England to Australia' daubed in large white letters across the centre of the fuselage.

And now they are just about ready, with the crowd and crew merry, the sun smiling upon them, and the engines sounding healthy. But Wilkins, as is ever his wont, insists on one final inspection – which . . . shows a loose water pipe. The mechanics give it a close look and assure him it will hold until they make Lyon.

All right then . . .

'Let's hop off,'[36] Wilkins declares, and so they do, waving out the cockpit window at the crowd who wave in return. This is it. After a shattering roar down the runway, the Kangaroo leaps into the air, and Lieutenant Williams and Lieutenant Potts continue to wave as Val takes her up. The crowd yell 'Cooo-eeeee'[37] as the Kangaroo circles *twice*. Why twice?

That would be Captain Wilkins! He is the only one not waving, as he is too busy filming the takeoff from inside the plane – and as an old professional, he knows a second take is always useful. Within minutes, the Kangaroo has disappeared over yonder horizon.

•

Good God, the memories. Roaring over Amiens, Villers-Bretonneux and Le Hamel late that afternoon Wilkins is quite stunned to see those places he had known as bloody battlefields, filled with death and destruction, blood and gore, fleeing French and rampaging Germans . . . now covered in a blanket of white snow, making it all seem so extraordinarily peaceful, as swirls of smoke spiral up from so many chimneys.

For Rendle and the others, it is just more pleasant French countryside beneath them, but for Wilkins it is somehow comforting affirmation that what so many good men had died for now has a tangible result: they are towns living in peace, free of Germans.

Onwards.

•

The one thing the crew of the Kangaroo can depend on? Trouble: 'In France we struck bad weather . . . We landed at an army station and stuck three days more. Across the Gulf of Lyon and all the way to Italy we had pouring rain. In Italy we landed at Rome and found the gasoline locked up and the man in charge away on a picnic. It was a saint's day. That delayed us a day.'[38]

You would need the patience of a saint to keep your temper after a problem-plagued trip filled with bad weather and long repairs that has taken them only a quarter of the distance to Australia in two-and-a-half weeks. George Wilkins and his Blackburn Kangaroo crew suddenly have the spectre of a grisly death riding along with them. Eighty miles southwest of Crete, co-pilot Reg Williams looks behind for one last glimpse at the receding coastline of Crete, when he sees it.

Christ!

'I noticed that the tail was all black on one side. I guessed immediately this was from oil leaking back in the slipstream from the port engine.'[39]

Urgently he passes a message to Val Rendle, at the controls beside him, while also alerting Wilkins and Potts in the rear cockpit. What can be done?

Nothing.

'Climb while you've got any oil! Get as much altitude as possible!'[40] Wilkins yells as Val nods.

It seems likely that somewhere in the engine an oil line has fractured and the lifeblood of the whole machine is slipping away in the slipstream, weakening with every passing minute, effectively pouring troubled oil on glistening waters.

Their chances, once that engine inevitably packs it in? Don't ask.

'The makers of the plane said it would only fly 30 miles on one engine from 2000 feet.'[41]

Still, reasoning that it is better to be 50 miles from land in the middle of the ocean than 80 miles, Val Rendle turns the plane around, and throttles back the stricken engine, hoping to nurse it through for as long as possible. But they are resigned to this likely ending ugly.

'There seemed to be no chance of getting out alive,' Wilkins will recount. 'Everybody believed it would be the end.'[42]

Wilkins is fascinated to observe the psychology of the situation; the certainty of their eventual death relaxes them. They will crash, when they do 'there was no chance to swim 50 miles or be picked up. Our thought was "Well here goes. That is the end of our adventure until we meet the big one."'[43]

Paradoxically, as they keep edging along past 50 miles, 40 miles, 30 miles, then 20 miles, the crew members now become ever more tense. Somehow, with a chance of survival, they all began to be *afraid* of dying. At last, at last, Crete comes into view, but still they are not safe. For, where could they land? Everywhere they look, all they can see are cliff faces, rugged rock, and yawning, hungry canyons, each one clearly happy to swallow them whole without burping, before, just clearing the cliffs they find themselves soon flying very low over some kind of town!

In vain do they look to the terrain beneath them for some flat land where they could possibly make a smooth landing . . . or at least some land flat enough on which they can get down intact. Finally, it is obvious that they are just going to have to take their chances, and – even as the last of the oil leaks from the port engine and the motor comes to a shrieking halt of tortured metal on metal and blown bearings – attempt a landing on the flattest handkerchief of land they can find.

But all they can see are houses that they are heading straight towards! At the last instant, with extraordinary skill, Rendle manages to get the last bit of lift the plane has in it to get over the houses, only now to be confronted with a large stone wall that seems to be racing towards them!

Brace! *Brace!* BRACE!

As the entire Kangaroo crew hold their head in their hands, in the instinctive belief that this would provide some protection, the plane rushes over a ditch, bursts its tyres upon such hard contact with the earth, bounces up a bank, and comes to rest very near that wall, with its tail held high, if not its head.

Now, given how many people think that only lunatics would have entered such a race in the first place, it is, perhaps appropriate that the wall that they have so narrowly avoided smashing into belongs to Crete's largest lunatic asylum.

Still, at least they are alive, and will live to fight another day!

Truth be told, George Wilkins is not at all perturbed about not winning the race and, though joyous that he personally has survived, running at a close second is the fact that his notebooks with all his meteorological data are secure, as is his varied equipment of barometers, thermometers, hygrometers, wet and dry bulbs and density meters.

(His bank balance is not so secure, even after the insurance money reluctantly appears, Wilkins still loses £2000 on the venture.)

Wilkins, with typical understatement, lets Charles Bean know via cable how the whole thing went:

```
FOR BEAN. MACHINE CRASHED RETURNING LONDON UNHURT IF DESIRE
MY IMMEDIATE RETURN AUSTRALIA CABLE ME AUSTRALIA HOUSE
LONDON [ . . . ] WILKINS44
```

May 1920, London, enough rope for Cope
And for his next adventure?

The New York Times is the paper that breaks the news to the world:

COPE PLANS TO FLY TO THE SOUTH POLE;
Head of British Antarctic Expedition Prepares for a Dash from the Great Ice Barrier. IS CONFIDENT OF SUCCESS

Special Airplane Will Be Used as a Sledge on the Ice in Event of an Accident in the Air.[45]

The paper informs its readers that the Englishman, Dr John Cope, who was a part of no less than the famous Shackleton Expedition, this time 'proposes, besides carrying on scientific investigations, to ascertain the extent of mineral and other valuable deposits on the Antarctic continent,'[46] all as part of what he calls, the British Imperial Antarctic Expedition of 1920.

Much of the funding and attention comes from his previous association with Sir Ernest Shackleton, and Wilkins readily agrees to join him, once he hears all Cope's plans about fulfilling his own dreams of flying in the Antarctic!

Cope hopes to have no fewer than a dozen planes, courtesy of $750,000 in funding from the governments of the UK and USA, to chart the entirety of the Antarctic continent. The plan is grand: the dozen planes to be purchased from surplus war stock, taken by ship to Graham Land on the furthermost part of that finger of land that stretches towards South America. From this base the unprecedented exploration will begin as the planes will fly the men to establish bases all over Antarctica; then half the planes will simply return to carry fuel for the second half to explore easily and endlessly. Wilkins is struck by the similarity to a dog sled exploration notion that Robert Peary once had; and is certain that it can work for aircraft, leading to a base within easy flying distance to the South Pole.

'I believed we could thus cover a much larger area of unknown territory than by other means of transportation, and I would have my chance to test my theories of the feasibility of flight in the polar regions.'[47]

After all, as he tells the Australian press on a quick trip home, 'I came across Australia on the trans-continental railway, and there are better landing places on the Arctic ices than I saw in my trip from Perth to Melbourne.'[48]

Oh, yes, everything is in hand.

'We will gather all the geological and biological information we can,' Wilkins goes on, 'and will keep complete meteorological records of the tour ... This expedition is only a preliminary to a larger one which will be organised by Dr. Cope and myself on our return to England, when we hope to fit out two fairly large vessels for the Antarctic, and will also take with us six aeroplanes. During the present tour I will be able to get reliable data as to how best to fit out aeroplanes for Antarctic exploration.'[49]

Alas for Wilkins, he will soon learn that Cope is 'just about the poorest executive in the world'.[50]

For it is while on that trip home to visit his aging mother and make arrangements for Bean to get separate Commonwealth space in Canberra for records, artefacts, photographs and footage of the Great War – the genesis of what will become the Australian War Memorial – that Wilkins is shocked to receive a telegram from Cope asking him if, well, if he wouldn't mind too much, going to Canada and buying sled dogs before coming to meet Cope in Montevideo?

Sleds?

Dogs?

Wilkins has already seen, and even starred in that movie.

But he has moved on now.

What of the 12 *planes*? The *scores* of men? The government *funding*?

Yes, well. Wilkins acidly records the result: 'We had to abandon that plan because Cope spent all the money on other affairs.'[51]

(It certainly isn't spent on accommodation because, as Wilkins will later find out, Cope's favourite method of finding lodgings is to invite himself to dinner and then fall asleep before his host can get him to leave.)

Wilkins now sends Cope a frank telegram:

UNDER CIRCUMSTANCES DO NOT CARE TO GO ON EXPEDITION. AM
TAKING ENGINEERING POSITION IN AUSTRALIA. WILKINS. [52]

The reply from Cope is as direct as it is desperate.

AM DEPENDING ON YOU. I HAVE ORGANISED THE EXPEDITION ON
UNDERSTANDING THAT YOU ARE PART OF IT. NO OTHER MAN HAS ANY
POLAR EXPERIENCE. I BEG YOU DO NOT LET ME DOWN.
COPE. [53]

As with all things Cope, the explanation contains parts of the truth but is inherently deceptive. As for 'organising', Cope placed an ad in the magazine *Nature* (on page 93 for those who get that far). The ad is a masterpiece of pomposity and grandiosity, and is, as ever, centred on John L. Cope:

A British Imperial Antarctic Expedition.

I, through the columns of NATURE, direct attention to the British expedition which I am at present organising and propose to lead to the Antarctic in June next year. The objects of the expedition are briefly as follows :- (1) To ascertain the position and extent of the mineral and other deposits of economic value . . . (2) To obtain further evidence of the localities of whales of economic value . . . The expedition proposes to leave England in June, 1920, and to be away for a period of five years. During this period important scientific research will be undertaken on the lines briefly given above.

Applications are invited from fully qualified men . . .

JOHN L. COPE. 66 Victoria Street, London.[54]

Only two men contact Mr Cope, but at least they meet the two key requirements: previous attendance at a public school and gullibility. Signed!

It is not quite two men and a dog, as the dog must have blanched at the last minute, but Cope insists they can get the dogs they need on the way, while also informing Wilkins that he is now no less than 'Chief of the scientific staff'.[55]

Right.

Whichever way you look at it, this Imperial Expedition must be one of the least Imperious Expeditions ever 'assembled', if you can even call it that, and the only thing that prevents Wilkins pulling out immediately is his sense of honour and the inescapable fact he has given his word.

And yet the more Wilkins learns about John L. Cope, the bigger a bounder he seems. That whole thing about having been part of Shackleton's great expedition? It is only tangentially true. The truth is, once the expedition began, Cope had never seen Shackleton again, as while 'the Boss' hurtled to fame on the *Endurance*, Cope – who might as well have been called 'the Junior Underling Whose Name Nobody

Can Remember' was merely part of the party charged with dropping supplies at the other end!

Shackleton's right-hand man, he wasn't!

Wilkins is still on board but not happy, writing to the Chief Officer on Shackleton's 1907 Nimrod Expedition, Captain John Davis, that, 'Cope, in my estimation, will never be fit to run an expedition.'[56] As for the reasons why, well, Cope is guilty of 'Unskillful, unnecessary and unwarranted lying, misappropriation of money [and] making payments with worthless cheques.'[57]

In the final scheme of things, these will prove to merely be his opening remarks. Nevertheless, he does finally agree to meet Cope in Montevideo, only to arrive after seven weeks travel by sea and overland – the Cope itinerary sees him journey via Sydney, Auckland, Wellington, San Francisco, New Orleans, Valparaiso and Bueno Aires – to find that the grandly named British Imperial Arctic Expedition actually consists of three men and a dog.

And now seven more dogs.

And no further funds.

Wilkins' alarm grows, from meeting Cope in Montevideo onwards.

Cope proves to be a shameless self-promoter, a man with a penchant for filling silences with stories of his own virtue and supposed genius, a man who has never heard a sweeter sound than the lilt of his own voice, who never saw an opportunity to hang his hat on the perpendicular pronoun that he didn't take.

Yes, Cope is a medical doctor and a biologist, with real qualifications from no less than the University of Cambridge, although Wilkins will ruefully declare: 'Practically the sum of his Cambridge education was the learning of many amusing anecdotes and stories.'[58]

Appalled does not sum up Wilkins' mood, which blackens even further when he grasps the backgrounds and roles of the other two men foolish enough to join Cope. Thomas Bagshawe, is a 19-year-old student from Cambridge University, and according to Cope a 'geologist', though just a few questions from Wilkins establishes that the boy 'had no real knowledge of geology'[59], and the closest he comes is to be goggle-eyed at everything he sees. The other man is Max Lester, 22, who until only weeks before 'had been second mate of a tramp steamer'[60] but is now given the title by Cope of, if you please 'Expedition Surveyor'.

Oh, really?

And what expedition would that be?

Powered by dogs?

By men?

No, by charity.

Cope explains: 'I am held up for debt. I don't have any money.'[61]

So he is adopting a Mr Micawber approach to travel in general – always of the Dickensian belief that 'something will turn up' – and Cope's current idea is that he ask some kind Norwegian whalers to give them free passage.

The Norwegian whalers prove to be amused but not terribly charitable and have no interest in Cope's proposals.

Wilkins steps in to ask them if an arrangement cannot be made?

It takes some doing but it is done: if Mr Wilkins will promise to take photos of their whaling work to be used in publicity for the company; passage will be given.

Done!

As good as his word, Wilkins dutifully photographs the Norwegians efficiently slaughtering whales even before departure, noting with interest their speed – nine a day are diced and sliced at once – and the fascinating reason: 'Whales must be got into the pot as quickly as possible; otherwise they cook themselves. This occurs because as soon as a whale is dead the ice waters chill the outside fat, which then holds in the animal heat, and within three hours the fat inside the body is ready to be cooked.'[62]

It is not the most elegant thing that Wilkins has ever seen, but it ranks somewhere among the most brutally efficient. The meat, to be cooked and eaten, is put aside. The strips of fat are thrown into boilers, to be reduced to whale oil that will be made into bars of soap and lamp fuel, perfumes and varnishes, nitroglycerin and . . . margarine. The bones will be cut and polished for brushes and finishes on luxury items. Even the tiniest bones will become the cinches in corsets. No part of the whale is wasted, every resource that can be harvested is made good use of. From the Antarctic to a dress shop on Oxford Street, parts of this whale will travel much further than Jonah ever did.

Once it is all over and the film secured, the whalers fulfil their side of the bargain and take Cope and co. with Wilkins to Graham Land, and Wilkins momentarily feels like an explorer again, as they slowly wend their way through the ice; towering black and green rocks cutting out of the relentless beauty of ice and snow. Yes, Wilkins' mood soars

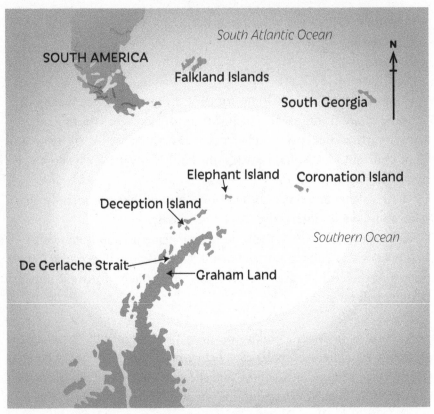

Expedition to Graham Land, Antarctica

once he sees the staggering pristine beauty of Antarctica – the whiteness, icebergs and whales – all of it so reminiscent of the Arctic that he loves so much, but the brightness does not last long.

For once the four are finally left alone on this northern finger of Antarctica known as Graham Land, making camp in a penguin rookery, and the 'expedition' proper begins, though, as Wilkins disgustedly notes, it is not actually 'much of an expedition at all – just four fellows camping in the Antarctic for a few months of summer'.[63]

Well, they must do something and Bagshawe and Lester are left at camp while Wilkins and Cope set out to explore the immediate area. Luck is not with them and the pair are dismayed to see that their way is 'blocked by an almost perpendicular cliff, with no apparent way around'.[64]

The only way to proceed, they work out is to 'cut steps for several hundred feet in the steep-walled snow surface'.[65]

Tethered together, they painstakingly cut their way up the cliff for four hours, each hand- or toehold having to be re-dug by the man following as drifting snow seems to pack each tiny crevice in an instant. Their reward for reaching the summit? Unpleasant knowledge, namely: 'the far side of the mountain was just as sheer as the one we had climbed'.[66]

What to do now?

Retracing their vanished steps in this weather seems a fast path to vanishing altogether. Descending the new cliff looks to be more dangerous still. And so a third option is taken. Very carefully they clamber along the narrow high ridge, to go higher up the mountain in the hope that an easier descent will be seen.

With oblivion beckoning on either side, Wilkins elects to crawl, on the reckoning that while four points of contact is undoubtedly undignified it is twice as safe as just two, especially given that he only has Cope as his safety man. The difficulty is working out what is solid and what is just loose snow, where to put his weight because with one error he will be . . . falling?

Falling.

F
  A
   L
   L
    I
     N
      G
       !

It is not unlike the feeling of raising your feet in anticipation of the final step at the top of a flight of stairs in the dark, only to find . . . nothing. Wilkins becomes a victim of the void, suddenly freefalling to his fate, which is likely . . . death. And so it would be, if not for his indomitable resolve and quick thinking. Swinging his arms out wide in search of purchase, some rock, some shard of ice, *anything*, he somehow manages to grasp the cliff face in his hand. Cope, formerly behind Wilkins, is now beneath him, falling and tumbling as ice and snow follow him as he was following Wilkins only moments ago. In a quick flash of steel, his ice-axe comes flying out and bites its cold teeth into the ice wall, rapidly slowing his fall.

But it is not enough, and still, they continue their descent:

> Down we went at great speed, first on our faces, then on our sides and on our backs – over and over, trying our utmost to control our downward slide with our ice axes. But our pace was too fast, and the slope was rough.[67]

All they can do is slow themselves a little.

By pure happenstance Wilkins finds that he is bumped 'into a sitting position',[68] which makes him clasp his arms around his knees, which thus gives him more speed, almost as if he is sitting up in a toboggan.

The two men keep roaring down the mountain, scraping, scratching and bumping further and further, feeling rather as though they have fallen into the Antarctic's answer to the rabbit hole in *Alice in Wonderland*, an endless falling that neither can quite believe: 'We careered on for what seemed an endless time. When our speed slackened somewhat, I gathered that we were either coming to the bottom or – and my heart almost stopped beating at the thought – we might be coming to . . . a crevasse.'[69]

At last Wilkins' trajectory reverses. Moments ago he had been falling further and further down, ever closer to a gaping maw of dark ice, but now he feels a new sensation. It is a tightening of his chest and stomach, of his breath leaving him, as his vision begins to darken, and fade to black, before . . . after a painful pause . . . he hears a distant din in the dim . . . the sound of a man on a faraway shore. It's Cope. And he still needs help!

'Wilkins, are you there? *What's happened?* Come and help me! Pull me out of here!'[70]

For a moment Wilkins thinks that he is dreaming, it takes another few moments for him to get enough breath to reply.

'Hang on!'

(It is, literally, excellent advice, but Cope has thought of this already, even if he is now right at the end of his tether.)

'I'm okay. I'll get to you in a minute.'[71]

Rudyard Kipling comes to mind: *If you can fill the unforgiving minute with sixty seconds worth of distance run . . . and so hold on when there is nothing in you, except the will which says to them: 'Hold on' . . .*

While Cope indeed holds on, Wilkins uses his unforgiving minute to dig his ice-axe into the snow, tie a rope to the axe and then carefully move forward, over to the edge of the snow ledge in front where Cope is now hanging upside down, about four feet in mid-air. His end of the rope is just sliding down his hips, and Wilkins regards it as a miracle that he has not already, 'slipped out of the loop and vanished into the deep, blue crevasse, the bottom of which I couldn't see'.[72]

If Cope does fall, he will tumble at least a hundred feet to his certain death. But, having lost his own axe in the slide down, the Englishman is completely helpless and entirely reliant on the exhausted Australian for his salvation. The chores, choices and chances are grim, dim and slim.

Wilkins furiously goes over his options. He could make an attempt to pull Cope up, hand over hand, but he is far from certain that his own ice-axe will hold its tenuous grasp if he does so, and the likelihood then is that they will both fall and die.

And if Cope tries to right himself and climb up – presuming he can do either – the same result may occur.

There is, of course, the infamous last resort of the mountain climber: cut the tethering rope, to release the weight of your partner and save yourself. In this case, Wilkins would definitely save himself, Cope would definitely die, and no-one would ever even know the tragedy of what had happened or the shocking hard decision that had to be made so that one of them would survive. But for Wilkins such an option is unthinkable. There must be another way.

Yes, that's it.

Cope, I am going to get the knife to you, and you must swing yourself to the mountain side then cut holds in the face, so you can climb up yourself. (Of course, if Cope drops the knife then Wilkins will be tied to him for good, and for eternity in a few hours – but there is a chance this will work, and Cope agrees.)

Thinking fast and acting faster, Wilkins tears a strip from his shirt and ties one end to the knife. The other end is looped onto itself and fastened to the rope.

Slowly, gently now, Wilkins pulls the rope so the knife edges its way towards Cope, held aloft by no more than torn and tattered cloth. But Wilkins knows his knots, and the knife remains steady, making it all the way to Cope without trouble. Just one problem; Cope is still upside down, struggling to right himself. But the arrival of the blade is hope enough, and before long the nominal expedition leader is back on his feet, so to speak, gingerly grabbing the steel in his cold fingers before driving it into the cliff face.

Again and again he stabs, each desperate thrust showering him in a spray of ice and rock as he painstakingly crafts his own footholds. Beads of ice-cold sweat drip slowly down Cope's face, only to freeze and fall off. Wilkins clutches his end of the rope as tightly as his hands will let him, his knuckles bright white with the effort. He prays the serrated teeth of his axe will hold their grip. It is not a long climb, but it is a hard one, every upward movement requiring a fresh blast of icy shrapnel as Cope gets closer and closer to the summit. The rope is as strained as the men's nerves, though thankfully not as frayed – for it holds. Finally, one of Cope's arms shoots over the ridge, though it may as well be a claw for the way it frantically scrabbles for purchase. Wilkins grabs it, the rope in one hand and Cope in the other, and *HEAVES*. The men collapse on their backs as their lungs burn and their fingers sting, their bodies aching for a moment's respite. It is the most physically draining ordeal they have ever been through, but it is over.

'Thank God! We've made it!' Cope yells, before sitting up and seeing exactly what Wilkins has been wanly gazing at for an hour; crevasse after crevasse after crevasse, onwards ever onwards, stretching out in an infinity of risk before smooth snow and safety can even be glimpsed.

'The sight was almost too much for him,'[73] Wilkins notes. The completely crumpled Cope has gone from elation to despair in a horrible second of clarity. Yes, they are both alive right now, but in some ways all they have done is an Antarctic version of climbing out of fire and into the frying pan. They have prevented their immediate rapid death, but quite possibly only to give them a long, agonising and freezing one. For now, the pair try not to despair and rest as best they can for three hours before making the next attempt to save themselves. At the end of that time they know they really must move. If they stay still they will freeze to death even before they have a chance to starve. Lester and Bagshawe have no idea where they are and no hope of reaching them even if they do.

They painstakingly crawl through the crevasse, a labyrinth that light doesn't quite penetrate. It is dark, and it is cold, and they are weary, but they must persist. The chill reaches their bones this close to the ice, but they are warmed by their certainty that a single mistake can be fatal. Wilkins has no energy for mishaps left, and no stomach for any further adventures on this day. Inch by inch they move forward, Wilkins leading and Cope coping. They grip nothing unless they know it will hold true, put no weight down on anything that risks collapse. It is slow going, but they go on all the same.

Navigating by Wilkins' usual method, a combination of pluck and luck, at long last they reach the top of a glacier from which point – praise the Lord! – a path to the promised *land* beckons. (Exhausted beyond measure, they nevertheless must proceed with extreme caution lest they lose their footing and slither to their doom. Finally, the icy steepness gives way to nearly horizontal ground – and the piles of rubble here and there say it really is exactly that, *ground*.)

With infinite relief, and their last reserves of energy, they make their way back to the harbour to soon be greeted by the positively delightful sight of the camp they were convinced they would never see again.

Their companions greet them in a straightforward manner as they are, of course, completely oblivious that, rather than merely two men before them, Wilkins and Cope are nothing less than two miracles. But Wilkins is only too aware of how close they had come, and will never forget this day. His verdict is final: 'I've never really enjoyed mountain climbing since.'[74]

The ensuing weeks are a slow agony for Wilkins with so precious little to do. Apart from mapping 30 miles of the coastline, his sole achievement is to find a succession of 500-foot-high sheer cliffs so severe that not only would no sane man, who cares for his life and limb, climb them but . . . nor even will Wilkins. And after his last experience, Cope declines to even look in their general direction.

The contrast between their essential inability to move far from their icebound actuality and the *12 planes* originally promised by Cope is not lost on Wilkins. More than ever he feels that planes are the future of the polar exploration. What they are doing now – cooling their heels on the edge of icebound Antarctica – belongs to days long gone.

Finally, after a long discussion, it is decided that Cope will return to Montevideo with the intention of securing a ship and additional help – his idea being to return the following season and transfer the base to a more suitable site. Wilkins is adamant in declaring that his intention is to return home on a whaler.[75]

In short, the merciful decision is taken to give up this farce and return home.

Bagshawe and Lester, however, are just as adamant that they wish to remain. They accept Cope's assurance that he will endeavour to come for them in February the following year, 11 months hence.

While Bagshawe is left to tend camp, Wilkins, Cope and Lester set out to the north in a small open boat looking for a boat – very likely a whaler – large and kind enough to take them slinking back home. Cope and Lester spend most of their time doing what they have been doing so much of in recent months, the thing they are most qualified for – sleeping – while an annoyed Wilkins has the watch and control of the boat.

And now what? In the polar twilight, Wilkins – a child of the desert, born to dust-devils of willy willies swirling all around – suddenly sees thin sprays of water rising all around him. And they are moving! As Cope and Lester snore, the spray increases more and more until Wilkins realises that it is not some hitherto unknown Antarctic phenomena but in fact is . . . a school of whales surrounding them! Initially, they seem more curious than anything, with one of them even gently nudging Wilkins' hand as it rests just over the stern. Recoiling, Wilkins turns to see that it is a killer whale, immediately identifiable by the sheen of its black top and the flash of white behind the dorsal fin.

Good God!

The killer whale in question seems to feel much the same, for after that first warm touch it swirls, whirls and tightly twirls beneath the fragile craft, rocking and knocking it so violently that water pours in from both sides in sudden succession. The most amazing thing? Cope and Lester don't wake! Yes, they are being monstered by a school of killer whales, for the turbulence soon spreads to the others, as angry snorts and thrashing whale tails are all around them, and the other two men are still in dreamland while he is living a nightmare.

'WAKE UP!' Wilkins yells and the startled Cope and Lester quickly do just that, appropriately alarmed to have gone from the Land of Nod to the Ocean of Menace in a split second. The minutes tick by like hours as these massive mammals continue to cavort all around them, rollicking and rolling the boat, snorting their contempt in clouds of steam; circling, churning and *toying* with them. With just one swish of a tail, their own tale will come to a tragic end, and there is nothing that they can do but wait for their doom or deliverance. At last, while they are all caked in a sweat of pure fear, they see the water ceases to swirl and the whales have vanished.

'Luckily for us they were not too set on an explorer dinner,'[76] Wilkins chronicles.

Badly shaken, all of them are relieved only a short time later to see a Norwegian sailing vessel come into view, some 40 miles from their original camp.

Wilkins climbs aboard the vessel, only to find that, while the captain is welcoming to him, there is one among them he cannot abide.

'You and Lester are welcome but I will *not* have Cope at my table,'[77] the skipper says flatly.

Wilkins groans inwardly, the best he can, not saying a word, but very clear in his thoughts.

Ah, you too? You must have met our skipper?

Not actually, no, but the story soon emerges.

The Norwegian whaler had been at a place called Deception Island some time ago, you know of that, Mr Wilkins?

Yes, Wilkins is more than familiar with it.

Well, several months ago an arrogant man called Cope had come aboard the vessel asking for the captain. Advised by the lowly steward that the captain and his senior men were not on board, no matter, Cope – with extraordinary presumption – had ordered the steward to cook him a meal and fetch him a bottle of whiskey.

And by the time the captain had returned, it was to find both Cope and the whiskey gone.

'I feel,' says the Norwegian in a way that lets Wilkins know this feeling is deep indeed, 'that Cope should not order my people about in my absence.'[78]

Wilkins entirely agrees, and tries his best to mollify the captain with some gentle words but the Norwegian has no interest. To have done this on his *own boat*! Taken his own *whiskey*! Ordered his own lowly steward about!

'On his previous visit, Mr Cope would have been offered all the hospitality in the world, if he had only waited a few minutes.'[79]

But he hadn't. So now he shall have none.

No dining at table for Cope. He can eat in the galley.

When Cope is told the news by Wilkins he is furious.

'I refuse to stay on this ship,'[80] he says, which suits the Norwegian captain just fine.

In the end there is nothing for it but for Wilkins and Lester to follow Cope and re-board their small boat and go in search of another ship, at Nansen Island, one that their leader has not yet insulted – which, under the circumstances, will be about as easy as finding an Antarctic bay without ice.

Alas, no sooner are they alone again in de Gerlache Strait than they are hit by a huge storm. Not four hours ago they were safe, secure, warm and dry in the belly of a comfortable Norwegian whaler, and they are now three men up to their waists in freezing water, ditching supplies and bailing as if their lives depend on it – because they do – all while trying to avoid jagged rocks that would cut them in twain if they fell at the wrong angle, and impale them if they fell at the right one.

The wind howls like a banshee, the waves rise and crash over the sides. Bail! BAIL!

Thoroughly soaked and shivering from the bitter cold, Cope and Lester bail even as Wilkins works the tiller and, after first sailing into the wind trying to get to the other side of the storm, and when that does not work, running before the storm – if you can't beat it, try and run away faster than it can chase you – hauls on the tiller right and left as rocks of doom loom. Over the next three to four hours they sail no less than 45 miles at an average rate of 17 knots – so fast they shoot past a small cove where they can see a whaling boat has wisely taken shelter.

In a smaller wind they might have tried manoeuvring into that cove themselves but it is impossible, and right now Wilkins has more urgent things to do.

For Wilkins must WATCH OUT FOR THE ICE CLIFF DEAD AHEAD!

It is no more than a hundred yards away and they will crash straight into it unless urgent action is taken. Wilkins thinks for all of five precious seconds before taking desperate action. There is no time to tack either way, or even take the sails in again, and so, taking his knife in hand, he leaps forward and starts slashing wildly, happily reducing the bottom part of the sails to mere ribbons. Parts of the sail now fall on him, blinding and binding him as the small boat 'rolled and twisted like a thing alive'[81] in the surging sea and wild wind.

At least they have slowed before hitting the cliff, which saves them for the moment, but in this wind and these waves, unable to proceed in any direction, it can only be a matter of time before they capsize and all will be lost.

And now they see it.

Perhaps there is a God after all.

It is a small motorboat with three crew on board, sent by the whaler they have just whistled past. The captain had seen them, and the law of the sea had applied – render assistance to all, whatever their flag.

Can we help?

Yes, you can!

A tow rope is thrown, and the three drenched, shivering men burst into a chilly laughter, each guffaw warming their souls if not yet their bodies. It has been another day of disaster averted; which seems like the only kind of achievement Cope's expedition will ever own, always carefully navigating their way from the fire back to the frying pan.

Once on board the whaler – it proves to be Norwegian, the *Solstreif* – Wilkins intently examines the captain's face, to see if he recognises either Cope's face or name.

Might they be cast off once more?

Happily, no, the captain proves to be humanity itself and even agrees that, once the storm has abated and they have been fed and warmed, his whaler will tow them all the way back to Bagshawe's camp and then take them all back to a port where passage back to England can be arranged. (They are in such a sorry non-ship shape the captain also has their humble lifeboat hauled up on deck to be re-caulked and given a coat of paint.)

It is a wonderful offer but to Wilkins' stupefaction Cope has other ideas. After all, he has reasoned, if Bagshawe and Lester and Wilkins stay in Antarctica, somewhere vaguely hospitable in Graham Land for the winter, the 'expedition' can be said to be *technically* alive. They would not be required to do anything, just sit there as Cope goes back to Europe or America or Britain and raises more funds.

Wilkins finds the proposal insulting and the very idea ridiculous and says so with some force.

'As far as I am concerned, the idea ends right here,'[82] he tells Cope. However, a charming and persuasive man when he wants to be, once they return to the camp Cope does manage to convince both Lester and Bagshawe to go along with it – neither man has anything better to do, which helps – and there is nothing that Wilkins himself, the captain of the whaler, and the whaler's doctor, can do to make them see the absolute folly of what is proposed.

'I thought it was only a waste of time. They had no instruments to carry out work through the winter . . . They had seven dogs to amuse them.'[83]

The only thing Wilkins does manage is to get them to agree to a partial compromise, in the hope that it will break the spell that Cope has clearly cast over them: 'You may remain two weeks and see how you like it. After the two weeks, if you still want to stay, I will withdraw my objection.'[84]

Done.

After they all shake hands on the deal, the two young Britishers are left on shore with ample supplies, ample dogs and ample penguins, as they have chosen to camp at the penguin rookery. Wilkins and Cope travel on to the nearest whaling station, on the promise made by the kindly captain that another of his company's boats will stop in on the deluded duo before they all head back to Norway. Perfect. In two weeks, Wilkins knows, that vessel will arrive with Lester and Bagshawe begging to come home. In fact, however, when the last Norwegian whaler returns to base two weeks later, they are not on board.

It is only a few days later when, well on their way back to England, Wilkins learns the horrifying truth. Cope had got to the original Norwegian captain and threatened that if any whaler visited the men, he would call his friends at the Colonial Office and have their whaling licence revoked in British waters for daring to interfere with a British expedition under his command. And so the duo had remained there, alone, with no visit from anyone. [85] Wilkins is so furious, and so appalled at Cope – what an inhumane and deceitful BASTARD he is – that he asks the captain to turn the vessel around to go and find them, but the captain in turn refuses. The Cope threat had been real, and it is a genuine risk to the entire Norwegian operation in these waters.

There is nothing Wilkins can do but to accept it, his only consolation being that he knows for a fact the two have more than enough supplies to survive for at least a year until the next whaling season begins, as well as a hut, warm clothes and even 'a Victrola with records'.[86] And there is one other thing in their favour when it comes to survival: they do not get on. For Wilkins has observed a peculiar truth of isolated life at the Poles; friends will be at each other's throats, while enemies will contentedly live with each other. Why? Well, he has a theory: 'Any two men who have an angry row in the morning and forget all about it by the afternoon can get along together anywhere for any length of time.'[87]

If they liked each other, well that would be a different story: 'If they had been boon companions, with never a cross word between them, I would have brought them away at whatever cost.'[88]

And yet, despite Wilkins strongly feeling that his own part in this expedition is over, the Governor of the Falkland Islands, when they arrive on that remote British outpost in the Atlantic Ocean – their first port of call after leaving Antarctica – takes an entirely different view. For no sooner has their boat docked than the Governor, His Excellency Sir John Middleton, demands to see Wilkins and Cope immediately.

What, pray tell, Mr Wilkins, is *going on*?

The Governor has been in touch with the father of one of the young men that you, Wilkins, have abandoned in Antarctica, despite the fact that the father had only consented to his son going in the first place if Wilkins stayed with him. This is news to Wilkins, and Cope insists – for what that is worth – that it is news to him, too, but the Governor takes a very dim view of men who would abandon their comrades in such a shoddy manner.

Agonisingly aware of the irony, and even his own craven hypocrisy – but it must be done – Wilkins now finds himself passionately arguing Cope's case and insisting that, as there is a whaling station only 30 miles away from the two young men, if rescue is required, there is no issue, Guv!

But it alters not a jot the Governor's view – in Antarctica 30 miles might as well be 300 miles. If anything went wrong there would be no help at hand, and no way of them letting the outside world know in any case. Very well then, might Wilkins gently point out that as Cope is the head of the expedition and is the one who made the decision, he is the one who the Governor should be talking to?

But His Excellency is way ahead of him.

'Although Cope was nominally the expedition leader, the officials knew I had more experience than Cope and was of a different temperament, so I should bear the blame.'[89]

Wilkins narrowly escapes arrest, but it is a close-run thing. Under the circumstances, Wilkins cannot get away from the Falklands in general, and Cope in particular, quickly enough. The whole thing has been a fiasco from first to last, and he is now, formally, finished with it. He advises Cope that he has changed plans, is resigning from any further involvement with this debacle, and will book his passage to America, where he intends to do precisely what Cope had promised to do in the first place – but had so comprehensively failed to attempt – and explore Antarctica by air.

Cope is sorry to hear it, but magnanimously accepts the decision, and grandly tells Wilkins that he shall go to England, obtain a steamer and remount the expedition.

'What he actually did,' Wilkins will subsequently report, 'was to hang around the Falklands until the British authorities grew impatient, got him a job peeling potatoes on a Scottish steamer, and sent him home. His latter career was in medicine, not exploring.'[90]

(As for Bagshawe and Lester, they spend a full year in the Antarctic, waiting contentedly. A ship dutifully called for them on 18 December 1921, but having many things to clear up, they did not wish to depart immediately. Arrangements were made, however, to remove the dogs to the ship.

Oh, there is one thing they would like from the whalers; forks. It seems Cope forgot to leave them and eating their meals has been rather a chore without them.

Finally, on 13 January 1922, the ship *Graham* had arrived to carry them to safety, and it became apparent that an odd thing had occurred in the course of their year in isolation together. They are now bosom buddies!

'My companion in this adventure,' Bagshawe will later recount, 'was one of the best and I shall always remember Lester with happy thoughts. As I write, memories are conjured up of the pleasant hours we had together, the laughter and happiness, the help and condolence in times of trial and tribulations.'[91])

Reaching New York in May 1921 Wilkins is as good as his word, and immediately starts negotiating with the German company Junkers for some of their planes to fly to the polar regions but, in a curious turn of events, it is Cope's former employer who derails his plans. For, out of the blue, Wilkins gets a cable from Sir Ernest Shackleton with two purposes: a warning and an offer.

YOU CANNOT AFFORD FOR YOUR OWN GOOD REPUTATION TO BE
ASSOCIATED WITH AN AGENT FOR GERMAN MACHINES. WILL BE RUINED
FOR LIFE IN THE BRITISH EMPIRE IF YOU USE GERMAN MONEY. AM

STARTING NEW EXPEDITION TO ANTARCTIC WILL GET YOU BRITISH
PLANE IF YOU JOIN. [92]

Sir Ernest himself makes Wilkins the offer of a position as naturalist and photographer for the expedition, no doubt because of the Australian's reputation, but also, Wilkins suspects, so the great British bulldog of a man can eliminate any competition in the Antarctic. The fact that the Germans now wish to delay any possible machines or funding for at least a year makes the decision easy. He agrees to Sir Ernest's terms on one condition, that, as well as photographer, the great man will regard him as an apprentice expedition leader; a condition Shackleton readily agrees to.

Shortly thereafter Wilkins arrives in London to join 'the Boss' – as all his men call Shackleton – barely daring to believe that he can be in the great man's presence. This is Sir Ernest *Shackleton*, the greatest living legend of the British Empire. He had joined the expedition of Scott of the Antarctic back in 1901 as, effectively, the 29th and last selection, more anonymous than a lost dog – but finished that expedition as one of the two men Scott had selected to venture into the Antarctic interior. Falling out with Scott, Shackleton had returned to Antarctica himself in 1907 at the head of his own expedition and got to within 99 nautical miles of the South Pole before – after doing his calculations that they did not have enough food for *all* of them to get to the Pole and back alive – he had turned back. It had been vintage Shackleton – always putting the lives of the men first. And then, returning to Antarctica in 1915 with 48 men, Sir Ernest's ship *Endurance* had become stuck in the ice. After extraordinary adventures involving leadership and derring-do unheard of in the annals of British history, Shackleton had guided his men from ice floe to ice floe to open boat to the remote Elephant Island, where he had left 22 of them, before taking five men and journeying 720 nautical miles across the fiercest body of water in the world, the notorious Southern Ocean. And then things had become difficult ... After successfully navigating to the tiny pinprick of an island which is South Georgia, they had, again against all odds, scaled cliffs and journeyed to the other side of the island, to get to a whaling station to get help for the men still marooned. In the end, Shackleton had saved every man-jack of them.

Knighted for his many efforts, Shackleton is as venerable as they come, and Wilkins is mightily impressed by him.

'Sir Ernest Shackleton,' he will later say, 'was the most beloved leader of men I ever knew. Even his enemies couldn't help liking him.'[93]

On this latest venture, Shackleton wishes to explore the coast west of Enderby Land and some of the little-known sub-Antarctic islands.[94]

'It was a hybrid affair,' one writer will characterise it, 'a strange medley of the quest of lost islands and of the exploration of once-glimpsed ice-infested coasts such as no other polar leader than Shackleton, with his Celtic imagination, sense of the dramatic, and faith in his star, would have conceived.'[95]

And yet it is also true that, these days, Sir Ernest has been so wearied by the treks, the blizzards, the sea journeys, the disappointments, the unrealised plans that, even though not yet 50 years old, he looks to be near the end of his tether even before this journey is begun. He is drinking heavily, seems uncharacteristically morose – when his hallmark had been an eternal optimism no matter the situation – and a sense of fatalism invades his actions and thoughts. Yes, he can still dazzle, and is able to regale the assembled company with long poems by Wordsworth, Shelley, Keats and Kipling, oh, all right William Wordsworth it is – '*Though nothing can bring back the hour, Of splendour in the grass, of glory in the flower; We will grieve not, rather find, Strength in what remains behind . . .*' – for hours at an end and for all occasions; but there is more than one such occasion when his words slur, he loses his way, and the eyes of crewmates meet.

As it happens, the expedition ship Sir Ernest has chosen for this venture is rather in his image: back in the day it was no doubt really something. But now the times seem to have rather passed *Quest* by. It is some 112 feet long for 132 tons, and Wilkins has been around the block often enough by now to know a dud when he sees one. Just one glance at her patchworked wooden hull, threadbare rigging and filthy, leaking, tired engines is enough for him to acidly note that, of the agent who purchased it, 'I believe he must have been drunk and seeing double when he bought her.'[96]

The ship is supposed to steam at seven knots, but on brief trips out into the English Channel it becomes apparent that its real limit is just four and a half knots, and even that would likely be shooting down a wave with the wind behind it. Meanwhile, endless additions and 'improvements' that Sir Ernest has had made seem to have made her top-heavy and barely able to stay right side up. Her green timber swells when wet, a serious misfortune in a vessel permanently at sea, the engine regularly burns out, the propeller has real problems in propelling and, as things turn out, the ship is so very *not* ship-shape it has to be towed into three separate ports . . . even before they have left England.

Shackleton seems unfazed, and the whole venture is launched in a blaze of glory only possible for England's favourite son, as the photographers rush, the press gush, and the whole thing includes a newspaper competition for one lucky Boy Scout to join the crew. Typical of Shackleton's generosity, he ends up taking two such scouts as the final contenders are so outstanding he can't bear to leave one behind. The Boss also does such things as promising to search for Captain Kidd's lost treasure in Trinidad on the way back from the Antarctic. The press and public lap up every bit, hungry for more.

Once launched properly and on their way to Antarctica – ostensibly via the Azores, Cape Town and Rio de Janeiro – *Quest* continues to misbehave, to the point that the daily mishaps soon become hourly mishaps, and Wilkins watches, fascinated, as Shackleton leads his men and builds their trust with each setback.

Shackleton's method is simple. Whenever a problem appears, he quietly takes each crew member into his confidence, one by one, and says: 'We're up against it. Now *you're* the one man I think has some *real* ideas about this job, and I wish you would tell me how to go about it.'[97]

And just as it was extraordinary some 150 years earlier how it was one of the most junior crew members of the *Endeavour* who, encouraged by Captain Cook, came up with the innovative solution of 'fothering' – using a cover of canvas pulled under the hull to block a hole – to get the ship off the reef it struck off the north coast of Australia, so too is it extraordinary how often one of Shackleton's newly empowered crew members makes a contribution to solving problems.

For, time and again, the honoured crew member then humbly offers an idea, any idea; whereupon Shackleton says, 'Good! That's a splendid idea and I'll use it. Thanks a lot.'[98]

Wilkins watches this performance repeated shamelessly, including with himself. Surely Shackleton can't forget that he has enacted this play before with the same man? One day Shackleton starts in on Wilkins with his, 'We're up against it' speech and Wilkins, amused, cuts him short:

'Boss, I am quite familiar with your scheme and it probably won't work quite so well this time!'

Shackleton has a different view, even though his bluff has been called. 'My boy, don't you make that mistake again!'

Mistake? Oh yes, the Boss explains: 'I'm going to tell you something. Remember this: There's an effect in the spoken word. You may think you won't be flattered, but you're wrong. Even though you know why I'm doing this, you will be flattered because there is magic in the spoken word.'[99]

The Boss is right. Even though they know the trick being pulled, and soon enough that truth dawns on even the dimmest crew member, they all find their spirits lifted by such a session with Shackleton.

Quest however resolutely refuses to have her spirits or speed lifted. They fall so far behind their schedule that they have to abandon the reason Wilkins had been so keen to join the expedition – to have a plane fly over Antarctica – as the plane and winter gear were to be picked up in Cape Town and, sadly, George, we simply don't have time to get there and still get to Antarctica for the summer months, I'm sure you understand?

As Wilkins despairingly notes: 'We staggered into Rio de Janeiro under sail, helped out by an occasional kick from the propeller.'[100]

Now, given that repairs to *Quest* are as urgent as they are substantial – it will take at least six weeks – Wilkins and the expedition's geologist George Douglas are quick to get permission from the Boss to go ahead to South Georgia on a whaler that is shortly departing so they can take photographs and collect specimens in a place far more interesting than Rio. Of course. The Boss farewells them with a warm handshake and a cheery wave of his hand. What he most admires in men under his command is the spirit of wanting *to get on with it*, and Wilkins and Douglas certainly qualify. Permission granted.

The two, thus, are able to spend six happy weeks in South Georgia, with Wilkins particularly delighting in exploring this windswept rock in the deep South Atlantic, capturing specimens, and even trying his hand as Antarctic gourmet. While his albatross egg omelette is a great success, his Christmas duck lunch is quite a lot less so. (Alas, cooking in poor light, Wilkins had mixed up his seasonings, and instead of salt had put in arsenic preservation powders. Two days of curses and oaths and stomach pains make for an unhappy Christmas and Boxing Day.)

Happily, the men are fully recovered by 4 January 1922 when the word comes that *Quest* has pulled into the port, and they can rejoin their comrades on the morrow.

And yet, as they row out towards the vessel mid-morning, moored just off-shore in Grytviken Harbour, could it be that something is amiss? For Wilkins notices a curious thing: the British Red Ensign on *Quest* is drooping sadly at half-mast, as are all the flags on the other boats in the harbour.

Quest is still and Wilkins yells out as they draw alongside.

'Hey there, on board! You all dead, or asleep?'[101]

A sole sailor appears on deck, the ship's navigator Douglas Jeffrey.

'Shackleton is dead,'[102] he tells them simply. A heart attack, striking as the Boss was settling in to sleep, having jestingly asked one of the men to play him Brahms' lullaby on the fiddle.

'This meant so much,' Wilkins will recount, 'that I could not at once grasp it; the dull shock numbed my senses and left my mind a void . . .'[103]

Dropping their kit, they hear the story.

His heart attack had been every bit as devastating as it was unsurprising. Even on his last two expeditions, the medical officer Dr James McIlroy had seen signs and wanted to put a stethoscope to his heart, but the Boss had growled his refusal.

And during this latest expedition, Sir Ernest had been so under the weather in Rio that the expedition doctor, Alexander Macklin, had been so bold as to suggest he stay behind in hospital, or even head back to Britain, but of course the great man wouldn't hear of it. He had insisted he would be right as rain and so Macklin had to content himself with keeping a close eye on him. And it had been in the early hours of this very morning that, while passing Shackleton's cabin, he heard a whistle from within. The good doctor had entered to find the leader looking gravely ill, and yet covered with but a single blanket. This lion among men, capable of taking on the worst weather the planet could throw at him, of traversing vast distances of the ocean in an open boat, and even greater distances pulling a sled across frozen wasteland, now did not have either the energy or the will to get another blanket from the bottom drawer, even if he'd wanted to.

Finding him in an unusually quiet mood, Macklin had judged the time was right to again suggest Shackleton take things more easily.

'You are always wanting me to give up something,' the Boss had responded, 'what do you want me to give up now?'[104]

At which point, Shackleton's own heart gave up. He closed his eyes, let out one last heavy breath, and was gone. Just 47, this giant of his generation, had effectively died in action.

At the request of his widow, the great man will be buried on this remote island at the foot of the mountains, by the harbour – it would not be right to consign him for eternity to some quiet suburban cemetery or even, perhaps, in Westminster Abbey. A funeral cairn is made, it is not easy, in more ways than one, his men 'quarried rocks out of the frozen soil with our own hands to build his monument'.[105]

With reverential care, keenly aware of the honour that rests upon them, the carpenters erect a large cross atop the cairn. With no man of the cloth among them, a simple service must suffice before the men gather as their photograph is taken, and for once Wilkins is able to be part of the picture instead of taking it, using one of those devilishly clever new inventions, the Kodak self-timer. He stands closest to the cross, stoic, as the hollow eyes of the men greet the camera.

Their leader is dead but their adventure is not over, as their orders from London are to continue.

And they try. And yet, the vacuum at their helm cannot be filled . . .

'There was no greater leader of men than Sir Ernest,' Wilkins will note, 'or a more sympathetic companion. Without his help we were like flint without steel – we had no fire.'[106]

In any case the ice has its own intentions, locking *Quest* into the Weddell Sea off South Georgia for the winter – Wilkins had not only seen this movie before, but shot a fair bit of it – before the ice relieves its deathly grip on the condition they concede defeat and sail for home.

Done and done.

Dispirited, dejected and, yes, defeated, all they need now is a deathly storm that nearly sinks them and, sure enough . . . here it is. *Quest* is in that singularly rough stretch of water between Elephant Island and South Georgia, when a hurricane hits very suddenly – *All hands on deck!* – and they must scramble to stow what can be stowed, tie down the rest, get the dogs safe, and batten down all the hatches. Typically, Wilkins does not let others do the hard work that needs to be done, even though he is surrounded by professional sailors, and his skill immediately impresses.

'He was an amazingly hardy sort of fellow,' one of them will recall, 'and could furl a topsail as well as any sailor by profession.'[107]

Which is as well, for now the wind howls, the waves get ever higher and start crashing over the bow, nearly swamping them time after time.

And watch now as one rogue wave, seemingly coming from nowhere, but in fact a surprise attack on their starboard flank, crashes over the side and ... throws one of their most highly regarded expedition members into the roiling ocean.

'Wilkie's gone!'[108] cries one sailor.

And just as they all look around to confirm the truth of that ghastly death sentence – for there is no possibility of rescue or even locating Wilkins in this tempest – yet another giant wave hits *Quest*.

And it deposits Wilkins back on the same part of the deck whence he was plucked, as the *New York Times* will record it, 'pretty wet but not at all discouraged'.[109]

(Not *quite* as miraculous as it first appears; Wilkins will later tell one of the expedition members that while in the process of being swept away he had managed to grab a rope.)

To anyone else it would be the story of a lifetime; Wilkins does not even bother to record it in his notes on the expedition.

Quest eventually limps up the Thames on 16 September 1922, and no-one is more pleased to be back from this wretched expedition than George Wilkins.

CHAPTER TEN

FROM THE KREMLIN TO KAKADU

Oh, the new-chum went to the back block run,
But he should have gone there last week.
He tramped ten miles with a loaded gun,
But of turkey or duck he saw never a one,
For he should have been there last week,
They said,
There were flocks of 'em there last week.

Last Week, Banjo Paterson, 1893

November 1922, London, I spy with my Quaker eye

The old rule applies. There can be no rest for the wicked, or the *wunderkind*.

For no sooner has George Wilkins arrived back in London than (after showing off his beard and proper explorer hirsute credentials to young Betty Dyson) he throws himself into writing a paper for the Royal Meteorological Society. It is on a subject he has been musing on for over 10 years and fiercely promoting: the possible establishment of occupied weather stations in a ring around both the North and South Poles, no fewer than 46 of them, supervised by 'an international bureau which would supervise the selections and training of observers for the outposts and to decide the nature of apparatus to be used at each station.'[1]

Briefly, his theory is that 'the weather in the southern hemisphere is largely determined by the weather conditions prevailing over the Antarctic continent, and if regular bulletins could be issued thence to the meteorologists in Melbourne, Cape Town and Buenos Aires, weather forecasts could be both lengthened and improved.'[2]

Given the benefits to those nations they could pay for these stations.

And yet, it is while his proposals are being mused over at appropriate bureaucratic length that Wilkins receives a most unexpected offer, signed off by a very particular organisation led by a remarkably singular man.

A decade ago the British had established the Secret Service Bureau – soon to be MI6 – devoted to gathering 'intelligence' on what was happening in the corridors of power as well as on the streets around the world. It is run out of Ashley Mansions in Vauxhall Bridge Road by Sir Mansfield Smith-Cumming, a monocled grandee like they just don't make them anymore – a man who signs all his letters with a 'C.' in green ink, has a wooden leg and a favourite prank, which is to stab said leg with a knife or fork and not react as fellow diners recoil. Such jolly japes aside, he is an astute judge of where danger lies, and has recently become concerned by the lack of information coming out of the newly established Soviet Union under Vladimir Lenin.

And so the call has gone out.

In his London hotel, George Wilkins picks up the phone to hear the voice of a friend high up in the British Government.

'Would you do a substantial favour for me, Wilkins?'

'Any time. What is it?'

'I want you to go to Russia for me.'[3]

Do tell?

'I . . . I'll consider it,' says Wilkins 'Can't you tell me more? Why do you want . . .'

But this Government man refuses to talk about the whys and wherefores of this highly sensitive matter over the phone.

'I want you there for six months. Pack a bag and come over to see me tomorrow. I want you to take pictures of course. You'll be off in no time.'[4]

Things start to fall into place and, as so often happens, while the risks of the venture are overwhelming, the potential rewards in terms of fascinating new experiences make it unrefusable. Broadly, if *extremely quietly* – for he will be nothing less than a spy – the British Government will pay him well to travel through Russia and report on just what is going on.

For cover, Wilkins will nominally be taking photographs and footage for the 'Society of Friends' Emergency and War Victims' Relief Committee' – a Quaker organisation of great repute – to highlight their famine relief operations. Unofficially, the government of Lloyd George wants Wilkins to travel to Bolshevik Russia, to photograph the reality of conditions, 'to make confidential reports of another nature for the government'[5] on the world's first communist regime. Most intriguingly it has been arranged for him to have an audience with Vladimir Lenin himself.

Now, as a 'blind', Wilkins is even to be given the services of a supposed Quaker lady companion and humanitarian observer, but who will be a fellow spy, an attractive woman by the name of Lucita Squier, who had started out as the shapely secretary of Cecil B. DeMille before doing a couple of minor films and becoming a stand-in for Mary Pickford on movie sets. In between such stints she had also become a substantial Hollywood scriptwriter, as the Lon Chaney movie *Bits of Life*, which she had written, had made her the youngest person ever to have scripted a feature film.

Wilkins comes away from a short meeting with her convinced she would be better advised to stick with acting.

'Women,' he notes, 'are unreasonable and apt to go off on tangents, especially when they are beautiful and charming.'[6]

And yet?

And yet, being beautiful and charming are compelling arguments in themselves for agreeing to her accompanying him. Wilkins remains blissfully unaware of the fact that Miss Squier is due to be wed to one Albert Rhys Williams, an American journalist who, much like the famed John Reed, has gone from covering the revolution to covertly pushing for it. Albert is no less than a personal friend of Lenin, a confidant.

And so the woman that Wilkins sees as a Hollywood socialite is, in fact, if very quietly, much closer to a Hollywood socialist, and she and Albert Rhys intend to make their home in Soviet Russia.

The Quaker 'mission' is rounded off in peculiar but charming fashion by Miss Laurette Citroën – yes, the daughter of *that* Citroën – who is perhaps, *perhaps*, the only one of the trio who is actually there with the sole aim of helping the Quakers.

It all happens so quickly, Wilkins leaves for the Soviet Union only 10 days after having arrived back in London.

•

It is the difference between night and day, drowsy and alert, closed and open.

It is the difference between the customs officers on the Polish side of the border, and those on the Soviet side – and Wilkins closely observes it all, already mentally composing the report he will soon write.

And then the train rumbles forth to first contact with the Soviet regime.

Looking out the window, the first thing the curious Wilkins sees is a company of young soldiers in singularly ill-fitting greatcoats. There are no officers apparent, and it is not quite clear what they are here to do other than make a show of force on the border, even as passport officials board the train to ensure that everyone's papers are in order.

The official who gets to the Wilkins party a long time later – no-one is in any hurry – is just 18 years old, with red cheeks, a spindly frame ... and the same drunken tailor as the soldiers outside. After taking the passports from all the passengers on the carriage he wanders off without explanation, as the train again rumbles forth. Some clue as to the deprivations in the new world come when an American traveller casually flicks the remaining butt of a cigarette onto the platform, whereupon half-a-dozen of the Soviet soldiers on duty scramble to be the first, or at least the strongest, to retrieve it and have the last few furious puffs on it.

This is your people's utopia of communism?

Fortunately, whatever the different agenda of the three companions, including the hidden ones, are, they arouse no suspicions from any of the officials and they are allowed to continue towards the heart of Soviet Russia through ever more check-points. Interestingly, there is frequent tut-tutting from concerned Russian officials that such beautiful women as Lucita and Laurette should be travelling about Russia with a man who is not the husband of either, or even fiancé! Yes, they are revolutionary communists, devoted to tearing down the conservative class system – no matter the bloodshed and general destruction – but only in a *proper* way. There are still certain standards of decorum to be met. Bolsheviks with Jane Austen etiquette standards are just one of the many contradictions of Russia.

Still, in a Russian effort to give the trio some suitable propriety, they often find their travel arrangements upgraded with somewhat embarrassing results. Instead of being classified as 'hard' Russian train travellers, meaning riding in an open carriage, the three are graded as 'soft' and so given a private berth, with four beds.

'The girls,' Wilkins notes approvingly, 'seemed no more embarrassed than I was,' and prove to be 'marvelous travelers'[7] – able to blend chicly into any new situation – while he attempts to fit in by growing out his beard, 'in Russian style'. And it is no bad thing, in equal turn, that the beard he grows, together with his rather stark physiognomy, give him a distinct resemblance to none other than Vladimir Lenin.

(This is perhaps another reason that they are treated well – just in case.)

•

Things become grimmer as the train stops at the Custom House at Nicgoroloje, where their bags are opened by a group of ruffians in rough uniform, and there is no doubt what interests them most – food, all of which 'is mauled by the dirty hands of the Customs Officers'.[8]

Things are no better when it comes time to get tickets for the next leg of the journey to Moscow, where *everyone* crowds around one beleaguered official, 'a greasy individual with hair and beard of equal length, some three weeks growth'.[9]

'No order prevailed. Here the idea of communism was that each one had an equal chance of thrusting the other aside and getting in first.'[10]

Arriving in Moscow itself, Wilkins also notes some extremely odd things, starting with the chaos apparent at the platform as, for reasons best known to themselves, all passengers appear to engage in 'a frenzied rush and struggle for the exit',[11] which continues unhindered until they get to the gates where everyone must present their papers, state their business, and undergo 'careful scrutiny and attention'.[12]

They are being watched, very, very carefully.

The city itself, if it is possible, proves to be even more downtrodden and chaotic than the station, and the cuisine is little more than slops.

'Their fare is rye bread, a little fat, sugar, and an occasional meal of cheapest meat. Very few vegetables are obtainable at reasonable prices.'

The truth is staggering.

> The peasants even with the help of the foreign missions, have to fall back on their supplies of edible grasses that are gathered during summer ... Preserved in bulk, as they are, these grass have an appearance of ordinary hay. They are pounded in wooden troughs and mixed with rough rye flour for making bread.[13]

Wilkins is shocked, as are his companions.

They had known things were grim under communism. But peasants eating grass to stay alive? It is unimaginable. But it is reality.

Wilkins and his companions at least receive permission to head out into the countryside south of Moscow, where, at first, things seem good, with plentiful grass growing.

But now Wilkins looks closer and the problems soon appear. A drought appears to have gripped the land, and something is going on, because he can see a great deal of grain left standing in the field.

'In many little supervised districts,' he will soon find out, 'the peasants, fearful of the tax imposed on grain in kind, buried their half-dried harvest underground and this is now a rotting mass and mildewed, useless for any purpose.'[14]

Can the Soviet officials so omnipresent in these villages – conspicuous for their poorly cut cheap uniforms, but infinitely superior attitude – do nothing to alleviate this desperate situation? Perhaps, but they do not.

To these officials, 'a tip or a bribe is never refused and its even asked for ... One wonders if, by the time the bulk of the population have reached the stage of learning of the present official class, whether the officials will not have evolved to the same state of mind as those of pre-revolutionary days.'[15]

As to the overall educational policy of the Soviets, there is an average of just one textbook for every 40 children. The explanation given at the Government district office was that they had no money for schools as yet.

When Wilkins asks where they expect to get the money from the answer is clear. 'From the export of grain in the future.'[16]

Really?

What makes you think you will be exporting grain in the near future when, just last year, millions of bushels of grain had to be sent to Russia by relief organisations like the Quakers, in order to prevent the people from starving? The answer is ... foreign capitalists! You see, under Lenin's plan, the capitalists will be allowed to exploit the agriculture of Russia, 100 per cent tax free, so long as they meet two conditions. 1) Each capitalist exploitative company that sets up operations in Russia must have a Government-appointed communist official to run things while they are here. 2) The Soviet government will keep 50 per cent of all the capitalists' profits. (Which is sort of like a 50 per cent tax, a fact Wilkins tactfully does not point out.)

Other ironies, however, are harder to resist highlighting ...

'So you must depend on the capitalists of other countries to help you,' Wilkins posits.

'Oh yes!' comes the reply. 'We need the help of the capitalists in other countries, but we always intend to be communists in Russia . . .'[17]

It is a contradiction that amuses Wilkins but annoys Lucita, who struggles with evidence which seems to point to the fact that communism is a theory that does not work in practice, or even a theory that needs a little work.

For yes, there is no doubt that Russia is a land of great beauty, but as the visitors cross the Volga there is equally no denying they have entered into a world of great misery, where famine rules, starvation stalks the land and disease and death are everywhere. Everywhere they look, they see emaciated ragged people and, worse still, mass graves where the bodies of naked children are about to be buried, even as the rags they had been wearing are being divided up for warmth by survivors. Of all things that Wilkins has seen on heaven and earth, few, if any, have shocked him more.

He and Lucita are united in their horror. Whether the cause is the economic blockade of the West or the economic madness of Russia is tragically irrelevant, the result lies before their eyes – mass death and misery.

The town of Buzuluk on the Samara River, some 600 miles southeast of Moscow, proves to be the centrepiece of such miseries and, after ensuring that the ladies are properly settled into an establishment where they are safe – for so desperate are these people, anything could happen – Wilkins ventures forth to properly document the grim horror. Although here as a spy, this does not require much covert intelligence on his part: these photos will be his proof. One look and anyone will be able to see that the Soviet Union is not a threat but a catastrophe in the making. And these people need help. The fact that he can also see that Quaker relief missions make the difference between life and death for thousands of people is a satisfaction and he fully intends his photos to also be used for their purposes to help bring urgent help.

It is while taking such pictures and lightly conversing with such few locals who can communicate through basic English, sign language, and his own smattering of a shattering of Russian, he hears rumours of something as extraordinary as it is horrifying: cannibalism. Surely it is too grisly to be true, something that might appear in a horror film penned by Lucita, but not in modern Russia? Wilkins hears of an elderly Russian woman who is in control of a gang of cannibals who are murdering people not for their valuables, but for the meat they provide. He finds out their location easily enough – a small village by the name of Palimovka.

The fellow who tells Wilkins this shudders as he speaks of Palimovka, just as a pious man might when invoking the name of Sodom or Gomorrah.

It is a trip that Wilkins would rather not do alone, and he seeks accompaniment from some of the good men of Buzuluk, offering a combination of money and vodka. But the response is swift.

Under.

No.

Circumstances.

We are NYET coming!

Still, as ever a believer in his own star, Wilkins packs a generous supply of flour and black sausage and heads off to photograph this criminal crone. He finds the village easily enough, but also finds that the first man he greets, 'without hesitation delivered me such a blow on the head that I was knocked to the ground'.[18]

Lights out.

Coming to slowly – for the hit was so hard he had entirely blacked out – Wilkins becomes aware of an angry squall of peasants all around him, engaged in furious argument. And yet now his weary eyes so bleary are drawn to a far more shocking sight.

'Imagine my horror when I saw on a bench the remains of several human arms and legs. The cannibal story was true after all, and I felt at last I had been too venturesome.'[19]

It seems obvious that the argument they are having is over who will get to eat him first, or even who will get his juicy thigh, and who will have to make do with his far more stringy biceps. Wilkins has faced some tough predicaments, but this one, providing dinner for strangers, is surely the toughest of the lot.

Before killing and eating him, however, they first go through his belongings, which brings a shout.

The stranger has brought food with him! He has flour, and meats!

Suddenly they realise: he has come to help them, and this food – for there is a lot of it – is for *their* consumption, not his. The change is instant, as now the peasants kneel about the prone Wilkins as if he were a visitation from heaven itself, as they weep and passionately pour out their fervent thanks in a tongue too quick to register.

'I could understand hardly anything of what they said,' Wilkins notes, 'but I managed to get them to understand with the few words that I knew that I was their friend.'[20]

And more importantly, not their dinner. Uncharacteristically shaky – for this had been a really close one, at the hands of people as desperate as he has ever seen – Wilkins beats a hasty retreat, after making a sincere promise, which he fully intends to keep, to send as much relief as he can, as soon as possible. The whole experience is enough to tightly inform his and Lucita's meeting, a few days later, with the man who has wrought this revolution on the promise of making a 'workers' paradise', Vladimir Lenin himself.

These days, as he is seriously ill – having suffered both a stroke and an assassination attempt, which had seen him hit by two bullets – Lenin divides his time between his accommodation in the Kremlin when he is feeling well and his *dacha*, summer house, in Gorki in the countryside just outside Moscow when he is feeling poorly.

Their own audience will be granted at the Kremlin. (And it is no small thing to get through each check-point with the same papers to be presented to an ever growing group of soldiers the closer they get to the annex where Lenin lives, past the base where a statue of Tsar Alexander I has been rudely ripped down. Hats and coats are hung in an antechamber, as they approach warmth, a rare thing in the Kremlin and a sign that the great man is near.)

Ushered into Lenin's presence, Wilkins is immediately struck by the leader's weakness.

(Not for nothing is he known, quietly, in official Soviet circles as Старый человек, 'the Old Man'.)[21]

The revolution may be still young, but he no longer is. And yet Lenin remains mentally sharp as he begins by asking if his visitors would care to speak in French, German or English? English it is, but can they speak slowly please, for the Russian leader's fourth language is not his strongest.

In halting but passable English, Lenin talks to Lucita of their mutual friend John Reed, the brilliant journalist turned revolutionary who captured the Bolshevik takeover with his masterpiece of reportage and romanticism *Ten Days That Shook the World*. Lenin lightens and brightens just to talk of that extraordinary time again! The glory of it! The Tsar and his corrupt court routed not by an army but by an *idea* whose time had finally come. Yes, they were great days, the greatest in the history of Russia, and a smile plays on Lenin's face as he relives it once more. It seems like decades ago, not just years, as Wilkins and Lucita see the remnant of the electrifying orator performing his magic for their benefit.

Most interestingly though, the Soviet remains absolutely resolute in the rightness of the revolution. For when Lenin asks Wilkins to recount his experiences in Russia and the sights he has seen, and the Australian tells him frankly more than a little of the horror of it all – the poverty, the sickness, the starvation so severe it has led to cannibalism – there is a pause.

It is a rare thing indeed for the dictator to be told such stark truths, and his prisons are filled with some who have tried.

'While I believe,' he says carefully, 'that I have done the right thing in bringing about revolution in Russia, after five years effort in Communism I have come to realize that I have made a mistake in regard to the rapidity of development possible within the Soviet Union.'[22]

(But perhaps that is enough, in the scheme of things? One of Lenin's favourite dictums was, 'Give me four years to teach the children and the seed I have sown will never be uprooted.')[23]

There is no fault in the *theory* of communism, you see, it is only the *reality* that might need some tweaking and twisting before we get it quite right. Yes, there are hardships and horrors, but no more or less than there were as mankind moved from feudalism to capitalism. Communism is simply the next step, the inevitable conclusion of the current system. Regardless, Lenin concedes that progress requires patience, and for the moment he accepts that even a mind such as his cannot, 'inject civilization into the minds of humans and get an immediate response'.[24] They farewell the ailing leader, aware they have witnessed a giant fading before them.

On leaving Russia at the Polish border, en route to London, Wilkins and his companions look back at the receding frontier gateway to see an enormous Soviet flag flapping proudly from a rough, improvised pole: a tree that has been hurriedly hacked and brutally battered, beaten and betrayed into a roughly suitable symmetry for a flagpole though quite unnatural for a tree. For Wilkins it is the perfect symbol for what he has personally witnessed for the last seven weeks.

The mighty tree, shorn of its limbs and roots, its vital organs, greedy and grasping though they might have been, cannot live by the aid of the fluttering blood-red flag that flies this way and that as the four winds blow. Even the trunk must rot and decay unless – but it is not my province to foretell the future.[25]

February 1923, London, return of the native

Only a short time before arriving back in London to present his report on the Soviet Union to the British Government – right now, the principal danger they present is to themselves – Wilkins is contacted by Sidney F. Harmer, the Director of the British Museum, the most venerable and oldest museum in the world, weighing in at 170 years.

Given how well you have done in collecting valuable specimens in your polar explorations, would you like to return to your native land, trek for a couple of years across its northern regions and do much the same?

We will pay you a salary of £600 a year – the same wage you received with Shackleton – and you will be free to hire expert help to run your own show to gather specimens of flora, fauna, fish, fossils, spiders, scorpions, minerals and even ticks, to send back to the Museum for storage and analysis – to explore your own country and document the 'unknown Australia'.

After all, for the last century and a half the British Museum has been relying on the copious collections of Joseph Banks and Matthew Flinders, and it has become aware of two things.

Firstly, there are glaring gaps and speculative spaces that they want Wilkins to fill. And secondly, it seems some Australian species are becoming extinct in the wake of environmental destruction, and they need to gather as many specimens as they can, while they can.

In the Arctic and in the Great War, Wilkins has captured history with his camera, but can he capture these Antipodean creatures and plants before they are lost to history? It is a prospect that intrigues Wilkins as he also hopes to assess the 'patchy' continent and see – much in the manner of the famous if cursed explorers of 60 years ago, Burke and Wills – what agricultural blessing may have been overlooked.

'There are, of course, many thousands of fertile acres in Australia which are as yet unoccupied,' Wilkins will note, 'but we cannot over-look the fact that there are also many millions of acres which present as little inducement to the settler as do the desolate valleys of the moon.'[26]

Somehow, despite the natural desolation of much of it, and the ravages of white settlement, some of the fauna clings on. Wilkins sets to; his principal task for the British Museum is capturing the creatures with his camera as well as garnering a record of their natural life before they are killed, skinned or preserved. Jack of all trades and crackerjack with any camera, Wilkins is bustlingly busy. He starts writing articles to be sent to local papers, drumming up interest in his upcoming quest for rare specimens and hopefully inspiring helpful locals who will read of his progress and impending presence to come forward and offer guidance as to where they will be able to find interesting specimens.

Wilkins pauses only long enough to cable his full acceptance of the proposal, on condition he have Australian men as his assistants and the positions be advertised widely, before setting sail for home. By the time he arrives in Fremantle, over 350 applications await.

Not all of them, in fact very few of them, have much in the way of credentials, but they are nothing if not amusing. To wit . . .

> *DEAR CAPTAIN,*
>
> *I see by the papers that you want some one to go with you on an expedition to Northern Australia.*
>
> *I have wanted to go to New Guinea for a long time, and I would like to go along with you; that's the life for me.*
>
> *I went to school and know all about our birds. I am not much of a scientist, but I suppose that does not matter. I can cook a damper but I don't want to teach my Grandmother to suck eggs. I suppose you will do that.*
>
> *I can handle blacks, there's been one old Binghi about our place for a long time.*
>
> *Of course I know that there will be a lot of roughing it, but I don't mind; I can put up with anything.*
>
> *I want £5 a week and keep; I can get that anywhere. Let me know where I can meet you.*
>
> *Yours truly.*
>
> *P.S. I am going to write a book about it if you are agreeable* [27]

Thank you, Bluey, it is mighty tempting, but Wilkins is not agreeable.

Another applicant informs Wilkins that 'I don't know much about mammals, but I can skin 'roos and have cyanided possums . . . Of course the salary is the first consideration.' [28]

Of course!

Look, if Wilkins could be any kind of word, he is much more a verb than a noun, but heading to Northern Australia with a man who uses cyanide as a verb?

No. Just, no. *Cyanara, Applicant No. 42.*

When it comes to salary, a further half-enthused applicant tells him that unless he gets at least £300 – £300, you hear, and not a penny less – he 'could do better on the Hookworm campaign'.[29]

So don't come to me with any half-arsed offers.

Wilkins is sure not to.

Meantime, he is infuriated to find out that many universities have been actively *discouraging* natural history students and graduates from joining the expedition (because, as one professor patronisingly informs him, 'They can get well-paid commercial jobs as soon as they leave the university; there is no need for them to gain experience in the field.')[30]

This consolidates Wilkins' view of his homeland and helps to provoke a blistering passage of his later writings that will come to damage his reputation in the land of his birth:

> Most Australians are well off in regard to creature comforts, and many of them soon reach independent means; yet the absence of the expressed desire for culture and for higher things, and their contentedness with the mediocre, make them perhaps the poorest rich people in the world today.[31]

For now, Wilkins heads to Brisbane's Queensland Museum to set up what is effectively his 'base camp' – at least in the sense of a storeroom for the specimens his expedition will collect – and quickly hires a qualified crew of assistants through his own inquiries.

Three men of promise and experience are selected: Mr Edgar Young, an old Brisbane botanist and ornithologist; and young Mr Oscar Cornwall, who has until recently been collecting natural history specimens in New Guinea. They are joined by an ex-pat Russian, Professor Vladimir Kotoff, who has spent two decades collecting specimens in Siberia, Japan and Manchuria while his own homeland was being collected by the Bolsheviks, and he is delighted to join Wilkins as a mammal specialist.[32]

Kotoff will be particularly useful should they find the famed 'Queensland tiger', which some say is myth, while others maintain they have seen it! Some serious scientists insist it is either the offspring of the infamous 'Tantanoola tiger', a mysterious animal that had terrorised farmers around the remote South Australian settlement three decades earlier or perhaps the 'marsupial feline'[33] the Sydney zoologist Reg Kendall had gone looking for in North Queensland a few years earlier.

'There is a lot of room in Queensland, for that type of animal to live,'[34] Wilkins tells the Brisbane press, before characteristically shutting down questions about his colourful past.

'Whilst Captain Wilkins would talk freely about the Queensland enterprise,' the journalist for *The Queenslander* would comment, 'it was like drawing teeth to try and get out of him a few details of his own remarkable career.'[35]

For transport, Wilkins chooses a brand new Ford van, specially built to carry endless expedition boxes and serve as caravan and tent as needs and climate require.

The van itself is transported at first by train, with Wilkins and co. enjoying the favour of the Railway Commissioner for Queensland, who 'very kindly placed at our disposal a special car in which we could sleep and have our meals'.[36]

At Talwood terminus (some 180 miles south-west of Brisbane, a little to the north of the Tweed River), the car is unloaded and their journey proper begins.

As they travel to their first cattle station homestead, Hollymount, and make a moonlight camp, Wilkins smells that unique, 'faint, clean, penetrating smell of the scorching eucalyptus leaves'[37] that so effortlessly takes him back to his boyhood. In this scent he is sent the truth.

Sitting beneath a gumtree, he writes something of an epistle.

> Twenty years ago, as a small boy and totally ignorant of city life or even the 'feel' of a large town, I left the backblocks of Australia. Since then I have visited most of the capitals, and have wandered around the world from east to west, and from 80 degrees north to 70 south. Today, as I sit beneath a tall blue gum, whose wind-stirred leaves seem to whisper a friendly greeting, the bridge of years is crossed, and from the archives of memory, scenes and incidents that have lain dormant for years come flooding. The rustling of the trees and the crackling of the sticks as they burn with bright red flames that lick the smoke-blacked 'billy,' stirs one more deeply than the shouts of welcome that one hears from strangers when returning from some much talked of expedition . . .[38]

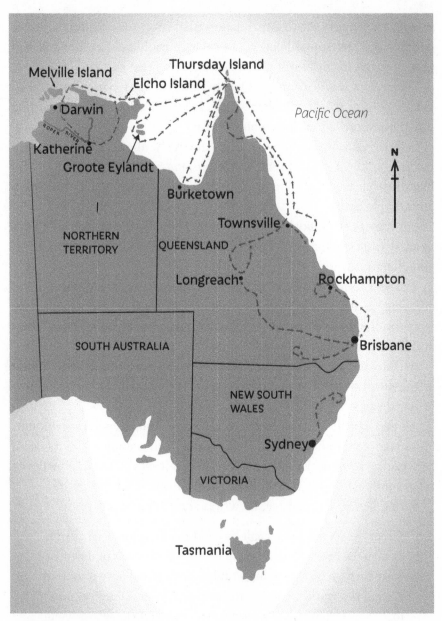

Wilkins' Australian expedition

That truth?

He is a boy of the bush and it will never leave his heart. He is truly home.

Yet another thing which reminds Wilkins of his childhood, though not with pleasure, is the fact that Queensland happens to be, in this summer of 1923, in the grip of crippling, sapping, soul-destroying drought and, just as he had in Russia, Wilkins sees in the trees a metaphor for life:

> The tall gum-trees and the dry red dust beneath them called out for moisture, but moving above were the leaves still bravely green. The roots of the trees are deep down, and with a firm, tight hold they live on and defy the drought conditions. When I see the stout hearted settler holding grimly on while the grass withers, and the stock get thin and die, I cannot but think that he must gain some courage and sympathy from these green-leaved trees.[39]

But to work. His key job in these parts is to capture and kill some kangaroos, a beast much harder to bag than he had anticipated. And where better to situate himself than – following ancient tradition – down by a billabong, to wait for some jump-bucks.

Sure enough, and soon enough, up jump the kangaroos, one, two, three; and three more and Wilkins is delighted to see they are 'beautifully furred'.[40]

Quickly now, quietly now, there is work to be done, deadly but necessary. Wilkins raises his gun closes one eye, and peers down the sights to see . . . to see another eye looking back at him. His timing is such that just as he is about to pull the trigger, 'a little "joey" peeped out from its mother's pouch'.[41]

The joey's mother turns her head now and, with twitching ears and lustrous eyes, mother and child stare silent at the weapon raised against them. Wilkins' finger tightens a little further and . . . stops. Smiling, Wilkins lowers his gun, and simply watches the mother kangaroo at play, for a full and fascinating 20 minutes as she hops about before the twilight starts to deepen. The head of the joey is glimpsed once more just as his mother starts to bound away to safety, and Wilkins cannot stop himself from waving the duo goodbye as they depart. Just this once, the hell with the British Museum. They already have plenty of mummies from Egypt and can live without the kangaroo one today.

'I had missed a splendid chance of starting the collection,' Wilkins says, 'but even in the cause of science one cannot always override one's feelings.'[42]

A kookaburra that Wilkins spies on his way back to camp is not so fortunate and, after the explorer shoots it dead, the bird is taken back to camp where it is quickly gutted and the preserving begins – with those specimens small enough being preserved in bottles of spirit. (All up, the expedition will collect a total of '697 [bird] skins, 658 specimens preserved in spirit, 28 eggs and one nest.'[43])

The first rays of dawn the next day find Wilkins and his men heading out to find wombats, and it is not long before they start to burn up in the morning sun. As it beats down, Wilkins cannot help but reflect upon how quickly his life changes from one year to the next. Last year in the Antarctic he had been trudging through snow, invariably seeing seals and occasionally wandering past whales self-cooking on the ice. Now he cooks in the Antipodean sun.

On this day, alas, wombats prove as elusive as the sun is scorching. Not a single one of those large ponderous animals is to be found, and even when they carefully lay traps, upon returning the next day, it is as much as they can do to find the traps, let alone actual wombats.

'There is remarkable sameness in this country,' Wilkins notes, 'and it was with difficulty that the place was recognized when approached from a different angle.'

Wilkins will soon conclude, crossly, that the wombat is 'almost, if not quite, extinct in this district'.[44]

The prickly pear proves somewhat easier to find, courtesy of Professor Kotoff. 'He brushed casually against it, and was soon in agony with sheaves of the little pointed barbs boring their way into his arms and legs.'[45]

He is not the only creature to suffer so, Wilkins being surprised to find the same bastardly barbs in the skin, feathers and even the bones of many of the native creatures they catch – creatures simply not equipped to withstand the horrors of the introduced plant species.

'Even the tough skinned goannas had not escaped, and the body of one huge lizard was riddled with the long spikes of the pestilential plant.'[46]

Pestilence it is: 'Thousands of acres in the State are so thickly covered with it that it is impossible to walk through them, and beneath the pear no vegetation grows.'[47]

As for goannas, Wilkins finds himself in much closer acquaintance with them than he would like. Some are six feet from top to tail and regard humans as no threat whatsoever, with ridiculous consequences:

> Sometimes when they are chased by the dogs and no trees are near,
> they will run up and take refuge on the shoulders of a man. It is

rather an unpleasant feeling to have a goanna dancing around one's neck and to see its black forked tongue darting about one's face.[48]

That evening they make camp by a creek, where they are able to shoot, gut and stuff black ducks, teal, heron and other birds they simply label as 'native companions'[49] – before dining on their flesh – knowing that it will please their employers at the British Museum to name them. On the creek bank they also find curious holes that contain – they find with a little careful digging – scorpions! For the purposes of science, Wilkins soberly details them as 'huge red scorpions, with claws as big as crawfish'.[50]

For their own purposes, they make a hurried investigation beneath and around the tents to ensure they have not pitched them on a bed of the brutes.

While preparing bird specimens inside his own tent, Wilkins is startled to hear a cry from Kotoff: 'Come quickly! Come quickly with a gun! There is some large animal here at the tent!'[51]

Wilkins – alarmed but also intrigued as to just what kind of large animal it might be – immediately rushes towards Kotoff but is only in time to catch a glimpse of some creature disappearing into the bush. *What the hell was that?*

'It is big as a dog, with a bushy tail and a round, flat head, and striped in colour.'[52]

Maybe, just maybe, it is the famed Queensland Tiger!

Or have they, perhaps, discovered a new Australian mammal? Not as startling as a kangaroo perhaps, but certainly startling enough to frighten Kotoff. With uncharacteristic excitement, Wilkins tells each of his men to keep a strong lookout and a gun handy in case the strange beast is seen again. And this time, we must kill or capture it! Sure enough, just the next day, there is a movement of a large creature in the bushes, Young fires off a shot and they all rush forward to see what it is.

It proves to be . . . a very large grey tabby cat. It is not quite the bunyip they were hoping for and Professor Kotoff's red face is not from sunburn alone.

Onwards.

The proliferation of such feral cats – for they soon see ever more – is one explanation for what generally seems to be the devastation wrought on the bird and small fauna population in so many areas they travel through.

'These cats have become an addition to Australia's native fauna, for they are found in numbers in many districts, and no doubt they prey heavily on the small birds and rodents.'[53]

So often there are stories from the old-timers, and sometimes even the younger ones, of flocks of birds so thick they could block the sun, of hairy-nosed wombats and reptiles so numerous you could barely spit across your shoulder without hitting one. But the rabbits had come, and the cats had come, not to mention the foxes, and everything had changed. And of course it was not just the feral species doing the damage as the settlers themselves had knocked over the natural vegetation for grazing and farming, at which point the hoofs of the sheep and cattle had further broken up the natural land covering – and then the settlers had to shoot all the native animals they could that, deprived of their own natural fauna to eat, have to live off the grass. It is nothing less than an ecological disaster happening before their eyes.

It is all rather depressing.

'When the trustees of the British Museum decided that they must have a collection of the fast-dwindling fauna of Australia,' Wilkins will note, 'they were not wrong in surmising that it was rare, and in fact, that it is even too late now to get some of the species.'[54]

There is no better example than possums and koala bears, which in the Talwood and Saint George area were so plentiful they were regarded as pests.

'But now,' Wilkins will report sorrowfully, 'they are practically extinct.'[55]

Yes, trappers and poisoners have undoubtedly killed many, but that still does not account for the stories he hears, of 'bears and possums ... found dead in heaps under the trees. The bears were usually found dead in a sitting position, with their front paws resting on the trunk of a tree as if in prayer. Part of their trouble seemed to be a disease similar to "mange", but this in itself was not enough to cause death.'[56]

And so to the sad conclusion. 'The longer we hunt and the more reports we hear, the greater is the evidence that Australian native life is becoming rare in many places, and extinct in parts.'[57]

At least at their next port of call, at Thomby Station, 'at a place where a bore drain fills a natural watercourse and maintains small lagoons the whole year through',[58] they do find that birds and fish abound all around, though of native animals there remains no sign at all.

Among those birds, glory be, is the very rare chestnut-shouldered parrot, *Euphema pulchella*,[59] long thought to have been extinct, apart from a small group discovered in New South Wales.

Bang!

The shot rings out, and it is added to their collection – Wilkins proudly proclaiming, 'the only one of its type that has been collected for museum purposes for the last 60 years'.[60]

Bang! Bang! Bang!

Down come some grey-crowned babblers followed by fine specimens of apostle birds.

There are lots of kangaroos and emus here, though station owners are doing their best to fix this.

'It seems a shame that so many of these great marsupials should be killed,' Wilkins writes, 'but when there are so many they become a serious menace to the station-owner, and soon deplete the food in the home paddock that is needed for the boundary riders' horses.'[61]

As to brush turkeys . . .

'They were plentiful in the district,' Wilkins writes, 'but the raids of the foxes and the cats have no doubt been responsible for the almost total absence of these birds at the present time.'[62]

In the end, all they can find left in the wilds in these parts, are two healthy specim . . . – *bang!* – make that one.

The plumage is delightful; the roast turkey it provides for their meals for the next few days and nights even better.

Still, they are lacking much of the fauna they had been hoping to capture.

It seems that everywhere Wilkins looks in Queensland he is told the same thing: 'It is no use looking here for specimens, you should go about twenty-five or thirty miles further west; and there you will find all sorts of interesting things.'[63]

Alas, just as 'tomorrow never comes', so too, when they travel the requisite 30 miles, they are told much the same thing. Finally, however, they arrive at the camp of a legend in these parts, a famed roo shooter by the name of Jack Gilbert, who lives by selling kangaroo hides. A grizzled bloke with a face of dark cracked leather – ever and always with the stub of a rolled cigarette hanging from the corner of his mouth, and that *never* goes out, emitting the occasional puff – he is himself an Australian original. Wilkins warms to him immediately for his talent and his tall tales alike, and he proves indeed to provide wonderful counsel on how to catch specimens whole in this area.

For what ya gotta understand, young bloke, is that the bloody roos are at their most bloody vulnerable at dawn and dusk. They get hungry hopping around all day and wanna garn getta good feed in the bloody twilight. So yer hafta get your eyes used to the gloom, and watch for the twitch of an ear or twist of a tail against the bloody red sky, always the fastest ways to spot a roo in the camouflage of the bush. These bastards give statues a run for their money they do, so good are they at remaining still as a widow's bed and silent as a pub for teetotallers. But all it takes, mate, is a tiny adjustment, the minutest movement and bang, you got him! And where there is one set of twitching ears, there'll be a herd that heard that shot and will start leaping. And then you've *got* the bastards.

Using Gilbert's expertise, Wilkins and his crew really are able to get closer than ever to the kangaroos, bag many of them and learn the fascinating habits that help them survive this habitat. I always like to understand something before I shoot it. Look now as the dogs attack the kangaroo. If the roo senses that the dogs have come with a man, then they will attack the greater danger, the man, first. But if the man is distant, well, the dogs may meet a quick and decisive fate as 'kangaroos are dangerous beasts when bailed up. They will sometimes seize a dog in their arms and disembowel it with a stroke from their strong hind-feet, or if water is near they will rush into the water and drown the dog by holding it underwater.'[64]

And the men so attacked?

Well, Mr Young is very unhappy to report that a kangaroo has ripped his trousers off, and he had only just got away. Maybe. Professor Kotoff barely manages to conceal his doubt and quietly tells Wilkins that he is sure that old Young has been infected by Gilbert's habit of telling tall stories around the campfire, and has come up with a tall story of his own.

Fine.

Which is why this next day Professor Kotoff acts as he does, while doing a spot of bird-watching at a nearby waterhole, when he spies a large kangaroo nearby. Taking up his gun, he fires, and drops the animal, cold as a spud. So much for Gilbert and Young and their stories of roo battles. Kotoff's only surprise is that Wilkins had actually believed such transparent nonsense. In fact, as he has now so comprehensively demonstrated, the kangaroo is not a terribly bright mammal at all, and is as easy to kill as any other. True, this one is not *quite* killed, as he can see the poor animal twitching. So Professor Kotoff approaches the roo with a club to finish the job.

But here is the strangest thing.

No sooner has the Professor's shadow fallen over the twitching roo than – comprehensively demonstrating that it had only been playing possum itself – it leaps to its feet as only a kangaroo can leap, and attacks him with swiping claws, massive blows to the mid-riff as the kangaroo leans back on its tail, and really puts some oomph into it.

Kotoff, no more than a lightweight nincompoop in the animal kingdom proves to be in the ring with a big boomer, the heavyweight champion of North Queensland, as the blows continue to rain down. At least his screams alert Jack Gilbert's dogs, which run for the roo, but again the roo has the better of them, by retreating to the waterhole as the canine hunters attack. Kotoff takes the opportunity to run for his gun and finally is able to put this Rasputin roo to death, but the effect on him is remarkable.

'Kotoff,' Wilkins dryly notes, 'would believe any story about a kangaroo after that.'[65]

Now full of renewed respect for the cleverness of the kangaroo, Jack Gilbert tells Wilkins that there is one animal in the Australian kingdom that is the cleverest of the lot, one that can do more damage than the kangaroo and comes with a higher bounty, for very good reason.

'A dingo,' his cigarette rising and falling with every word, 'is more cunning than a fox and is trapped with the greatest difficulty.'[66]

And make no mistake – a 'bad' dingo in a paddock filled with sheep, can do more damage than a fox in a henhouse, for it can kill sheep worth hundreds of pounds in a single night. That's why Jack can get as much as £40 for a 'bad' dog proved to have done damage, while an ordinary dingo can still bring £3 or £4 if a farmer can be shown the dog was killed on his land.

As for getting emu specimens there is nothing to it due to a peculiar quirk:

'They are curiously attracted to anything red in colour,' Jack Gilbert recounts, 'and an old trapper told me that he had once shot five emus with one discharge from a gun. They had approached a red cloth which he waved, and when they were close to him they put their heads together as if to discuss the strange object. He killed them all with one cartridge.'[67]

Sure enough, the emus prove to be as easy to capture as the kangaroos and dingos were difficult, as the long-legged flightless birds simply wander into their camp – curious about this new thing, little knowing they will soon be a new thing in a British Museum glass case.

Moving on, and now looking for wallabies to add to his growing collection, Wilkins soon comes across an old pipe-smoking stockman called Bill, who agrees to give Wilkins a ride on his cart to the gully where wallabies will certainly be found, though as Wilkins carefully puts it: 'Bill was the type of stockman who, having spent many years in the saddle with no other company other than his horse, had developed a habit of talking to his animal in general terms.'[68]

Generally, very blue that is, mixed in with what Bill assumed is a normal chat. An amused Wilkins notes down his peculiar monologue, to be later cleaned up a little for publication, but this is what he says: 'So, you come from the British Museum, do you? You bloody bastard. I'll learn you to look round and wander off the track, you fucking cow.[69] I'll lay along your bloody ribs and cut the fucking hide clean off your bones, you bastard.'[70]

Bill continues without pause, talking to Wilkins and his horse, and in no need of reply from either.

'S'pose you are pretty good at shootin'? You cock-eyed son of a fucking bitch; if I was to set about you your own mother-fucking mother wouldn't know you from a bloody fucking bunch of raw hide.'[71]

Wilkins is just about to answer the shooting question, and opens his mouth to do so, when Bill cuts him off.

'Don't you try to fool me, you fucking bastard brumpy*, I know all your fucking bloody tricks!'[72]

'The sunshine has a pleasant effect on the leaves ...'[73] interjects Wilkins.

'Too fucking right, my hearty!' replies Bill. 'Step up now, damn your arse! Smart today, aren't you? You can be a bloody good old bastard, can't you, if you want to be?'[74]

---

* 'a stumpy, scruffy little horse'.

When it comes time to hunt birds and wallabies, Bill proves equally loud and scatter-gun in his approach as they have no sooner started out than his gun roars, spraying gravel right into the back of Wilkins' trouser legs.

'After that we moved forward side by side.'[75]

No wallabies are shot, but against that neither is Wilkins.

Finding that his fame precedes him, thanks to the bush telegraph, Wilkins fields endless offers of country hospitality, leading to too many supremely long, dull evenings, and he declines when he can manage it without offence. One who does not fit into that category is a wealthy homestead owner with a very large station, where Wilkins is eager to bag many specimens, so he accepts their invitation and turns up with a swag over his shoulder ready to camp outside should it be too late to return to his own camp. Sure enough, dinner is only an entrée into a long evening of earnest conversation and paralysing stories that go until after midnight, but when Wilkins tries to make his excuses and sleep outside, his hostess will not hear of it.

'Of course, you must use the spare room; there's a bed and everything there all ready.'[76]

'It would be a shame,' tries Wilkins, 'to give you the extra work of making up a bed on my account, for I have my blankets and am accustomed to sleeping out.'[77]

'Oh Lor', 'taint no trouble!' she cheerily replies. 'Why I put clean clothes on it on Sunday and we've only had the mailman and my old man's two nieces sleeping in it since then.'[78]

Do tell? Given that it is now Friday, Wilkins manages to arrange a compromise whereby he will indeed sleep in the spare room but will use his own blankets. The following days are spent hunting in the Talwood and Saint George districts with very little success. Not to worry, Wilkins is told, one thing this whole area is famous for is small marsupial mice, sure to be of interest to the British Museum. Yes, they would be, but where exactly are they to be found?

'Why, you'll find them under every old dry tree,'[79] Wilkins is told. He and his men search under every old dry tree they can see with no success. Hundreds of trees, no mice. Wilkins then goes back to all the helpful people who gave him this advice, and gets exactly the same explanation.

'I haven't seen them things *myself*, y'know. Old Jim So-and-so, he saw 'em and he told me.'[80]

Wilkins succeeds in finding Old Jim, who has a familiar refrain. 'Never seen 'em *myself*, but there must be lots, everybody says so.'[81]

No, everybody says *you* said so. Old Jim considers this and comes up with the definitive answer: 'A bloke up at the next station, 'e saw one under a log.'[82]

It takes a while, true, but eventually Wilkins finds the bloke up at the next station.

'Yus, I sawed one,' says the old fella ruminatively.

Thank God!

Where? Where can I find this natural wonder?

'It was – let me see – yus, it must be some fourteen years ago. I was raking up some old logs to make a yard when one of them little things 'ops out. Just like a roo 'e was. That's the first and last one that I ever seed or 'eard of.'[83]

And here endeth both the great marsupial mouse hunt and the last shred of faith that Wilkins has in local tips.

Moving on, as has ever been his wont, two months of hard slog through the stony-hilled country in the west of Queensland – all red rocks, dry watercourses and a hundred hazy horizons stretching before them – the explorer decides to head north to the Cape York Peninsula, that great finger of land in Australia's north-east that points straight to the Equator and gets itself singed for its trouble.

Upon reaching Townsville, the party is welcomed aboard the steamship, *Kallatina*, which will ferry them on the northern leg of this journey.

CHAPTER ELEVEN

TO THE FAR NORTH

*I discovered that adventures such as boys hanker after are gener-
ally the outcome of ignorance or incompetence and there is more
hard work than adventures on expeditions. An adventure often
means a task not done, a condition unforeseen, and in these days,
provided always that you know of others' adventures, adventures
may be avoided.*[1]

Sir Hubert Wilkins, 1928

June 1923, Cape York, secret Wilkins' business

The Far North? Wilkins is distinctly unimpressed, starting with the people
who live here as he finds that the indolence-inducing tropics produce folk
who fritter away their precious time on earth by leading inactive lives.

It is while at Temple Bay, in the far, Far North, right on the Coral
Sea, that Wilkins has his first encounter with members of the local tribes.
The scarcity of encounters to this point has two causes. Local Indigenous
people who don't want to be seen – which is most of them – are generally not seen. And the other, tragic, truth is that the Indigenous people
in these parts have suffered terribly since the days when the first settlers
had come, taken their land, their waterholes, and their whole way of
life. (In the lifetime of surviving 60-year-olds there have been no fewer
than three local massacres of Indigenous people, and two of them were
done officially, by the police.)

'Most of the telegraph stations on the Cape York . . .' Wilkins will
note of his hosts, '[are] seven-roomed house, in the form of a hollow
square. A veranda runs completely round the inside and on two sides
of the outer and on the corners are iron battlements with loopholes.
On the walls hang old Martini-Henry rifles and heavy revolvers, but
these weapons are not needed in the service now, and are only moved
for periodical cleanings. The natives are more easily and successfully

subdued today by the threat to cut off their supply of tobacco, than by the use of firearms or force.'[2]

Still, one evening after they have spent a day collecting and have returned to camp, Wilkins sees two Indigenous women, one old and venerable, the other young and beautiful.

But careful now.

From long experience, Wilkins knows 'something of their customs',[3] which is why, as they approach, he does not even look up. As silent as the grave, as stealthy as if they are of the land and not simply moving across it, the two women keep coming and now sit soundless by his tent.

Again, Wilkins understands.

It seems they are likely envoys from their tribe, and are waiting for him to deign to notice them. Wilkins takes his time, as is the fashion, but finally looks up and says, 'Welcome.'

They smile briefly in reply, but hang their heads.

More silence.

Very well then.

Wilkins must push a little.

'I am pleased you have come,' he says. 'For I wish to get in touch with your menfolk. I hope they will be able to bring me some specimens?'[4]

'This is not possible,' they reply in pidgin English. 'Our men are far away.'[5]

And yet it is not what they say, but the way they say it that counts.

He doesn't believe them and senses the women are afraid of telling him anything but have come on a mission to find out something from him. Sure enough, finally, the question comes.

'Are you a policeman?'

'No.'

'Are you a Government man?'

'No.'

'Are you a missionary?'

'No.'[6]

With those big three dangers denied, the women smile a little more but soon take their leave. Trusting a white man can be dangerous, and they are not yet prepared to give him that trust. They already know he is not a trader or a fisherman, for he has been silently observed over the last few days and it is obvious that he is engaged in neither activity. But they need to take their time to work out what he *is* doing here.

The next day they return in similar manner, and again put the question:
'Are you policeman, missionary or Government man?'

Wilkins shakes his head. The two women eye him warily while having
a long discussion in their own tongue. Then, shyly, the young one asks:
'You want black woman come live longa you?'

'No, but I want your men to come and help me hunt for animals.'[7]

His reply surprises the women greatly.

You don't want black woman come live longa you?

Why not?

'Other white men want women in their camps,' they say. 'They give
the women lots of flour and tobacco.'

'That is not my custom,'[8] replies Wilkins.

Well, then.

He really is a strange one.

An hour later, they have returned, though this time accompanied by
several of their men. Like the women, the men are all but naked, their
modesty protected only by a bunch of leaves ... at best. But they are
missing front teeth and have deep scars on their torso, signs they have
been initiated into manhood, Wilkins knows.

The men have questions of their own. Who is Wilkins? What is he?
What is he doing here?

Again Wilkins explains – he is here collecting specimens of local
animals and plants so they can study them at the Big Smoke – but he is
curious as to why the black fellas are *so* suspicious of his party.

'Why do you think we are policemen?' he asks.

'We are much afraid of policeman, they take us away and lock us
up,'[9] replies one man.

Understandable. Moving on.

'What is the matter with missionaries?'

'We have not seen a missionary,' the man answers, 'but we have heard
they make the black people do many things they do not like and they
take the children away from their parents.'[10]

And the children do not come back. The missionaries come, the children disappear as the missionaries leave, and are never seen again, leaving
wailing parents who mourn them for the rest of their lives.

Wilkins is curious about why women are offered to a white man.
The answer is a rather brutal pragmatism: 'Suppose that man take the

women, then we would no longer trust you people, but would steal all we could from you and then run away . . .'[11]

As Wilkins has refused such offers, it makes all the difference.

'Now we will stop and help you and we wish to be entirely friendly.'[12]

This friendliness runs to teaching the visitors how to enjoy one of their favourite local delicacies: tree-ants, the eating of which is a bit . . . delicate in itself. First you must grip them so quickly and firmly by the head that they can't bite you, and then bite off their large green bottom.

To Wilkins' surprise, they are delicious, with a sweet tangy taste. Given these ants build large balloon-shaped nests in bushes and trees, they are easy to find.

There is even a medical angle, with the Yidiny people insisting that mixing the ant juice with water and ashes can be a cure-all for most ailments.

On the other hand?

'In fact, so many had attacked and nipped pieces from us that to eat them would have almost amounted to cannibalism.'[13]

Wilkins is also fascinated to note that this whole exercise is not just a matter of him observing the Yidiny – for they, too, are closely observing him, and sometimes are more than a little disappointed that he is not reaching their expectations.

One day, for example, he is having his usual lunch from a tin can, using a sheath knife as his only utensil, when he notes one of his Aboriginal guides looking at him with what is clearly disdain.

Why?

The answer, Wilkins finds out after some insistent inquiry, is that the man has turned up his nose at the white fella's dropped standards.

'For a white man to be without a knife, fork, plate and spoon at all meals in front of a native is to lose all dignity.'[14]

Oh yes, make no mistake. The black fellas and white fellas are studying each other, and the black fellas have now learnt enough to know the signs of who are important white fellas, who are middling, and who are scum. And as a general rule, woe betide one with poor table manners, for 'the white man is treated thereafter with studied contempt or ignored'.[15]

Happily, in this instance, this loss of face works out well for Wilkins: 'After our first meal together he showed plainly that he did not consider me to be a respectable white man, he treated me like a brother.'[16] This is precisely what Wilkins wants, to be treated as a fellow, not as an other, with both parties adopting 'acquired company manners'[17] as they usually do.

Every day and many an evening, midst the hard slog, there are new and extraordinary things to see and chronicle. Some 80 miles south of the tip of Cape York, and some 40 miles from the coast to both the east and west, they make camp at the crossing of Lydia Creek. Wilkins is firmly of the view it should be renamed 'Frog Swamp'. For never in all his born days has he seen anything like the number of small green frogs as he sees here. One step in any direction and no fewer than half-a-dozen must croak their last. They're in the trees, by the shores of the creek, in the long grass, everywhere. Without fear or favour they jump everywhere with suicidal eagerness.

> Scores went into the fire, and our billycan and cups of tea received their share. It would have been useless to throw the tea away, because the next brew would have been served in a similar manner, so we sipped our soupy beverage as stoically as possible. I have eaten frogs' legs in France and have found them tasty, but the thought of drinking frog was certainly not pleasant and we went to bed rather thirsty that night.[18]

There will be no easier taking of specimens on the entire trip. All Wilkins has to do is leave a jar with its lid off, and within three seconds as many frogs have jumped in.

It is with relief they leave Lydia Creek to head for the hospitality of nearby Merluna Station, which Wilkins intends to use as his new base for specimen collecting. The hosts are friendly and the accommodation comfortable, but there remains a problem. Though he had already secured one roughly local Indigenous man as a guide, all the others – the ones who can be counted on to know about every living thing in the area – avoid his camp like the plague.

It takes a little while, but Wilkins finally gets his guide to tell him why the others are staying away.

'The bush natives won't hunt for you, not even if you offer them any amount of money or tobacco.'[19]

Why not? Well, they have heard that Wilkins collects skin and skeletons for Britain; after he finishes with the wallabies and the possums, he will start with the local people. It takes Wilkins a little time, but he is finally able to convince his guide that there is no chance of that occurring, whereupon the word spreads to the local Indigenous people, and the power of money and tobacco resumes as normal, with Wilkins' specimen collection rapidly expanding accordingly. Still his pursuits remain a source of merriment for Australia's original inhabitants, no matter how he explains it: 'I told them I was making a collection. That amused them very much. They thought I must be perfectly crazy because I didn't eat these things, but put them into tin cans.'[20]

And yet, other locals in these parts up around the Wenlock River of Cape York remain suspicious of him. The gnarled old prospectors of the Batavia goldfields – all bushy and 'baccy-stained beards, battered visages, and wild squinty eyes – have seen a few things come and go in their time. Why, some of them have been on site since the early days when the local Aboriginal woman Kitty Pluto, you know the one, had discovered a large vein of gold by the river.[21] She'd worked her finds at Pluto's Gully and Camp Choc-a-Block with that bloke 'Friday' Wilson – so-called because as a baby he had been found on a Friday after a massacre of Aboriginal people, and adopted by the Wilson family – before they'd been forcibly removed by officers operating under the *Queensland Protection Act.*

Broadly? Gold was for white people, not people such as them, and they had to be protected from that.

Yes, these old bastards have seen a few things in their time, but a cove putting frogs in jars? Bullshit. That is obviously just a cover for what he is truly here for, no doubt sniffing around to jump their claims.

Again, it takes some time, and strains the visitors' reserves of charm, but finally Wilkins is able to more or less gain their confidences. Each one of them makes Wilkins swear on the head of his mother he will keep secret both the location and the amount of their discoveries, and the newcomer is happy to do so – confident that it makes no difference in any case. He has no doubt that he could yell everything he knows on the corner of George and King streets and it would make not a jot of difference to the net 'wealth', or actual poverty, of these old diggers.

'There is much that I could tell,' he will diplomatically note, 'but the most that I can say is that the men are satisfied to work them. They were each quite pleased to give me specimens for the Museum, and for these gifts I was duly grateful.'[22]

Wilkins is less grateful about the behaviour of his next hire, Bill McLennan. After cheerfully agreeing to collect cave animal specimens,

McLennan is left with guns and equipment for preservation. The result? 'McLennan evidently got bitten by the gold-bug after I left the district, for I never received a single specimen from him, neither was my equipment returned, and all my letters and telegrams were unanswered.'[23]

(Worse still? Wilkins will later be furious to hear that McLennan opened up a 'rich gold claim' and was a magnate in the making, made unwittingly by the explorer.)

In the meantime, returning to his main focus, Wilkins has heard tell of a particular wallaby no bigger than a watermelon, a reclusive marsupial that hides in caves and shies away from human contact, which he expects will, 'likely be new to science'.[24]

Through an arduous day on arid land, Wilkins clambers and climbs over root and rock in search of the blasted little beast. Finally in one of the low caves, peering through the dark he – THERE! *Do you see it?* Twin pricks of light in the musty murk betray the wallaby's hiding spot. Wilkins has seen, 'a glimpse of an eye and the faint outline of a wallaby'.[25]

True, in firing off the shot that kills it he brings down dust and debris on his own noggin, but when all settles down and he is able to drag the wallaby into the light, he feels a surge of excitement. Is it . . . ? It IS! It is, his expert, Professor Kotoff tells him, a brand-new species, never before captured and catalogued by Western eyes.

Several days later they are in the Far North, moving along by the Flinders River heading west into the Gulf Country, when they see something far more exciting.

For, look there!

It is the clearly defined remains of a kind of dinosaur – or at least half fish, half lizard – plainly visible in a nearby hillock, exactly where it got stuck likely 100 million or so years ago.

'We traced the outline of a huge ichthyosaur, about thirty-five feet in length, the head and shoulders and hips of which were plainly visible.'[26]

Wilkins is excited. This is an ancient creature, and you can clearly see it, right there! But the specialists with him, men who have devoted their entire lives to searching for and studying this very kind of thing, are *beside themselves*.

Excitedly, they talk of the world that this creature had been a part of back then, when the Australian continent was covered with oceans where deserts now lie, where the entire land was one vast tropic roamed by giant beasts akin to this.

And yet what competes with the majesty of the find is the irony of it all. Wilkins has been tasked by the British Museum to find and document bird and animal life in the Antipodes before it disappears; and here is a remarkably well-preserved relic of life that predates Britain and any human being that ever walked the earth. An ancient creature of an ancient land, its form has somehow survived all the way into the modern world.

Wilkins carefully shoots the site and the sight with his camera; the twentieth century capturing pre-history in an instant.

These large fossils weigh tons and are too big to carry, but Wilkins does carefully note the location so those who are truly expert in the field can come and see this land so steeped in its extraordinary past that you can still see ancient bones poking out of cliffs and hilltops, pre-historic remnants jutting out like signposts.

Wilkins and his companions keep covering the ground with remarkable ease. All up, he cannot sing the praises of Henry Ford enough as he is able to travel with a speed and ease unimaginable for previous explorers.

'One never ceased to marvel at the ease with which a Ford car can negotiate rough roads; with the aid of pick and shovel work and four or five lusty individuals pushing behind, we could cross most creeks and black soil patches.'[27]

At Mount Wheeler, near Rockhampton, Wilkins was told a story by a local that he finds deeply troubling, one that nevertheless perhaps explains why they have seen so few local Indigenous people in these parts.

> Years ago, a sergeant of the police enthusiastic in his duty of quelling Aboriginal disturbances, developed a blood lust and sought to carry on a wholesale slaughter in support of the theory that no matter how good a blackfellow may be, he is better dead. During a raid on a tribe, and to impress others in the neighbourhood, the sergeant and his constables of the Native Police drove two or three hundred – it is curious to notice that a hundred or so does not seem to matter in the estimate of numbers of victims in a tragedy, when they are only 'blacks' – men, women and children, up the sloping sides of Mt. Wheeler and over the precipitous sides to be dashed to pieces on the rocks below.[28]

His name – Inspector Frederick Wheeler.

Travelling on to the Crocodile Islands, and one in particular, Milingimbi, home of the Yolngu people, Wilkins writes an account that will be sensationalised by others, much to his chagrin. The inhabitants of this island have a peculiarity:

> A curious physiological feature was noticed in many of these unclothed people. They exhibited two or three joints of the lower vertebrae in the form of a tail. The tail was not so noticeable on these people as it is on some of the African tribes, but the tails were quite conspicuous enough to show in a photograph.[29]

Wilkins, with some sensitivity, refuses to reprint such a photo in his book and is infuriated when tabloid newspaper sub-editors put shocking and unfair headlines like 'MEN WITH TAILS' to the extracts that will eventually be published from his book *Undiscovered Australia*. As it turns out, these 'tails' are in fact a lingering symptom of leprosy, but the tales of the tails prove hard to extinguish.

Other things fascinate him about tribes in this area, including one north of the Roper River, 'who, during the mosquito season, live in mud houses. Hitherto it was believed that no Australian Aborigine built houses.'[30]

Extraordinarily sophisticated structures, using solid principles of engineering and construction, albeit with very basic materials, their strength comes from a dome-like frame of heavy tree limbs, upon which a grass thatch is laid to help with waterproofing, and upon that a final layer of mud with a great deal of clay for further waterproofing and insulation.

Much further north, he is fascinated to come across a tribe of 'some 400 people of two distinct types – one tall, the other small'.[31]

He does not understand how this can be, but they have a very clear explanation. Back in the Dreamtime, 'a race of black men lived in the sky. Then a flood came and washed them away. The Milky Way still showed the track of the flood waters. Thus, the black men were washed to earth. Those who fell on soft ground (the swampy country near the Goyder River) remained tall and sound: those who fell on hard ground (the rocky country round the King River) suffered from the contact. So, they are stunted in growth ever since!'[32]

Wilkins enjoys being around this tribe, and admires the simplicity of their lives, but will go on to recount one troubling episode where he and his crew are in their camp when they are approached by a young

Aboriginal woman, again caught between that world and the white world she has half gone to.

'[This] young girl,' Wilkins will recount in one of the newspaper columns he is regularly sending back, 'had for a time worked on the nearest cattle station [and] was a pathetic figure as she approached our outfit. She had saved from her earnings a few rags of calico and was endeavouring to cover as much of her body with these as she could in the presence of white men. With a knowledge of immodesty and immorality born of "civilised" association, she craved for a dress and tobacco, and, in understandable English, voluntarily offered her all in exchange for these things, especially tobacco.'[33]

Wilkins is disgusted. Not by her, but of the forces that have caused this.

And so to Darwin where Wilkins has very little good to say about his countrymen of the Far North:

> It was in this straggling town that we first noticed the meanness which seems to permeate the northern part of the Northern Territory. There was a conspicuous absence of that . . . large and generous spirit which exists elsewhere in Australia.[34]

Others might glory in Australians' propensity for saying 'near enough is good enough' and 'she'll be right, mate', but Wilkins deplores the whole attitude.

Wilkins relates an anecdote about how Darwin fares in the service and hospitality departments. He arrives at the leading hotel in Darwin, the Victoria Hotel, 'the Vic' on Smith Street. Made of stone, it was one of the few buildings in Darwin that survived the Great Hurricane of 1897, and the locals reckon the coolest place in the city on a hot summer's day is on its second-floor verandah, with a beer in your hand. The publican? One of the toughest of all, by local legend. May Brown is a ripper, who learnt how to box from her first husband, and uses those very skills to clear the bar of drunks. Just so long as they don't bleed on that part of the wall where Ross and Keith Smith – who were the winners of the England to Australia 1919 race, that Wilkins had crashed out of in Crete – had signed their names, when they had stayed there. On the day that Wilkins arrives, there is no porter, no clerk, no concierge. He goes to the hotel's bar and approaches the barman.

'Can I obtain accommodation?'[35]

'I s'pose you can,'[36] replies the barman, who keeps pulling beers for the 30 longshoremen, who keep drinking them to the point that the barman only just stays ahead. Wilkins waits for a few minutes and then realises he will have to take the initiative.

'Can I have the number of my room?'[37]

The bartender sighs. These out-of-town dandies are always demanding like this. He has never been able to work them out.

'If you want a room *right now*,' he says, 'just go to the top of the stairs and shout for Johnny.'[38]

The beer-drinking customers look at Wilkins accusingly with what he describes as a 'What's the bloody hurry'[39] look.

As Wilkins will tell the story, he moves away from 'eyes that glared above froth flecked moustaches'[40] and climbs two flights of stairs. 'Johnny?' he calls hopefully.

No reply.

'JOHNNY!' he shouts. Eventually a Japanese boy, rubbing his eyes while yawning, appears.

'What's matter?'[41] he asks, clearly a little offended at being woken.

'I want to engage a room for the next few nights,' replies Wilkins. Johnny looks suspiciously at Wilkins' beard.

'No more room!'[42] he says flatly.

Wilkins is about to explode when, mercifully, Johnny cocks his head and takes another look at the stranger. Why, this is not a crook, he's a star. Johnny has seen his face in the paper! A room is found at once – the best room in the hotel, with a window that actually works – and as soon as Wilkins is settled, Johnny insists he comes downstairs to shake the hand of every man who glared at him, whereupon he is shouted a beer by each man.

It all comes, however, with one coda . . .

One of the drinkers, Bill, offers to send his 'boy' to fetch Wilkins' bags and, sure enough, just a few minutes later, Bill's Indigenous servant appears with the bags at the pub's entrance. Seeing him, Wilkins gratefully invites him to come inside. But the young fellow won't budge.

'He beckoned me to come out, and this I had to do, for there is a law in the Northern Territory which forbids a black man to enter an hotel.'[43]

Just to add a final jab, Wilkins notes of Darwin that: 'The local townspeople were most hospitable to the stranger within their gates; but behind it all was a background that was not pleasing.'[44]

Perhaps Katherine, 200 miles to the south, might be different? Not that it will be easy to get there. The train is known as *Leaping Lena* and leaves Darwin every Wednesday at 8 am. While waiting for it to leave, Wilkins is approached by a local journalist who clearly comes away impressed.

'Captain Wilkins,' the journalist for the *Northern Territory Times and Gazette* notes, 'is a remarkable man . . . An outsize among men, a very modest man (who hates newspaper reporters), but whose wonderful exploits are well known . . . and according to ordinary chances should have been killed about 50 times, but he had wonderful luck . . . It can be said of Captain Wilkins, in the same way as it was said of Cecil Rhodes, that he thinks in continents, especially those continents which very few people think about, viz., the Arctic and Antarctic. He is hoping that next year he will be able to prove that the aeroplane is quite competent to be used in polar exploration and hopes by that means to add another 1500 miles of coastline to the known territory of the Antarctic.'[45]

For yes, travelling through drought-stricken parts of Australia has indeed increased his passion to unlock the weather secrets of the polar regions and he is determined that will indeed be his next project.

For now he must catch the train to Katherine, where the track proves to be so poor the train frequently slows to walking speed and the passengers get out and walk alongside.

Upon arrival in Katherine, Wilkins endeavours to engage guides and donkeys to use as transport and so be on his way, only to find that his timing is out. Any other day of the year, including Christmas Day and New Year's Day that might be possible, but today . . . ?

Today, mate, is the day of the Katherine Races, and the *only* thing that will be moving any distance are the horses and jockeys themselves, for everything and everyone else is at the track, don't you get it, mate? As it happens the race occurs at the town of Emungalen, just a mile north of Katherine.

A strange name, yes?

Yes.

Wilkins is bemused to note that the land was supposed to be called 'Mungalen', as the Aboriginal people had called it for millennia, but no sooner had a 'high official' of England arrived on site than he had, with the confidence of those born to the Order of the Silver Spoon, declared the name should and shall be . . . 'Emungalen'.

'No,' he had been told by the locals, 'there is no "E" in the native pronunciation.'

'Never mind,' he answered. 'We must have it a bit English-like. Emungalen is near enough.'[46]

And Emungalen it had been ever after.

Disembarking from the train, Wilkins is surprised to see an enormous crowd heading to the track, even though the racing itself will not start for many hours. Again, it takes a little while for him to understand. For today is not just about a few horseraces in the afternoon. It is about drinking that started four days ago for a few people, two days ago for most, and this morning for everyone else. It is why everyone he talks to is completely sloshed, slurring their words and occasionally reaching out to grip him by the shoulders – not by way of welcome, but so they can stay upright.

While they're there, George, let them breathily tell you how long their journey has been, how some have ridden 130 miles to watch these races – at least they *hope* they'll still be conscious when the race starts. (After all, just a couple of years earlier this race had made the news in the *Northern Territory Times and Gazette* for the fact that there hadn't been any riotous atrocities of note.)

Looking at the crowd now, Wilkins can't help but notice that many of them are already swaying like straw in a breeze of booze that can be smelt and felt by any within six feet.

Most amazingly, it is not just the spectators who are, to use the vernacular, as pissed as newts, spifflicated from the shivoo, whiffled and jiggered, Aussies ossified and petrified, ploughed on plonk. In other words . . . drunk. As skunks with a bad drinking problem.

For, at the track itself, carried along by the tide of drunken humanity, he is stunned to see that many of the would-be jockeys are also passed out drunk in the gutters, waking occasionally to be barracked as they 'pranced their hacks up and down the stump-ridden lanes which did service for streets'.[47]

At least the racecourse itself is in reasonable shape, *reasonable* being a relative term. Keen patrons will note that several trees have been freshly chopped down on the track itself, which is a good sign. There *is* some organisation here. Wilkins has arrived just in time for the betting to get into stride as one likely lad yells out, 'Ten bob on So-and so!'[48]

This opener does not impress a walrus-moustached drover who answers, 'Ten *fucking* bob! What yer comin' at!? Ten *quid* on So-and-so for me!'[49]

Despite the fact there are only four horses in the race, the drunken drover backs two of them so heavily that Wilkins works out he will only just make a few shillings if either of them wins.

'But what did he care? This was one day in three hundred and sixty-five, and he would show them that he, on that day at least, was no piker.'[50]

God forbid.

But pleashe hush now, for itsh time for the firsht race!

Some races around the country use horns to begin a race, a few use opened gates, still others use a fired pistol.

Here in Katherine, as it turns out, the starter – the one fellow in town with a coat – simply shouts 'Go!' and off they do.

Racing now in the Katherine Cup . . .

The horses take off around the track, their order occasionally obscured as they might pass a thick grove of bush but soon enough reappearing and, amidst enormous cheering, it proves indeed to be one of the drover's horses who gets to the line first but, alas, it has passed 'on the wrong side of a tree stump'[51], a Hoyle's technicality which means the drover has now lost £20, while many others have equally lost. But others who have put their money on the one that came second have won big.

Those who have won, drink heavily to celebrate. Those who have lost, drink heavily to console themselves. But the drover in question – the latest in a long line of chancers since time immemorial – remains undeterred by his loss, and with the simple logic and grasp of maths of the true gambler, asks his wife to give him £40 to make it up at the next race. That good woman, equally the latest in a long line of women of long-suffering sanity, noting that this amount is half their yearly earnings, declines to give it to him.

Wilkins leaves them to argue as he goes for the race 'luncheon' which consists of cheese and beer. A ploughman's lunch for the ploughed . . .

As it happens, and perhaps not surprisingly, Wilkins finds that it is at the bar that the *real* money is being spent. And it is far from just the locals knocking it back. Any stranger, like him, is met with the half-threatening welcome, 'Here you, will you fucking drink with me or won't yer?'[52]

Got it? You can either knock back a beer with me, or get a smack in the mouth for your trouble. So, what will it be?

Wilkins chooses to have one beer and then several, with as many men who started as strangers but who he is soon sure feel so warmly towards him they would happily rename their firstborn after him. In the wildness of it all, as drink after drink is knocked back again and again – *Barman! Give me two more!* – Wilkins is bemused to note how, while speech becomes harder for some, it has the opposite effect on others. Truly, previously taciturn bushmen with thousand-yard stares and mouths that looked as if they had been sewn shut, are now regaling all and sundry with tales of stampedes and other adventures. Elsewhere in the bar – in fact, a rough enclosure, a bark hut with a tin roof – warm greetings turn into fist fights in a couple of blurry instants as the amazed Wilkins records:

> From friendly handshakes they would come to grips and tussles, and then hard resounding blows, 'just friendly like' and others would join the melee. Half a dozen arms and legs would winnow the dry leaves and dust from the trampled ground, and then, when each had received his brand of gravel-rash, the social bout would be adjourned to drinks.[53]

The drinking is then adjourned for more fighting, which not only interests Wilkins as a cultural phenomenon. For as a photographer he also thinks it will make an interesting series for the papers and so sets up his camera accordingly. The men fighting don't know or care.

But the 'semi-sober'[54] certainly do and are appalled – *appalled*, do you hear us? – that anyone would seek to damage our reputation in this manner, making us look like a bunch of ill-educated drunken yobbos who would sooner belt someone than shake his hand. There may be some irony in the fact that to make their point they simply grab the camera and stomp it 'into the dust'[55] but they don't see it like that.

As to the snapper, when he takes a dim view of this approach, he is rather firmly told: 'No fear, you fucking bastard. No monkey tricks today.'[56] Wilkins poetically notes that the would-be chronicler of his times is now treated to a 'continuity of language [that] paled the bright blue sky and echoed to the hills'.[57]

For the unfortunate fellow must be made to understand: This is race day! It is a glorious 'day of freedom to be talked about and laughed about throughout the year, but no permanent record of it was to be placed on photographic paper'.[58]

For his part, Wilkins wisely makes his notes in his head rather than on paper, observing that while, 'the brawling was continuous, the racing was spasmodic'.[59]

It is the tradition in these parts, just as another interesting tradition is how carefully those actual races are run, with the result determined in advance by the bookies.

'The favourite seldom won, unless the side bets made it profitable, but in one case the favourite won by accident.'[60]

What to do?

Lodge a protest!

One jockey obliges, his official complaint dutifully recorded: 'The fucking starter shouted "Go" when I wasn't looking.'[61]

Indeed? And did this result in an adverse outcome? It certainly did, the jockey concludes: 'I couldn't hold the cow.'[62]

The 'stewards' inquiry' – three blokes having a bit of an animated debate away from the others, with just the odd *Fuck 'im!*' escaping – sees the result overturned, which results in wild rejoicing from all bar the jockey who actually won and those who backed him. The re-run sees the jockey who had been protesting take off even before the starter yells go, which secures victory, only for the previous winner to violently protest.

'I don't care if he is my boss!' the jockey roars. 'He chipped the starter and got off before I was ready. He can sack me or I'll sack myself, but my horse will beat his any day!'[63]

Again the stewards consult, and come back with the same verdict: *Fuck 'im!* The verdict remains unchanged, Wilkins notes, despite one unhappy punter threatening to beat one of them to death even while swirling a large club about his head.

This proves to be no more than the opening salvo of those who are aggrieved.

'A drink-demented station hand seized an axe and hurled it with all his might at an officer. The officer dodged skillfully, walked calmly up and took the would-be murderer to the shade of a near-by tree, where he advised him to have a sleep.'[64]

Now, now. You have to understand, there is no sense in pressing charges of attempted murder on Katherine Race Day, as 'on days like this a good deal of latitude and judgement must be exercised'.[65]

(And besides all that, think of the paperwork.)

With the racing finally over – let the record show, 'Burgundy [beat] Rambler'[66] by a head in the Katherine Cup – there is time for just one last tradition to be observed. That is for one last giant punch-up, seemingly involving most male attendees, while their wives, daughters and sweethearts enjoy tea and scones with strawberry jam under the trees not 200 yards away, insistently not noticing. The same cannot be said for the Aboriginal trackers who are accompanying Wilkins and cannot take their eyes off this mass bloody brawl. Their verdict?

'All about that fellow him bin poison longa *debil-debil*.'[67]

Translation: the fighters are drunk as skunks on the devil water. As Wilkins watches, quite stupefied, a solitary Chinese prospector – who comes replete with a long stick over shoulders bearing a wrapped load on both ends, long plaits in his hair and a curiously pointy hat – wanders into town, just in time to stand silently watching the end of the battle.

'What do you think of that?'[68] asks Wilkins, curious.

'No savvee,'[69] he replies.

Wilkins couldn't agree more, and yet his final verdict on this strange tribe of whites he has come across here in the Northern Territory also notes there is another side to them once they sober up.

'And yet,' he notes with wonder, 'these very men on the morrow organised a benefit and raised some thirty pounds for a sick "cobber" up at Darwin hospital. They may not have been able to put up the money, but they mortgaged their services to meet the debt.'[70]

True, there is a problem when the said benefit on the following Saturday becomes a drunken mess that takes up most of the day, but at least their instincts are good. The problem for Wilkins is that given the next day is a Sunday, which is drinking day, and Monday is reserved for hangover and recovery, before Tuesday results in a mass 'hair-o'-the dog' cure, the Friday races take up five full days before anyone is sober enough to offer Wilkins any assistance. In the meantime, Wilkins engages an Aboriginal guide, Charlie – who may well be the only sober man in the district – and camps by the Katherine River.

There is an aching and timeless beauty to the whole area that captivates; the sense that as flows the river, so flows the millennia, onwards forever, and they are little more than specks of the here and now ... privileged to be here. To gaze up through a clear night sky to see the Southern Cross in this land is to be linked to the ages, the lightly lapping waters of the river echoing a very present past. It is a magical place to collect specimens using a flat boat directed by Charlie, and yet far more interesting than the local flora and fauna is having the time and opportunity to talk to Charlie about his extraordinary life, the things he has seen and done, the cataclysmic changes his people have suffered since the white man came, his casual stories of a history filled with hidden bravery and tragedy.

> ... sometimes he would tell me tales of the early days when the white men first visited his part of the country. He had witnessed more than one deliberate murder of a black man and had assisted in the killing of several white men himself.[71]

(Wilkins takes a mental note: be good to Charlie.)

On the other hand, and there really is one, Charlie has also saved the life of several white men, including on one occasion when he had swum through swamps infested with crocodiles to get medical supplies to a dying man – an act all the more courageous because Charlie had a more than usual detestation of crocodiles, because of previous grim experiences with them.

As a matter of fact, one evening as Wilkins and his men smoke their pipes, Charlie is suddenly anxious that the day is not done, saying leadingly that he is sure if they concentrate they can kill plenty of crocodiles this evening. Always interested in getting more and better specimens, Wilkins obliges, and yet there is a problem. The crocodiles' massive jaws and teeth present a remarkably strong argument that you take them on at your peril. And like the roos they also possess an extraordinary ability to 'play possum', to feign death until a better moment comes to strike back.

On this night, Wilkins is excited to shoot a superb specimen which Charlie throws in the bottom of the boat. Later they can gut it, and get it ready for transportation to later be stuffed. Heading back to camp Wilkins spies an elderly lady with two attractive young lasses who want to cross the river. Ever gallant, and interested in attractive young lasses,

Wilkins offers his services, though is careful to explain that, before they accept his offer, they should know that there is a dead crocodile in the boat. Of course, the ladies don't mind. They are Territorians and otherwise extraordinary things simply go with the . . . territory. Alas, no sooner have they stepped into the boat, than one of them actually steps on the crocodile.

'What's that?'[72] comes a sudden shrill female cry.

What's what?

'I felt something move on the bottom on the boat!'[73]

Good God! She's right. The lady's heel has awoken the 'dead' crocodile and, in an instant, instead of the carcass of an enormous 'possum' the bottom of the boat is filled with a wildly thrashing beast and screaming women. The boat rocks so wildly that it threatens to tip them all into the water where they would surely come across *many* thrashing crocodiles, each more aggrieved than the last.

With no other choice, for otherwise all is lost, Charlie and Wilkins both jump on the enraged beast. Wilkins manages to get thick rope between the crocodile's jaws and his foot on the back of its neck. The rope acts as a kind of leash that keeps the crocodile under control just long enough to enable Charlie to steer the boat for the bank. The instant the boat hits the shore the women jump off and, still screaming, flee into the night. The only thing faster than them is Charlie, closely followed by Wilkins, who decides discretion is the better part of valour and – not wanting to look a gift-horse in *this* mouth – it is time to race, Ace.

•

With Charlie by his side, Wilkins' exploration of his own native land continues and, as they push their way up the river, one night they camp by an Aboriginal camp boasting extraordinary bark *mia-mias*, the roofs of which bear beautiful drawings of men and animals, with one peculiarity. The drawings show the internal organs, hearts, lungs and stomachs with remarkable accuracy.

The most fascinating thing of all though is that all the drawings of the men show them with no hearts. What is going on? Wilkins' guide has an intriguing answer: 'Men have peculiar hearts that are not always in the body.'[74]

Wilkins recognises the central truth of this observation, and so poetry trumps physiognomy.

In the meantime, from the Dreamtime, Wilkins is told a story by an Aboriginal guide that he will never forget.

> During the rainy season when the floods were at a great height an old man wanted to send a message to a friend who lived on the other side of a river. Several messengers tried to swim the river, but could not get across owing to the strength of the flood. The old man then made a rainbow and sent his message by that means. The message is still to be found at the end of the rainbow, but no one now living has been able to find out what the message is.[75]

Wilkins loves such stories and so much enjoys spending time with these original inhabitants of Arnhem Land, he promises them a Christmas treat at camp in Milingimbi. Alas, the white man he sends to get supplies and presents for them from Darwin, Mr Watson, is delayed on the boat voyage back and an embarrassed and crestfallen Wilkins has to announce that Christmas will be cancelled. And yet, not long afterwards, a surprise ... A group of Aboriginal men quietly approach his tent. There is hesitation from them all, until the oldest and the boldest says quietly, 'We bin make this big one corroboree all for you. You no come look?'[76]

Wilkins is truly touched.

As Wilkins approaches their corroboree ground, a flat circular space well tamped down, and stamped down from eons of pounding feet, he can see this is no casual gathering. For it is clear all of them daubed with white markings have 'spent considerable time and trouble in arranging a display for me. I had not asked them to do any such thing and, like the average white man, did not credit them with much understanding.'[77]

He is to learn the error of his ways. If they don't quite understand the whys and wherefores of Christmas – beyond the strange story about the birth of a special child in some kind of a large humpy – they do know it is a special day for white people. For *this* white man, they will make an extra effort to make sure it still is.

Delighted yells and capering from the participants give way to an elaborate mock battle, with eight warriors taking up arms on each side and facing off at a distance of 100 yards. These eight stalk and whirl until two men are left, each has four spears at his side and a throwing stick. Wilkins watches as one warrior stands still, defiant and waiting as his opponent selects his spear. The first spear is launched and it flies over the human target's head. Having been missed, the stationary warrior now paces back and forth 'like a tiger in a cage'[78] before settling, and waiting. The next spear comes, it falls just short and the tension rises as the warrior paces back and forth once more. Too far, too short, all know the next spear will be just right. Now the attacking warrior throws one spear fast and straight at his target and Wilkins is amazed to see a blur of motion and second spear launched while the first is still in the air.

With a level of sang-froid that even Wilkins finds breath-taking the motionless warrior watches nervelessly as the first spear hurtles straight towards his chest. The white man gasps, certain he is about to see a grisly death – nothing less than a murder – when, at the very last instant, with a lightning flick of his wrist the warrior brings his own throwing stick up to divert the spear, which sees it glance off and miss his shoulder by three inches. The second spear is also heading straight towards his heart, another flick and that spear is launched straight up into the air, accompanied by a roar from the crowd at such death-defying skill. Although the crowd yell their approval and Wilkins and his men applaud and cheer; the two warriors accept no applause, as they 'with conspicuous bashfulness and modesty'[79] run into the crowd so that they can be lost within it. Their bravery and skill have been demonstrated and that is enough. It would be nothing less than unseemly and certainly un-warrior-like to claim acclaim for their skill and bravery.

(This makes them Wilkins' kind of men, and the Yolngu his kind of people. If only he could do the same after his own feats, and just melt back into the crowd his life would be so much simpler.)

The ritualised dancing now begins – a comprehensive performance involving the imitation of animals and waving of bunches of eucalyptus leaves, all to the beat of bilma sticks – and Wilkins receives his unique Christmas present with delight. (Later in the evening, one youth tries to emulate the warriors and learns how hard it is; his spear goes right into a crowd of observers, missing all but their glares. He too melts back into the crowd, though with a little more haste than the first warriors.)

A fascinated Wilkins chronicles some of the Yolngu's beliefs about 'spirit children', beliefs that have endured since the Dreamtime for the people of Milingimbi and others in this region:

> Several of the Arnhem land natives told it to me as being true.
> They said that all children are first of all controlled by spirits which
> roam the bush, and that they are under the guidance of various

controls, such as emu, crows, pandanus, turtle, etc. Women are not able to see these spirit children, but men can see them, and when a married man sees the spirit of a child under control of a suitable totemic guide, be it bird, tree, or fish, he will send his wife to the place where the spirit child was seen and the child will enter the woman, to be born in due time. Because the father was the first to see the child he is in a position to know its totem, and he alone has the right to name the child after it is born.[80]

Mr Watson and the Christmas presents turn up in time for New Year's Eve and Wilkins is able to hold his own gathering – a 'tournament' of cricket, football and sprint races – before 1925 arrives. Organising a barefoot 200 yards race, Wilkins notes that by his stopwatch the competitors seem to have nearly equalled the national record of 22.2 seconds! When it comes to football, the Yolngu have extraordinary athleticism but, Wilkins records, 'they found it difficult to keep to the rules of football, and liked to kick the ball all over the place and keep it from one another, and they took no notice of "off-sides" or goal-posts'.[81]

Fortunately, the rules of Aussie Rules seem so nebulous anyway, no-one seems to mind.

That evening however, is Wilkins' proper gift to them: a 'lantern show' with magical slides and sights to the amusement of the locals. Only 10 years ago Wilkins had spent a Christmas showing movies to the Eskimos; now he gives a lantern show in Arnhem Land on New Year's Eve, showing slides 'mostly of local scenery and native life in other parts of Australia'.[82]

The Yolngu find Wilkins such a refreshing change. He is genuinely interested in them, he sits with them, eats with them, learns from them, even learning some of their language, and gives to them.

'You proper white man,' Wilkins chronicles them telling him. 'You come sit down "longa camp; no humbug longa women". You eat tucker allasame black people. You no more make 'em allabout work for give 'em tucker, no more make 'em allabout listen when you talk; you sit down quiet and listen allatime and eyes belong you look-about, see everything. Allabout [everybody] feel quiet inside when with you and allabout want to touch you.'[83]

As touching a verbal tribute as it is, far more moving to Wilkins is the intimacy of their collective actions, as time and again, he will be doing work around the camp when he will often find an arm suddenly linked through his, or a child placing their arms around his neck as he works. Most interesting is that, despite the ill-treatment they have too often received from white men, they have such little fear of him. For many he is the first white man they have got close enough to, to explore, and many of the men simply want to touch the oddity of his white flesh, to poke and prod and rub it, to see if it feels the same as theirs. And yet, as ever, Wilkins is careful not to try to touch any Aboriginal women; as such they decide that he is not 'a real man' and so they can be comfortable around him.

He finds the women 'had many secrets of their own that they could not tell a "real" man, some of these they told to me'.[84]

Indeed. There is a sensitivity to Wilkins – perhaps even a curious vulnerability?

This happy bubble is in strict contrast to his next ports of call, as he moves through Elcho Island, Melville Island and Caledon Bay, where he hears a story about an entire Japanese sailing crew – bar the two men who were spared – slaughtered after making advances on Indigenous women. The two surviving men, shattered, made their way to Elcho Island, only to find the local authorities had little interest in vigorously pursuing the matter. This is a frontier land. Things happen. Yes, they make some inquiries, but no Aboriginal people offer them any information at all, and there is no sign of any bodies. Justice, it seems, has been carried out and the matter is concluded by artful neglect.

Wilkins spends the next three months with the remote mission outpost of Groote Eylandt as his base.

Most distressing for a man of his sensitivity is when he is approached by some 'half-caste'[85] teenage schoolgirls who confide in him that they feel they have no future. Neither fully black nor fully white, they find themselves not as the bridge between the two societies but exemplars of the chasm, as they are accepted by neither blacks nor whites. Can this kind man help them escape, does he have some advice, can he point to a path they might follow?

> These girls could not see through the dark days ahead to a clear
> light beyond. Were they to grow up and marry half-caste boys
> – their only choice – and face a life of drudgery? They could
> not legally or contentedly marry a full blood native. Would any

322 • THE INCREDIBLE LIFE OF HUBERT WILKINS

respectable white man marry them and take them to his family? They knew enough of civilisation not to hope for that, although many men might do worse than marry them. I could give them very little consolation, for I could not solve their problem, but tears of compassion were often in my eyes when these girls sobbed in the anguish of their fate.[86]

Finishing his assignment for the British Museum at last, Wilkins sails for Thursday Island, 400 miles east at the tip of Cape York, and has no sooner docked and made his way into the town proper than he sees a casual acquaintance rushing towards him, hat in hand, across the street. The man shakes his hand and gasps breathlessly: 'So, you're safe after all! How glad I am to see you again! We all thought you had gone for good this time.'[87]

'Of course, I am safe,' answers Wilkins 'Why not? What do you mean – gone for good this time?'[88]

'Why you have been lost for months! They were just going to send an expedition to find you when you were seen at Townsville!'[89]

Wilkins, not especially concerned about other people's fears, summarises it thus: 'Those that knew me felt sure that I was safe; those that did not know me felt sure that I was dead. There were many that did not worry either way.'[90]

Wilkins is amused, more so when a flood of cables arrives celebrating his still being alive and found. 'For my part I have never been lost; I have been as dead as newspapers can make me several times, but with Mark Twain I can only say that these reports were somewhat exaggerated.'[91]

The press is nothing if not laudatory of the man himself, and Australia gets to know him a little better, with *Truth* one of the papers that brings him to the attention of its readers in late June 1925, together with the first signs of the next project he has in mind.

CAPTAIN GEO. HUBERT WILKINS, the explorer, who is anxious to raise £15,000 for an Antarctic expedition, is the strangest man in Australia today, and the most travelled. He has not been longer than six weeks in any one place during the last 16 years, and never longer than 14 days in any one house during the last seven years – and this unusual man who has thoroughly explored every country in the world, including both the North and South Poles, and excluding only China and Japan, is only 36 years of age. He has been six times reported dead, once for nine months, but as he says in a cultured accent and a soft, melodious voice, 'My friends don't really worry very much about me now.'[92]

Quite. The *Truth* continues:

> In fact, he has made of the world a chessboard, and has moved on to every square ... He has but one recreation – making the world seem smaller every day, and growing each day more modest about it.[93]

Whatever else, there is no doubt he has done a wonderful job on this last project, as his collections are complete and he has ample samples and specific specimens all ready for the British Museum.

The final tally of two years of toil? Wilkins and his men have collected 1514 ornithological specimens, 861 mammalogical specimens, 1631 entomological specimens, 231 botanical specimens, 160 zoological specimens, 225 reptile specimens, 417 geological specimens, and 76 ethnological specimens.[94] (Of course, there was that female roo and joey left alive, but it is a pretty good tally, all things considered.)

The most astonishing discovery for the British Museum is that the expedition has come in under budget; only costing them £3,010! The Director of the British Museum, Sidney F. Harmer, is delighted.

> Dear Captain Wilkins ... I may perhaps be permitted to allude to the very successful efforts you made to find ways of diminishing the costs of the expedition to the Museum, which you have always treated with conspicuous fairness.[95]

It is an unprecedented situation, so much so that the Museum gratefully gives Wilkins the £3,000 they had already budgeted for extra expense that never eventuated.

In Wilkins' travels and trials, he has also found the time to pen a book about his experiences, titled *Undiscovered Australia*, labouring little to put any positive sheen on things. Although the book is full of his flashing wit, its verdict is gloomy. 'There is no doubt,' he wrote, 'that we are witnessing the passing of these mammals, and that as far as indigenous life is concerned, Australia is in the death-throes.'[96] Wilkins portrays his fellow Australians as brash, drunken, lazy and all too often racist and callous when it comes to the plight of the Indigenous people they displaced. It is far from a fond farewell, and Wilkins leaves for England more a foe than a friend to his own country.

Australians are often amused to be criticised by foreigners but . . .

To be criticised by one of their own, a farm boy from South Australia no less, or rather no more, is an unforgivable offence. Wilkins will never again spend more than a couple of months in Australia; the Arctic and the Antarctic are once more to be his obsession and his true home.

When it comes to true love; many of Wilkins' friends and all of his family are most surprised to read the following announcement in the press:

> VICTIM TO CUPID: Capt. G H. Wilkins Falls . . . the Australian explorer, has become engaged to Miss Lorna Maitland, says a cable message to 'The Herald,' Melbourne. Miss Maitland is a daughter of the late Col. Maitland, a doctor of the British army, who was stationed at. Madras, India.[97]

A few weeks later Wilkins' bemused mother receives a letter from her son, already speeding to America by boat: 'Perhaps you have seen some things in the paper about my engagement to be married. That is so. I have known Lorna Maitland the girl I'm engaged to for seven years.'[98] *Well, if the boy is in love* . . . but as his mother reads on, these words cast doubt on that notion: 'I do not know when we will get married. Perhaps not at all, for I must go on with my polar work and while I am moving about so continuously I do not intend to get married.'[99]

CHAPTER TWELVE

ARCTIC BY AIR

If you can . . . watch the things you gave your life to, broken,
And stoop and build 'em up with worn-out tools . . .

Rudyard Kipling, 'If'

With so much achieved at 37 years of age, the name of Captain
George Hubert Wilkins has become one of the most honoured in
Australian annals and world famous, adding to Australia's status
among the nations.[1]

Daily Advertiser, February 1926

Wilkins is the bravest man I have ever met.[2]

Vilhjálmur Stefansson, 1928

Early November 1925, London, no way Norway

Captain E.R.G.R. Evans, the one-time comrade of Scott of the Antarctic – second-in-command of his last expedition – momentarily has the floor.

'Gentlemen, please put your hands together, for the most remarkable and most distinguished Arctic and Antarctic explorer of the day, a worthy descendant of the Vikings, one of whom landed on the American coast more than 500 years before Columbus sighted it . . .'[3]

And now, here is the Vikings' modern heir, Roald Amundsen!

In this private London club, for this private swish lunch, the said gentlemen applaud warmly to welcome the great Norwegian – the first man to force the fabled Northwest Passage from the Atlantic to the Pacific Ocean, across the top of the Asian and European continents, the first man to reach the South Pole, just four weeks before the doomed Scott.

These days, Amundsen is getting on. Though only 53, he looks well over 60, with chopped white hair, craggy features and a weathered face that would do a rock-face in Siberia proud. Despite that, he stands tall and handsome still, the piercing blue eyes of a Viking gaze intently; and he dresses more like a dandy (fawn gloves, no less!) than an explorer.

Today, however, he is not here to talk of his triumphs, but his recent near-death experience, the story of how just last year he had tried to match his South Pole feat by being the first to *fly* to the North Pole and back. What a tale he tells, in his deep, gruff, heavily accented version of English, complete with lantern slides.

Starting from the Norwegian island of Spitsbergen – halfway between the Norway mainland and the North Pole – he and his crew had set off on 21 May 1925 in their two German-designed, Italian-made Dornier Wal flying boats. They had been funded with $85,000 by Lincoln Ellsworth, son of an American millionaire, on the condition he could come too. Their plan was to cross the ice pack and reach the North Pole some 750 miles away, then land on the ice, and transfer the remaining fuel in one plane to the other before flying back to Spitsbergen, leaving the plane with the empty fuel tank abandoned.

But, my friends, what a near-disaster.

About 150 miles away from the North Pole, Amundsen's plane had developed engine trouble and they decided to land on what looked from on high to be clear ice. Only when they got closer to landing did they see it: mountainous pressure-ridges of ice, snowbanks, open water, icy sludge, icebergs.

'We were caught like rats in a trap,'[4] Amundsen tells them. Nevertheless, at the last moment, both planes manage to get down intact on small patches of slush and find themselves about four miles apart, just visible to each other. On Amundsen's plane – with its nose right up against an iceberg – they must jump out quickly to prevent the slush freezing around the plane's floats. And this, friends, is where their whole dreadful saga began . . . as for the next *four weeks*, half-starving, the two crews, six men in all, had to first get to each other, abandoning one plane and then working to fashion a runway out of the ice-ridges long enough, wide enough, and stable enough to get off again. This was complicated by the fact they had no tools for the task and had to use everything from ski tips to tin cans to form the 1500 feet of flat, firm ice required. And the whole thing had been hopelessly complicated by the shifting ice, and the extreme weather conditions which kept destroying their runway. Time and again they thought they were ready, only for the ice to break up.

'Captain Amundsen,' a correspondent at the lunch will report, 'gave a vivid and detailed account of the difficulties experienced by the party when fighting the ice in their attempts to free the aeroplane. Time after time they levelled the surface for hundreds of feet, only to find their labours in vain owing to the sudden change in the formation of the ice.'[5]

For, make no mistake . . .

'It was,' the explorer tells the enthralled audience, 'a battle for life.'[6]

As they kept working, getting ever weaker every day on half-rations, a huge mass of ice resembling the Sphinx kept getting closer, forcing its way through the main body of pack ice.

'We imagined we could see the Sphinx nodding its head and chuckling with joy; now it was going to get us.'[7]

After four failed attempts, they had one last chance on the afternoon of 15 June. After discarding everything from the plane bar the bare essentials, to allow it to lift more easily, they climb on board. The engines are put to full throttle at 2000 revolutions a minute. Vibrating heavily, the plane starts to move forward and then starts to jolt and scudder across the mostly levelled ice-ridges, and just manages to jump several small chasms that had opened up on the runway that very morning. And finally the blessed moment.

The sound of the dreadful scraping on the ice ceases, and they can only hear the engines humming.

They had lifted off!

Their last look back had revealed the Sphinx, gazing forlornly at the spot from where its prey had so suddenly escaped. A few anxious hours later they had landed back at Spitsbergen, a seeming visitation from the dead.

'What was the result of the expedition?' asks the explorer rhetorically, before giving his own answer. 'Two hundred thousand square kilometres of new territory.'[8]

Having flown where no-one had flown before, they had been able to determine a few islands, but no mountain, and certainly no vast tracts of land in that area. Against that, however, and on this point Amundsen is particularly firm, they had also proved the complete hopelessness of planes for polar exploration. The ice floes were simply not fit for purpose, and no plane, no matter how modern or well-designed, could hope to land and take off from that ice without trouble.

'Aeroplanes must always be prepared for landing,' he says, 'and landing upon polar ice is not practicable.'[9]

And that is that. An extraordinary story, wonderfully told. Everyone applauds warmly, even those who do not agree with his conclusion about the impossibility of using planes for polar exploration. One of those is George Wilkins, who had been sitting quietly up the back, and is convinced from his own knowledge of the polar regions that not only is it possible, but that he personally is the very man who can do it successfully. He departs silently, with much food for thought, but as determined as ever to explore Antarctica by air. Only a short time later, however, he has what might well be described as a case of '*vu déjà*' if one may so describe something feeling familiar as one heads in a polar opposite direction to that originally intended.

Over a decade earlier, while in the Caribbean, Wilkins had thought he was heading to the Arctic, only to quickly head to the polar opposite. Now, despite his Antarctic intent, a cable arrives from none other than Stef urging him to come to America at once, as a project is forming to fly across the frozen ground they know so well, the Arctic, and he is telling everyone that you, George, are the only man for the job. (Stef remains one of Wilkins' most enthusiastic boosters, for, as he frequently declares, Wilkins is one of two men who learnt and followed his lessons properly: 'Wilkins and Storker T. Storkerson are the only other Arctic explorers who practise the Stefansson method.')[10]

Wilkins is on the next ship across the Atlantic, and disembarks in New York a fortnight later to find that the project is well advanced and Stef's good word has advanced him into being the man for the job.

'There have been controversies as to whether this region is land or sea,' *The New York Times* notes. 'Tides, winds, movements of birds and other phenomena indicate the existence there of a polar continent. Captain Wilkins should solve this problem.'[11]

True, there are many worthy experts who say the whole concept of flying in the polar regions is madness from the first and, of these, the great Norwegian explorer Roald Amundsen continues to be the loudest and most forthright, even writing a note of warning and reproach to Wilkins for entertaining such a folly: 'What you are trying to do, is beyond the possibility of human endeavour.'[12]

But Wilkins will endeavour despite all naysayers – he knows planes, he knows the Arctic ice; it can be done or he will die trying. Amundsen is more succinct: *to* try, is to die.

Wilkins' goal is to be the first man to explore the Arctic by plane and to reach the 'Pole of Inaccessibility' or 'Ice Pole' – a concept invented by Stef, which is essentially the point on the Arctic ice mass furthest from all land masses, 800 miles from the nearest land (Henrietta Island in the East Siberian Sea), and 500 miles from the North Pole itself.

How to get such a project underway?

There is only one place to launch from – the USA, a place awash with both money and interest in the northern climes, not to mention owning the perfect staging point to begin the journey, which is the Territory of Alaska. What Wilkins needs to do is to find potential sponsors through his many friends there, men who can write big cheques in return for the publicity that will come their way, once he succeeds.

Unfortunately for him, he is too convincing. In New York, he asks for funds for one plane, one pilot and a mechanic. Smaller is better in his view: 'When all is said and done, the navigator and pilot would have to do the work. But they don't do things that way in America'.[13] No, they do not. The North American Newspaper Alliance, a syndicate of over 50 papers who pool their resources and coverage, agrees to give Wilkins $25,000 just to get the ball rolling. Announcing the news to their readers, they define him as 'the greatest aviator of all world explorers, and the greatest explorer of all aviators'.[14]

Once the publicity drive starts rolling it does not stop. Henry Ford consents to meet Wilkins and shake his hand.

'Within the means of millions' is Ford's latest slogan to sell the Model T, and his blessing means millions will take Wilkins' schemes and dreams seriously. Wilkins knows that to interest Mr Ford is in his interest and proceeds to dazzle him. As his old friend Frank Hurley reports, with a combination of admiration and envy, Wilkins can charm any titan when he sets his mind to it: 'He can enter the sanctum of a captain of industry like Henry Ford and thoroughly impress his person-ality on the great man – and, again, get what he wants (or something like it!). One of his most formidable weapons, apart from his peerless powers of persuasion, was his fund of stories. At telling stories he can beat newspaper men – which is saying much – and he can divert Henry Ford himself.'[15]

And so a transfixed Ford is treated to Wilkins' highlights reel: the story of how the simple ability to fix an engine had led to his life of adventure, including Balkan flights over the foe while juggling a camera, being on the spot when the Red Baron was shot down, dangling from a balloon while some German Fokker fired at him, flying into a lunatic asylum in Crete on the way to Australia, and the stunning reliability of the Ford van he had driven all through northern Australia. Each yarn leads to laughter and questions from Ford, who wants to know all about this fellow's nine lives and what he can do to help with the tenth.

The Detroit News (a single paper but a mighty one as the Fords had forged Detroit into a world power instead of a mere city) also offers funds for exclusive coverage – even more exclusive than the 'pool' coverage already promised by the Alliance – and the millionaires of the Detroit automotive industry line up to see their engines tested in the air.

Engines? Oh yes, the Detroit men do not believe that less is more, they believe that more is more, and the only thing better than more is *even more*. Wilkins will not be allowed to creep out of America with just one plane.

'We must,' he records his quandary, 'have two if not three planes, mechanics, superintendents, photographers, newspaper-correspondents, wireless operators, and assistants, a supporting party to depot our supplies, spare pilots and a host of other things, to my mind quite unnecessary.'[16]

And speaking of unnecessary, a Board of Control is appointed to, well, be in control of all approvals of plans and management. Wilkins goes along with it reluctantly, later noting, 'A Committee is a group of incompetents, appointed by the unthinking to determine the non-essential.'[17]

Still, it is nothing if not high-powered, and is headed by William B. Mayo of Detroit, general manager of the Ford Motor Company. It also includes Dr Isaiah Bowman, director of the American Geographical Society, and none other than Wilkins' old compadre, Vilhjálmur Stefansson.

A committee of aeronautical engineers is assembled to choose the best equipment, a finance committee to manage the funds; not to be confused with the official manager, the assistant manager, a secretary, a treasurer, transportation staff and trained associates. All Wilkins had wanted was one plane, one pilot and maybe one mechanic.

The Board of Control produce many candidates, compulsory and otherwise, for the expedition, including a favoured pilot: the eminent and eminently qualified Major Lanphier, the 'dashing, brilliant commander of the First Pursuit Squadron of the United States Air Force'.[18] However,

Wilkins is not pursuing glaciers, he hopes to rise above them. Lanphier can command a squadron, but Wilkins is after a solitary man to follow his own command. Nevertheless, Lanphier joins the expedition as the Board's favoured fellow. (The US Air Force gives him four months leave, for a key reason. His American presence will strengthen American claims to any new lands discovered.)

The Board has strong views on the aim of the exercise, having seized upon Wilkins' proclamation that he hopes to see no less than a million square miles of territory previously unseen. For the Board this is one million square miles that might be claimed for the United States. Of equal interest is establishing a commercial flying route over the top of the world, capitalising on the fact that the shortest route between America and Europe is not around the globe but over the top, and, if this works, an aerial Suez canal might be theirs.

Wilkins' aims are more modest, perhaps a base might be established for . . . scientific research?

He is duly humoured that this may well be the case. (They will be sure to find a nice square space for his base, somewhere in the million square miles.)

Amidst the whole circus, Wilkins follows the recommendation of his old colleague of icy climes, Stef, and arranges a meeting in New York with Lieutenant Carl Ben Eielson, an extremely serious and seriously gifted pilot.

Short and darkly handsome, with a rather hawk-like visage, Eielson genuinely looks a little like the American bald eagle. He is a man who ideally needs a plane to fly but, if it came to it, could probably take flight himself. There is something about Eielson's completely self-contained nature – he is in it for the flying, he has come to New York on another matter with no interest in Wilkins' expedition or its attendant publicity – and this particularly appeals to Wilkins. The more Wilkins gets to know him, and his background, the more he becomes convinced that it is Eielson who should be his principal partner in the northern skies. True, the North Dakota–born man, nudging 30 years old, has no experience flying in the Arctic, but his experience in other realms, including extensive flying in Alaska, is particularly compelling.

Eielson was born in the fly-speck of Hatton, North Dakota, and spent his boyhood outdoors – camping, tramping and hunting – before enlisting in the US Army Air Service during the Great War, only to narrowly miss out on seeing action. The war ended, but Eielson's passion for flying grew, most particularly after he got the job of delivering mail in Alaska. His first delivery was a disassembled plane he boxed in Chicago and accompanied to Fairbanks, where – in the absence of air mechanics of any description – he re-assembled and mounted the engine himself, at which point the real fun had begun. There were very few landing fields anywhere in Alaska, least of all in the towns he would be visiting to deliver mail, so Eielson pioneered the use of ice instead, landing on frozen rivers, bays and harbours.

The mail must go through, and Eielson – who now calls himself an 'air gypsy'[19] – got it through. And now he is in New York, looking for extra finance to form a flying company to bid on the Government's Alaskan mail contracts.

But couldn't he put that off for the moment, and come and join our expedition? Wilkins queries.

Well, despite his reluctance to have anything to do with all the hoopla of expedition, for an air gypsy of his nature the opportunity to fly the Arctic is seriously tempting and, as he likes Wilkins as much as Wilkins likes him – they are a match in character, approach, personality and understated manner – he agrees to terms.

Wilkins feels blessed to have engaged such a good man and would happily have made him his No. 1 pilot. Nevertheless, another man he respects, Dr Isaiah Bowman, proposes that Wilkins offers the role of deputy in his expedition to Richard E. Byrd, a wealthy former navy pilot who has announced his intent to be the first man to fly to the North Pole.

Byrd quickly accepts with a grand and gracious letter. 'I will be delighted and honoured to serve as second-in command. I do this unconditionally as I do not hesitate to follow the leadership and judgement of Captain Wilkins.'[20]

Two weeks later another letter arrives from Byrd. He now has funding from the US Navy and is leading his own expedition. 'I must say to you very sincerely, that I am distressed that I won't have the opportunity to serve under you.'[21]

This distress is, true, a little surprising given that it could be so easily remedied by not breaking his previous commitment, but Wilkins is nevertheless happy to accept the paper resignation of his paper deputy.

And yet, the fact that Byrd wants to be the first to fly to the Pole goes with the times. For no fewer than three other explorers have also announced such plans in what is, in many ways, a revisiting of the famous race to the South Pole in which Roald Amundsen had triumphed, and Scott of the Antarctic had perished. Amundsen himself is now going to try to cross the Arctic in an airship, which is currently being reconditioned by the chief of the Italian Air Service, Colonel Nobile. The millionaire explorer Lincoln Ellsworth is to act as his navigator, supplying direction and funds as required. Meanwhile, the older Norwegian legend, Fridtjof Nansen – no less than the doyen of polar explorers – will also try, using a German plane. The French, unwilling to be left behind by the Germans (*sniff*), have also announced they will be making an attempt under the auspices of the Ministry of Marine.

A race!

The press loves it, and Wilkins is dragged into it, his lack of interest no match for his press men's ceaseless attention to this angle, for the interest it generates. None other than the American president, Calvin Coolidge, affirms his interest, writing to the chairman of the Board of Control: 'The flight has aroused the keenest personal interest, and it is fitting that we should strive to be the first to open these unknown lands to the knowledge of the world. The importance to commercial aviation in the possible development of air routes across the Arctic region makes the proposed enterprise of particular value.'[22]

But just as Wilkins' South Australian predecessor in polar exploration, Sir Douglas Mawson, looked with disdain on the carry-on theatrics of being 'the first' to plant flags – preferring to explore the new territory of Antarctica and expand the boundaries of scientific knowledge, so too with Wilkins. From the beginning the actual crossing of the North Pole is only tangential to what he has always wanted to look for – *new* land, most particularly land where they might be able to build the myriad meteorological stations he dreamed of, stations that may help the world understand the weather and predict the crippling droughts of his youth. And this trip is only a preliminary to the big one he has planned for later in the year, flying along the coast of Antarctica to explore from the air what has previously only been done by hard slogging. If all goes well, they could perhaps begin building the meteorological stations in a few years' time.

But what route? What goal?

'The Board of Control and my friends,' Wilkins will recount, 'were anxious that we should attempt a trip right across the Arctic Ocean. I was not keenly interested in this because such a trip would mean crossing many miles of territory already explored. It was my desire to spend all my energies and take whatever risks were necessary to cover only unknown conditions.'[23]

Broadly, he feels himself in the science business, not the hero business. 'However, it was agreed that when we had concluded as much real exploration as possible, we should, as a culmination to our efforts, attempt a flight from Barrow to Spitsbergen.'[24]

It makes sense. While flying from Alaska all the way to continental Europe would be an impossibility, the remote island of Spitsbergen halfway between mainland Norway and the North Pole would be the perfect spot for an incredible improbability, a target to aim for when flying over the roof of the world.

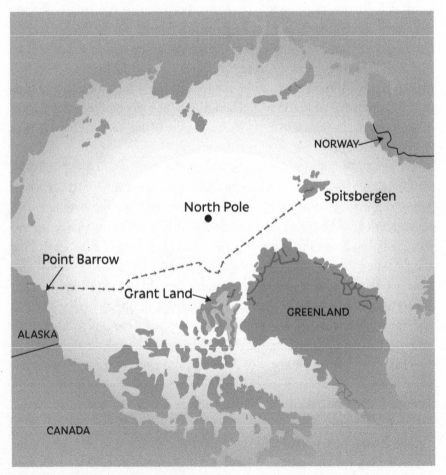

Flight from Alaska to Spitsbergen, 1928

Immediately taken with the idea, Wilkins moves with particular speed. Even though the Select Committee of Detroit Engineers is still debating which planes might be best for the expedition, Wilkins knows precisely what he wants, and delights in discussing it with no less than Anthony Fokker, the most famed aircraft designer in the world, who he meets in Detroit not long after arriving. Wilkins confesses his admiration and desire for a Fokker aircraft to do the job and the softly spoken Dutchman – whose planes had become famed in the Great War courtesy of the Germans embracing his designs, and the fact that the like of the Red Baron had made his planes legendary – immediately offers to take the Australian up for his first go in his newly designed plane with *three* engines – driving propellers on each wing and on the nose.

When?

Right now.

Within two hours of shaking hands, Wilkins is sitting beside Fokker as the Dutchman puts the craft through its paces, even while they talk about Wilkins' plans. Fokker is impressed and, that evening as they journey back to New York together, Fokker tells him of the stunning new large-wing monoplane currently being built in his Holland factory.

'With that machine,' he said, 'I believe you can do all you want to in the Arctic Ocean and then fly to Spitsbergen. If you can find your way, and if you do get there, I myself will come and fly the machine from Europe to America. We will bring the plane at once from Holland, and if those Detroit men won't stand behind you I will provide the machine.'[25]

Wilkins, of course, leaps at the offer, convinced that there is not 'another machine of its size in the world so well made, or so efficient as a large Fokker monoplane'.[26]

With friends like this Fokker, Wilkins can achieve anything. His American backers, however, are more happy to plan than to *do*. When Wilkins asks about a date this year when he can leave he receives a worrying reply from a cavalier Detroit supporter: 'What difference if we don't do the job this year? You go and gain experience, test the equipment, and the work can be done next year or the year after if necessary.'[27]

Gain experience?

Gain experience?

Is that a serious comment or meant in jest, Wilkins wonders. He responds with an uncharacteristic venom, pointing out that he *already* gained that experience after spending three years in the Arctic, particularly given that they're including the experience to 'walk home if need be'.[28]

As it happens, however, the brief impasse between Wilkins and the Board is broken by the public excitement that starts to build over this exciting idea, this 'spectacular dash from Barrow to Spitsbergen over the top of the world'.[29] While Wilkins might roll his eyes at such a 'stunt', it grips people's imaginations – not just to reach the North Pole but to have it as a halfway point as the frozen world is conquered from the sky – the sort of feat that gets people talking and the Board back on board to make it happen in the coming season. Whether it is possible is another matter entirely, but it is the seeming near impossibility that makes it so captivating in the first place. What matters right now is that the press has a headline and the public has a hope, and that is more than enough to get the ball rolling and get the propeller turning.

With a plan in place, and on the path to public promotion, the expedition is inundated with funds, never mind what Wilkins thinks of where it came from. Some of the wealthiest men in America are now behind this expedition as well as 80,000 private subscribers and thousands of school children who have taken a portion of their lunch money and donated it to the cause, five cents each. The children do so with the promise that their names will be inscribed on a monument to commemorate the expedition, but such an offering is not enough to deflect the detractors who criticise Wilkins and co. for 'extracting for our personal use money from defenceless infants'.[30] His old friend Frank Hurley will later note just how much Wilkins hated having to agree to it. 'There was another side to the money-getting which George detested, but which American psychology made it necessary for him to follow; that is, getting money from the schools. His backers told him it had to be done – he must go to the schools individually, talk to the children and "inspire" them to contribute. This Wilkins called "petty larceny" from the kids.'[31] When he also has to literally march in front of a brass band for publicity, Wilkins tells Hurley he felt like a fool and a knave at once. (Then again, the combination of Wilkins and music has never been an overwhelming success.)

As it turns out, Wilkins is now able to learn the key difference between people who pledge to be subscribers and the people who actually pony up real money. It is a large chasm, but at least, after Wilkins has worked on some of the biggest millionaires in the city, large cheques are indeed written, enabling him to make plans to mount the expedition for *this* year.

The expedition has even taken on a press man in the form of Palmer Hutchinson – a special correspondent for their sponsor, the *Detroit News* – in spite of Wilkins' qualms. But what a press man he proves to be, distilling his own excitement into palatable print that the public lap up, hungry for more. It's no wonder, Palmer started out as a sports writer and if there's one thing he knows, it's getting his readers as invested as he is. Palmer is determined to transform Wilkins into a star, a public darling whose name will linger on the lips of everyone who speaks it. On top of that, the young Detroit man happens to be a cheerful and charming character that the Australian greatly warms to.

Yes, a press man to beat your drum is all so very American, but . . . on the other hand?

On the other hand, so is all of their money, so grin and bear it, Captain Wilkins.

Besides, apart from everything else, Palmer writes well, pointing out to his readers, among other things, that this whole venture is so much more than 'a race', and that it is beyond everything else a highly complex exercise . . .

> SEWARD (Alaska), March 6. – The scientific equipment of the Wilkins' Polar Expedition has excited the admiration of the experts attached to the civil and military administration here . . .
>
> An 'Earth Inductor Compass' enables the pilot to determine the direction in which he is flying. It is the only compass which is satisfactory for use in the Arctic regions. The ordinary type of magnetic compass is quite useless, because the earth's magnetism is not horizontal in the polar regions, but is very nearly vertical thus exerting a strong force tending to tilt the compass card and causing very uncertain functioning.
>
> When Captain Wilkins flies northward over the uncharted polar wastes, he will seek to trace his course by a new short method of navigation . . . It deliberately sacrifices accuracy for speed of calculation. It gives ground to the fact that the Detroit expedition navigators must work rapidly And in bitter cold. But it [should] tell Captain Wilkins his true position within 30 miles.[32]

One man continuing to pay very careful attention to everything written about the Wilkins expedition is Amundsen, and his public criticism becomes ever more shrill as the plan gains momentum.

'He will never make it in a plane,' says Amundsen to the press in Vancouver.

Why, the lad proposes, no, *announces*, that he will land a plane *on Arctic ice*! Amundsen knows better: 'I learned a lesson on my last voyage. It is almost certain death or ultimate destruction to land on the ice floes.'[33]

Fine. Wilkins is not bothered, and informs a friend in Melbourne of his plans.

> My first goal is the ice pole, but many geographers believe that I shall discover new lands before reaching the Pole. If so, I shall drop an American flag on it, claim it for the United States, take photographs and return to Point Barrow, my starting place.[34]

It will be risky, but he is fully cognisant of those risks.

> When we start, it will be with the understanding that if our plane fails to reach Spitsbergen or return, no rescue expedition will be sent out for two years.[35]

But not to worry. So long as they survive the landing, all is not lost.

> We can live on the ice indefinitely. Seals and polar bears will supply food; their hides will be material for boots and clothing, and their blubber will supply fuel. Amundsen succeeded in getting his plane away from the ice after a forced landing and we see no reason why we cannot do the same.[36]

One of the key things to be successful, Wilkins knows, will be timing. They will need to make the attempt in that brief window of time when there is enough light to see for most of the day, while it is also still so cold that the ice will be thick and able to serve as their landing field. Consolidating March–April as that window is that, as Wilkins notes, 'later on, obscuring fogs would rise above the Arctic snow, making accurate navigation impossible and absolutely preventing exploration.'[37]

(As Byrd is using only solid land for landing in his own plans he is not as time restricted, while Amundsen, Nobile and Ellsworth's *Norge* Zeppelin can also freely float above such considerations.)

Now, while no public announcement has been made on which man will actually act as pilot for the Arctic flights, Wilkins has already made a quiet decision: 'I had early made up my mind that Eielson was the best fitted for that work.'[38] With that in mind, Eielson is duly sent to

Fairbanks, Alaska, to get ready for departure while Wilkins remains with his ever growing crew, and does the publicity rounds at the behest of his sponsors, to keep the money flowing, much as he detests it.

> I would not have believed it possible that I could withstand the humiliation of accepting a luncheon in my honour and there and then plead for money. The greatest hardship I have ever suffered, and the most trying ordeal ever undertaken, was to address from time to time an audience of harassed business men on a subject in which they had not the slightest interest nor understanding, and then stand while some raucous-voiced, high-pressure salesman sold my photograph for a hundred, fifty, twenty, five dollars, fifty cents or what have you. I still shudder when I think of it.[39]

Before leaving Detroit for Seattle on the cross-continental train, Wilkins and his companions must weather the 'blare of trumpets',[40] as they are formally farewelled at City Hall by the Mayor, backed by a big brass band which marches with them to the railway station and must compete for volume as thousands line the streets and cheer.

Arriving in Seattle – where a steamer awaits, already packed with supplies and their two disassembled Fokkers, which have just arrived fresh from the New Jersey factory – the press is quick to push the whole idea of this being a race.

> CONTEST BETWEEN CAPTAIN WILKINS AND AMUNDSEN (Australian Cable Service.) SEATTLE, 14-2-26. – A perilous and thrilling race for the honour of being the first to fly over the North Pole will engage the two noted explorers Captain G. H. Wilkins, the Australian, and Captain Roald Amundsen, the polar veteran, in aeroplanes. Captain Wilkins will start from Alaska, with Spitsbergen as his objective, while Captain Amundsen, who is flying in the opposite direction, will make his first attempt by airship.[41]

When the question arises as to the sovereignty over any land that may be discovered, Captain Wilkins tells the press, just before getting on the steamer:

> I am a British citizen, but this is an American expedition, so the United States should have the first opportunity to claim any land that may be found in the blind spot, of a million square miles,

between Alaska and the Pole. My own nationality will not be involved. Anyway, if all goes well, the theory of an Arctic continent should be proved or disproved.[42]

As Wilkins and his crew begin to settle into the steamer, they finally have the time to get to know each other, whereupon Wilkins can't help but notice that *some* of them are . . . borderline useless, walking liabilities who would be better left behind. Just one problem: Palmer and the rest of the press have already proclaimed them *savants*. It is too late, there is no way to cut them loose without starting a row.

'It was not the first nor the last time I regretted the unwieldiness of our organization,'[43] Wilkins notes.

The main thing, though?

The main thing is he has the one man he truly wants, Ben Eielson, already on site at Fairbanks waiting for them in the town that is 200 miles south of the Arctic Circle and 550 miles south of Point Barrow. If, after flying north from Point Barrow, they sight new land, Eielson will be the co-discoverer. If no land is discovered, the plan is to keep going north until they cross the North Pole to land in Norway's Spitsbergen – something that will satisfy their backers in the North American Newspaper Alliance.

In the lead-up to their arrival, Wilkins is careful to build the profile of his co-pilot.

'Lieutenant Eielson has flown 60,000 miles in the Arctic and near-Arctic,' Wilkins tells the press. 'No other pilot has flown a third as far under like conditions. He has faced the fiercest weather the north has to offer, has made landings on rough, unlighted fields during the Arctic night and has served as his own mechanic and rigger when a snowbank was his hangar and the temperature was far below zero.'[44]

Finally, alighting from the steamer at Seward on the southern shores of Alaska – if cumbersome plane parts can ever do something so graceful as 'alight' – the heavy boxes are now loaded on to a freight train and two days and nights later they arrive at Fairbanks in the middle of Alaska, where Ben Eielson is waiting with a posse of heavily rugged up men – for winter has not remotely let go its grip – to guide them to the hangar at the Fairbanks Airplane Corporation.

The streets are icy and it is near impossible to get traction, even as the wind howls and blasts snow into their faces, but they are helped by many of the gold miners around Fairbanks who retain memories of the times Ben Eielson had flown through storms to get medical help to injured miners. It means that, in short order, the expeditioners are able to haul the bare torsos of the two wingless planes down the streets and through sparse but cheering locals all the way to the hangar of the Fairbanks Airplane Corporation.

After a morning of heaving and hauling in now bright sunshine, the men set off for lunch in town. It is only at this point that anyone looks at a thermometer. Good God almighty, it is 52 degrees Fahrenheit below zero!

At least the assembly of the Fokkers goes extraordinarily well. Wilkins is not only impressed by the Fokker factory's workmanship, but more than delighted that their Detroit 'expert' in charge of assembly has no interest in directing or participating in anything so coarse as manual labour, meaning that Wilkins can get the job done the best way he knows how – entirely under his own direction, and to his own satisfaction, in just under eight hours. One of the more tricky fixtures is a permanent aerial camera, the 'Fairchild Aerial' which weighs in at 46 pounds. Whether it will capture anything of worth, at such a height, with such blindingly white topography, remains to be seen, but Wilkins knows the drill in taking photographs under adverse conditions: snaps will be taken and developments awaited.

Another unusual piece of equipment will also find its place in the air: a fishing rod. Every detail about Wilkins is now judged newsworthy and readers are breathlessly informed that the use of the rod will involve: 'a casting reel, band lines, and a plug trimmed with the heaviest of triple-gang hooks'.[45] Even with the heaviest of such hooks, Wilkins will not attempt to fish while in the air, but content himself to some nice ice fishing should a crash occur.

At last, all is in readiness, and Wilkins – not wishing to waste any time – gives the word. On the morrow, 11 March 1926, they will take it up for a test flight . . .

But wait!

One of the most popular members of the expedition to date, Palmer Hutchinson, has a typical press man's idea. This is the perfect moment to generate a bit of publicity for the sponsors. Captain Wilkins, the planes must be christened! And he doesn't just mean by the Cap'n making a speech before cracking a bottle of champagne against their sides. No, he means a real christening, as if they are actual babies.

Wilkins pauses.

Such malarkey is against everything he believes in. But, against that?

Against that he really does like Palmer for his energy and earnestness – and he does make a fair point. This is precisely the kind of pre-attempt publicity they want back in Detroit. And so Wilkins agrees. With typical thoroughness, Palmer arranges for no less than a Catholic Priest, a Presbyterian Minister *and* an Episcopal Parson to all be present for the christening, not to mention a layman preacher, just to make sure that all bases are covered, even as the cameras roll and click.

Through the good graces of both the wife of the Fairbanks Mayor and a local beauty, a curious ceremony devised by Palmer himself takes place.

The two planes are wheeled out of their hangars while an enormous crowd gathers.

Once the speeches are made and the blessings bestowed from the Lord for both planes to always have the Almighty looking over them, each woman pulls an American flag from the nose of, first, the three-engine dashing *Detroiter* and now the single-engine *Alaskan*, before indeed smashing a bottle of . . . gasoline against the propellers as everyone claps vigorously in the freezing air.

Look, it is all a bit odd, but it is also such a novelty, and so exciting, that Wilkins has no doubt that every man, woman and child in Fairbanks – all 2101 of them – has taken the day off to come and witness this historic, just invented, publicity ceremony.

Most importantly, no fewer than 11 'special cameramen'[46] take photos before racing away to send those photos with the reports to newspapers around the world. As everyone disperses on the calculated false announcement that no flying is imminent, they all get back to work and, only an hour later, all is in readiness. Wilkins gives the order for the first test flight to commence. But hang on, someone is missing. Where is young Palmer?

Oh. Writing up dispatches on the christening? Look, it would be such a pity for him to miss out on the 'scoop' of this *actually* historic moment not the contrived one, this first test flight, and he is such a fine man, Wilkins himself tracks Palmer down in town.

'Come to the field, inconspicuously,'[47] he breathes. We are about to go up. Palmer nods, thrilled. He understands.

Captain Wilkins really is giving him a major scoop, as he alone of the press men has been given the word.

Just one thing though, Captain.

Can I dash off a short dispatch to tell our readers that the first test flight is about to take place, get it away discreetly, and see you there?

Very well then. It only takes a few minutes and Wilkins and Palmer head to the plane, enjoying their easy companionship, the sense that they are on the edge of history.

They arrive at the flat field on the edge of town serving as an airstrip to see the newly christened and now fully tanked up *Detroiter* being rolled out of the hangar, and she is indeed an extraordinary sight. With wings spanning 75 feet, it is the biggest Fokker ever constructed, as Wilkins' good friend Anthony Fokker himself had proudly informed him. A quick inspection walking around it reveals that all appears to be in order, with all of the ailerons, tyres and rudder looking perfect, with no leaks apparent from any of the engines or the petrol tanks. And so, with a cheery pat on the back to Palmer, Captain Wilkins climbs on board beside Major Lanphier, who will be the pilot for the test, and sits at his side in the cockpit.

There are many, including many in Detroit, who think that the battle-tested Lanphier should be the permanent pilot; and while Wilkins is indeed impressed by Lanphier's darting skills in the air it is Ben Eielson's cool that has the biggest effect on him. For the moment though, let us stay with the senior man for this test to make sure that Detroit stays cool. (Here in a nutshell is the problem with taking big money from sponsors – they think it entitles them to pull levers and press buttons from afar, whereas Wilkins knows that the key to the operation being successful is for him to call the shots as and when he sees them. He has nothing against Lanphier, and if this were a dog-fight in the Great War he would indeed be the first chosen. But this is not combat – it is the Arctic. You need to outmanoeuvre the weather, not another pilot trying to kill you.)

'All clear!'[48] yell the mechanics, and the dazzling *Detroiter* starts to slowly taxi towards the runway, only to stop a hundred yards later, thanks to a small snowbank. Of course, the mechanics and crew men run to their aid, stamping the snow down in front of the wheels. Now, strictly speaking, Palmer Hutchinson is there to chronicle that history, not be a part of it, but he can't help himself. For now, momentarily stopping his eternal scribbling, Palmer runs with them, and enthusiastically starts stomping the snow with the best of them, before all of them, bar one, try to give the plane a push.

But no-one had ever told Palmer that when using your collective strength to get a plane with propellers whirring the only proper way is to push from *behind*. For while he had been pushing from behind, he now decides he could get a better grip by pulling on the wheel. This places him right between the wheel and the whirring, invisible, propeller of the starboard engine, just inches behind him.

George Wilkins is in the cockpit with his co-pilot, and knows nothing of what is going on below.

Straining, *heaving*, to get the plane moving the other workers spot Palmer's danger, just as the plane breaks free and once again lurches forward.

Once more the cry comes: 'All clear!'

The men spring clear of the plane, and Major Lanphier thrusts the throttles open. It is only a second later when they hear it: 'a dull, heart-sickening thud'.[49]

Aghast, fearing something has gone badly wrong – perhaps an engine mount has suddenly come loose, or a wing joist has broken – Wilkins looks through the window to where the sound has come from only to see . . .

Only to see . . .

Palmer lying in the snow . . . with the only thing connecting his head and his torso being a river of blood.

The instant they turn off the engines, they can hear screams and howls from outside, getting progressively louder as the engine noise lessens. A newsreel man who has captured the whole thing on his camera is even now pulling open the back of his camera and exposing the film to the light, certain that he never wants to see it again, and nor should anyone.

When 'All clear!' was yelled, the mechanics and professionals leapt back; the eager amateur leapt forward. It is a shocking irony that Palmer Hutchinson has created a story that will make the pages of many newspapers around the world and the front pages of most in America: his own death.

Beyond the obvious tragedy to Hutchinson and his family, it is a brutal beginning for the whole venture.

'There was a hex on that machine,' Wilkins will write of *Detroiter*'s beginnings. 'Everyone felt it.'[50]

There is no way around it. One second into the first test flight, and they have one man dead. The only consolation the men have is that Palmer's death was at least instantaneous but, of course, so is their shattering sorrow.

'Everyone with the expedition,' Wilkins will note, 'and I think every man and woman in Fairbanks, was more than usually grieved at this fatal accident. Palmer Hutchinson had been so well-liked. He was always with us, boisterously happy, joking, laughing, teasing or helping and comforting someone throughout the day. His career as a journalist had promised well.'[51]

In the wake of the widespread grief that crushes them all, Wilkins makes his despairing crew a promise which – as difficult as it will be to live up to – he hopes will recharge their spirits.

'Our operations have brought about the death of one of our number,' he tells them. 'We must atone by making our success a monument to his memory.'[52]

And we also must ensure that no-one else dies in this venture. From now on, Wilkins – who, despite his own many miraculous escapes, has always been assiduous when it comes to the safety of others on his team – is hyper-vigilant in reducing risks. For one thing, he orders that the airfield be prepared for taking off in the snow, with tracks pre-dug for the wheels rather than a hurried stamping down of any icy hillocks that present themselves.

Still, with a man dead before *Detroiter* has even left the ground, Wilkins decides – after poor Palmer has been removed from the scene – to leave the plane on the ground and take *Alaskan* on a trial flight at the next available opportunity. It is in the grand tradition – *flying will soon resume as normal, and we apologise for this short delay*, with the only delay being, in fact, that heavy snow now sets in for a week, before they are able to fly again.

For this test flight Wilkins employs the services of his favourite pilot, the unflappable Ben Eielson. This time the takeoff is thankfully normal – with everyone certainly standing well back from the single propeller – and with the single motor purring like a panther, Eielson takes her up smoothly, 1000 feet then 2000 as the patchwork of polar panorama unfolds beneath them, whole geographical features gradually being revealed as they continue to gain height. Eielson turns and yells to Wilkins: 'She is a mighty fine machine!'[53]

That she is. She also has dual controls, and Wilkins controls the flight now, finding that *Alaskan* is the best plane he has ever been in. 'I have never, in my limited experience,' he will recount, 'flown a machine so exquisite to handle . . . I lost my heart to the *Alaskan* on that first flight.'[54]

As Eielson takes the controls once more, Wilkins leans back to admire once again his easy calm style. From long experience, the South Australian is able to divide the best flyers into two categories: 'as a rule, a pilot will be either a slap dash, brilliant dare-devil whom Lady Luck has seen fit to favour in a miraculous way; or a steady discerning cool-minded man who flies steadily'.[55] Ben is the latter. Then again, never let it be said that the American cannot call on at least his fair share of luck on the odd, crucial occasion.

Speaking of which . . .

Eielson lines up *Alaskan* for a practice descent, or 'landing in the air'[56] as Wilkins describes it, as they approach the waiting airfield and are 200 feet above the snow and closing fast. Ben eases back the throttle on this unfamiliar machine in difficult conditions, only for . . . the single engine to . . . stop. They drop like a stone, and, though Eielson shoves the engine throttle open, it is too late to have any effect, the practice descent becomes terrifyingly real in a moment. There will be no 'landing in the air', the air ripping past them now as they plummet.

Ben and Wilkins are completely powerless to do anything other than brace themselves as the runway springs up and they – *brace, brace, brace!* – SMASH into the ice! They tear and rip into the ice just 50 feet from the end of the 2000-foot runway. *Alaskan*, or what remains of it, slides and slithers right through the fence that marks the end of the runway and keeps going for another 20 yards. The two men sit in a ruined plane as their crew runs the 2000 feet to see if they are alive.

Wilkins is silent.

'I had the throttle full open,' Ben remarks, mildly, 'but she did not have time to pick up.'[57]

Do say?

Yes, Wilkins had noticed that.

Ah well, there is still time for a homily and, as the horrified crew reaches them, the leader of the whole operation remarks to ease their pain a little, 'There is no use crying over spilt milk.'[58]

Quite. The truth is, the two men are lucky to be alive, no matter what damage has been done to the plane.

Still, the mechanics might be excused for shedding a tear when they see what has happened to their brand-new plane: 'The under carriage was wiped clean off, and badly smashed and twisted. The propeller was twisted and turned, each end in the shape of a ram's horn ... We could see that in time we could weld all the broken parts together and set the machine once more on its landing gear, but it would be a long job'.[59]

With no choice in the matter, they leave the crumpled plane precisely where it had come to rest, if such a gentle word can be used for such a crash, while Wilkins and Eielson go into town with the crew for several – *brace, brace, brace!* – bracing drinks.

The next morning, Wilkins finds his crew remarkably forthright in their opinion as to who is to blame for the crash.

'Well,' he will recount them saying, 'you have still another machine and two good pilots, you can carry on with those.'

In no way does Wilkins agree that Eielson had been to blame for what had occurred. The engine had malfunctioned at precisely the wrong moment, that is all. It happens. Against that, Eielson is indeed the third pilot with the expedition, while Major Lanphier and Sergeant Charles F. Wisely are the other two pilots, and the only two who still have the faith of the crew. Most cogently in this situation, Wilkins knows the crew will all be reporting back to their backers in Detroit so – after a quiet and consoling word to Ben to explain the situation and make clear that Eielson still enjoys his support – he reluctantly chooses Lanphier to pilot *Detroiter* today.

But history will record his view.

'If I had not been responsible to anyone else, depending entirely on my own resources, I would have had Eielson fly.'[60]

Ben's time will come.

For the moment, *Detroiter* is carefully wheeled out with Lanphier in the pilot's seat, and this time the plane has cases of oil shoved in its tail to make the balance perfect for the day's conditions.

Lanphier pulls on the throttle and *Detroiter* slowly gathers speed ... alas, to the left and heading straight for a snowbank at speed. Entirely unused to the three-throttle control system, Major Lanphier is now taking the plane straight for the high snowbank on the left side of the runway. It is not so much Wilkins' life that flashes before his eyes as its end, for, instead of braking, Lanphier is gunning the plane madly. With not a split second to spare, the plane miraculously *liifffts*, and actually bounces into the air, its wheels skimming the top of the snow as they climb. Then, once up in the air, one of the engines is vibrating so badly that Lanphier must pull back on the throttle in an effort to stop it. That done, once they have gained sufficient altitude, Lanphier tries to put the machine through its mild paces – banking, turning and deliberately stalling – but whatever he does the vibration remains. Perhaps, Wilkins suggests via a hastily written note, do some runs directly over the landing field just to get a feel for how the air currents flow in that area, and to gain familiarity with the space. Lanphier agrees and lines the plane up for the first run, dead in line with the runway. Wilkins notes they are directly, exactly, above the spot where *Alaskan* crashed yesterday, when: 'to my utter astonishment, Lanphier cut the two side engines'.[61] Lanphier is expecting the plane to approach on the power of the central engine.

Instead, Wilkins gets a shocking case of deadly *déjà vu* as, to his horror, 'suddenly, we dropped straight to the ground. It was a bad stall, from almost a hundred feet, exactly the same mistake that Eielson had made.'[62]

An 'uncanny feeling'[63] grips Wilkins; he is living the same nightmare in less than 24 hours.

So hard does *Detroiter* hit the ground that the landing gear pancakes beneath it as they continue to plough forward, gouging their way along the runway as they go. Yes, it is much the same as had happened with *Alaskan*, except this time, as Wilkins slides towards his unknown fate, he is keenly aware that there is an 800 gallon tank of petrol behind him and a 500 pound engine in front of him. What will happen when they inevitably hit the snowbank at the end of the runway?

It all happens so quickly.

As they hit the bank the weight in the tail takes the path of least resistance and rises. In seconds the plane is practically vertical, nose-first into the runway, and it is surely just a second until the plane falls backwards and 'we would probably be as flat as pancakes'.[64]

For a couple of seconds the plane quivers uncertainly, in a strange battle between snow, air and gravity, but finally decides it is undecided, and so remains ... as is, nose-down, tail-up. Neither Lanphier nor Wilkins has a scratch on him, despite another plane being scratched from the running. Wilkins can barely believe it, as for the second time in 24 hours he has dropped from the sky like a rock, and lived. Now, what's the next line in this play he is so familiar with?

Ah, yes, now the pilot turns to him, as they watch their crew run towards them and offers him an earnest explanation.

'I fully intended to circle the field again, but when we came this far we were in such a fine position for a landing that I thought I had better come down.'[65]

You don't say?

Well, they are down all right. The engine is poking through the fire-wall into their cockpit with the first saving grace being that it is not actually on fire. The second one is that, while the plane is not in one piece, they both are.

. . .

. . .

But, Christ Almighty. The damage is extensive.

While *Alaskan* will need a lengthy repair working with the parts they have, that is not the case with *Detroiter*, as so many of its parts are simply beyond repair. It will be many weeks to even get the materials needed to begin a repair.

In sum?

In sum, within the space of a week, they had lost a press man dead before the first test flight had begun, followed by both test flights seeing both newly assembled Fokkers crashing.

It all leaves them with an exceedingly ordinary report card. In Wilkins' words: 'Failure was plainly evident.'[66]

Quite.

What on earth will everyone think? What else can they think? Wilkins knows it only too well, and can't blame them. It is a low point in his adventuring life.

'I had accepted the help of over eighty thousand people and these people looked to me to succeed and justify their confidence. I could hardly be blamed for the crashes, yet I would have to bear it.'[67]

After all, the world has been waiting to hear of the great air race to the North Pole between Amundsen, Byrd and Wilkins and the first bit of news they will likely hear is that the Australian, Wilkins, couldn't even properly get off the runway, having crashed two – count 'em TWO – planes! And killed a fellow to boot!

Worse still, however, than damage to his reputation is the knowledge that, 'it was utterly impossible we could complete our program that year or even really begin it'.[68]

Now, at this point, with $US50,000 worth of planes smashed, and a man already killed, a lesser explorer than George Wilkins would have abandoned the project in tears. But Wilkins is not such a man. He continued to believe that it was possible to do what he had set out to do, and so methodically sets about repairing both planes. Though there is no way he will be able to fly over the top of the world that season, he is at least determined to get both planes up in the air long enough before the flying season closes that he will be able to ferry supplies of petrol to Point Barrow, using its frozen lagoon for a runway.

Once Wilkins gets his cable through to Detroit with his plans, the response is not long in coming. They are surprisingly understanding that accidents do happen and are fully prepared to continue funding the expedition on one condition: Wilkins must dismiss the pilots he has, the ones who crashed the two planes, and get new pilots. As Wilkins reads the telegram of dismissal to the men, he smiles, as they surely know he would never agree to such a condition. It is the job of Detroit to provide the money, but that is where it ends. He will not take direction from them, and certainly not in something as crucial as pilots.

But the pilots do not understand that at all, and two days later Wilkins is sullenly asked, 'When will the new pilots arrive?'[69]

What?

Could they really think so little of him to imagine even for a moment that he would countenance such an order?

No, lads, we are in this together, for good or ill, until the end.

'We have planned a job which we said could be done, and we must do it!'[70] After all, the pilots and the rest of the crew are 3000 miles away from those in Detroit and they have no *actual* control of us. We are here on site, with the capacity to fix the planes – and we will!

Alas, not all of the men of his crew can maintain a stiff upper lip.

'The task to them now seemed hopeless,' Wilkins notes, 'and truly it was, but we could not afford to sit and weep.'[71]

CHAPTER THIRTEEN

SOARING

We begged for money, bought machines, flew them and smashed them, rebuilt them and smashed them, rebuilt them and smashed ourselves ... But we saw the job to be done. No man drove us to it, but there is no harder taskmaster than the 'will to do it'.[1]

Sir Hubert Wilkins, 1928

It was Wilkins' restless inquisitiveness about the weather that set him apart. It was a goal far more ambitious than the mere attainment of some meaningless geographic spot and it made him very different from the better-known explorers. While they searched breathlessly for location, he sought revelation.[2]

Simon Nasht, The Last Explorer

March 1926, Fairbanks, third time's the charm

Wilkins rallies the men to keep working to fix the planes, come what may, tightening what can be tightened; re-shaping that which has been crumpled, fashioning solutions from the material available; adding fencing wire and elbow grease where desperate and, when all else fails, using a bigger hammer. Sure enough, in just three weeks, *Alaskan* is again ready for flight, but this time Wilkins refuses to tempt fate with more 'tests'. He and Eielson – he announces – will fly to Point Barrow, now, 'endeavouring to do something before there was a chance of another accident'.[3]

And yes, no doubt eyebrows will be raised when Detroit hears that he has used the No. 3 pilot, and the one who had crashed *Alaskan*, but what does it matter?

This is Wilkins' last throw of the dice for this season, and for him the important thing is he takes the pilot he believes in, and that is Ben.

In the meantime, Wilkins & co. are not the only ones struggling to get north. A couple of weeks earlier their Advance Party had left from the settlement of Nenana – taking 15 tons of supplies overland to meet them at Point Barrow. The Advance Party consists of five men and two 'snow motors' pulling ten sleds in a motor-train. The Eskimos had snorted, saying motors would never work for that kind of work and only dogs could be relied on, an impossibly old-fashioned view that has proved . . . correct. High in the ranges the Advance Party is beset by many problems, not least of which is that the ice refuses to remain solid and the snow to stop . . . snowing, all while the machines keep breaking down. And the other problem is the consumption of fuel, with Advance Party member Gordon Scott reporting over the radio: 'We could reach Point Barrow, but we fear we will burn up the entire petrol supply in making the trip.'[4]

Ah. Given that it had taken 14 days to travel 65 miles, it is unlikely the Advance Party will make enough progress to achieve this Pyrrhic triumph, but they join the many parts of this Arctic exercise that Wilkins cannot exactly be sure of. The Advance Party retreats to take on a squadron of dogs to make the best of both worlds.

31 March 1926, Fairbanks, off the charts

Up, up, *up* . . . just . . . and away. On this bright morning, *Alaskan* manages to get off the icy runway, despite carrying a 3000-pound load of equipment and fuel, plus Wilkins and pilot Eielson. Both men are gratified to find that the plane feels nearly as good in the air as the first time they had taken it up, even if it feels a little less powerful as it is now fitted with the back-up engine personally given to Wilkins by Henry Ford in Detroit. (Whisper it only, but the first engine, built in New York, was much better.)

Settling at an altitude of 6000 feet they fly on through the fog, with no choice but to trust their maps, which show no peak reaching this height. And when the cloud breaks, so it proves, as not only are there no mountains for them to crash into, but before them lies the northern vastness of Alaskan grandeur expanding, and demanding awe: stretching grassland and tundra, and the Yukon River, which is so crooked it reminds Wilkins of the River Jordan. As Eielson takes them up to 9000 feet, Wilkins sights mountain peaks in the distance and takes a reading, using landmarks on his map and the time it takes to pass them to estimate their ground speed and comes up with – *dot three, carry one, subtract two* – 140 miles per hour, which is unlikely for a plane designed to go

132 miles per hour at top speed, and usually trundles along at 110 miles per hour for the purpose of fuel economy.

Something is wrong here; and when a mountain top looms out of nowhere – 3000 feet higher than the terrain should be in this spot, according to the *map*, the problem becomes obvious. These maps might be more use for Wilkins to blow his nose with than to give an accurate guide to the landscape beneath them.

A short time later Eielson threads between two mountains.

'At this point our map was a blank,' Wilkins notes. 'On it nothing appeared between the mountains and the coast.'[5]

But they are flying between the mountains on the map and the coast.

Wilkins and Eielson are caught amidst consternation, elation, error and terror.

On the one hand they are exploring land unknown to the Western world, flying where no-one has ever flown before and seeing things unknown to all but the Eskimos. On the other hand, it means mountains really could be anywhere, including in those clouds up yonder. Mercifully the cloud ahead clears to show them . . . more cloud below, endless and unbroken. They dare not try to get beneath it, as they fly on and on and on, two intrepid souls in a roaring machine, into the unknown.

> It was a weird and uncanny sight. We seemed to be the only speck in a boundless world. There was nothing for contrast and from which to judge space or distance; nothing in front of our eyes except the tapering bonnet of our engine; nothing below us to be seen but the same grey, grey mass. I am sure that we could find no situation more weird if we were to travel through space to the moon. The monotony and uncertainty of it would drive any man crazy if endured for long.[6]

But now here is something Wilkins has never seen before: the image of their own plane, so extraordinarily framed.

> The sun shone dimly, about on level with our wing. On the opposite side to the sun there shortly appeared two complete rainbow circles and in the center a ghost-like shadow of our plane.[7]

The sensation is extraordinary, a beautiful if troubling optical phenomenon that raises the hair on the back of their necks.

> It seemed to me that it mocked us as we speeded on. The very
> shape of the shadow gave it a sinister appearance. I have seldom,
> in a variety of experiences, been so awed.[8]

The living dream continues, unchanging, and it feels they are 'doomed forever to fly into an endless grey mist as punishment for having dared to venture over the Arctic wastes on wings of fragile wood'.[9]

What can they do, except what intrepid explorers have done since the dawn of time, when to turn back is at least as dangerous as going forward?

On they go, Wilkins busying himself operating the gasoline hand pump to fill the upper tank from the reservoir, as they sail on, a ghost plane of the northern skies, a tiny speck in the grey eternity . . .

At last, after long hours that feel like weeks, the cloud breaks and far ahead he sees . . . *what is it?* . . . 'water-sky, which I knew from experience would only be seen over open water'.[10]

As strange as it seems, it is like the water is in the sky itself, reflecting from the clouds. It is a beguiling phenomenon, a blurring of the lines between dream and reality experienced only in the Arctic and Antarctic, which allows you to see in which direction the open water lies. The stark white nature of the environment leads to a bizarre lighting effect in which the water below is reflected, truly mirrored, in the clouds up above. Likewise, it can create a darkness beneath particular clouds that assists in navigation, highlighting the direction of the open water based on what is shadowed and what is lit. It is an occurrence that was noted some 150 years earlier by no less than Captain Cook atop a ship, and is now reflected in Wilkins' experience aboard an aircraft.

What is going on?

Eielson is clearly wondering the same thing, for he now turns to his navigator, and raises his eyebrows.

Wilkins decides that masterful certainty is the right pose for now.

'I nodded wisely and pointed.'[11]

Very well. But what *direction* should we proceed?

'Keep going,' instructs Wilkins, pointing straight ahead. 'We should take at least five hours to get to Barrow and we have been in the air not much more than four. We can't be there yet.'[12]

(The explanation for the water-sky is not apparent right now, but it will surely emerge. At least it indicates that there are unlikely to be mountains ahead.)

But has he miscalculated?

For only a few minutes later, Wilkins is delighted to realise that the ice beneath them is land ice, as the 'small cakes of rubble', the tell-tale piles of rocks that poke through, are the same distinctive patterns that he had navigated on foot years before. But where in the Arctic are they?

They must be just crossing the coast right now, and likely have Point Barrow more or less beneath them? No. Based on Wilkins' calculations ... Point Barrow is in fact now well and truly behind them. Incredibly, their flight has been so rapid they have overshot their goal, and Wilkins now realises that their correct position in just a few minutes will be over 100 miles from shore.

> So far as we knew, no man had been so far in that direction. I was jubilant! We had actually started our work of exploration! Quite unwittingly at first, but now we were on the way, it was our chance to make a little headway.[13]

As navigator and commander, he probably should break the news to the pilot. Wilkins leans over and yells: 'If you look ahead you will see a hundred miles further north than any man has seen until today!'[14]

Ben Eielson's face says it all. It was only a few weeks ago that he had been able to fly no further than the end of the runway. But look at him now. He has flown his plane where no-one has gone before, over uncharted land, further than any explorer has ever been, seeing things never seen.

Captain Christopher Columbus and Captain James Cook can eat their hearts out. It is heady stuff but they must both keep their heads.

They really should turn back to Point Barrow soon, for safety's sake, but ... Wilkins is the one who puts into roared words their previously unspoken wish: 'We are a hundred miles out over the Arctic Sea. What do you say to going a half an hour longer – just to make it a good measure?'[15]

(For a moment Ben appears uncertain. But quick enough Wilkins' faith in the skills of the North Dakotan is matched by Ben's faith in the South Australian.)

'Whatever you think best,'[16] he roars back.

So on they go for another 30 minutes, hurtling at over 100 miles an hour into the unknown. Wilkins is sure where he is, though the real risk will come when they turn around. Will they be able to find Point Barrow again through grey clouds that mask their view? Should Barrow prove to be unreachable it will give them the opportunity to try something never even attempted before: to land on Arctic ice and, ideally, take off again. Nobody is even sure if it is possible, but Wilkins is convinced that it is and, if needs must, this will be the time to try it.

Finally, they turn, a roaring dot of navy blue in a blur of grey and white eternity, and head back to the Alaskan coast for the next two hours as fierce black clouds appear on the southern horizon.

Christ.

Following Wilkins' hand signals, Eielson drops the plane from 4000 to 2000 feet, whereupon the clouds give way to a brawling blizzard that buffets and blanks them out, just about where Wilkins hopes Point Barrow *is*. They climb to 4000 feet and try to fly around the blizzard.

'It was now a case of figuring and dead reckoning. We were undoubtedly lost in the sense we did not know where we were. There was nothing to identify.'[17]

Turning south to at least make sure that they are over land, they continue for 50 miles inland, before turning west and descending a little, with Wilkins hoping that the light of hazy sun will reveal the steep coast around Point Barrow. The atmosphere is tense, as their fuel supplies run as low as their chances of finding and reaching their desired goal, and the sky gets ever darker as the long polar twilight deepens.

Under such circumstances, it is more urgent than ever that they make contact with the world – to let them know their approximate position, and it is for this reason they have carried on board a 50-pound light radio-sending set and rigged its antenna in one wing. Cranking the set up, Wilkins starts tapping out the message. He and Eielson are still in the air, the plane is in good shape, but they can't find Point Barrow.

•

What's that?

At settlements and towns all over central Alaska, there is a surge of excitement as their radio receivers just barely crackle into life with a series of beeps – some short, some long. The *relief*. They're alive!

That is the good news. The bad news is that the signal is very weak, and clearly very distant, and trails off before any intelligible message can be taken down. Wherever they are, it certainly doesn't sound like they are over Port Barrow or anywhere near there.

At least some reports come into Fairbanks not long afterwards, Eskimos had sighted a plane 140 miles south-west of Point Barrow, while another message follows that a plane was seen by Eskimo hunters flying westward towards Point Barrow. They can't both be true, can they? Unless, perhaps they are lost in *Alaskan*, going in circles, looking for Point Barrow. It is all so confusing.

But now comes a solid report, relayed to them from – of all people – Robert Waskey, the radio operator of the Advance Party making its grinding way to Point Barrow. High in the mountains, south of Point Barrow, Wilkins' message has come through loud and clear.

They're still in the air! Still going!

•

In *Alaskan*, as their fuel continues to dwindle and there is still no clear sign of where they are, or how close to Point Barrow they might be, things are becoming tense.

Suddenly, however, Ben sees something.

'What's that over there on the left?' he yells excitedly. 'Looks like houses!'[18]

Wilkins looks and indeed sees something – maybe houses? – but not in the direction that Ben had been pointing.

> Then I carefully looked with binoculars in the direction Eielson was pointing. I could see nothing where he thought he saw houses, and he could see nothing where I thought I saw houses. We were beginning to get over-anxious and our minds, as do the minds of men weary and wandering, began to conjure up all sorts of visions.[19]

With both of them seeing houses and villages everywhere, castles in the air that are simply not there, it means they are nowhere.

Stay.

Calm.

One's mind must be clear, so you can see what is there, not what you hope is there.

> The only thing to do in such a case, as I had learned in early youth, was to concentrate on some particular thing. Do one thing at a time and make sure of it, depending on some solid fact.[20]

In extremis, Wilkins focuses on what he *knows* to be real: the compass. He can see it clearly. It is real. It will guide his dead reckoning, not the wild reckoning of his mirage-filled eyes. Settling down, Wilkins calculates they must be within 15 miles of Point Barrow, and guides Ben accordingly, hoping for a break in the clouds, so he can . . . wait? What's that?

Given the previous mirage of villages, this time Wilkins can barely bring himself to believe he is seeing what he is seeing, but it looks like . . . a cliff. Surely, the coast? Frantically, Wilkins tugs on Eielson's arm and points down, *down, now*! The American hauls smoothly on the throttle, gently losing height as the speed slows, but neither of them can see anything and Wilkins has that sinking feeling that his eyes have tricked him again when . . . there it is. It really is the coast, the blessed coast, all of its cliffs coated in tips of white.

'The snow could be seen drifting and pouring over the tops of the cliffs like sugar from a barrel.'[21]

Many times before, Wilkins has seen such a sight, but never has he seen it from the air, and never has it looked so sweet. They have made it. There is the lagoon he once walked around. There is the trading post of Charlie Brower, who had so wonderfully fed and warmed Stefansson's men when they shivered into sight so many years ago now. There is the church, with its spire pointing to the heavens. It is all so familiar, so wonderful and such a delight to swoop and sweep over the paths he once trudged. With each sweep across Point Barrow back and forth, Wilkins and Eielson look at each other, and nod.

Are you seeing what I am seeing?

I am.

We agree. This is real. We are saved. And the fact that it is real is confirmed as, down below, the Eskimos come out of their habitations and gaze up at them, waving. It is one thing to imagine villages, but quite another to imagine villagers waving furiously. As if by way of ceding defeat, the blizzard has now died down from its howling fury, though still leaves a wind whipping so vigorously that 'the drifting snow looked like huge writhing snakes squirming across the [frozen] lagoon'.[22]

And yet the vision of those snakes is crucial, as their broad direction tells which way the wind is blowing, and in turn the direction the plane must land, for it is also crucial – on Arctic runways particularly, which are generally much shorter than in warmer climes – to land into the wind, not with the wind behind you. Struggling to steady his hand

The single-engine Fokker *Alaskan* – the second plane to be used by Wilkins in the *Detroit News*–Wilkins Arctic Expedition.
(University of Alaska Fairbanks)

Wilkins' expert pilot, Ben Eielson. Wilkins and Eielson would achieve several history-making flights together, both in the Arctic and the Antarctic.
(University of Alaska Fairbanks)

Wilkins and Eielson preparing for their trans-Arctic flight in the Lockheed Vega plane in March 1928. The *New York Times* would call it 'an amazing victory of human determination' and the journey would earn Wilkins a knighthood.

(University of Alaska Fairbanks)

The end of a charmed life. In November 1929, Wilkins' brave and trusted pilot and great friend Ben Eielson disappeared in the Arctic while on a rescue mission. His smashed plane and his body were found weeks later on the Siberian coast of the Bering Strait. (University of Alaska Fairbanks)

The airship *Graf Zeppelin* landing at Lakehurst, New Jersey on 29 August 1929, after its epic around-the-world flight. Wilkins was an enthusiastic participant in the journey and corresponded on the venture from the 'technical viewpoint' for William Randolph Hearst's newspapers. (OSU)

View of the crowd that attended the reception given to Dr Hugo Eckener, commander of the *Graf Zeppelin* in New York City. (OSU)

A studio portrait of Wilkins and Australian actress Suzanne Bennett, who, on 30 August 1929 in the back room of the Ohio Justice of the Peace, became Lady Suzanne Wilkins. (OSU)

Nautilus in dry dock, undergoing repairs. (OSU)

Lincoln Ellsworth, renowned multi-millionaire and adventurer. In 1931 he was a principal backer of Wilkins' *Nautilus* expedition. (OSU)

Nautilus in Plymouth, 26 June 1931. (OSU)

On board *Nautilus*, Wilkins and chief scientist Harald Sverdrup conferring over calculations. (OSU)

Crew of *Nautilus* stretch their legs on an ice floe during a pause in the trip, a welcome respite from the cramped quarters of the submarine. (OSU)

Making radio contact from *Nautilus*. Wilkins and chief radio operator Ray Meyers set up a radio on the Arctic ice. (OSU)

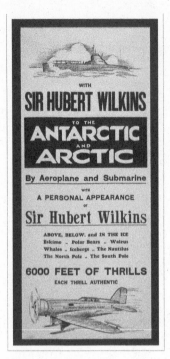

Poster advertising a Wilkins lecture tour. Wilkins became one of the US's busiest and highest-paid lecturers. (National Library of New Zealand)

Sir Hubert Wilkins in later life. Data gathered on his *Nautilus* expedition proved vital to the US Navy. In 1958, the nuclear-powered Navy submarines USS *Nautilus* and *Skate* achieved what Wilkins had always believed possible: a crossing of the North Pole under the ice. (OSU)

The Navy submarine USS *Skate* breaks through the ice at the North Pole to scatter Wilkins' ashes, 1959. (Naval History and Heritage Command)

in the back of the cramped cabin, as loose gear jostles and juggles itself around him, Wilkins pens a sketch for Eielson of where they must land. They are soon to disturb virgin snow, attempting to land a plane on Arctic ground, a foolhardy endeavour to say the least as there is no established runway. The spirit of Roald Amundsen is watching closely.

Not knowing quite what to expect, they sweep in over the ice and 'touched the snow as lightly as if touching a cloud'[23] coming to an elegant rest.

When Wilkins opens the side door of the plane he is immediately hit by a crippling blast. It is 40 below zero, with high wind and drifting snow, and though no brass monkeys are in evidence, a group of Eskimo lads do rush towards them, followed by an enthusiastic if cold crowd.

The Eskimo lads, mystified, look the plane over and ask, 'How can it fly? It has no feathers.' Others say it looks 'like a duck when overhead, but on the ground it looks like a whale with wings'.[24]

One of the first to greet them is an old friend, the Eskimo Pyruak who Wilkins introduces to Ben as 'a fat, lazy but likeable old beggar'.[25]

How many years ago, Pyruak, that we explored this land by sled?

Nine or ten summers?

And now here I am, dropping in from the sky.

This landing on ice really is the first sense Wilkins has of fulfilling the dreams he had had all those years ago. But he now wants so much more.

For his part, Pyruak, nodding appreciatively, is clearly in no doubt as to which is the better method of travel. The next man to arrive is Charlie Brower himself – a little older and stouter, admittedly – but still the same joyous figure who greets his old friend George Wilkins with a whoop of jubilation and a hearty slap on the back. It is truly big news, and significant for Charlie beyond the reunion with an old friend: Point Barrow can be reached by air! And speaking of air, it is freezing out here! Can we not retreat to your trading post, Charlie, and catch up in there? Damn cold, and wouldn't they all prefer to swap tales inside? Done.

After overseeing the covering of the plane with canvas to protect it from the elements, and tying it down to prevent it blowing away, Wilkins and Eielson are soon inside the Point Barrow trading post, drinking a steaming cup of coffee and chatting to 'two handsome young ladies',[26] one of whom is hoping to write a book about Arctic life. A few chapters will be written tonight as all clamour to hear about what Wilkins and Eielson have done and how on earth they did it.

Wilkins knows, of course, just how extraordinary the plane is as a means of transport in these parts, but his point is rather emphasised when, just a day after they arrive, the three-man crew of the Advance Party hauling the planned radio arrive, 'after their terrible trip on the over-land trail'.[27]

Battered, frozen, notably thinner – their leader, Malcolm Smith, has lost 20 pounds – they can barely see from snow blindness, they can barely stand and 53 of the 68 dogs they started with have died. They have taken six weeks to travel the distance overland that the plane has covered in just eight hours. The only thing among the freshly arrived crew capable of doing any work is . . . the radio they have been hauling all this way, which proves to function perfectly. Again, the Eskimos are stunned as they gather round and the box starts talking to them.

'Waskey,' Wilkins will report, of the newly arrived radio expert, 'is regarded as a great magician, because of his ability to pick messages out of the air. Awed silence prevails while we communicate with the out-side world. The natives follow Ben . . . all around the village. They regard him as a messenger from the gods.'[28]

(One female Eskimo elder seems to be not nearly as impressed as the others and, after poking the fabric of the plane with her pudgy finger, announces that she is certain that with the right sealskins she would be able to sew one for herself. Which is as may be . . .)

For his part, an old Eskimo man tells Wilkins he is sure it was a ghost he saw – a noisy one, yes, but a ghost all the same – a vision that kept flickering and vanishing in the sky.

This resonates with Wilkins and Eielson, who have so recently seen themselves as a 'ghost plane'[29] in their own windscreen.

The man circles the plane warily; it is such a heavy thing, did it really go up in the sky?

It really did, Wilkins assures him, and what is more they were inside it at the time.

'You are here, I can see that,' says the Eskimo. 'I suppose it must be the fortune of the white men now to travel this way. In my early days it was only the *angakoks* (the old medicine men) who were able to fly through the skies.'[30]

But, finally, you white men have caught up and so you must be congratulated then.

Welcome to the twentieth century.

Thank you, and a pleasure to be here.

Still, Wilkins questions the man closely and establishes he lives just seven miles from Point Barrow and had first seen the noisy ghost go over his ice quarters and disappear, before returning three or four hours later. It means that, despite the cloud, Wilkins only missed Barrow by seven miles heading north and on his 'return' course missed it by just two miles; flying in blind conditions both times. It restores his confidence at once; they had luck, yes, but even a blizzard, a wrong map and endless clouds only resulted in a tiny leeway after 550 miles of flying.

True, navigation in these climes is no easy task, as a magnetic compass is about as useful as a wooden one this close to the North Pole – the strange thing being that determining which way is north or south becomes ever more problematic the closer you get to the most northerly or southerly point.

'The compass error of declination,' Wilkins notes, 'would vary from 30 degrees east to 180 degrees when the north end of the compass would be pointing south.'[31]

The only way to counter it is to consult and calculate using precise tables from the American Geographical Society, which show the level of magnetic variation for every latitude.

It would be hard enough atop the roil of the ocean, and the speed and altitude of an aeroplane only make it harder. But there is no other option, no alternative that Wilkins can see; and so, it must be done, checked and re-checked, a constant calculation born of the novelty of his craft.

•

Ben is searching for agreement from Wilkins, for the American opines that they should set about preparing the field for takeoff, with the aid of the locals, but Wilkins insists: no. This is the opportunity to see if the plane can take off *au naturel*, just as they would have to if an emergency landing had taken place somewhere. Although Wilkins does make a request the Eskimos find bizarre indeed. He would like nothing less than a lit kitchen stove, please, to be placed under the canvas, right near the engine of the plane, in order to warm it, with an Eskimo watchman appointed to keep the fire going at all times.

This Wilkins! He is a strange man, yes?

Yes, but, very well, the ghost shall be warmed, ready to leave when the weather breaks, which could happen any time. Wilkins is pleased.

As the freezing point of oil is considerably higher than water it is impera-tive that the engine warm up so the whole thing does not become so sluggishly viscous that it will not start.

Now, as it happens, it is five long days of waiting before the snows stop and the sky is clear for the first time. Wilkins tarries not, and the plane is quickly loaded.

Once the propeller is swung and the whole plane roars into life the crowd quickly divides into onlookers and workers, with the former falling back and the latter moving to the front to put their shoulders to the wing to help push it forward. Once out on the frozen lagoon, Ben guns it, and after a wobbling and bouncing start, *Alaskan* is soon airborne, to the applause and hoots of those on the ground at Point Barrow.

They are away.

Settling in at an altitude of 11,000 feet, to be clear of those oh so recently mapped mountains, Wilkins quickly notes how much the smooth white blanket of snow covering everything has disguised the topography, so well that once more they must rely on their instruments and not their eyes to tell them where they are.

At least Ben and Wilkins can agree that the dark blue thing ribboning out beneath them is the Yukon River, and simply by following it – or at least going straight towards the furthest point of it they can see on the horizon – they must soon be circling over the town of Circle. Sure enough, one hour later, there it is!

They know Circle is a place as good as any for a quick refuel, a quicker rest, and a weather report. Wilkins has the plane land atop a frozen river, the rubber of the tyres colliding with the ice in a curious sodden squeal. Such an audacious entrance causes as much delight as their arrival in Point Barrow did, and locals run from far and wide to greet them. Wilkins is particularly enchanted with a young woman who made it halfway across the ice to meet him before turning heel upon the realis-ation that he and his men might be hungry:

> Putting aside her personal interest in the machine she ran the half mile back to her house, snatched from the stove the dinner she was cooking for her husband and herself, placed it in one can and put that can in another can full of hot water, then hurried back over the ice to the plane and offered us her hot meal before gratifying her curiosity.[32]

It is quite simply the kindest gesture Wilkins has ever seen and he thanks her sincerely for her 'thoughtfulness and energy'[33] as he and Ben devour the meal. (Somewhere in the middle of the white Circle, there is likely a highly disgruntled and hungry husband a little less enthusiastic than most at the new arrivals, but no matter.)

In the meantime, the welcome news goes out: they are alive, and the press rejoices: 'All Alaska is rejoicing tonight at the return of the fearless aviators, for many had given them up for lost.'[34]

But now in contact with the world they also receive a weather report from Fairbanks. They are beset by heavy storms – do not try to fly here. And so the night is spent sleeping in the cabin of their own plane; leaping out now and again to check that the oil stoves they have been lent by the locals are stopping the engine from freezing solid. In the morning Wilkins accedes to Ben's previous request, engaging a local to sweep the snow from a flat area to create a makeshift runway. It works perfectly, and after a friendly wave of warm thanks to these good people, they are able to take off smoothly and land at Fairbanks just 90 minutes later to an ovation from friends and crew.

They have made the 'longest aeroplane flight of Arctic history'[35] – 1000 miles against the odds and reason, and travelled further north by air than anyone yet.

In the meantime, as repairs to that hopelessly ground-bound plane, the doomed *Detroiter*, are still being made in Fairbanks, Wilkins and Eielson take the opportunity to make another flight to Point Barrow in *Alaskan* – to drop off more fuel in order to build up a cache that will be able to sustain a flight across the entire Arctic Ocean. What had been historic and record-breaking to the residents of Point Barrow just ten days ago seems now, if not commonplace, at least not particularly remarkable. Now they have proven it *can* be done, there is little fanfare left.

'They are beginning to realise its possibilities though,' Wilkins notes, 'and much of the awe of the first visit is vanishing.'[36]

It is only on their return flight that some drama enters the frame. Yet again, their map proves inaccurate enough to cause trouble, and they become disoriented when a village clearly marked on the map proves not to be there in actuality, which raises the question – just where the hell are they, apart from in a roaring machine with diminishing fuel that will soon need to get to Fairbanks? Finally, though, they see on the horizon a single plume of smoke. A village!

If only Wilkins could find out which village it is – although *village* proves to be generous on closer inspection.

Hamlet, is closer. To be or not to be – *certain which hamlet it is, well, that is the question.*

Rather than land and take off again, with all the familiar risks and fuel required for such an undertaking, Wilkins yells to Ben: 'Fly low over the houses!'[37] As the pilot does so, Wilkins drops a note, weighted by a spanner and wrapped in a long cloth with the tail of the cloth streaming out flapping behind, so it's easy to see, and it lands next to the people gathering below. The note asks if the people wouldn't mind writing the name of their village in the snow. They obediently do so, stomping out the name A l l a k a k e t[38] clearly, with the only problem being that name is not on any map Wilkins has. Very well, then. As Ben swoops low again, Wilkins drops another note, asking them to please stamp out an arrow pointing in the direction of Fairbanks.

The locals go one better, as they quickly form up into a human arrow, clearly pointing to the south-east. Sure enough, less than two hours later, Wilkins and Eielson land at Fairbanks with a little less than a gallon of gas left in their tanks. It is just another narrow escape for George Wilkins. There are close shaves and there are close shaves, but even the Man from Ironbark never went as close as this. Still, it provides yet another lesson to Wilkins. Never be blasé about risk; they are still attempting the improbable with each flight.

Still, risks must be run if a fuel depot is to be built up; and the fewer flights and the more fuel each flight carries to Barrow the better. (One peculiarity they must factor into their flight: there are four extra hours of light in the far north. While Fairbanks at this time of year has 14 hours of light a day, Point Barrow, which is over 500 miles closer to the North Pole, has 18. In such conditions, 'it would be quite possible to start from Point Barrow early in the morning, return to Fairbanks, and, without stopping the engine, load another cargo, reaching Barrow again before dusk'.[39]

And so like a frantic moth fluttering to a dying light, they fly hither and thither pursuing the sun's sinking sparkle to the end.

On their third flight between Fairbanks and Point Barrow, their plane is carrying 4750 pounds, which is nearly double the amount it was designed to carry, and they only just manage to get off the ground. While approaching the Endicott Range, it becomes clear that their load is too heavy for the plane to reach an altitude that will be enough to clear the mountains. The solution is obvious; dump some fuel and lighten their load. Wilkins writes a note ordering this but Ben yells back, 'Let's try for five more minutes and see if we can't get through.'[40]

For you see, there is no need to go over if you can fly around. With clouds now competing with fog to limit their vision, Wilkins is not sure if this is the wisest choice, but, as he nearly always does, he is prepared to back his life on Ben's judgement.

'He was not a daredevil who would dash into anything,' Wilkins writes. 'He summed up every situation and looked at it from all angles and then whenever there was a possible chance he went ahead.'[41]

(He is, in short, exactly as Wilkins had always picked him – just the kind of pilot he was looking for. Many pilots will enumerate the reasons why something can't be done. Ben's starting point is how it might be done better.)

Still, while Ben looks to see if he can thread the camel through the eye of the needle, Wilkins heads back to the cargo bay and starts to shift the cases of gasoline around in the hope of aiding ascent: 'There were many cases of gasoline distributed on the floor and to give the machine every advantage for climbing I placed all these cases up near the pilot's cockpit.'[42]

Look, it's his instinct, rather than physics, but the very action at least gives a sense that something is being done. But now the plane hits a sudden air trough and throws him so heavily against the cabin wall he fractures his arm. That crippling pain, however, will have to wait, for they now have a problem as big as a mountain.

'Just then I noticed on the left side of us a sharp mountain peak. It seemed as if we must smash into it. I hurriedly motioned to Eielson to keep to the right . . .'[43]

No, as big as two mountains.

'. . . but he indicated that there was a mountain peak on his side also.'[44]

Would that all decisions in life could be so easily made.

Straight ahead it is then.

Sweat drips from both men as they zoom ahead, climbing ever so slowly, thwacking through the clouds into what they pray will be more air, not a fatal mountain. The clouds clear and they see . . . a mountain top . . . dead ahead!

One way or another, this is going to be a bloody close shave.

> *(Then slashed the red-hot razor-back across his victim's throat;*
> *Upon the newly shaven skin it made a livid mark—*
> *No doubt it fairly took him in — the man from Ironbark).*

Lift, Ben, *lifffft!* Ben hauls on the joystick, and the plane starts to rise. It is going to be close.

> *He fetched a wild up-country yell might wake the dead to hear,*
> *And though his throat, he knew full well, was cut from ear to ear,*
> *He struggled gamely to his feet, and faced the murd'rous foe:*
> *'You've done for me! you dog, I'm beat! one hit before I go!*

There seems no possible way they can miss it. Wilkins is certain of it.

> *And all the while his throat he held to save his vital spark,*
> *And 'Murder! Bloody Murder!' yelled the man from Ironbark.*

In fact, they just make it. After they pass, Wilkins looks over the side to see the wheels of the plane spinning incredibly fast, and he realises they have kissed the snow ridge as they flew between two peaks and over a third. Another 'narrow shave, too narrow for comfort'.[45]

All up, when they make it to Point Barrow it is with more relief than when Charlie Brower subsequently dresses Wilkins' fractured arm – it still hurts like the Dickens. The main thing is they are safely on the ground, intact, warm and now with a hot cup of coffee and . . .

And, what?

Shouts in the distance. An Eskimo runs in with some shocking news. 'Your plane is burning!'[46]

Running out of the store, they can indeed see the fire leaping up above the snowdrift, and their hearts sink as their alarm rises – that is precisely where they left *Alaskan*. But as they run to the top of the drift, they now see that what is ablaze is in fact the canvas 'tent' that they have placed over the engine cowling to protect the engine and stop it freezing. It burns so quickly that, beyond the propeller's varnish vanishing, little damage appears to have been done.

'Too much blubber and too much fire,'[47] explains their interpreter, referring to the fact that the Eskimo watchman left in charge to make sure the oil didn't coagulate had put too much seal blubber in the kitchen stove, and the flames had billowed forth.

There are some repairs to do, but it is manageable. It has to be manageable, as a message from Major Lanphier informs them that a replacement propeller cannot be flown to them because *Detroiter* is still not repaired. (They are not surprised, a genie would give up trying to make that plane fly.) It means their expedition is now left without a functioning plane and, though a valiant attempt is made to repair *Alaskan*'s propeller, it shakes so much it ends up vibrating the entire plane; forcing Eielson to land on a lagoon where he and Wilkins have to bind the blade with brass to give it improved, if improvised, balance.

Once back at Fairbanks, the two decide to take *Alaskan* up once more for a test flight, only to find something else is wrong. Yes, the plane is no longer vibrating terribly, but it refuses to move.

'Many willing hands helped us drag the machine back to the field and we decided to reinstall the old brass bound propeller and make it do.'[48]

When all is in readiness, loaded down with tins full of petrol, the two pilots in *Alaskan* are hurtling down the airstrip at Fairbanks when it is clear that yet another thing is wrong, as the plane simply refuses to lift.

Crash positions! (Basically, tense up and pray.) While stopping a fast-moving and heavy plane is problematic at the best of times, achieving that feat on snow and ice is something else again . . . and now the massive Fokker begins to slide from side to side, careering towards disaster. At last, the plane comes to a crunching rest against a snowbank, though the main thing, as ever, is that neither Wilkins nor Eielson is hurt. The mystery remains, however: why have they crashed? They had loaded the plane like this before and done this takeoff many times. There has never been a problem. They know the plane is capable of taking the load, mainly a quarter ton of fuel cans in the fuselage. This time, however, the plane simply refused to leave the ground, as if it was too heavy. But how could that be?

Well, for the moment, there is nothing for it but to entirely unload the plane to lighten it enough that they can drag it free and begin repairs. It is while doing exactly that, however, that Eielson sees something move in the cargo hold. What the devil . . . ?

Reaching a hand into the semi-darkness he grabs at a shape, to pull it out into the light, not knowing what to expect and . . .

And he is suddenly confronted with a very beautiful young woman, flashing angry eyes at him as she tries to wrench her arm free. A stow-away! George Wilkins recognises her immediately. She is a gypsy, a rather modern independent one who, instead of moving around with a band of other gypsies, is moving around on her own, exploring the world. As a matter of fact, he had danced with her just the night before, during which time she had told him she was a bit of everything – a musician, artist, writer and explorer – and wanted to know whether she could accompany him on this flight? It had been with some reluctance that he had declined – for fear that her weight would make them crash – but it now turned out she had taken matters into her own hands. Mindful of his own past, where stowing away had started him off on world adventure, Wilkins is not too angry despite the wrecked plane, and the fact that she is now swearing at them in a manner that would have made an Australian wharfie blush. He understands only too well her compulsion, and besides, she is too beautiful for him to be angry with her for long anyway. Wilkins ignores the damage to the plane and arranges for her to slip away into a friend's car, so she would not be embarrassed in the Fairbanks community. He, Eielson and their mechanics then do the best they can to repair the plane; a new propeller does the trick and that same day they are out on the landing ground again.

But wait, what's this? They are no sooner out on the runway and getting the revs up as they start to taxi forward in preparation for taking off, faster now, seconds away from being up in the air when they hit a small depression, followed by their own great depression as . . . the right wing of the plane falls off, and they slam and slide along the runway, the contents of the plane shaking and shifting around them as they glide on the ground, tail in the air and fuselage shuddering to a halt.

Look, if you are to have a disaster in a plane, the best place to have it is on the ground, which is precisely where they are. Far more problematic is that Wilkins is now covered by 30 cases of gasoline, while from the upturned tail of the plane, gas from the wing tanks is pouring onto the hot exhaust pipe.

'Hurry up and get out, she will surely catch fire!'[49] shouts Eielson with uncharacteristic urgency. Alas, Wilkins is completely submerged buried by the cases and cannot move. He is also in the way of Eielson's exit. Still their luck, if men who have just had a wing drop off their plane can be called lucky, holds and the plane does not catch fire as their brave crew men rush forward and risk their own lives – for *Alaskan* really could explode at any moment – to haul the cases off Wilkins and get both men out.

Once out and recovered, they examine the plane and find that they are two of the luckiest men in the world. For it 'seemed that the wing had been broken for some considerable time'.[50] Which means for some considerable time they had been flying in a plane that could have dropped out of the air at any moment.

Given that *Alaskan* is now certainly beyond repair, they must try the same exercise with the (yes, Lord be praised, Lazarus be invoked, it has risen) just repaired *Detroiter*. And so Wilkins takes off to Point Barrow days later, with Wisely as pilot and Major Lanphier as mechanic; getting their crack at last as Ben rests.

Alas, alas, only a few hours after taking off, *Detroiter* starts to weave and wind in a most erratic fashion, like a car with a drunken driver all over an icy road. Wilkins approaches the cockpit, only to see that Wisely and Lanphier are physically fighting with each other over the controls. The curse of *Detroiter* has struck again; when the damn thing finally flies the pilots strike each other.

It turns out that after Wisely had lost his nerve completely and decided to return to Fairbanks immediately, Lanphier had started by urging him to stay the course, *Cap'n, my Cap'n*, and finished by trying to seize control, and most pertinently the controls. And that's when the trouble had started . . . Gobsmacked, their astonished commander watches as they punch and pull at each other, all while roaring insults and threats. It would be farcical if the stakes were not so high and Wilkins barks at both men to get hold of themselves and the plane. Mercifully, his authority and presence is enough to restore order and the journey to Point Barrow is from this point uneventful, but it is enough for Wilkins to vow that he will never go up again on this expedition without Ben Eielson as pilot.

Under the circumstances, landing safely on the ground at Point Barrow is no small blessing. This plane really is *cursed*, or worse! For rather than a mere hex, it might be that *Detroiter* is just badly constructed from the first with Major Lanphier claiming that, 'the centre engine was out of alignment . . . the controls were awkward and uncertain, and that even with the three engines at full speed, the machine could scarcely climb to an altitude of 5000 feet'.[51]

Whatever the case, they will be trapped at Point Barrow for days as a blizzard closes in and all they can do is hunker down.

At least there is now a permanent radio here – the modern world has arrived – and so it is they hear the extraordinary news of a feat that occurred on 10 May 1926.

Ex United States Navy aviator, Richard Byrd, has flown to the North Pole with his co-pilot Floyd Bennett. In a Fokker tri-motor christened *Josephine Ford*, after the daughter of Edsel Ford, who had helped finance the expedition. Taking off from Spitsbergen in Norway, Byrd's claim was that they had flown for eight hours, reached the North Pole and circled it for 13 minutes, before returning to land safely at Spitsbergen – a round trip distance of 1535 miles.

A very fine effort. (And before long, seriously doubted as to whether they had, in fact, got within 120 miles of the North Pole.)

Wilkins is interested to hear of Byrd's feat, but no more than that. The press might carry on about 'races' and desperately care about who is first to certain places but the Australian is first and foremost interested in expanding the frontiers of science, and eschews such exploratory circus acts. Far more interesting, and impressive, is that the same bulletin that reports Byrd's feat carries the news that the airship, *Norge*, conceived and organised by Amundsen, had started from Ny-Ålesund, Spitsbergen, reached the North Pole on 12 May and continued on its way, and is currently coming their way.

This, Wilkins is sure, is the future of air-travel. A few quick calculations and the Australian works out that *Norge* should pass over their present position at Point Barrow on the evening of 13 May. Wilkins and Lanphier wait all evening, hoping that persistent clouds will break and they will see history. At last, the clouds clear and in the distance they see it, a small dark object that they know at once must be *Norge*.

'That was perhaps the greatest thrill of my life,' Wilkins notes, 'because in 1919 I had planned to use an airship to make an Arctic flight . . . It mattered little to me who was at the control or who had organized the expedition, the fact that the machine had safely crossed the Arctic was sufficient.'[52]

Despite that, the success of *Norge* actually portends a cold wind blowing for Wilkins when it comes to raising funds. For when you are in the business of the exploration of polar regions – selling the risk, the romance, the adventure to readers – the fact that the Pole has not only now been reached by plane, but also crossed by people eating caviar and smoking cigars in an airship, perhaps means there is little left to sell?

It is with that in mind that, still stuck at Point Barrow, Wilkins cables the Board of Control on 4 June to inform them that, with no planes, he cannot do any more useful work this season but he hopes he has their support to return to the Arctic next year? The good news is that the response sounds positive when it comes: DETROIT WILL STAY WITH YOU UNTIL THE JOB IS DONE.[53]

The bad news is, Wilkins has no sooner returned to Detroit than it is a very cold wind blowing indeed. All of the men who have been involved in the project are summarily dismissed, and Wilkins is told that the expedition is $30,000 in debt.

As Wilkins writes to his mother – who is indiscreet enough to pass it on to the Adelaide *Advertiser* – he 'is disappointed at the attitude of the Detroit [sponsors], as they had in most cases proved to be interested in my last flight solely from the publicity they got out of it. Americans are prepared to pay anything for publicity, [even though Ben and I] are accomplishing things that are big enough to speak for themselves without the need for any boasting.'[54]

And yet?

And yet it is not long before a certain buzz attaches to the Wilkins name and his proposed next venture: the completion of the task to fly over the roof of the world and cross the entire Arctic by air, from one continent to another, Barrow to Spitsbergen in one leap. Despite the North American Newspaper Alliance withdrawing all future funding, the *Detroit News* agrees to pay all the running expenses of the new expedition in exchange for exclusives – something which prompts the Detroit Aviation Society to give Wilkins two Stinson SB1 planes, *DN-1* and *DN-2*, to fly with, at which point the North American Newspaper Alliance changes its position and donates funding for all wireless communication on the understanding that all news conveyed using this wireless is exclusive to them.

It means that Wilkins goes from famine to feast in the space of just a few weeks, and he is able to start hiring again, with Captain Ben Eielson naturally the first hire as chief pilot.

After everything is in place, with other hirings organised and supplies purchased, they leave Seattle on 11 February 1927, with Wilkins finding himself with the unusual luxury of three planes, two Stinsons ready to be assembled and one 63-foot winged single-engine Fokker F.VII. Alas, on assembling the planes in Fairbanks, the Stinsons feel like two jigsaw puzzles with many pieces missing and the result is that, 'One machine was assembled from the parts that fitted best.'[55]

And yet, a very serviceable machine it proves to be, powered by a state-of-the-art Wright Whirlwind J5 220 horsepower engine noted for its lightness and reliability, and an extremely Spartan interior to save weight. There are not even seats, with the pilots simply sitting on cushions atop the fuel tank.

Wilkins is confident that, once they have sent for the missing parts, his crew will eventually be able to form the remaining Stinson parts into a whole plane. In the meantime, he and Ben get to work on another Arctic innovation: skis, in the place of wheels. For the kind of surfaces they intend to use as their runways, it makes sense.

After much inquiry Ben Eielson finds the wood he needs – light, but strong, and absorbent of the many layers of varnish required to be perpetually slippery – and who else but Wilkins himself starts to construct them.

Once the second Stinson has been successfully built, Wilkins – with an eye for coverage and for gratitude – christens the planes, without fanfare this time, *Detroit News No. 1* and *Detroit News No. 2*. When it comes to their colours, the planes are striking indeed, painted orange to be spotted with ease in the white of the Arctic; their names are emblazoned in a deep blue, clear to any view.

After the Fokker bends its new skis while taxiing, Wilkins orders the two Detroiters to Point Barrow, a first dual flight. At least it starts off as a dual flight. Eielson and Wilkins, in *Detroit News No. 1*, arrive on schedule but *Detroit News No. 2* takes another nervous day to arrive. Wilkins' plan is to fly a course towards 78 degrees north latitude, 180 degrees longitude; to land on the ice and take 'soundings'. As orderly as it sounds ... the fact is the sounding will be achieved through ... explosives. By measuring the time it takes for the echo of the explosion to come back to them, they should be able to gain knowledge about the depth of the Arctic Ocean and the extent of the continental shelf. If Wilkins can show it is possible to land and take off from solid ice floes, the possibilities for exploration are endless. If he cannot take off, well, the possibility of a long and uncertain walk back is the best case scenario. The worst scenario? Don't ask.

Ideally the planes could fly as a duo, which would provide some comfort in that, when landing, at least someone would know where they are. But as *Detroit News No. 2* could not even reach Point Barrow in tandem, Wilkins is under no illusions: it is highly unlikely two planes could stay in contact for long in Arctic conditions.

At midnight before the dawn of the day of their departure – 29 March 1927 – all are asleep in Charlie Brower's warm but ramshackle trading post and home, bar Wilkins, who, in the absence of sleep, spends his time making more and more detailed notes on possible routes. The quickest and safest route is not over the North Pole, but skirting it, keeping closer to the known northernmost land masses – which, ideally, will give them a couple of regular navigational check-points to determine precisely where they are on the map, before adjusting their course accordingly.

Between a couple more visits to the Land of Nod, Wilkins twice rises to check the barometer in the kitchen of the Barrow store, and is relieved both times to see that it looks promising for good weather! And yet, in the gamut of things he is about to attempt, it is not the flight that concerns him most. He and Ben had proved last year that using the ice as a runway was feasible. This year will be different. Not only will they deliberately be seeking out and landing on detached ice – ice islands, if you will – but they will then be setting off an explosive to measure the time the reverberations take to bounce back from the seabed, as a modern way of taking depth sounding. The good news is, it should result in an enormous expansion in human knowledge. The bad news is, it risks causing the death of two fine humans, him and Ben. All he can do is trust his instinct, borne of his vast experience on the ice, that it will be all right.

With the first glow of that dawn at last, Wilkins is instantly up and about, soon joined by Ben and the rest of the crew, making final preparations, which will include Wilkins' demonstration of his belief in Kingsford Smith's attributed dictum: 'A plane only ever has too much fuel on board when it is on fire.'

Their big bird will carry 172 gallons of fuel in her belly, the fuel tank acting as both a seat and backrest in the cabin. A secondary tank lying next to Ben's feet, will carry 30 gallons. Both are connected via a wobble pump to additional tanks in the centre and starboard wing of the plane. As the level of the main tank goes down, Wilkins will use the pump to fill 'er up, from the auxiliary tanks.

To sustain themselves, they are taking 'ten pounds of Norwegian biscuits, twenty pounds of Norwegian chocolate, five pounds of Powell's

Army Emergency Ration, three pounds of mixed pemmican – a highly concentrated mix of meat and fat, delivering the most nourishment, for the least weight – chocolate and biscuit prepared for consumption during the flight'.[56] Oh and 'one thermos full of black coffee – a gift of Fred Hopson'.[57]

While the local Eskimo community work like navvies clearing away the last of the snow on the frozen lagoon, making a 5000 feet long and 14 feet wide 'runway' – ideally just enough to get such a heavily laden plane in the air at full throttle – Wilkins oversees the loading of the last thing: the sonic sounding gear. It is a very odd thing to carefully stow the very thing that is mostly likely to bring you down, but Wilkins only does so having examined it from every angle.

'For detonating there was a battery and switch connected to a long length of insulated cable at the end of which was attached, as often as required, a detonator . . .'[58] That detonator will set off the dynamite, which is particularly carefully stowed up the back. Well up the back . . .

All stowed.

It is by now, six o'clock in the morning.

All that remains are handshakes all round, wishes for *bon voyage*, a safe flight.

With a final wave, the two aviator–explorers climb through the side hatch as those Eskimos and citizens of Point Barrow who have come out to see them off, stand back. Now, one of their ground crew members, following the thumbs-up signal from Wilkins, grips the propeller, and gives it a mighty swing. The engine purrs into life, the plane vibrates happily, a bird ruffling its wings before setting off. Two other crew members push hard on the end of each wing to get them through that first bit of snow that has fallen overnight, to get them *moving*, and are soon happily left behind as the plane proceeds under its own propulsion . . .

Within seconds they are skimming down the runway on their skis, getting faster and faster until, just as the snowbank at the end of the runway looms, they are in the air.

Wilkins makes a simple note: 'At six a.m. we struck out into the unknown.'[59]

At Point Barrow, wireless operator Howard Mason knows they are fine for the time being. 'They were soon lost to sight, but the "OK" signal cranked out.'[60]

Heading out over the ice, Wilkins occupies himself, as ever, by recording detailed observations along with short notes chronicling their basic situation.

At 8 am he notes: 'All O.K. Eielson seems happy and I have every confidence in him and everything else on board.'[61]

What is not okay is the radio signal, which is maintained for no more than two hours as Mason notes: 'It gave me a sinking feeling as I knew the old set was not kicking out like it should and we would soon lose the signal altogether, which we did.'[62]

In the cockpit, all is fine apart from the sputtering and backfiring from the engine which also comes and goes . . . the tension tightening as it comes, and dissipating as it goes. Unaware that they are now beyond range, Wilkins continues to send a wireless message to Point Barrow every half hour.

At 10.55 am, the engine moves from sputtering and backfiring to a loud knocking that persists, the devil pounding on steel. Something is wrong – mostly likely 'carburetor ice', as it is known – and might be badly wrong. The good news is that the clouds have cleared, the vision is good, and up ahead Wilkins spies a good 'young' ice patch – relatively smooth and therefore good for landing on.

Wilkins hands Ben a note.

That's good ice to land on. Think we had better land and fix it?

Ben nods, cuts the throttle back as they circle lower and at 11.15 am they land, alone and adrift, on ice. Wilkins watches with bated breath as the 'beautiful, flower-like crystals'[63] of ice, two inches high and sharp as steel in the 30 below temperatures, cut into their skis. The skis survive and so do they, completing the first plane landing on Arctic pack ice in history. Both men leap from the plane; Eielson to the engine, Wilkins to prepare his sounding to determine how deep the ocean is where they have landed.

With enormous determination Wilkins whacks away at the ice with a pick, and with a huge effort it takes a full eight minutes of smashing at the ice before something breaks – sadly, it is the ice-pick. As unperturbed as ever, Wilkins mends the pick with screws from the cabin and starts again, going more slowly this time, digging another hole 30 yards away.

Eielson is still tending to the running engine; but Wilkins will need silence to set the depth charges off and take accurate sounding.

'Ben, can you stop the engine?'

'Go ahead and take the sounding,' Ben replies, 'if we stop the engine we will never get it started again and nobody but God, you and me will ever know what the sounding is.'[64]

But Wilkins insists.

As the engine dies – hopefully soon to rise again, or it will be them who die – the silence of the tomb descends, with the rustling of a light breeze over the snow and ice to keep them company.

Ben and Wilkins exchange a nod.

Ready?

Ready.

Ben has the receiver to his ear, and a stopwatch in his hand, watching Wilkins closely. Wilkins nods as he presses the plunger to set off the charge he has suspended in the hole. An extraordinary 7.3 seconds later Ben hears the echo and yells the time to Wilkins. Rapid calculation results in the answer: they are above a depth of 5000 metres of water, using the metric system because it is easier to calculate. It is an enormously significant discovery, providing precise knowledge where there had previously been none at all. Clearly, they are nowhere near land, or at least not a continental shelf of any description.

With the first priority of the service of science dealt with, it is time to focus on the second order of proceedings – surviving. Working with freezing hands they drain the oil, heat it up, pour it in again, disassemble, clean and oil crucial parts. They work in the freezing cold on the engine, even as the wind rises, the clouds close over and the temperature plummets further. With each part oiled and reinstalled, they must regularly restart the engine to prevent freezing, and at each pause, before disassembling again, manage to get 1400 revolutions per minute out of it, registered on the tachometer. That may be enough to take off, with the emphasis being on *may*.

Finally, they feel they have oiled and cleaned every part of the engine that needed attention. Eielson and Wilkins exchange looks.

It is time to give it a go, with Eielson taking his place in the cockpit and Wilkins giving the propeller a whirl in the hope of starting it.

Ready?

Ready.

Wilkins gives it a massive swing.

Nothing.

Stone-cold nothing.

Again, they exchange looks.

They're not dead yet.

But it is certainly a knock on the door.

Wilkins tries again . . . and again and again.

No better, no better, no better.

The temperature and their hopes continue to drop, but on the fifth attempt, there is a cough, a sputter, an utter sputter, another cough and now . . . the engine roars into life. Better still, even if the engine still sounds a bit rough, they are soon airborne once more, flying off into the wild grey yond . . .

Oh. Oh dear.

Only 10 minutes after taking off at around 1.30 pm, the engine kicks and stops. Clearly, they must land once more, and while Ben takes it lower and Wilkins shakes his head at the first two possible landing spots, he gives a thumbs-up for the third. Again, their luck holds and they land intact.

What can they do but the same again? The engine must be overhauled, the cowling must come off and their freezing hands once again battle to fix the engine before the oil freezes so thick and gluggy that it will no longer properly lubricate at 32 degrees Fahrenheit, and they are trapped. Despite the freezing temperatures, and now heavily falling snow, Eielson is a stoic champion. Of course he knows his fingers are literally freezing, that in an effort to protect itself his body is shutting down circulation to his exposed extremities, starting with his digits. But he equally has no doubt that they will die if this engine is not fixed by his bare fingers. Despite being in a desperate hurry against the flurry of slurry, only after an hour is the shaking Ben satisfied. Wilkins now gets on the propeller and hauls on it as if their lives depend on it – because they do.

Once, twice, thrice it fails. But on the fourth attempt, it catches, coughs, spits out a puff of smoke and takes, this time with the roar of an angry lion unleashed! But, after Wilkins clambers back on board, taking off is not as easy as last time as so much snow has fallen. Once again Eielson has the answer. After the first attempt to take off fails, the skilled pilot turns the plane around and is able to ingeniously use his own freshly made tracks as the smallest possible runway in the world, at the end of the world, to indeed . . . rise again!

Wilkins records the time: 2.20 pm.

They have been out for nearly eight hours and have eight hours' worth of gas left if they remain in flight. *If.*

True, their speed remains alarmingly slow as they battle headwinds, but compared to where they have come from this not only feels like the heavens above, it feels like heaven itself, most particularly because the engine is purring perfectly, with no knocking on heaven's door and no stalling to stop their hearts and flight once more.

Given how exhausted, freezing and hungry they are, Wilkins is quick to dole out the pemmican and hot coffee.

'The food tastes good. Everything O.K. but speed over the ice painfully slow,'[65] he notes.

Eielson holds the coffee mug fiercely, wanting to get warmth back into his frozen fingers but there is little feeling from the worst affected. It does not look good. The weather thickens at 6 pm and just 15 minutes later Wilkins gloomily records that both the sun and ice are now almost completely obscured. They are flying blind, with only their instruments to tell them their direction, altitude and how much fuel they have left: about three hours.

They could, of course, try for a landing and take off again when the light is better but given their most recent experience neither wants to take the dual risk of landing in the dark or getting the engine started again after many hours on the ground. No, they are intent on riding this beast till she drops.

Onwards, two souls in the northern skies – now heading south again – with Wilkins using a torchlight to see the compass, tapping Ben on the left or the right shoulder to spare shouting and indicate which turn to take to keep them on course. This is hands-on navigation of a most unnerving kind. At 8.40 pm visibility is nil, only a dull glow through the clouds[66] indicates where the sun is skimming the horizon. They are at 5000 feet and time and fuel are running out.

Wilkins passes a note to Ben: 'What do you think; let her go as long as she can, then drop straight down ahead?'[67]

Slowly, Ben nods. Neither speak. Both know they are pretending a necessity is a plan, the risk of landing in such light rotates in their minds as the engine . . . stops. Not a stutter or a splutter, it just cuts dead, as the propeller windmills, propelled by the wind rather than propelling the same. There is a terrible, horrible silence.

The truth is obvious: 'We had run out of gasoline. There was no sputter or gasp from the starved carburetor; it was sudden silence, except for the hum of the wind vibrating the wing wires.'[68]

Ben snaps the magneto switches left and right to no avail. There is spark, but now, definitively, no fuel. Both feel the plane sag as they rapidly start to lose altitude. By the reflected light from his torch onto the altimeter Wilkins watches with admiration as Ben simply steadies the dead machine, 'righting her to an even keel and an easy glide'.[69] But this glide will not be easy and they both know it, and it is unlikely to have a graceful end. Ben's eyes stay fixed dead ahead, while Wilkins takes a quick look at the compass, a last reading noted, for themselves and for those who may find their smashed craft and lifeless bodies.

Wilkins has no doubt that is the most likely outcome. As they swoop down to just a few hundred feet from the surface, by the dim glow of the falling sun they can now see ice-ridges in the distance dimly, rushing towards them as they sink ever lower.

'We floated down through utter darkness, a grey forbidding darkness,' Wilkins will recount, hauntingly. 'Not black like a winter's night, but a nerve-wracking, sense-dulling density. Beneath us lay what? Rough ice we knew and perhaps a lane of open water. Injury, minor, or fatal seemed imminent but we were resigned helpless in the hands of our Maker. His to dispose of without effort on our part. There was nothing we might do to help ourselves.'[70]

Closer to the ice, the plane swerves and pitches with the terrestrial winds, but with masterful expertise Ben holds steady as they come in and down and . . . hit a snowdrift at speed.

'The left wing and the skis struck simultaneously. We bounced and alighted as smoothly as if on the best prepared landing field. I gripped Eielson's shoulder and slipped through the door of the machine to the ice. Wind and driving snow filled my eyes.'[71]

The lower wing has been torn in the landing. They will not be flying home to Point Barrow. 'The plane still was on its skis, but they had turned on their sides, with stanchions twisted and broken.'[72] It will not be taking off again anytime soon. But they are alive. The two men climb back into the cabin and swap estimates as to where the hell they are. Before sleep takes them, Wilkins taps out a message in Morse to whoever might be receiving:

.-- . -. – / --- ..- – / . . .   . . . ----- / -- .. .- . .   . . . -.-.-

Went out 550 miles. Engine trouble. Forced landing three hours. Sounded 5,000 meters. Landed out of gas sixty-five miles N.W. Barrow.[73]

At Point Barrow, the eyes of wireless operator Howard Mason flick open as his ears prick up. Long hours of silence and dashed hopes had ended when dashes were heard for a few seconds, not enough to discern anything but it was on Wilkins' wavelength and now . . . --- . . . At last Mason has something to make a record of: 'A pretty good signal . . . two letter combinations that stood for "forced landing on sea ice", "out of gas" and "plane damaged". The whole message was repeated several times and that was all.'[74] The rest is silence. Back to waiting now, helpless but not hopeless.

The survival of Eielson and Wilkins will now depend upon the skills that the Australian had learnt all those years ago on the Stefansson expedition.

Unable to keep his eyes open a moment longer, Eielson has taken the top of the empty gas tank as his bed, a sleeping bag the only padding between himself and the hard metal. Wilkins merely curls into a ball in the corner of the cabin and lets his lids fall, waiting for the peace of unconsciousness, which arrives in an instant.

They wake in the morning to a banshee's howl, the sound of a brewing storm's emissary, a violent gale that threatens to tear the cabin apart.

'We could see through the windows of the cabin, where we had slept,' Wilkins will recount, 'that conditions made outdoor activity most inadvisable if not impossible.'[75]

Still, given they will need to walk home from here, he knows he needs to climb out and have a cold recce of their surroundings. When he does so, the cold sweat on his forehead instantly freezes – a more than curious sensation.

You have to see it to believe it! For they have landed on a smooth patch of ice a little less than 30 yards by 15 yards. On three sides are high rough ridges; on the fourth, one three-feet high ridge. It is chilling to see in daylight that 'our safe landing the night before was miraculous'.[76]

Mark it down as yet one more time in Wilkins' life when he had survived against all odds.

Still, given that that is situation normal for Wilkins, he continues to behave in a manner which will heighten his chances of continuing to survive. After taking a large ice-pick and digging a hole through the six-feet deep ice – no small thing, and not easy to break through the bottom without falling in himself – he is at last able to drop a tow line to calculate that the ice pack they rest on is drifting north of east at six miles per hour, taking them even further away from Point Barrow. That afternoon, Wilkins takes two separate observations using his sextant to measure the angle of the sun at midday and calculate their current position, in the current, to be at 72 degrees latitude, 155 degrees longitude. Wilkins taps out what this means in Morse to any who might be listening: 'Now about one hundred miles northeast of Barrow. Position tomorrow.'[77]

Immediate duties now done, the madness of what they have just survived, and the lunacy of what lies ahead, strikes them both, at such a combined angle it produces a surprising reaction.

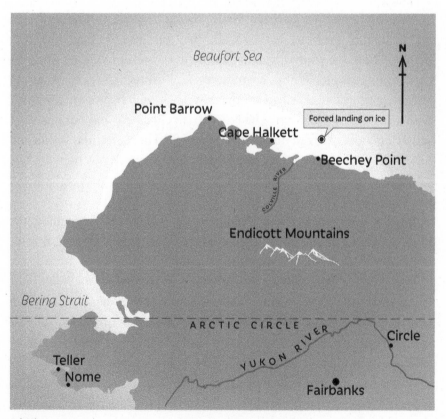

Alaska

'Ben looked at me and I looked at him. Then we began to laugh nervously, and could not stop for about five minutes.'[78]

With the storm still howling, they rest through the night to build their strength. On the morrow Wilkins again takes their position to find that they are broadly stationary, if adrift. It is not until 3 April 1927, three days later, that Wilkins judges conditions good enough to walk in; which is fortunate as the drift by now has turned and carried them 80 miles from solid land.

At 8.15 am on a snowy Sunday they farewell their stricken craft with some emotion – it has served them well in the air, and provided key shelter on the ground – and begin to drag their sleds to the south-east. The sleds are improvised, newly crafted from the wooden parts of their now useless aircraft: 'one from the lower part of the cowling and the other from the tail-ski – to which was attached a section of corrugated duralumin from the cabin wall'.[79] What they were flying they are now dragging into the far from Friendly Arctic.

The wind cuts like a frozen blade, slicing through clothes and skin and chilling to the bone as they walk. By 1 pm they are trudging more than walking, the snowdrift reaching their waists and making progress close to impossible. Within 15 minutes Wilkins recognises they can no longer continue and calls halt for the day. Their tired bodies are rewarded with a treat from Norway: 'The Norwegian chocolate and biscuits were satisfying foods. We did not stint ourselves.'[80] There are signs of life and other feet that have trodden their path, but they are not human: 'We saw many fresh bear and fox tracks and many seal breathing holes in the young ice.'[81]

It is the young ice that has Wilkins concerned about dying young; the illusion of solidity surrounds them and that it is the real danger. But for now, the solace of shelter should be sought.

It is time for Ben Eielson to learn the art of building an *Iglu*. True, it is no less than 12 years since Wilkins has last built one, but happily it is a skill that is never forgotten. And yet it is only now that Wilkins realises just how bad Ben's frostbite is: 'He could not hold a knife or saw and was scarcely able to carry the snow blocks. Four fingers on the right hand were badly nipped; the little finger and the one next to it badly blistered and blackened . . . he must have suffered excruciating pain which he bore heroically without a murmur.'[82]

At least they are soon sheltering inside the *Iglu* that Wilkins has constructed with record-breaking speed. It is done so well that Ben is not only soon inside, but confesses surprise at how warm it is. Drawing further on his own deep knowledge of how to survive in the Friendly Arctic – and here's to you, Stef – Wilkins tells Ben to throw away his 'expensive riding breeches, woolen underwear and elaborate sheepskin jacket'[83] and try Eskimo furs instead. Though reluctant, Ben does as he is bid, and is soon so warm and comfortable that, in this amazing new abode, it is soon time to put the rubbish out – as the pilot's clothes are simply thrown away in the snow.

The following morning, Ben's understanding of how to survive and prosper in these climes goes up another notch as he watches in wonder as Wilkins negotiates the floes. With extraordinary agility and a nimble touch, it is like he is playing a deadly game of chess with the ice – Knight to Queen 4, and then two steps forward and four to the right by hook or by Rook – as he darts forward, then left and right, from floe to floe, as they drift past each other, crash into each other, and grind past each other. Do as he does, Ben, and stay close, for he understands how to *read* the ice. Look for the rough ice for it spells old ice and is likely solid. Otherwise, *watch for steam, Ben*, for that hot instant mist spells danger – betraying the fact that the underlying ice is melting and with just one wrong foot on that fatal fragility you will be making the breakthrough of your worst nightmares, through the ice and into the freezing water, dragged down by your pack. Certainty does not exist here. Safety is impossible, chances must be taken; but unless they are educated chances survival is impossible.

What is particularly driving Wilkins' need for speed is his deep concern for Ben's frostbitten fingers.

Ben is as stoic as ever, but a single faltering confession finally escapes his lips. 'My fingers ache,' he says, which Wilkins instantly translates to 'I am in agony,'[84] for otherwise he would never mention it.

Wonderful!

'That is a good sign!' the older man exults. 'They are regaining life!'[85]

Instantly relieved and feeling more cheerful Ben tells Wilkins that one finger feels just fine though, and in an instant Wilkins knows that he can no longer be candid with his pilot.

Which one, Ben?

'The little finger on the right,' Ben informs him, and Wilkins' heart sinks further. 'That promises trouble,' Wilkins thinks to himself.[86] As counter-intuitive as the whole thing is, no pain on a frostbitten finger means it will likely have to be amputated, should they live to see a doctor. And if any more of Ben's fingers start 'feeling fine', then Wilkins will have to become a brutal makeshift doctor – or at least bloody surgeon – himself. They have a surgical kit with them – Wilkins has used it before – but if they are fast enough across the ice, Wilkins can fulfil his silent hope to 'save Ben's hand without mutilation'.[87]

The key thing now is to get Ben more ice-savvy, to give him a crash-course that ideally will not see him crash through the ice to his death, but survive all the way back to Point Barrow, where he can get medical help. As the days pass he indeed becomes more proficient, able to choose where to put his feet with much more success, and each night Wilkins is scrupulous in making sure Ben looks after his feet by going through a nightly ritual of cleaning his boots and socks.

Be *rigorous*, Ben, and clean it of the very last icicle!

> Constant care is necessary to prevent them from becoming a mass of ice. The boots must be turned inside out, beaten and scraped; socks beaten, turned and rubbed, then slung about the chest and under the arms to dry. Cleaning boots and insoles is a cold-fingered job, and drying socks on one's bosom is not exactly a pleasant pastime, but care of clothing is the most essential part of the day's work during Arctic travel.[88]

While it is all as familiar a chore to Wilkins as brushing his teeth after a meal in normal circumstance, Eielson finds it more than just burden-some and confesses to Wilkins that for the last nine years he had done no manual work apart from handling the controls of an aeroplane. Still, every night he gets a bit better at it, and understands that not freezing to death is laborious work, but someone has to do it – and that someone is him.

At last, after five days of pushing hard, the lanes of moving ice give way to a stout-looking solid floe of ice which, to judge from its lack of movement to either east or west, is very likely landlocked. Wilkins recognises it as their chance.

It is time to dump all heavy equipment and move, quickly. With land so close they must cross this ice before it inevitably breaks once more.

Putting the bare essentials they need in their backpacks, they abandon the sleds one after the other and their toughest push across the ice begins. This close to land, the ice is by nature uncertain as it pushes up against an immovable object, becoming such a treacherous mixture of melting blocks covered by deceptively full-looking snow that only two points of contact with the ground is too risky.

'We had hoped to walk,' Wilkins notes, 'but much of the way was so rough – upended ice blocks surrounded with soft snow in which we sank to the waist – that it was necessary to crawl slowly ahead on hands and knees.'[89]

Wilkins is aware of how absurd they must look – 'A moving picture of our floundering efforts would be considered much over-acted'[90] – but the danger outweighs any dignity as their every step forward must be first tested as to whether it can take the weight. There is literally no going back, they crawl, walk and clamber their way forward, they hope towards land, land, land, blessed land.

Stopping for a quick lunch of pemmican, with extra lashings of anxiety for dessert, it is more than a little disconcerting for them to see seals suddenly popping their heads through the ice to have a stickybeak at them – a sure sign that the ice is thinning around them at a steady pace. There can be no sleep now, only rest, then onwards.

On 9 April 1927, 11 days after their journey had begun, Wilkins spots a thicker pack of ice beckoning, which is to the good. What is to the bad is that it is effectively protected by a half-frozen *cordon sanitaire* of spongy ice around it. Wilkins, the old-stager, cautiously explores the solidity on hands and knees; the impatient Ben steps on the ice with one foot to test its strength, only to slip and fall in up to his knees and then scramble back like a mad thing – once frostbitten, *thrice* shy.

Lesson learnt, Eielson now gets on hands and knees himself and slowly follows Wilkins as he painstakingly determines a likely path and they slowly cross, inch by inch. Both men spread their legs as wide as possible, holding their ice-picks out as another possibly buoyant limb should the ice betray them. And sure enough, after some six hours of this agonising work, they get to within just three yards of the clearly solid pack, which sees the jubilant Wilkins stand and turn to happily yell encouragement to the lagging Ben, 'Come on!' whereupon, as he turns, the ice breaks beneath his feet and he sinks. Partly through instinct and partly through experience, Wilkins throws his body forward, back to the more solid ice he has come from, and manages to get his arms on it, even as – as fast as he can, without breaking through himself – the deeply alarmed Eielson pushes forward to help. With one mighty heave, knowing he has only minutes before the freezing waters will render him incapable of movement and he will die – Wilkins manages to swing his right leg up and onto the more solid ice. It takes some doing, including one failed attempt, but he is mercifully able to jam his ice-pick into the ice pack and give himself a solid enough purchase to drag his whole body up and out.

After Ben rips Wilkins' boots off, 'they stood up stiff and solid as he tossed them aside'.[91]

Wilkins boots are stationary, he decidedly is not.

For to Eielson's stunned amazement, Wilkins now strips stark naked and starts dancing about. Has the man lost his frozen mind?

Not quite.

A veteran in these matters, he is doing what he had first learnt from the Eskimos all those years ago, and now starts rubbing his wet clothes in the soft, dry, snow to blot up the water, while also dancing around and flailing his arms to keep up the circulation.

'Afterwards Eielson told me that at that moment he felt certain that there was no hope for either one of us. When he saw me stripping stark naked and dancing like a madman, he thought I had gone crazy from the shock of falling into the water.'[92]

In what the shivering Wilkins knows is the time-honoured fashion, he now tightly squeezes the water out of his pants and shirt – wringing them till they can be wrung no more – so that if he stays alive his body heat must eventually dry the rest once he puts them back on.

That night they both slumber in their clothes, with legs and feet warm at least, through their curious sleeping configuration.

'We had abandoned one sleeping bag and one half of the other,' Wilkins will recount. 'At night we slept in our parkas and in our canvas packs, both of us with our feet in one bag. We found this warmer and more comfortable than an individual bag.'[93]

And now both their bottom halves have the half that remains. After everything they have been through, now they can also say they have slept together and be half right. It is 20 toes together at 10 degrees below zero.

In the morning they continue and it is far from easy, Wilkins dourly summarising the day as 'ten steps, or one step, then a tumble',[94] but at least they do not fall into water. Once again they resort to travelling by hand and knee over uncertain ridges, and a 'cautious slithering dash across young ice that sagged as would a stretched blanket, if you stepped on it'.[95]

Hour after hour, day after day, they slog on, Eielson with his head down and only occasionally lifting it – lately to groan at the sight of high ridges of snow. Wilkins, on the other hand, the old hand, always has his head up and is constantly scanning the icescape before them, *reading* it – and he loves nothing better than seeing such high ridges. For he knows that only shore ice is stable enough for snow to build up on, and where there is shore ice, there must be a shore not far away. And at last, it happens, getting to the top of one such ridge, far to the south, Wilkins sees something that is man-made – two high, vertical straight lines. Taking his Mirakel five-power field glasses in hand, Wilkins confirms it: two poles. This in itself is a marvel of navigation: 'After thirteen days of zigzagging across the ice-fields, finding our course by the aid of the pocket compass and two watches I carried around my neck, we had come out within three miles of our intended destination.'[96]

Delightedly they scramble ahead, and soon cross a sled dog trail, which confirms that some kind of civilisation is near, and following the trail it soon proves to be – he can recognise its key outlines even from a great distance – the trading post at Beechey Point.

> In sight of relief and human habitation, our tongues began to wag. Whom might we find? What had our friends thought of our delay? Had our wireless messages been received? Would we find transportation at the house or would we have to walk to Point Barrow?[97]

And now look.

There!

Coming their way at full pelt is a dog team pulling a sled upon which is Alfred Hopson, the son of Fred Hopson of Point Barrow – the cook who had delighted in swearing at his uncomprehending wife. Within just two hours, instead of fighting for their lives in the icy wilderness, they are hale and hearty in the trader's house at Beechey Point, coffee alternating with alcohol to warm them as toasts are drunk. The trader, Anton Edwardson, shows them, Wilkins reports gratefully, 'every hospitality'.[98]

Here's to life!

And to friends!

And to being safe at last!

Wilkins' old Eskimo friend Tapuk volunteers to race to Point Barrow with a dog team, where he will get the remaining pilot, Alger Graham, to fly *DN-2* to Wilkins and their walking days will be done. Tapuk leaves less than three hours after Wilkins and Eielson arrived, and the two tired men now eat the first hot meal they have had in 18 days – roasted seal.

At Point Barrow search flights have already given up looking for them. As Howard Mason knows: 'The proverbial needle in the haystack is but a drop in the bucket compared to the chances of finding two men and an airplane somewhere on the Arctic Ocean.'[99]

But as they sleep soundly, about 1 am the dogs begin to yowl. The door is flung open and it is the Eskimo Tapuk, who tells them he has travelled six days to bring them a letter from Wilkins. Forget the exact wording, the gist is this: 'They are safe!'[100]

If Graham wouldn't mind hopping over to Beechey Point and picking them up in *DN-2*.

Back at Beechey Point, after Tapuk has departed, Ben's fingers are carefully inspected and with care; all save the little finger on the right hand are showing signs of life. As for Wilkins, an extraordinary thing happens – now that they are safe, his body seems to take the opportunity to collapse. Within hours, his feet and legs swell so badly that he can barely hobble from the couch to the table and back again. Yet the Eskimos, of course, have seen it all before. Application of a hot liniment by the old women, and the passage of time sees such a quick recovery that by the time Graham touches down a week later Wilkins is able to walk to the plane quite normally. And what a joy to once again be airborne, to look down on all the ceaseless snowdrifts, fearsome floes and rigid ice-ridges that, if not for this glorious plane, they would have had to struggle over, but are now traversing at the rate of 115 miles per hour.

They land in Point Barrow for a joyous reunion with their crew, tempered only by grim news from the local physician, Dr Newhall. There is nothing for it, he advises after quick examination, but for two joints of Ben's little finger on his right hand to be amputated. Eielson receives the news stoically and Wilkins assures him that, after his recovery, he will fly with no other. By cable, Wilkins summarises their achievements for the press and his backers in Detroit (there is a fair overlap between the two).

Their flight, and return, has proved, he says:

(1) the approximate limit of the Continental Shelf north of Siberia

(2) that landings and forced landings on the Arctic pack-ice are possible

(3) that it is possible for experienced men, without dog teams, to walk to safety over the Arctic pack-ice.[101]

Certainly, ahem, the loss of Stinson *DN-1* is to be regretted. But look to the upside! They have not only proved 'the ability of the healthy-bodied, healthy-minded man to rise to all emergencies,'[102] but also provided a thrilling story guaranteed to sell papers and provide a return on their sponsorship.

As to the thrilling story there will be no argument from the Eskimos, who regard Wilkins' return as almost supernatural. They know, better than anyone, the risks that Wilkins has taken and somehow overcome by crossing shifting, shallow ice in such a manner.

The notable exception to those impressed is the remaining pilot, Alger Graham, who is clearly shaken by the whole affair, and seems to have come to the conclusion that flying in this part of the world is madness. That much is evident by his return of *DN-2* back to Fairbanks in a shaky progression of forced landings and frequent repairs. Once there, Wilkins will recount, Graham 'was not anxious to leave,'[103] and he must directly order him to take off for Point Barrow once more to transfer supplies. Well, Graham will go that far, but no further. When Wilkins suggests a trip from Point Barrow to Greenland, Graham's response is sharp.

'I was engaged with definite and clear understanding that I could refuse to go on any flight over ice.'[104]

But Wilkins insists, whereupon Graham indeed makes a half-hearted attempt, only to return a few hours later, insisting – to Wilkins' fury – he had encountered heavy fog.

With Ben still out of action recovering from his operation, Wilkins is dependent on a pilot who is determined to remain on land. Bitterly, Wilkins writes his verdict of Graham: 'Neither constitutionally nor mentally was he a suitable pilot for flight over the ice.'[105]

It leaves him with, effectively, only one other option to consider – himself.

> By going alone I could reduce weight of the equipment and food;
> I could carry more gas, probably sufficient to actually reach Etah
> [in Greenland], in the machine and I could – even alone – with

the Sonic gear take a sounding if a landing on the ice was made. If I were forced down, I could walk out alone with less difficulty than with a crippled companion. I could also more cheerfully face greater risks when none other than my own life was at stake.[106]

Against that?

Well, against that, if he pilots *DN-2* alone, it will leave the rest of the crew stranded at Point Barrow until Eielson is well enough to fly the Fokker from Fairbanks to pick them up – which might well be problematic, given that summer is not far away, and there may be a real problem landing the heavy Fokker on ice in such weather. There is also the factor that, flying alone, he would have no witness for his achievements. As he puts it delicately: 'There was also the ever present possibility of a lone traveler's word being doubted by some individuals and societies and any observations taken on a lone flight might prove valueless.'[107]

Truth be told there is already controversy attached to the claimed achievements of some men who actually have a witness. There is no better example than what had happened to Richard Byrd who, only last year, had been widely celebrated as being the first person to fly over the North Pole before serious questions had been asked after his pilot had refused to confirm whether or not they made it over the Pole itself – or, as seems more likely, just got fairly near, before turning around and leaving Byrd to 'prove' they did it on paper. Wilkins knows his own achievements will be real, but that is not enough – they must be undeniable, which is no easy thing, particularly when it is the nature of many of his feats to be fantastical and unbelievable.

You did what?

You faced a firing squad in Turkey, three times, and were never actually shot?

You went to the Arctic as a photographer, and finished as a commander, rescuing the leader?

You were at the Western Front and constantly went over the top ahead of the troops, armed with nothing more than a camera, and never got shot?

One way or another it is time to get back to America and once more get funds together, this time with the goal of flying across the entire Arctic.

After securing their stores and supplies in a Point Barrow shed, they leave on 5 June 1927, with the seven-and-a-half-hour flight to Fairbanks made by Alger Graham, his swansong of flying for Wilkins, for obvious reasons. After some more aerial hops, skips and jumps, the expedition arrives at Seattle by boat on 18 June 1927. Once more, all men are paid off and formally dismissed, at least being told that Wilkins will be in touch directly if they are required to return in preparation for the leap across the entire Arctic.

What they don't know – for he simply doesn't have the heart to tell them – is that Wilkins is already aware that it is highly unlikely that he will be needing them any time soon, and it certainly won't be on the coin of the *Detroit News*, which had cabled him while sailing in notably stark terms:

> DETROIT NEWS NO LONGER CONCERNED IN YOUR WORK. CLOSE
> EXPEDITION WITHOUT FURTHER EXPENSE. [108]

Charmed, he is sure.

CHAPTER FOURTEEN

ACROSS THE ROOF
OF THE WORLD

One man in a thousand, Solomon says,
Will stick more close than a brother.
And it's worth while seeking him half your days
If you find him before the other.
Nine hundred and ninety-nine depend
On what the world sees in you,
But the Thousandth man will stand your friend
With the whole round world agin you.

Rudyard Kipling, 'The Thousandth Man'

July 1927, Detroit, the presses stop

Wilkins has failed. They wanted him to cross the Arctic and not only has he failed to do so, but a journalist has been killed in the process, planes have fallen out of the sky, they have been beset by constant problems – and all while Robert Byrd had reached the North Pole first, plus a flying mailman from Missouri by the name of Charles Lindbergh had flown all the way from New York to Paris! And what does Wilkins have to show as actual achievements for all their money? He has landed on ice, dropped some depth charges, and taken off again. Congratulations, but it is not enough to interest our readers, or justify further expense on our part.

We will pay all outstanding bills, and you may keep *DN-2*, but the expedition is over, as far as we are concerned. Wilkins' best reckoning is that he might be able to sell the Fokker for spare parts to raise $15,000, but it is beyond him to work out where he might find a buyer.

•

Travelling to San Francisco, Wilkins does the rounds of aviation and adventure sponsors, presses for press interest and looks for funds, finding only a predilection for ejection and rejection, leading to uncharacteristic dejection.

On the day he is due to leave for Los Angeles, he is sitting despondently in his hotel room (brooding on the possibility, perhaps the necessity, of being the only man remaining in the expedition and flying *DN-2* solo), when he looks up and sees a vision. Shooting past the window of his room is a plane, the like of which he has never seen before.

> I marked its beauty of streamline, angle of incidence of the wing in level flight . . . As it turned towards me I realised the full beauty of its design. It apparently offered no head resistance except for the engine, leading edge and a slim landing gear. It had no flying wires; no controls exposed – nothing but a flying wing.[1]

What on earth is this creature? Wilkins is besotted at once: 'It gave me a thrill that another might experience if he saw his ideal woman in the flesh.'[2]

Shortly thereafter he is ringing all the local airfields, describing this beauty that had literally flashed before his eyes, only to be told time and again that no-one has seen anything like the picture he draws. Drat it. Wilkins heads off to Oakland Airport to meet up with an old friend from adventuring circles, Ray Shrek, and . . . there she is!

It is, he quickly finds out, a Vega, and by inspecting the name on the rudder he is delighted to find out it is made by the Lockheed mob in Los Angeles. Just two days later he is sitting before Mr Allan Lockheed of that very company, pouring out his admiration, and his need to have one of those planes so he can fly over the roof of the world and make Lockheed even more famous!

Alas, alas . . .

'I was somewhat dismayed to discover that the machine I had seen was the first and only one of its kind.'[3]

She is a beauty, no doubt about that, with a 225 HP Wright Whirlwind J5[4] horsepower engine capable of propelling her at 180 miles per hour and so light that she could be pushed around by two men – extremely useful, should they have to make an emergency landing! It is precisely the kind of plane he needs for his next attempt at a polar flight (as he, and his non-sponsors, have determined that this time he will be running a much more streamlined and cheaper operation). But no, Lockheed has no interest in offering him one, the way Anthony Fokker had. If he wants one, he will have to order one outright. Very well then, Wilkins does exactly that.

He signs the orders for the second one to be made, and is informed it will take four months – hopefully giving him the time he needs to raise the necessary funds. How, exactly?

Very good question. Sponsors are still rare – the likes of the Detroit Aviation Society and the *Detroit News* had both run screaming from the room when he had last approached them – and straight out benefactors rarer still. The amount of money in his bank account, plus a dime, could just about get him a cup of coffee.

Assets?

Two wrecked Fokkers. His best chance is to sell them for whatever he can get and use that money for at least a down-payment for the Lockheed, which will be $12,500 in total.

As it happens, while in Seattle, he read an intriguing story in the local syndicated newspaper. One of his fellow Australians, Charles Kingsford Smith – the very fellow Wilkins had replaced on the Billy Hughes Great Race to Australia – is in San Francisco with a couple of other Australian pilots, and they are looking for a plane to fly across the Pacific. Their plan is to fly from San Francisco to that tiny speck in the blue eternity of the Pacific called Hawaii, on to another speck called Fiji, before going to the barn door of the Australian east coast, ideally landing in Brisbane.

Wilkins immediately dispatches a cable to Kingsford Smith . . .

HAVE FOKKER I CAN SELL YOU, WITHOUT ENGINES OR INSTRUMENTS.

The return cable from Kingsford Smith is not long in coming.

RE YOUR WIRE. COME DOWN TO 'FRISCO AND TALK IT OVER.[5]

George Wilkins does exactly that, meeting Kingsford Smith, and his fellow pilots Charles Ulm and Keith Anderson, a few days later in the Roosevelt Hotel. Kingsford Smith and Wilkins have an easy affinity, and Wilkins warms at once to the other two pilots as well. They are all Australian aviators in America, and they are all pursuing a dream, all chancers – and in the case of Wilkins and Kingsford Smith, they are both war veterans with slight limps.

None of them, however, have too much time for chit-chat. Wilkins gets straight to the point. 'I think I have the machine you require for your

Pacific flight,' he says simply. 'It is a tri-motored Fokker. No engines or instruments, but the wings and airframe are in excellent condition.'[6]

Charles Kingsford Smith likes the sound of it.

'Well, George, I'd very much like to see the machine. Where is it?'[7]

Arrangements are made, tickets booked, and within the week the covers are removed from the plane in its Seattle hangar as they all gaze upon it.

Wilkins looks upon the cursed *Detroiter* with a creeping horror that he struggles to disguise. There is nothing wrong with the thing *per se*, but from the moment it had killed poor Palmer Hutchinson and everything had gone wrong thereafter, he has found it hard to escape the feeling that it has nothing less than a *hex* on it. Maybe Kingsford Smith will have better luck with it, but frankly, he doubts it. In any case, that will be for him to work out. The wonderful thing is, Smithy takes to *Detroiter* immediately, while discarding the name. *Southern Cross* will sound better.

The only sticking point is price. Although £3000 is a more than fair sum for Wilkins to be asking – as it was only about a third what he had paid for it[8] – the simple reality is that his fellow Australians don't have remotely that kind of money for a Fokker without engines or instruments.

As to sponsors and benefactors, Kingsford Smith, Ulm and Anderson are in exactly the same position as Wilkins. These are grim times. They, too, have written to many companies in both the USA and Australia, for, let's see . . . *zero* interest. After cabling the one benefactor they do have, the NSW Government, the best they are able to offer is £1500 immediately, with the remainder to come . . . when they can raise it.

Done. It is something, anyway, for Wilkins. And just as there is honour among thieves, so too is there collegiality among flying adventurers – for if they won't help each other, who could?

Meanwhile a nascent aeroplane company at Fairbanks makes an offer for *Alaskan* and other bits and pieces that Wilkins has, and he quickly accepts.

All put together he can just scratch together enough money to buy the Vega. But for a crew?

Well, there is one man he knows will likely come for no pay at all, and that is Ben Eielson. He writes to Ben about the Vega and Eielson replies that he is currently working as an inspector for the Department of Commerce, Aeronautical Division but he will apply for leave immediately on Wilkins' word.

As the new Vega is built, Wilkins must decide on a name. His choice? *Detroit News*. It is a naked gesture – designed to get that newspaper organisation to come up with some more money through either shame or gratitude, he doesn't care – but although he receives a polite thanks from his former sponsor, they add that they will be in no way responsible for any further expedition and also that they would rather not have their name associated with a failure, which they clearly feel (*sniff*) this expedition will be. Ah well, Wilkins is always confident his actions can change people's minds and as he is sure of success '*Detroit News-Wilkins Arctic Expedition*' is 'painted in dark blue letters on each side of the deep-orange coloured fuselage'.[9] The plane, at an added cost of $1000, will have a short-wave wireless system installed to broadcast triumph or disaster as it occurs in real time, via Morse code. Such a system has already been tested on the recent Dole flight competition from California to Hawaii and it worked; the listener could follow William P. Erwin and A.M. Eichwaldt's death throes, live, 'as they tail-spinned into the Pacific Ocean and to death'.[10] Drama aside, Wilkins has little faith in the reliability of short-wave alone, he decides that Morse will provide a less graphic and more certain way of communicating their own fate, for good or ill, its dots and dashes transmitting at least 500 miles, courtesy of a large aerial 'permanently slung during flight'.[11]

For their part, those at Lockheed are fascinated with what Wilkins is going to attempt and whether their machine can pull it off.

Captain George H Wilkins   *July 16 1928*
Waldorf Astoria Hotel
New York City

Dear Captain Wilkins,

You have by this time received the blueprint of the plane that we are building for you . . . we want to build this sea plane for you, not for the profit that we could make out of it, but because of our interest in you, and we want you to be thoroughly satisfied with the plane . . . If the plane for any reason does not meet with your entire approval, you need feel under no obligation to take it.

Yours Very Truly
    Lockheed Aircraft Company
    Ben S Hunter[12]

As it happens, Wilkins is there on site as the Government Official Pilot arrives to take a test flight with still the only Vega in existence to determine her airworthiness or otherwise. He stands with Allan Lockheed as the pilot takes her through her paces. There is no doubt, this Lockheed model 'Vega I X3903' with the 'X' standing for 'experimental', handles beautifully, banking and turning with ease. And yet either because of a failing of the design or more likely the Official Pilot himself, she stalls as she comes in to land and hits the runway so heavily that the landing gear breaks and it is only by a seeming miracle that the plane comes to a halt, swinging in 'a sickening circle' on just one wheel! Lockheed, whose approach is just like his eggs – always sunny side up – is seemingly not worried that his only Vega plane has just suffered big damage. For now he approaches Wilkins and another shocked watcher with a broad grin, and an enthusiastic shout.

'Boys! Did you see how that fellow held her up after the landing gear parted?'[13]

The following day, when Ben Eielson arrives at Los Angeles' Union Station after his long cross-country haul from his home state of North Dakota it is to find Wilkins impatiently waiting for him, and he cannot even drop his bags at his hotel before they have caught a cab to the Lockheed factory to get a close-up look at the new plane being made for them.

Ben's opinion?

'She looks all right,'[14] he says carefully, though he is clearly unimpressed, and all the more so, after chatting with the officer who had flown and crashed the Vega yesterday, whereupon he becomes 'even less enthusiastic'.[15] No matter, Wilkins knows just what is needed and quickly arranges for Ben to become a 'demonstration flyer' to Lockheed so he can give the one plane they do have endless hours of testing, with a guarantee of another one gratis if this one crashes. Sure enough, just as Wilkins had hoped, Eielson grows to love the Vega in flying her – her speed, her manoeuvrability, her light touch! – and even forgives her occasional eccentricities, such as the engine cutting out sometimes in difficult moments. All up, Wilkins and Eielson both sense that, with this plane, nothing less than *history* is already in the bag!

It is a strange thing but they both feel it: destiny, inevitability. Wilkins writes of the curious sense of *preja-vu*:

> There was an odd sense of annoyance in the back of my head that it should be necessary to spend the time to go to Barrow and actually do the flight. It was like setting out to run a race after the prize was won. But this strange foreordination did not prevent me from going ahead most carefully in the arrangement of every detail.[16]

That detail includes sending 1500 gallons of gas to Point Barrow by ship in order that that side of things can be taken care of from the beginning.

Wilkins and Eielson themselves, meanwhile, travelling by boat and train, through snow and flood, arrive at Fairbanks on 26 February 1928. They cannot help but notice that the greetings from their old friends are a little perfunctory: 'Their manner of saying, "We know you will make it this year", implies there was nothing else to be said.'[17]

After all, these two have already passed this way at much the same time with their disassembled plane for the last two years, before retreating in defeat about five months later. Why should this time be any different?

Good to see you, of course, but let's not get carried away, yet.

(Among the very small number of people who actually think they might make it? Stef! But even he is giving odds: 'I think Captain Wilkins has at least three chances out of four of success,'[18] he tells the press. It is the fourth chance that others are betting on.)

Wilkins notes other rather depressing changes. When he first arrived in Alaska, flush with money and publicity, he was given free use of the airfield by its shrewd manager, Bob Lavery. The second time, still in funds, Wilkins was only charged a minimal fee. Now, with no publicity and no funds, Lavery charges him four times as much! As Wilkins mockingly declares: 'Unto him that hath much shall be given, but to him that hath little much shall be charged.'[19] This time, travelling light with no crew, Wilkins must hire two day labourers to provide the grunt to help him and Ben assemble the Vega. Here, at last and at least, they actually do find some charity. For while they are assembling the plane, other men working at the airfield become fascinated with the unusual plane and soon lend a hand and a monkey wrench – pass the screwdriver – as before long a small swarm of them are setting to with a will.

It is all quite jolly, albeit with one problem that Wilkins had not foreseen, the sudden and overwhelming popularity of Ben Eielson.

Why? Because the good people of this frozen burg have become convinced that this time, this time, the mad Wilkins really is going to kill him.

> Eielson, who is Alaska's popular hero, has in Fairbanks, as every-where, many friends. Many of them, I know, thought that he was heading for his doom. They admired his pluck and determined that he should have a good time before he departed. His round of social obligations gave him no rest.[20]

The exhausted Eielson soon feels the same.

Get me out of here, Wilkins!

On the morning of 19 March 1928, with the temperature hovering at 24 degrees below zero, Wilkins does exactly that as the Vega hurtles down the runway, skis showering snow, and is soon soaring away, leaving only a crowd of sombre spectators in its wake, sure that an actual wake for dear Ben cannot be far away. But the two in the Vega are not sombre at all. They have no less than 130 gallons of gas in their tanks, in a new and wonderful plane, and food to last 30 days travel should they crash and survive. Not only that, but for once, the weather is perfect.

'This trip over the Arctic slope,' Wilkins will recall, 'was one of the rare occasions when we could see every detail on the snow-covered tundra.'[21]

Indeed, as far as the eye can see – which is literally hundreds of miles in every direction, it is a picture postcard, with even the taciturn Ben grinning in delight as he points out a herd of thousands of caribou moving towards the mountains near the Bering Sea. Does life get much better than this?

Bliss was it in that dawn to be alive, but to be young was very heaven.

And to be in a plane like this, on a venture like this, is nothing less than seventh heaven.

Still, with everything going so well, and easily for the moment, the truth is that the two men feel more like tourists than explorers; the fact that they constitute the entirety of the expedition makes this seem like a private trip rather than an historic endeavour. When they land in Point Barrow, a happy group of old friends appear, with Charlie Brower and Fred Hopson praising the new machine as the Eskimo children – seem-ingly oblivious that it is minus 48 degrees Fahrenheit – compete for 'the privilege of carrying our gear to the station warehouse'.[22]

And yet, not all children are so enamoured of the Vega. As Wilkins and Ben gulp coffee at the trading post, one of the villagers comes rushing forth: 'That crazy Eskimo boy took a gun and shot into the wing of your plane!'[23] They rush outside to find to their relief that the boy is about eight years old and his 'gun' is a slingshot! Wilkins muses on the story that might have been made of this by the press men that once surrounded him: '*Eskimo, Mistaking Airplane for Evil Spirit, Brings Down Wilkins and Eielson with a Discharge from His Gun*'.[24]

Wilkins and Ben kit themselves out with new fur shirts, and an innovation to write home about: silk shirts and trousers. It is 'Pongee silk', expensive but worthwhile for its non-absorption of moisture, a critical advantage when it comes to not being slowly drenched by sweat or snow.

In the meantime, the Eskimo women are pleased to see Wilkins back, and now call him 'Tnakuta'[25] meaning 'otherwise strong, wise man'. One day, perhaps, he might prove completely wise if he stopped arriving from a tottering craft in the sky, but they are happy to help him in any way they can.

On 3 April 1928 Wilkins is sure that he is finally ready. He records his aim simply: 'The course I proposed to follow this year was from Barrow to Latitude 84 N., Longitude 75 W. and from there to the northwest corner of Spitsbergen.'[26] That course, if successfully followed, will be the first trans-Arctic flight in history, crossing two countries and to two continents with a little matter of the North Pole in between them. The navigational difficulty of flying by magnetic compass so close to the North Pole, on a route never before taken let alone completed?

Well, there is no need to make a fuss about it: 'I was agreeably surprised to find, when marking each leg of the journey involving at least twenty-two changes in as many hours, that the compass changes would not nearly be so great as it first appeared from a casual glance.'[27]

Of course Wilkins knows that the shortest route of all is to fly directly over the North Pole and then head straight for Spitsbergen, but two things count against this path. First, the *Norge* had already covered that part of the globe and Wilkins cannot resist the lure of new land. There is also the vital anchoring of known markers; if his calculations are correct both Ellesmere Island and Greenland may be sighted and serve as signposts of accuracy on this nightmare of navigational difficulty. It echoes Lindbergh's strategy when he diverted his course to fly over Ireland before getting to Paris: length promises certainty; and that certainty may save he and Ben from certain death if Wilkins' maths goes awry.

Still, Wilkins is as taut as he is tense on the night before flight, going over and over his plans, their preparations, and rising at 3 am to check the barometer before going out to the Vega to light the kitchen stove that will heat her engine for the flight this day. That kitchen stove is not the problem – it is the one back in the store that proves to be causing a spot of bother when he returns. It is billowing flames from some cooking oil that has caught fire, which now sees the effing and blinding cook attempting to put the flames out with one hand, even while continuing to cook breakfast with the other.

At least all the swearing wakes Ben and, after a quick burnt breakfast, the two are walking out to the plane, only to find out that somebody has thoughtfully 'removed the stove fearing the engine would get too hot'.[28] The sooner they are up in the air and in control of their fate, the better.

Alas, getting the achingly heavy, fuel-packed plane up in the air, in what proves to be a ferocious wind, proves problematic and in the process one of the Vega's beautifully made skis buckles and breaks. There will be no attempt made today, and while they wait for their next chance, Wilkins enlists dozens of Eskimos, at $6 a day, to smooth and cut a much longer runway for them amidst now heavy snow. The Eskimos are hard workers but late starters. They refuse to work early in the morning but will keep going as long as Wilkins likes, stopping at about 7 pm. It takes a *week* before the runway is completed to his satisfaction.

If the wind holds, Wilkins is sure that they will be able to take off tomorrow morning on Sunday, 15 April 1928. And the Point Barrow community feels it too. As they make final preparations, some of the old Eskimo women carefully go over the men's reindeer skin clothing to make some last-minute repairs, while some of the young Eskimo women use their teeth and arms to stretch seal skin mittens until they fitted Wilkins and Eielson perfectly. (One particularly beautiful young woman takes *extra* care of Ben's mittens, smitten as they both are with each other.) Not to be outdone, an experienced Eskimo man sharpens their seal spears and their ice-picks, against the possibility that they will have to live off the ice.

What else?

Our Arctic food supply consisted of chocolate, 5 lb. of biscuits, 20 lb. of pemmican, 20 lb. of malted milk, 24 lb. of raisins. Other things included were medicines, a flask of ether, surgical instruments, stoves for heating over the plane's engines, a tent, a saw, an axe – to cut our way out of the plane in the event of emergency. A forced landing might have meant 18 months stay in the Arctic before reaching civilisation.[29]

Wilkins wakes as usual at 3 am to check the weather, and again at 5 am, this time for a 'splendid breakfast' cooked by Fred Hopson. Eielson will sleep for two more hours as Wilkins checks over the runway once more. The weather clears and a late start of 10 am provides perfect conditions. Now or never.

Eielson sits in the plane as Wilkins stands on the ice, going through the final checklist hopefully for the last time.

'Switch off?' Wilkins asks.

'Switch off,' is the reply.

'Gas on?'

'Gas on.'[30]

Wilkins swings the propeller once, twice, three, four times.

'All clear?'

'All clear,'[31] answers Ben, as the engine coughs, kicks, coughs once more then begins that magical purr that only the Vega makes.

Wilkins can see some ice stained by leaked petrol during the night, they will be taking off about five gallons light, but they will be taking off.

Wilkins climbs into the navigator's cabin.

'Let's go!'[32]

It is the moment. And Ben takes one, to reflect . . .

'There are not many men with whom I'd be willing to set forth on such a trip, where there is always the chance of a year's walk back.'[33]

Wilkins is such a man. Ben has seen Wilkins' uncanny ability on the ice and has never seen such an accurate navigator; if you do have to follow a chap for a year's Arctic walk, Wilkins would be the one.

The Eskimos begin nudging the plane gently, wary lest in these extreme conditions – it is 20 degrees below – the bottom layer of the Vega's skis have frozen stuck to the ice. But all is well, and the mighty bird starts to move smoothly across the frozen snow until it is in position – at the beginning of the cleared runway a mile long, with only small bumps where protruding ridges had been mostly smashed down by shovels, just as big troughs had become small troughs by being mostly filled in. All up, it is a better runway than the one on Barrow Lagoon – and the only thing that remains is for Wilkins and Ben to make the Eskimos' typhoon of labour worthwhile. True, there is only a very narrow leeway on either side between the snowbanks, which means they will have disaster riding shotgun on both sides as they attempt takeoff, but Wilkins, as ever, has faith in Ben's consummate skill.

Wilkins clocks their building speed – 30, 40, 50 miles an hour, 'the machine behaving like a proud horse tugging at a load'[34] – even as he nervously keeps glancing out both side windows, monitoring the proximity of the twin disasters that beckon from left and right. But Ben barely blinks, the very picture of concentration as the Vega hurtles down the narrow corridor:

> I marveled at Eielson's skill and courage. An error of a few pounds pressure on the rudder, a swing of a few inches one way or the other and we would have hurtled into the snowbank, our skis would have buried themselves beneath the blocks we had thrown from the runway and disaster would surely have followed. Eielson kept his nerve. I prayed.[35]

Faster and faster, 60 miles an hour, now 70, God is indeed in his heaven and, after a couple of preliminary practice bounces, they genuinely *liftttttt* . . . the wings bend upward as they take on the weight of the fuselage and now . . . soar. They are away, into the free air at last.

'Never has there been a more fervent prayer of thanksgiving than the one I uttered.'[36]

Surely, the toughest part of this journey is now over. As for the Vega, she is clearly a machine no longer.

'No. No! She was a thing of life; buoyant, swinging out with a manner true and straight, seemingly matching her will to ours.'[37]

Wilkins writes a note to Ben at once: 'Wonderful take-off. How's everything?'[38]

'Everything great!'[39] Ben shouts in reply, over the roar of the Pratt and Whitney nine-cylinder engine driving the propeller round and their slender strong craft onwards.

Holy Christ.

When Ben says things are great, you know they are at a level of perfection rarely reached by the Gods themselves. It frees Wilkins to turn his head to his calculations.

Ground speed 108 miles per hour, wind slightly north of east, weather fine with no sign of storms in their path. They are 30 miles over the Arctic Ocean and they are greeted by a marvellous thing: a clear horizon.

'But soon,' Wilkins will recount, '[the] icepack showed rough and jagged. There would have been no salvation if the engine had failed.'[40]

Carefully looking for 'land-fast ice'[41] which would indicate a spot where one, one day, might build meteorological stations, Wilkins initially thinks he sees some, but soon comes to the conclusion he is mistaken, just before the cloud cover closes in and they cannot see what lies beneath at all.

But Ben has it right: everything really is great, the flying is easy, steady and certain.

'We flew on and on,' Wilkins will recount, 'and I was busy as a bird-dog for the first few hours.'[42]

Indeed, as Ben flies, Wilkins puts out his charts, tests the sextants, clears the gas pump, checks the dump valves, sets the wireless aerial, then checks the drift indicator, his stopwatch, the turn and bank indicator, the oil pressure and the airspeed indicator again and again. With ice beneath them and certainly no land – nothing but open ocean for a hundred horizons – there is nowhere to put down the Vega now as they rise slowly from 500 feet up to 800 feet.

Looking down upon it, hurtling north at the rate of 120 miles per hour, he cannot help but compare his situation with what it was a decade earlier.

'I had stood upon the ice and watched it squirm and pile in jumbled masses. I had listened to it groan under pressure of high winds and swift moving currents.'[43]

Now *they* are the swift and the currents seem glacial, turning in to slow motion as the Vega flows through the air.

Ironically, in the midst of incredible pressure and unbelievable risk, they must fight that strangely lilting lullaby of light the Arctic serves the unwary pilot: 'Driving steadily into the bewildering whiteness where there is no earth, no sky, no horizon, brings a semi-consciousness that one must continually fight off.'[44]

Wilkins and Eielson are both conscious of this trap and check on one another constantly to ensure they remain alert.

For another 200 miles they cross over the old floes, even as the sky starts to darken as the clouds thicken, and for the next 120 miles the earth beneath them disappears, something Wilkins finds deeply frustrating.

'What lies beneath that hundred and twenty mile belt of cloud is still a secret.'[45]

Have they missed whole tracts of land, perhaps a whole – still – undiscovered people?

It is unlikely. Occasional breaks in the cloud reveal only ever more ice, and when the cloud lifts they are over an ice floe once more. It is true that while the romantic in him would be quite gratified to be the last man to discover a new land – Christopher Columbus, eat your heart out – Wilkins is at least content in that he is in all likelihood the *first* man to confirm that the continental maps really are complete. And yet still they must be careful for there remains a chance that not only will there be land ahead, but it may have a mountain upon it that they could fly into! In all seriousness, they must be careful not to finish this journey on the face of an undiscovered Everest between them and Norway.

They have now been in the air for 11 hours: 'It was past midnight local time but the sun – now almost due north – had been well above the horizon all of the time, yet for an hour or so it had been almost impossible to get even a reasonably accurate sextant observation, probably because of the refraction.'[46]

An acceptable difference in usual measurements is to vary by fractions of a degree. Right now, successive measurements are varying by as much as two degrees, making them useless. Nobody had ever been faced with this problem before; to travel at such rapidity that it is barely possible for Wilkins to 'keep the sun centred on the bubble of my sextant'.[47]

If that centre is glimpsed then Wilkins can look at a chronometer strapped on his arm, which in turn allows him to make his calculations based on Greenwich Mean Time to give them their longitude – with corresponding figures and calculations carefully marked on all maps to reveal the meridian to be followed. But if the centre cannot hold for more than a split instant and he cannot get a bearing, well, where does that leave them?

It leaves them for the moment with little more than Wilkins' dead reckoning which, if he is wrong, could very well see them going not only to the ends of the earth, but over it, overshooting any prospect of land and seeing them finish in the drink. One thing Wilkins has learnt over the years, however, is that roaring anxiety does not help. All he can do is concentrate, plotting short course after short course to protect against error, and adjusting Ben's course accordingly. These 'adjustments' will change their course 22 times in less than a day, a dazzling display of nerve and navigation, using 1100 pounds of petrol as they are propelled hither and yon, no, yon a bit more to the left, Ben.

Endlessly making his observations, consulting his tables and doing his calculations, Wilkins continues to pass Ben notes.

> Turn South
> More Southward
> North
> Stay Course[48]

Stay the course Ben does, while both men remain vigilant, straining their eyes for what lies up ahead. For back in 1909, the American explorer Dr Frederick Cook had claimed to have seen mountains – big ones – between Grant Land, the northernmost point of Canada and the North Pole, the precise area they are flying over now.

Wilkins thinks it unlikely, because Cook could not be trusted. His claim to have reached the North Pole had been exposed as a lie, but nevertheless Wilkins keeps his eyes peeled as much as possible in the tight, swirling conditions. Whenever the clouds do clear, all he can see is ice, and slight cloud 'stretched like a silken scarf in the wind, but beneath it I could see a clear-cut horizon . . . We felt certain then and believe now that no mountainous land exists between Grant Land, and the North Pole.'[49]

A slight deviation in their course provides no confirmation for Cook's claim; another mirage revealed in its true light. There will be no continent revealed, Spitsbergen takes its place once more as the sole focus of their thoughts and hopes.

With that worry receding both men can focus on a more immediate one, as Eielson passes Wilkins a note: 'Engine been turning about sixteen fifty. We are using approximately eighteen gallons an hour.'[50]

Oh.

Dear.

Following quick calculations, Wilkins knows that for the rest of the journey they will need to average 12 gallons an hour to complete their

20-hour flight. It is not completely impossible and, as it has always been in those very realms Wilkins has done his best work, he is quick to find reason for hope: 'We had see-sawed over the clouds and we would probably see-saw again. We had also, from my observations, swung a little to the right on one leg of our journey and a little too much to the left on the next leg . . . I felt certain if my interpretation of the meteorological conditions were right that we should have no violent head winds and that we could make it.'[51]

That is, unless they have to go through . . . *storms*.

Precisely like the one he can suddenly see looming now?

Yes.

'Far, far away in the very distant eastern sky, pillars of high storm clouds hung like wraiths under the pale blue zenith.'[52]

On a quick estimate of their altitude of roughly 3000 feet, and the fact that these clouds are on the limits of their horizon, Wilkins estimates they are 'perhaps 150 or 200 miles away'.[53]

When Eielson tries to change route to avoid the whole black mess – 'wishing that it had been my lot to be a chicken farmer and not an aviator'[54] – Wilkins corrects him swiftly by note: 'You must not go west or south, turn east or north if necessary to avoid cloud.'[55]

The key will be if they can get a glimpse of Grant Land, off to their right, which, most importantly for their survival, will give them a sure fix of where they are on the map. Some 13 hours into their flight now, and completely lost in calculation, Wilkins is startled by a shout.

THERE IT IS!

Ben is pointing: 'slightly to the right and not far – perhaps less than twenty miles away – the rugged mountain tops of Grant Land piercing the clouds. It was a fleeting glimpse we had but served to stir deep emotion in our hearts.'[56]

It is a significant moment.

They are halfway across the Arctic.

'The real flight of exploration was now over, but we still had 900 miles to reach our goal. We headed straight for Spitsbergen by the sun's position and the compass.'[57]

From now, every mile they travel is a mile closer to safety, not a mile further away from it, and they relax just a little – at least as much as one can while attempting the impossibly improbable – uniquely alone on the earth with no chance of rescue or respite. 'We were cheered immeasurably by the fact that our machine had performed wonders, our engine faithful every moment. Calmly content, I reached for the thermos bottle, poured myself a cup of coffee, munched pemmican and biscuit and quickly summed up the situation.'[58] The situation is simple but even writing it down cannot make it seem quite real: 'It was 38 degrees below zero in the open air, and we were three hundred miles from the Pole.'[59]

And at least they are now through the worst of the storm!

'I celebrated the return of the sunshine,' Eielson will recount, 'by having a stick of chewing gum.'[60]

They fly on calmly, Wilkins focusing on their course, Ben's eyes on the prize of the far horizon and his ears keenly attuned to the hum of the motor, even as he regularly responds to a gentle tap on the left or right shoulder from Wilkins before more detailed adjustments to their path are handed over on a hastily scrawled scrap of paper.

Just three hours later, there is another breakthrough, for again out to their right they see, on the far hazy horizon, what can only be the snow-capped mountains of North Greenland! With that sure fix, Wilkins adjusts their position on the map accordingly – not far, for he is nearly bang on – and they keep going.

All else being equal, they now have just 600 miles and six hours to go, before they should be in the area of Spitsbergen.

But will all else be equal?

When they are still some 220 miles away from Spitsbergen the high curling cloud masses ahead rise to a height that they cannot get above. And so Ben now starts to dart along 'selected lanes between the feathery masses'[61], while Wilkins takes observations from the brief snatches of sun to determine their position. By estimating their speed and direction, with constant references to his watch, he is able to estimate their longitude. Now and then, far below, they can see, 'a dark stream of water almost free of ice beneath us'.[62]

Travelling in this manner for an hour and a half the good news is that they know they must now be in the vicinity of Spitsbergen! The bad news, however, is that with that proximity comes the possibility of running into that island's mountains, obscured by the cloud they are still flying through. And if they fly over those mountains, the other danger is, 'we might be going too far inland, and would therefore probably have to journey over the mountains on foot to reach the settlement'.[63]

The results of that summation are passed in a note for Ben:

There are two courses open. We are above storm now. Down there we can land and wait until it's over. Can we get off again? If we go on we will meet storm at Spitsbergen and perhaps never find the land. Do you wish to land now?[64]

Ben thinks for a full minute before yelling his reply: 'I'm willing to go on and chance it.'[65]

Wilkins smiles. Of course you are, Ben. We are both born chancers.

And so they fly on, their light craft now drained of nearly all of its petrol, flung about by the storm.

'She leaped and bucked like a vicious horse, and to add to it all, fine snow and the wind made everything invisible.'[66]

Everything loose in the cabin begins to tumble and rattle.

Rattled they may be, but nerves give way to excitement when the clouds break and they see them!

'Suddenly two sharp peaks, almost needle-pointed, appeared beneath us.'[67]

They must be . . . the mountain tops of Spitsbergen! They have *done* it! Yes, those same mountain tops are in imminent danger of becoming their gravesite but for the moment let them savour the flavour of their triumph. After all, it is no small thing to have made the impossible now possible and both men are near overcome with an elation tempered only by the terror that their reward might be a grisly death, just minutes from now. Yes, they have done what no human being has ever done in recorded history, and near completed the hardest, most dangerous, most improbable flight in history, but the final challenge might be the toughest of the lot – to get this bird safely on the ground, or at least on the ice. Their task is clear: they must now make history and not disappear into it permanently.

Nose down with the engine roaring as they descend, the storm continues to slap and thwack the Vega. The air above the cloud was wild but beneath it is hardly tame and the Vega continues to buck like a bronco.

'Eielson, never losing the upper hand, held and guided her splendidly around the rugged mountain tops.'[68]

For all that, things remain ferociously difficult.

The wind is so furious and freezing that it is not only turning sea-spray to ice but actually breaking up some of the icy surface below and sending small pieces of it smashing into their windscreen. It doesn't break, but visibility is shocking and –

CHRIST! LOOK OUT! – it is only through Ben's furious concentration and quick reflexes that they miss another mountain peak, which suddenly appears, by what surely must have been inches!

This is madness, and Wilkins is quick to signal his pilot to immediately head back out to sea at once. Landing and gas be damned, they must let the dog see the rabbit. It is only minutes later, at sea and turned around once more they can see the mountain they missed. Heading in once more, towards the island, and into the storm, much the same thing happens with another mountain that had been obscured by the storm!

'We were like an imprisoned bird beating against a window pane.'[69] The risks, great as they are, must be run. They head into the storm and towards land, Eielson flying entirely at Wilkins' direction.

The windscreen of the Vega is almost 'totally obscured with snow and frozen oil'[70], but through the small patches of visibility they do have, the ice, land and sea, all look . . . angry.

'The ice-strewn water and the wind were furious, while spray was whipped from the sea and filled the air. Over the land the snow drifted high and thick, and it was therefore impossible to judge distance.'[71]

One way or another they have to get down, and quickly. Desperately looking left and right out the cabin windows, Wilkins sees it.

It is a single smooth patch of white snow.

It is their one chance.

How does one steer a blind pilot?

By instinct of precisely where he thinks that flat white snow is, and with note and note after note, frantically written in just legible haste:

> Turn right.
>
> Now to the left.
>
> A bit more.
>
> No, we have passed it.
>
> Turn back.
>
> Keep as close to the land as possible.
>
> There it is on the right.[72]

Eielson takes just a glance at each note before instantly responding, throwing the plane through trough after trough of heavy wind, even as he brings her down and turns her into the wind to attempt a landing on that white handkerchief of snowy ice. It looks tiny, but ideally both the snow and the wind will bring her to a quick, tight halt. If, that is, the surface allows for a landing in the first place. Wilkins presses his face hard against the window pane, aching to see the first sign of safety or danger, will the surface be jagged ice or smooth? Will they crash or land?

'It was impossible for Eielson to see but with steady nerve braced for all eventualities he levelled the ship and lowered her gently until lost in the swirling snow.'[73]

There it is!

And it looks . . . smooth enough.

Unbelievably, Ben feathers her down for *the* landing of his career. Both the thick snow and the wind coming straight at them indeed stop her cold before she can break against the mountain or a snowbank.

'Such was Lieutenant Eielson's skill that the machine stopped 30 yards after the skis touched the snow.'[74]

And yet, despite being at blessed rest, they cannot rest for a blessed second because, as Wilkins is all too aware, the tanks will freeze in minutes if not drained immediately. He leaps out of the plane, yelling, 'Open the tap!'

Ben, in a seeming daze, does not move. For one thing, as Wilkins soon works out, the pilot is, 'temporarily stone deaf as the result of the constant throb of the engine. He could not hear a word and only when by signs was I able to make him understand, the oil flowed.'[75]

Wilkins continues regardless, stamping the snow down around the skis, hoping to ensure no frozen walls will form to impede their future takeoff, if one occurs. For once he does not lope, he leaps, his stride disappearing into a dash as he further quickly puts up canvases and waterproof covers around the engine, all while still incredulous that they have landed in one piece, with a better landing than they have made on clear runways in good weather!

'THANK GOD, THE MACHINE'S SAFE!'[76] Wilkins yells to Ben, who nods solemnly, although Wilkins thinks he may have nodded solemnly at anything at that moment such is the silent fog he is lost in.

'The minds and hearts of both of us were too full of thankfulness for conversation even if hearing had been easy.'[77]

Simply radioing their position to the waiting world is not possible as, alas, when landing *in extremis*, Wilkins had forgotten to wind in the aerial cord and it had sheared off. Apart from that, when it comes to the radio itself, one of its valves has burnt out, thus further ruining any chance of communication. He and Ben know they are alive, but what the rest of the world must think, who knows?

Now, after carefully brushing all the snow off their clothes, they can rest, eat and drink at leisure even while the wind continues to buffet them with every gust, and they gratefully chomp down some dry biscuit, chocolate and pemmican, chased down by one gulp each of the still hot coffee in Wilkins' thermos. Together with a smoke to calm their nerves it gives both men a chance to reflect on their extraordinary achievement.

They have flown 22 hours and 20 minutes.

Coast to coast, non-stop, over the Arctic, 2200 miles, the first men in history to do it. And, yes, it had taken it out of them.

'We had to fight,' he will recall. 'Fight every inch of the way, anxious, uncertain, never quite helpless but ever against tremendous odds. We had, as we sat in the plane, reached a position of safety not only for ourselves but for our plane.'[78]

As Ben's hearing gradually returns, the two discuss where exactly they might be. Dead reckoning suggests they may be in the neighbourhood of Kings Bay in the mid north-west of Spitsbergen, but the map they have doesn't record any mountain like the one they almost crashed into anywhere near that bay. Still, it is possible, as they know more than most – and have the scars to prove it – that Arctic maps are little more than rough guides.

On the crucial subject of fuel, Ben is sure they have about 20 gallons left. If they are indeed near Kings Bay that should be enough to allow them to fly the last leg to Spitsbergen in triumph rather than risk the agony of another two-man trudge through treacherous ice.

And helping to buoy them further, 'Kings Bay or Green Harbour could not be far away, and we might walk or construct a boat and reach some habitation.'[79]

These are all worthy subjects of discussion, and other key subjects come up – how long this storm may last, where their next meal might be coming from – but again and again the wonderful realisation springs anew: they have DONE IT! Ben's wide grin 'neath his hawk-like nose mirrors Wilkins' own delight.

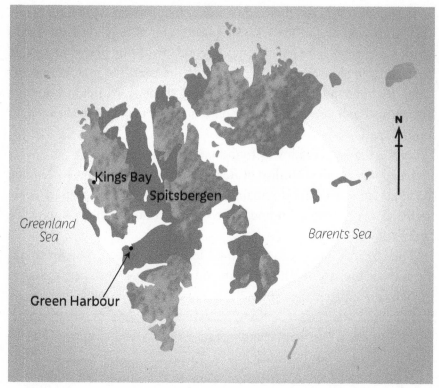

Spitsbergen

'We had flown just about half way around the world in one hop.'[80]

And that hop was over unknown, frozen territory, never crossed by anyone, ever!

As well as sky travellers they feel like time travellers:

> We tried to make it seem real that at Barrow, Fred Hopson and Charlie Brower, whom we had left twenty hours before, would be eating breakfast while the inhabitants of Spitsbergen would be having their evening meal.[81]

The excitement ebbs enough for sleep to overtake them, a 'Rip Van Winkle sleep but our watches showed that scarcely five hours had elapsed since our landing'.[82]

After some blessed slumber, they wake to find the storm has abated and they tentatively set out to explore their immediate surrounds by ski, hoping to see some recognisable landmark, something that will allow them to get a bearing on their *exact* whereabouts.

In the end, it is the damnedest thing, like a cosmic joke, for while they do indeed find in the snow, a quarter of a mile away, a surveyor's mark – a triangular beacon of sawed wood, 'evidently placed as a guide to masters of ships that might pass along the coast'[83] – it lacks one key thing for their purposes. The masters of those ships might know its significance but there is nothing on it to help people like them, right on the spot.

'Not a line or scratch was found on any of the timber.'[84]

Laughing at the absurdity of it, for given their salvation by landing on solid land their mood remains light, they try their binoculars again in the clearing weather to find some further, more useful beacon of life and man.

There! Far in the distance, to their left, across a bay, Wilkins thinks he may see houses, or a house. Can you see it, Ben? With perhaps some smoke coming from a chimney?

Ben looks and can see nothing . . . actually, no, he sees *something* . . . but not houses. Is it happening again? Are they seeing images formed by their desires, not by reality?

Wilkins looks again and feels certain it is a house, but under such circumstances as this, what is such a feeling worth?

'Too often had I been fooled by mirages in the Arctic and other places,' Wilkins will note. 'Under just such conditions I had seen palaces, ships, palm trees and giant cities outlined above treeless, sandy plains or Arctic wastes. A mirage will throw up shadows resembling any known physical thing depending upon the imagination of the observer . . . We turned away from the mocking shadows.'[85]

They can certainly see mountains and small islands, but the guesses of Wilkins as to what they are, on the sparse map he has, and his resulting calculations contradict rather than clarify. The best he can work out, they might be near Danes Island, 60 miles north of Kings Bay, the settlement on the west coast of Spitsbergen. If so they should have just enough fuel to make Kings Bay. If not, well another miracle landing and a march is in store.

An attempt to melt snow by burning some driftwood is made, but the 'fire' is so reluctant and miserable that they succeed only in having two thermos bottles of lukewarm water and sludge. Well, there is an alternative: 'For further liquid we must resort to our small store of alcohol.

The knowledge of this aggravated our thirst, but expended little energy, lying cosy in the cabin.'[86]

Slightly less cosy is the shock they get when they measure the amount of fuel left. They don't have 20 gallons in the tank as Ben had estimated just before they landed, but only 10. With just that amount, all they could be sure of in taking off would be that they would have to land again very soon. Their only recourse is to rip the seals from the 'dump valves' of the cabin tanks – a small pocket within the fuel system, which just might deliver a gallon or two. A fuel can is carefully placed beneath the seal and Wilkins pulls a cord to set the seal open.

'Much to our astonishment a full stream of gasoline came forth.'[87]

So much bursts forth, in fact, that it keeps going even when their five gallon can is full! How to stop the flow while Ben grabs another canister? It needs 'some soft obstruction fitting snugly against the pipe. [Alas,] the only soft thing available was my hand.'[88]

As soon as Wilkins has taken his glove off in his best attempt to seal it, and forms an imperfect seal with his hand, but as he does so the gasoline runs down his shirt sleeve, evaporating at a painfully low temperature.

At an estimated temperature of 14 degrees below zero, Wilkins can feel his hand freezing in seconds, the exact reverse of putting your hand on a hot stove except he cannot remove his hand. 'We could not afford to lose the gasoline. Eielson made all possible speed, but I suffered excruciating pain and frost bite before the can was replaced. Less than half a gallon more came from the tank. Still, it was worth it.'[89] The same method is used on the other cabin tank after Wilkins' hand thaws; another five gallons is captured, making 20 in all, which, as Wilkins dryly notes, 'would take us anywhere we wanted to go in Spitsbergen'.[90]

Now all they need is some clear weather to take off. With that in mind, Wilkins awakes every night at 3 am to check conditions. On Saturday, 21 April, at 3 am, five days after they first landed; those conditions are right! There is little wind, and the sky is relatively clear.

Wilkins rouses Ben from his slumber and the two comrades work furiously for the next six hours, shovelling snow as the 'runway', such as it is, is cleared – their urgency pushed by the desire to get up and away before the weather can close in once more. As the machine is starting on a downhill slope, they at least should be able to get speed up fairly quickly if they can just get it moving in the first place.

In addition to its own power, a push will be required to slip the sticky bonds of snow. Wilkins hauls on the propeller to start the engine and then races behind to push the tail. Success!

The plane surges forward and breaks contact with the ground – up, up and awaaaay – but, alas, Wilkins is not in it! Banking the plane in a tight turn to go back over the runway, Ben sees Wilkins frantically waving to him from the ice! Again Ben lands and they try again, this time with a back-up plan if the plane rises too quickly: a rope ladder is slung out for Wilkins to catch and climb aboard.

Once more Wilkins swings the propeller then runs to push, as the Vega gathers speed so does Wilkins, but only enough speed to grab the rope ladder. To grip it properly he rips his gloves off, feeling his hands numbing as he does so. His grip is slipping and, in desperation, Wilkins bites a rung of the rope with his teeth! 'A foolish thing to do, perhaps, but it seemed imperative that I cling to the machine.'[91] Cling he does until the madness of what is about to happen sinks in and Wilkins lets go with teeth and hands and is hit by the tail of the plane as it ascends.

Wilkins descends to the snow at speed.

'It was fortunate for me that the snow was soft. I was half buried and partially stunned from the fall.'[92] The part that is not stunned gingerly touches his mouth and finds, not to his surprise, that all of his front teeth are loose: 'Whether from the impact of the fall or from the grip on the rope I was not sure, but I think now it was the result of holding onto the rope.'[93]

Six of one, half-a-dozen loose of the other.

Ben comes in to land once more, but this time it is no easy landing. Wilkins watches as the Vega descends into ridged snow. 'I was directly before the machine as she landed. The sight appalled me. It struck the snow and bounded over ridges like a frightened deer. The skis, slung and kept in tension with rubber cords, bounced and wobbled almost like the arms of an octopus.'[94]

Mercifully, however, they hold.

Third time lucky?

They'd better be, for the first two attempts have seen their engine running for nearly an hour, and no less than half their fuel is gone. Perhaps, Ben opines delicately, we should take out a tent and a rifle, so that if Wilkins does not make it aboard next time Ben can fly on and Wilkins will walk?

No, upon discussion, it is decided that by splitting up they place both of them at risk, while together, whatever happens, they can hopefully nut something out. Besides, Wilkins has an idea. Instead of pushing from a wing, he will push from the cockpit. It will work like this, the Vega's tail will be lifted onto a block of snow, a driftwood log will be used as a sort of stilt for Wilkins to push at with his arms while one of his legs is in the cockpit and the other pressed against the fuselage.

Yes, not ideal, but on the upside it is marginally safer than gripping onto a rope ladder with your *teeth*.

Wilkins hauls on the propeller once more and the motor starts easily. But now, though Wilkins strains and pushes, and the engine roars, the Vega remains stubbornly still. For a full minute they are frozen, Wilkins straining with all his might, Ben willing himself not to turn around but to focus on the runway that he hopes to God will see them *run*. Still there is nothing, until, with everything he has in him, Wilkins grits what is left of his teeth, puts the last reserves of his strength into it and is rewarded with a small, slight lurch. She's free! Wilkins drops the log, and drags himself tumbling and bruised into the cabin. A whoop and a roar comes from Ben. *Are you on board?* Wilkins is too exhausted to answer for a moment, but recovers and lets the relieved Ben know that, yes, this time two of them are in the air at the same time.

Now, to find out where they are.

The windows of the plane have hoar-frost upon them, caused by Wilkins' body heat after falling into place, meaning he will only be able to see by going forward to the cockpit, which he now scrambles to do.

Just as Wilkins pokes his head through he hears Ben shout: 'What's that over there in the bay to the left?'[95]

Wilkins can see very clearly what it is: two radio masts in a distant group of houses! His vision now clouded by tears, he collects himself and writes a note to Ben: 'Must be Green Harbour. Go over and land where you think best.'[96]

Without a word they fly over five miles of open water; the distant houses coming nearer; the end literally in sight, if blurred by more tears at the vision splendid.

'Our first sight of Green Harbour from the air was a golden city paved with gold for us.'[97]

Ben swoops over a coal mine, over the houses, circling back and comes gently in to land on the ice at the foot of the radio masts. A yell from both men erupts as the engine dies, not six seconds after they have come to earth. They had landed on fumes alone.

'Our flight from our base in Alaska to a town in Spitsbergen was ended.'[98]

What is also ended is anonymity; they are about to become two of the most famous people in the world. A group of eager Scandinavians greet them in excellent English, only to have 'a look of pleasant astonishment when Eielson, from the cockpit, greeted them in an old Norse phrasing and accent with a slight American influence'.[99]

George Wilkins' first words will soon be reported in the press: 'I am tremendously hungry.'[100]

One of the gentlemen here to meet them is a fat fellow with an 'old-world grace and charm of manner',[101] who – though they don't know it – has spent a good deal of the last week making them famous. His name is Thoralf Bowitz-Ihlen, the manager of Svalbard Government Radio. Wilkins had felt as though he was tapping into the void with every Morse message he sent out in the ether, but Bowitz-Ihlen was always listening, watching and waiting and transcribing every message and turning them into news for the waiting world.

He shows Wilkins the two final messages, sent five days ago:

> Now within a hundred miles of Spitsbergen. We are in a bad situation. Heavy clouds about us. Have two hours' gas but will not be able to see the land. All open water below.[102]

> Spitsbergen in sight. Spitsbergen in sight. We will make it.[103]

But did they make it? *The New York Times* has been screaming to find out for the last five days! The King of England wants to know. The King of Norway has asked for updates. The world waits to hear from Captain Wilkins!

Why yes, he has made it and is pleased for Mr Bowitz-Ihlen to announce that he and Mr Eielson have arrived.

'Three hours after our arrival, a wire from our friends came to us. They and the world had learned of our safety and they and the world were pleased to let us know that they were pleased.'[104]

Kings, Queens, Ambassadors, Premiers, High Commissioners, the Secretary of State and the Secretary of War, Nansen, Stefansson – and yes, Roald Amundsen! – not to mention seemingly every scientific body known to man cable their delight and congratulations to the two men who have just 'covered twenty two hundred miles of Arctic snow and ice; thirteen hundred miles of which had never before been seen by man'.[105]

Amundsen's cable is especially prized by Wilkins, it reads: 'Please forgive a stupid old man for thinking he knew all about it. When you come to Oslo will you stay with me at my home so that I may apologise.'[106]

Oh, how the mighty . . . are gracious!

The newspapers around the world exult in reporting, 'as a result of the flight it is thought that geographers will now be able to fill in maps of the Great Arctic Circle – that great white space between the north of Canada and the North of Siberia'.[107]

Leave it to Stef to sum up what elusive task his apprentice has just mastered. 'The dream of ages, a north-west passage across the world,' he tells the international press, 'has been brought nearer to practical achievement through the pioneering of Captain Wilkins.'[108]

Yes, the quest that defeated Captain Cook after two years of trying has been achieved in a single flight by Captain Wilkins and Ben Eielson in two days. Stef explores the future now with some shrewd speculation: 'I am picturing polar cross roads, where future Continent to Continent caravans will meet, shortening the distances by hundreds of miles.'[109]

And you, Ben Eielson, what do you have to say for yourself? What were you thinking as you crossed the roof of the world?

'My thoughts were hundreds of miles behind the fastmoving plane, with an Esquimau girl on the coast of Alaska,' the American says disarmingly. 'I am proud to have accompanied Wilkins, I do not accept any credit, for when Wilkins clamps his Australian jaw and says he is going somewhere, I dare not say I am not going.'[110]

RAH!

Still the cables keep coming, so many of them requiring responses – and energy the men just don't have – that Wilkins authorises 'the wireless operator to say that he is unable to answer further questions'.[111]

(He does, however, hand the operator a cable to be sent to the American Geographical Society: 'No foxes seen.'[112] It is the agreed code, to signal that they have discovered no new land.)

What he can do is put out his own article, to be reprinted around the world, with many of those questions answered.

'With good weather, good luck, and by careful navigation,' Wilkins writes modestly, 'we travelled 2200 miles above the Arctic ice from Point

Barrow to Green Harbor, Svalbard, three-quarters of which area has never before been seen by man. We are thankful that aeronautical and engineering skill of conscientious workmen and our accumulated Arctic experience has brought us through in comparative comfort and safety.'[113]

No less than *The New York Times* runs their story on the front page, for three days straight.

WILKINS FLIES FROM ALASKA TO SPITSBERGEN IN 20 HOURS; COVERS POLAR SEAS IN TINY PLANE AND FINDS NO LAND; ISOLATED FIVE DAYS ON UNINHABITED ISLAND NEAR GOAL; TRIP ENDED BEFORE START WAS KNOWN

Wilkins and Eielson, an Alaskan Flier, Left Point Barrow Unannounced – Flew Route Below the Pole.[114]

CALLED THE GREATEST FEAT OF ALL AVIATION

Amundsen Hails the Flight as Far Exceeding Any Hitherto Accomplished – Success Crowns Wilkins on Third Attempt After Many Discouragements.[115]

And the winner of 'the race'?

The Herald of Melbourne has no doubt, declaring:

WILKINS BEATS ITALIANS
Great Polar Race
WAY NOW CLEAR FOR ANTARCTIC PLANS
Explorer's Big Ambition

Captain G. H. Wilkins, the heroic Australian explorer, airman and photographer, and his American companion, Ben Eielson, in crossing the North Polar region from Pt. Barrow, Alaska, to Spitsbergen have beaten their Italian rivals, led by General Nobile, who are about to fly from Spitsbergen in the opposite direction in the airship *Italia*. Wilkins thus has his revenge on Nobile . . .[116]

CHAPTER FIFTEEN

O'ER THE OCEAN,
AND UNDER THE SEA

As the captain was finishing his sentence, I said to myself: 'The pole! Is this brazen individual claiming he'll take us even to that location?' ...

'Yes, sir, it will go to the pole.'

'To the pole!' I exclaimed, unable to keep back a movement of disbelief.

'Yes,' the captain replied coolly, 'the ... pole, that unknown spot crossed by every meridian on the globe. As you know, I do whatever I like with my Nautilus.'

'I'd like to believe you, captain,' I went on in a tone of some sarcasm. 'Oh, I do believe you! Let's forge ahead! There are no obstacles for us! Let's shatter this Ice Bank! Let's blow it up, and if it still resists, let's put wings on the Nautilus and fly over it!'

'Over it, professor?' Captain Nemo replied serenely. 'No, not over it, but under it ... He who enters the Nautilus is destined never to leave again.'

Jules Verne, *Twenty Thousand Leagues Under The Sea*, 1870

11 May 1928, Green Harbour, Lord of the Dance

No matter that Green Harbour is not even close to being a town, Wilkins and Eielson are the toast of it all the same and, among other things, are honoured with a gala dinner before leaving.

'Captain Wilkins' farewell party before sailing,' one of the news correspondents will report to the world, '[was] a strange affair. The only dinner jacket worn was by Ihlen, the wireless superintendent. The attire of the rest was a laughable mixture. Captain Wilkins wore a fur skin, in which he fascinated the Esquimaux by dancing the Charleston and the Black Bottom, the natives replying with a display of their own dances.

The news that the steamer *Hobby* was approaching abruptly ended the festivities, Wilkins donning his snowshoes and beating all native performances by covering the seven miles in 50 minutes after which he and Eielson embarked on the *Hobby*.'[1]

The said vessel, with the Vega lashed down on the deck, smashes through a hundred miles of pack ice before reaching clear waters and then the Norwegian port of Tromsø on the European continent, in the wee hours of 15 May.

One would think the humble residents of Tromsø would be fast asleep at such an unholy hour, and yet Wilkins and Eielson are again greeted with a throng of cheers from those gathered in anticipation of the boat. The two men do not set foot on the port, but appreciate the sentiment none the less. They carry on towards the more southerly port of Trondhjem, only to realise that, once again, the people have gathered *en masse* to greet them, and this time it seems as though the entire town has turned out for their arrival.

Did anyone, ever, make such an entrance to this seaside town? The excitement of the *Trøndere* locals is palpable as large crowds gather by the pier where they are due to dock, many of them bearing Norwegian and American flags.

'At 12 o'clock precisely,' the local paper will report, 'there was a big train of singers, athletes, students and scouts marching. The music corps took position on the pier, close to the reception committee and a number of the city authorities . . .

'At 1.30 pm the ship came covered in flags through the river inlet during salute, and was escorted by student rowers who bravely defied wind, rain and rough sea, during greetings and cheers from the many thousand people on the pier. Wilkins and Eielson who stood by the row on the promenade deck, nodded and smiled.'[2]

Cue the band!

With some enthusiasm the band plays all of the Norwegian, American and British National Anthems – the last to honour Wilkins who, as the newspaper explains, because, 'as an Australian citizen [he] is a British subject'.[3]

To cheers and furious flag-waving from the crowd, the two great men come down the gangplank and are soon on the podium where the mayor offers the official welcome.

We are sorry, that you do not have more than a few hours at your disposal, but we hope that your stay in our city will be among your best memories from the trip in Norway. I warmly welcome you to Trondheim![4]

Again the Norwegian National Anthem is sung, and Wilkins is invited to say a few words.

I just want to emphasize the skill of my Norwegian–American pilot. He brought the machine from Alaska to the coasts of Norway. We were lucky that our journey brought us here to this country, which has always been at the forefront of polar research and in other tracks we follow. We are in a debt of gratitude to the Norwegians and this reception you have given us has formed an impression that will last our whole life.[5]

And so to the climax of the occasion.

For now, Ben Eielson steps forward and, as the crowd falls quiet, speaks to them in their own language!

'I'm glad to have the opportunity to speak. We arrived in Tromsø and were received at 2.30 at night and nobody was in bed. I am proud to be of Norwegian descent.'

Cheers!

'There is no . . .'

Cheers! (Again and again Ben tries to speak, only to be interrupted by the wild enthusiasm of the *Trøndere*.)

'I say a big thank you for your greeting!'[6]

And again, the crowd starts cheering!

And so it goes. The two intrepid aviators are finally able to make their way into the town proper to do a quick loop through the streets lined with the cheering crowds.

When asked to sum up what he has learnt on this expedition, Wilkins gives an answer that is as simple as it is moving: 'Eielson and I have learned, at all events, the sincerity of friendship.'[7]

Even as they are accepting their wreaths and laurels, others are still in the Arctic, searching for more glittering prizes. With some fanfare the *Italia* has just become the third airborne vehicle to reach the North Pole (and very likely the second, if those who doubt Byrd are to believed). Commander Nobile of the *Italia* even reveals through the wireless some details that the press laps up. 'We reached the Pole and dropped the Italian flag, also the Cross which the Pope gave me for the purpose. Am sending reverent message to the King of Italy, one to Signor Mussolini informing him that the Italian flag flew over the Pole, and one to the Pope announcing that the Cross has been dropped.'[8]

Yes, Santa is not the only figure of magic and mystery to stake ownership of the North Pole. Now the Vatican has walloped its own claim through the ice. Christ alone knows what Christ himself would have thought of all this, but for *Il Duce*'s part, he is delighted. When explorers explore, countries claim – it is the way of such things.

•

Wilkins and Eielson return to the boat drenched to the bone, wondering what the next stop has in store for them. To their surprise, the answer is dinner jackets. They reach Bergen days later – to be greeted by the now usual thousands of Norwegians waving flags, while dignitaries make speeches before applauding their every word – and the explorers are immediately afterwards told they must buy dinner jackets. The reason? They are dining that evening with King Haakon VII of Norway, in Gamlehaugen, effectively his western palace.

The next, when they arrive in Oslo, is, if it is possible, even more grand in its manner of reception. For after settling in to the Royal Suites of the Grand Hotel, they are placed in Royal Carriages and drawn through streets packed with cheering crowds – all the way to the home of that prince of exploration, Roald Amundsen.

As they get within a mile of Amundsen's house, planes from the Norwegian Navy come in low and drop bouquets of flowers upon them, just before they see . . . Roald Amundsen himself, running to his gate to greet them – *and . . . can it be true?* . . . – he is weeping! Throwing an arm around Wilkins, the very man who has spent years ridiculing the very notion of polar flight – polite in public but pugilistic in private – now embraces him and whispers: 'You must forgive me for trying to prevent you from carrying on. I thought the job was impossible with an airplane, but you have shown I was mistaken.'[9]

(As a matter of fact, Wilkins is entirely unaware of exactly how much Amundsen had done in attempting to prevent the whole exercise, only learning later that Amundsen had gone out of his way to warn off many potential financial backers.) And yet let bygones be gone, as now

Amundsen looks to laud, as they are greeted like conquering kings with a reception in his garden, peopled by local notables and replete with a cake he has had baked for them, 'decorated with a map of the Arctic on which our Vega's course was marked in red'.[10]

Delicious!

That evening Amundsen also hosts them in town in his role as President of the newly founded Norwegian Aero Club, where the two are presented with the Norwegian Air Club Gold Medal of Honour, and appointed honorary members of the Norwegian Aeronautical Association. No less than Major Tryggve Gran, the Norwegian who had been a member of Scott's South Polar expedition, proposes the toast of 'Australia, Captain Wilkins' Native Land.'[11]

It is, of course, Amundsen himself who leads the ovation.

Ben beams and Wilkins returns a gracious nod of respect from Amundsen. It is a shining moment. While millions of people around the world have read all about what they have done, the number of men who *actually* know what they have achieved because they operated in the sphere themselves are no more than a handful, and of these Amundsen is right up there with Charles Lindbergh, the American who just a year before had stunned the world by flying non-stop from New York to Paris. So, to have such acclaim from such a man is nothing less than extraordinary. And again, Amundsen is more than generous in his remarks, telling the waiting press that the blushing double act are no less than, 'two of the greatest flyers of our time'.[12]

He means it all right.

'They covered more unknown territory than any other airmen. No flight has ever been made anywhere, at any time,' he enthuses, 'which could be compared with it.'[13]

Wilkins and Eielson respond graciously, with Wilkins noting that, 'Mr Amundsen was the hero of my youth.'[14]

(To be a hero of your hero – now that is truly something.)

•

Coal miners know the sensation, as do mariners, as do . . . flyers. The word travels hot, only to give chills on reception: men are trapped, lost, missing. It's not that you hear the news . . . you *feel* it. There but for the grace of God . . .

And if you possibly can, you must *help*, just as you know that if the situation could be reversed they would help you. It is the code they live by. News of men in peril is a call to action.

This word comes through on the afternoon of 28 May 1928, and Roald Amundsen hears it just as he is farewelling Wilkins and Eielson from his fair town. The airship *Italia* has gone missing on their way back from the North Pole. A rescue mission must be mounted, and Amundsen tells the two intrepid aviators with him of his intention to go himself, whereupon the duo offer to go in his place. (They are delicate, but perhaps the great man should recognise he is also now an old man?) No, Amundsen assures them, they have performed quite enough heroics for the moment; he shall answer the call. They shake hands.

God speed.

Germany has sent its largest plane, as a courtesy, to transport the two heroes from Norway to Berlin; accompanied by two squadrons of fighting planes as a ceremonial escort. Looking out at his escort on their way to Berlin, a thought occurs to Wilkins.

'I couldn't help remembering that day some ten years before when I had dangled by my hands from that balloon basket, with a German fighter plane circling around me and letting rip its machine gun.'[15]

Oh, how the wheel turns.

Touching down in Berlin, they are greeted on the tarmac by the beaming ambassadors of both Great Britain and the USA, with, of course, an aide of German President, Field Marshal Paul von Hindenburg, who, naturally enough, wishes to dine with them that evening and hear of their astounding journey.

And so it goes.

Thanks to the proof established by Wilkins that flight over the Arctic is feasible, the future travel patterns of the world might have been rearranged. 'The Polar Regions would probably become the principal West–East air route. It was 6500 miles from England to Japan via the Pole whereas it was 11,000 miles by ordinary ship flying routes, which indicated that these regions might, with the progress of aviation, become the regular route between Europe and the Far East.'[16]

A pipe dream? Perhaps, but the future of travel is air, not ship, Wilkins is certain of it; and the world has had its trade and travel route reshaped in a day by the Vega. The Silk Road took a thousand years to evolve;

with a snap of their fingers, Eielson and Wilkins have carved a road in the air that millions will follow, one day.

Cables with offers keep coming and, among other things, Wilkins is commissioned to write an instant book, *Flying the Arctic*, which will essentially be a collection of the acclaimed articles he had penned for *The New York Times* on Ben and his travails and triumph. Typically, Wilkins gets straight to it when he is not being honoured, but for the moment that means there are only fleeting opportunities for writing.

In Paris, they are greeted once more by roaring crowds who turn out just as they had for Charles Lindbergh two years earlier, and preening politicians anxious to be photographed with these newly crowned aristocrats of the air. Wilkins is particularly amused to note the formal enthusiasm of the French for the rather good-looking Eielson. He and Ben are talking to the American Ambassador, Myron T. Herrick, when a well-dressed young French officer approaches them and speaks.

'You are Monsieur Eielson?'

Ben answers in the affirmative and the officer nods and then summons a very beautiful young woman. 'Monsieur, permit me to present my wife. She wishes to kiss you.'[17]

And what is George Wilkins? Chopped liver? It would seem so!

Ben obliges as the astonished Ambassador and the delighted Wilkins watch this very personal award ceremony.

'For some reason,' Wilkins notes, 'I was not so lucky.'[18]

When it comes to precious honours, however, the situation is reversed as, to Wilkins stunned amazement, it is announced that he is to be ... knighted, an honour unavailable to the American, Eielson ... so, there!

For Wilkins it seems extraordinary that a mere decade after chronicling, as a humble cameraman, the knighthood conferred on General John Monash, he is to be equally honoured, and in no less a location than Buckingham Palace! Oh, how proud his father would have been to see him arrive at the Palace in his morning dress – all too tight shoes, rented sheen and a hint of handkerchief from his breast pocket, just before noon on 14 June 1928 for the ceremony. Wilkins is surprised to find that King George V is an enthusiast when it comes to all matters aviation, while the King himself is more than a little surprised that the famous Captain 'George Wilkins' has elected to have himself knighted as 'Sir Hubert' Wilkins.

'Well, George, why Hubert?'[19] asks the curious Monarch. Wilkins jokingly replies that he would not be so presumptuous as to use the King's own name.

King George rises to the occasion, even as George Hubert Wilkins kneels before him.

'Perhaps your parents named you "George", because they expected you to do things like flying the Arctic and leading expeditions into the unknown,'[20] the Monarch drolly suggests.

In any case . . .

'Arise, Sir Hubert.'

Oh, the glory of it all!

It is not long before the journalists come knocking on his aged mother's door in Adelaide.

'It is delightful to hear that George has been so honored by the King,' she tells the journalist from the Adelaide *News*. 'I have seen little of him during recent years, and have not been able to keep in touch with his movements. It was a fine performance on his part to fly to the North Pole, and I hope that some good will result from it. From a child he has been possessed of an indomitable spirit. His success is not due to education received when a child. We lived in the country, and the children had to walk six miles to school. They did not receive the same instruction as children do at present. I doubt whether I will ever see him again, but I am certainly proud of his success.'[21]

This English knighthood, while a great honour, comes with some crippling expectations of its own: the way one dressed and transported oneself, the servants you were meant to have, the clubs you belonged to and the hotels you stayed at.

This knighthood business, Sir Hubert tells Frank Hurley, 'is a positive drawback. It takes so much living up to, even on the financial side, that it is like trying to live the life of the normal subaltern in the Brigade of Guards without a private fortune: "It can't be done".'[22]

For this reason, he often travels incognito.

'On one occasion,' Hurley will recount: 'I visited George at a well-known London hotel and asked if Sir Hubert Wilkins were stopping there.

'"No," replied the booking clerk, "but there is a Captain Wilkins on our books!"'[23]

Among the many telegrams coming for Sir Hubert congratulating him on his knighthood is one from Charles Kingsford Smith and Charles Ulm:

```
WE BOTH SINCERELY AND HEARTILY CONGRATULATE YOU ON
THE SUCCESS OF YOUR FLIGHT AND THE SUBSEQUENT WELL-
DESERVED KNIGHTHOOD. WE WISH YOU ALL THE SUCCESS YOU
DESERVE, AND WILL, WE KNOW ACHIEVE.
    KINGSFORD SMITH AND ULM. 24
```

The most staggering thing? They write from Brisbane where, just two days earlier, they had landed after an historic three-hop crossing of the mighty Pacific Ocean from San Francisco via Hawaii and Fiji, in the newly named *Southern Cross* which had been the *hexed* – Wilkins was convinced – *Detroiter*. Clearly the hex only worked on ice . . .

•

Amundsen is to depart in search of *Italia*. His offer to take his French-built Latham 47 flying boat out has been gratefully accepted. Before taking off with his co-pilot, an Italian journalist posits that it might be too risky to take such a slight craft to such climes after such little preparation?

'Ah,' Amundsen replies, 'if only you knew how splendid it is up there, that's where I want to die. And I wish only that death will . . . overtake me in the fulfillment of a high mission, quickly, without suffering.'25

Which is fine for him. The Swedish pilot who will actually be in control of the craft, Leif Dietrichsen, would rather wait for a day before attempting the first leg, across the Barents Sea to Spitsbergen, but Amundsen insists. It is urgent. Radio contact has now been established with the captain of the airship – they have crashed, and the few survivors are on an ice floe, north of Spitsbergen.

We go *now*.

At four o'clock on the afternoon of 18 June, they take off from Tromsø and head towards Kings Bay, Spitsbergen. Several hours later, a fisherman off the coast sees her flying low.

'A bank of fog rose up over the horizon . . . she ran into the fog and disappeared before our eyes.'26

No-one on the Latham 47 flying boat will ever be seen again.

'He vanished into the Polar Sea which had been his only true home on earth.'27

(As to the *Italia*, it had indeed crashed on 25 May, with most on board killed immediately. Seven survivors including Captain Nobile were found 48 days later, having camped out on the ice, and living off supplies from the airship.)

2 July 1928, New York, Broadway, star-crossed lovers

In her small dressing room backstage of the Frolic Theatre on Broadway – a minor theatre, true, but at least it's *on* Broadway! – the beautiful young actress reads and then re-reads the review from *The New York Times*, about the rather uninspired melodrama she is starring in, *The Cyclone Lover*. Alas, no matter which way she tries to spin it, it is less a review than a public slapping.

While the first act is 'promising'[28], the following acts are distinguished chiefly by the fact they are not *quite* as bad as the actors themselves. The worst thing about this review? While it actively pans two of the actors, it cannot even be *bothered* to mention the name of the starring actress. The ignominy of it! The sheer indignity and *injustice*! Her name is Suzanne Bennett, and she has already reached a certain level of press fame for being, 'one of the dancing partners of the Prince of Wales during one of his visits here'[29], as well as starring in the productions '*Vanities*' and '*Guns*'.[30] But *this* reviewer, did she mention, doesn't even give her a passing reference!

And now what? Suzanne's next reading matter does not improve her mood. It is some odd cable from her agent, wanting her to call him about 'a ceremony'.

A ceremony? Is it a speaking part?

Not quite. It is a grand welcoming ceremony for Sir Hubert Wilkins and Mr Ben Eielson.

'Why would anyone want me to meet them? What have they done?' Suzanne asks.

'Have you ever read the *front* page of *The New York Times*?' asks her agent.

No, she doesn't make a habit of it; she only reads the reviews! Patiently, if slightly exasperated, the agent explains who this Sir Hubert Wilkins is and what he has just done. 'You've been asked to welcome him because you are a fellow Australian. If you don't want to, doubtless there are other Australian girls around who would seize the opportunity . . .'[31]

Consider it seized. *The New York Times* is going to notice her this time. Despite the fact that she has to arrive at the ungodly hour of 5 am, which to a 'Broadway Baby' is usually the time you are just about to settle in to sleep, Suzanne gets a taxi to the Battery, hops on a well-worn harbour tug and experiences the novelty of, 'fresh morning air to which we of the theatre are not accustomed'.[32]

She alights on the *Stavangerfiord*, the ship that has brought Wilkins and Eielson across the Atlantic. The Australian Consul is there and passes to Suzanne the flowers he wishes her to give to Wilkins, which she does as the cameras flash and roll. Neither is impressed with the other. In fact, Miss Bennett – as she styles herself despite now being divorced from Mr Bennett, an Australian soldier of the Great War to whom she'd been briefly married a decade earlier – notes that this so-called 'hero paid hardly any attention to me'.[33] Sensing some awkwardness, the Consul steps in with some diplomacy.

'Miss Bennett is an Australian too,' he notes hopefully.

'Yes,' replies Wilkins dryly, 'I meet them all over the world.'[34]

Perhaps, the photographer asks in a thick New York accent, Sir Hubert would like to have his photo taken with Miss Bennett?

Not really.

But the photographer is having none of it. For suddenly the snapper drops his professionally feigned accent and uses his *natural* accent.

'*Come on, Aussie*! 'Ave your pitcher taken!'[35]

Yes, the photographer is Australian too! Wilkins and Suzanne both roar with laughter and at last, 'the ice was broken'.[36]

They are all Aussies here, all children of the Southern Cross, so enough with fake airs and graces, and, with that bond established, Wilkins quickly finds himself in conversation with the most charming and beguiling woman he has ever met. She knows nothing about him; he must know everything about her.

But to the speeches . . .

After the New York Mayor welcomes Wilkins and Eielson to their fair city and congratulates them on their achievement, the Australian makes reply.

'I can never express my gratitude to the American people for the splendid cooperation they gave us,' he says. 'It was one of the greatest privileges of my life to carry the Stars and Stripes to the Arctic, and I appreciate the confidence which the United States placed in me.'[37]

Upon the shore a roaring crowd awaits, with no less than a ticker-tape parade so as many New Yorkers as possible can see them, before Miss

Bennett tries to steal away to get some sleep. But the Australian Consul catches her before she escapes, *can she not say goodbye to Sir Hubert?* She does so and realises he does not wish to say goodbye to her.

'I'm sorry you are not staying for lunch, I have an official dinner tonight or I would ask you to dine with me,'[38] he says plaintively.

'Call me at midnight,'[39] she says, and then blushes as she realises how that sounds and Wilkins smiles in delight at her embarrassment. She is an actress, and rarely home before midnight, she explains. *Of course.* 'What is your number?' asks Wilkins, but Suzanne is not sure whether he is just being polite as she gives it to him.

Suzanne is intrigued and reads *The New York Times* to find out just who it is that she has met and why all those people are cheering for him. That night, after the theatre, she is nervous, an emotion she rarely experiences. She bathes and changes, waiting by the phone, holding some orchids which she has felt a compulsion to buy for herself that night. The phone is silent and she feels like a fool, but then, as the clock strikes 12, her phone rings. It is Wilkins, he has been waiting until exactly midnight to call and ask her out to dance.

They go to the St Regis Hotel and dance until three o'clock in the morning. Suzanne is surprised that Wilkins is a graceful and talented dancer. When on earth did he learn?

'Surely not in the Arctic? The Great War? The Balkan War?'[40]

No, not quite.

'I brushed up by taking a lesson,'[41] he replies.

When?

Today.

And that is Wilkins all over.

Between the time she handed over her number and the time he worked up the nerve to call, he had taken a lesson, improved his skills and is the better man for it. One way or another, they connect, and one thing they have in common, oddly enough, is admiration for Sir John Monash. Her former brother-in-law, Captain Gershon Bennett, had married Bertha, General Monash's only child, not long after the war and she had got to know him then. On and on they talk, into the night.

Theirs is less a whirlwind romance than a cyclone, as after intense activity Wilkins is whisked away after just a few days, for he and Ben must be off on a lecture tour. Upon departure, Wilkins shyly gives Suzanne a book, his own, *Undiscovered Australia*. When she reads the inscription, she knows that she has his heart:

To a discovered Australian, the lovely Suzanne, with love, GHW.[42]

Wilkins is already at work polishing *Flying the Arctic*, recounting the long road to triumph that he and Ben have had; but as he writes his mind turns again and again to the woman to whom he just gave his previous book. When they next meet, Wilkins has only two days until he will leave for the Antarctic; a matching Pole waits conquest from the air. He sits through a performance of Suzanne's new play which, she reasons, he 'must have liked, because he invited me to supper – and proposed to me. I said yes.'[43]

Though thrilled with her answer, Wilkins has no interest in confirming it to the press when the story gets out, even resorting to that most hackneyed of all responses: 'We are very good friends, but I cannot say anything at this time.'

It is no-one else's business, after all.

Miss Bennett, however, with a métier that lives on encouraging personal publicity not trying to limit it, is more than happy to tell the press the truth. 'We became engaged three days ago. I met Sir George for the first time when he arrived in New York from Spitsbergen after flying over the North Pole. No date has been fixed for our wedding.'[44]

And where is Wilkins' nominal fiancée Lorna Maitland in all this?

Exactly.

They have been courting for all of a decade now, and engaged for half that time, but it is one of those things. 'Settling down' had just never been possible for a man on the move like Wilkins and they had drifted apart – or at least he had drifted from Lorna, even while she had continued her devotion to him, filling endless scrapbooks with every fresh newspaper article about him, and in recent times having to buy ever more scrapbooks!

This time, though, Wilkins is committed to his new love, he is nearly sure of it. The pair are delighted but Suzanne reminds him of the book he first gave her.

'Let me know why you shouldn't stay in New York and "discover me" some more?'[45] she teases.

There is no need for that, Wilkins tells her. 'We'll still be married in thirty years,'[46] he says, certain of it.

But to work, always to work!

Just 12 months ago, Wilkins was more a beggar than a chooser when it came to funding for his adventures, but everything has changed as commercial suitors elbow each other out of the way in their eagerness to be associated with him. Inevitably it is the biggest man with the sharpest elbows who gets to the front of that queue, and is no less than the most famous press baron of the last half century, William Randolph Hearst. A genuinely enormous man of well over six feet tall, weighing in at at least 15 stone, he looks like he could break you with his hands not just his power and influence. To Wilkins, however, he is nothing if not generous from first to last, giving the Australian $25,000 on the simple promise of exclusive press reports for his next venture, whatever it is. Grateful, Wilkins tells him that he will go to the Antarctic, to explore the continent where no plane has ever flown.

Excellent! Hearst effectively reaches down the back of the couch for some loose change and throws in a $10,000 bonus if Wilkins actually reaches the South Pole – a secret arrangement that is to be the first of many between the two.

The Wilkins–Hearst Antarctic Expedition is exclusively announced to the 20 million readers of the Hearst empire. Having learnt lessons from previous cumbersome outings, Wilkins' expedition will consist only of five men: he and Ben, Ben's pilot chum Joe Crosson – better known as 'Fearless Joe Alaska'[47] – who is also an experienced hand, plus a radio operator and a single mechanic.

Wilkins dryly notes that the Lockheed Vega, technically previously nameless, has been renamed *Los Angeles* quickly, 'since she had become such a famous plane'.[48]

Again, Hearst reaches down the back of the couch for the loose change to buy another Vega, this one named *San Francisco*, and both planes are soon disassembled and secured to the deck of the 16,000-ton steamer *Hektoria*, on their way to Deception Island, the only land worth naming that lies in the Southern Ocean between Cape Horn and the finger of Antarctica that reaches the most northerly point, Graham Land.

The New York Times is there at the prestigious City Club to chronicle Captain Wilkins' speech to the members just before departure, as he spells out his plans to move forward the science of meteorological prediction by intensively 'studying Antarctic conditions to determine their influence on weather throughout the world',[49] while also finding bases for the many meteorological stations he wishes to see built.

'With such stations established in Arctic and Antarctic regions it would be possible in the future to forecast seasonal conditions with considerably greater accuracy,' he says. 'Seven years ago meteorologists told me this idea was fantastic. But then such things couldn't be instituted without the development of reliable communication between long distances such as we have in fast planes and wireless today. Now the time has come when we can expect to take advantage of such information as can be collected in the Arctic and Antarctic. The Aero-Arctic Society in Berlin is to establish thirty-five meteorological stations in the Arctic and we hope twelve can be opened in the Antarctic, established and maintained by Governments of the Southern Hemisphere.'[50]

Everything is soon in place and Wilkins and his crew are on their way to Deception Island, which will be their base before transporting their other supplies by the planes to Graham Land further to the south. It is, of course, largely a revisitation of the Cope plan of eight years before, with the difference that this time there are real plans, actual planes, and Wilkins in charge, which means it is actually organised and backed by real money.

A large part of the funding comes from a deal Wilkins signs with a new sponsor.

> If in the Antarctic flights Mobil Oil is used exclusively for the lubrication of the air-plane and Plume Aviation Gasoline used exclusively for fuel, we will pay to you, or to whom you may designate, the sum of Ten Thousand ($10,000) Dollars.[51]

On the way they stop at the Falkland Islands, where a new Governor has thankfully taken over, and this time instead of being in trouble as he had been coming back from the Cope Expedition, Wilkins is asked to help the British Empire on a rather delicate matter.

For you see, Sir Hubert, we have become quite . . . concerned that the Americans are going to try to claim large swathes of the Antarctic for themselves. Their man Richard Byrd has already declared that he wants to be the first man to reach the South Pole by air, and who knows what subsequent claims they will make if that succeeds? We would like to back up their claim to sovereignty by placing, or dropping, Union Jacks all over the frozen continent! And we think you would be just the man to carry out such a delicate mission to help keep other countries, and most particularly the Americans, at bay.

Now, we are aware of course that you are in partnership with Mr Hearst, and so we are unsure whether you are travelling under an American or Australian flag for this venture?

Wilkins gives his answer, and Governor Ellis delightedly cables the results to London.

> Sir Hubert stated emphatically that while he received generous support for that venture from friends and sympathisers in the United States he is beholden to no one . . . He added that although an internationalist in his ideas at heart he remained a Britisher and would therefore be glad to assist so far as he could in furtherance in the cause of Empire by dropping or planting British flags in the manner suggested.[52]

With Wilkins' acquiescence, Governor Ellis puts the wheels in motion and the notion is granted Royal Approval. Yes, a bemused Wilkins now receives an official blessing from King George, appointing him (with rather bizarre boilerplate) to the task of claiming a continent by flag drop.

> **George R.I.**
> George, by the Grace of God, King of Great Britain, Ireland and the British Dominions Beyond the Seas, Defender of the faith, Emperor of India &C, &C, &C . . . Greeting!
>
> Whereas our trusty and well beloved Sir Hubert Wilkins Knight Bachelor is presently to conduct an expedition . . . And whereas we have judged it expedient to appoint a fit person to take possession in our name of such territories now unknown . . .
>
> Now Know ye, that We, reposing especial trust and confidence in the discretion and faithfulness of our Trusty and Well-beloved Sir Hubert Wilkins . . . hereby appoint him to be our commissioner for the purpose aforesaid . . . [53]

And so it is that, together with food and fuel, Wilkins is also provided with a secret supply of Union Jacks, cotton claims to sovereignty intended to hold back Yankee carpet-baggers heading to the New Deep South. With his blood brother Ben Eielson by his side – the two have become extra-ordinarily close – Wilkins celebrates his 40th birthday on the Falklands, with toasts to his longevity despite all odds. Nearly all present – especially the Norwegian whalers, for whom it appears to be something of a daily tradition – get dutifully drunk to mark the occasion. The exception is Wilkins, who drinks the toast and makes his excuses shortly afterwards, much more interested in planning his future survival than celebrating his present.

Just a week later, *Hektoria* steams into the harbour to be greeted by thousands of waddling penguins, each one of whom appears to have graduated from the Charlie Chaplin School of Walking, causing high hilarity among all members of the crew, particularly those who are seeing them for the first time.

It is no small exercise to unload the planes near the whaling station – first getting them into the water on their pontoons, before dragging them to the shore, and putting on wheels for the pontoons, but when it is done Wilkins is able to send out a joyous radio report that will be the foundation stone of newspaper reports around the world:

> **MESSAGE FROM LEADER.**
> LONDON, Monday. – Sir Hubert Wilkins, the Australian explorer, in a wireless message from Deception Island via Port Stanley, says: – 'Our American monoplane, the *Los Angeles*, which Lieutenant Carl Eielson flew over the North Pole, was today the first complete aeroplane to touch Antarctic soil.[54]

Inspecting their proposed 'runway', the ice on the harbour, Wilkins notes something more than passing strange. The ice is so much thinner than the last time he was here. Back in 1920, when he had first landed with that impostor, Cope, he had measured the harbour ice as being six feet thick, but it is now not even half that! Having, typically, a contingency plan for the possibility that the ice would be too thin in parts – he has brought the pontoons along so that the Vegas can be transformed into sea planes – ensuring that the unexpected warmth does not immediately place all their plans or their planes at risk. It is just very odd. (And perhaps the first time in history that instead of a polar explorer worrying about being crushed by ice, his ship being trapped, freezing to death etc., he is worried about there simply being not enough ice to be safe.)

As it happens, although the pontoons prove to work perfectly, a new and entirely unexpected difficulty soon presents itself: suicidal water birds. In a poetic irony that Coleridge would appreciate, the birds are a little more than just omens of bad luck.

'They were not at all afraid of the plane,' Wilkins writes, 'but had an immense curiosity about it. Whenever it got up to speed on the water, clouds of them flew with it, dashed into the propeller and made it impossible to take off.'[55]

At last, a satisfactory cold snap comes, which sees the water transformed into ice solid enough that it will suffice for a runway. They hope. At the least it is strong enough for Eielson and Crosson to be able to take off with ease on their respective test flights, one craft soaring after the other off the ice, but . . .

But taking off is the easy part. It is the landing on ice which is most difficult, and most dangerous when it comes to breaking through that ice.

Wilkins is watching as nervously as everyone, thus, as Ben brings *Los Angeles* in for landing on the area that Wilkins has marked out as having ice sufficiently thick to take the weight. Alas, one of the features of new ice is its extreme slipperiness, and though the first part of the landing is without incident, those on land must watch in horror as *Los Angeles* keeps going, slipping, sliding, slithering now from side to side, onwards ever onwards to the thinner ice! Finally, it comes to a shuddering halt on thin ice . . . at least laterally . . . and not a little literally.

For, the moment it stops, the weight is too much for the ice to hold it and before their very eyes, the front wheels break through, putting the plane's nose and engine beneath the ice, with only its tail in the air!

No-one move!

Despite their obvious urgent instinct to rush forward and get Ben out of there, they have to restrain themselves as their added weight would almost certainly crack the ice more and the whole plane might disappear.

'We stood horrified and held our breaths as we saw the plane hanging precariously over the hole in the ice.'[56]

Inside the plane, slowly, oh so slowly, realising that he might be just seconds away from sinking to a frozen and watery grave, Ben Eielson has unstrapped himself and is very carefully climbing his way towards the rear hatch.

There he is!

A great gasp of relief is let out as Ben successfully climbs down and – as a fully grown man trying to tread as lightly as a sparrow – steps with infinite care to thicker ice. Now that he is safe, the rest of them try to get to the plane but the situation is every bit as precarious as had been feared. For even as they approach, there is another fearful cracking, a lurching forward, and now *Los Angeles* sinks even deeper! Just one wing and the broad fuselage remain visible. This harbour, as Wilkins knows too well, is 1200 feet deep. Just one more move and *Los Angeles* will never be seen again.

Above, Crosson is still flying *San Francisco*, and has seen the whole thing. But what can he do? With fuel running low, he circles back to land on the runway he took off from.

Sure enough, he brings the plane down to a sliding stop and . . . it holds! The back door bursts open and Crosson rushes out to some nearby whalers he has spotted, begging them to help at once. The whalers rush towards the *Los Angeles* with rope and planks to secure it – though just how a rope and planks might prevent a plane from sinking is not clear to anyone at the moment.

At least for the moment they are able to get ropes on the *Los Angeles* and begin pulling, with the occasional man falling through the ice as a useful guide as to where it is thinning. With the temperature of the day rising with the sun to well above freezing, the ice is thinning by the minute and the men playing tug-of-war at least know not to bind themselves with rope; for they will *not* be joining the craft if a deeper plunge happens. Finally, however, by putting wooden slats cross-ways under the skis, to ensure the weight of the plane is more widely spread on the ice, the job is done, and the *Los Angeles* makes its way back onto solid ice!

It has taken . . . 18 hours, but it is done. Wilkins is overjoyed, thanking each whaler in turn for their incredible efforts. Amazingly, *Los Angeles* seems fine, with Wilkins noting the sole black mark: 'salt water had got into the cylinders'.[57] Salt water may well have got the whole bloody plane forever and, once again, Wilkins delights in turning a disaster into a triumph. There remains the matter of finding ice solid enough to venture more landings, and the whalers lend Wilkins a 'catcher' vessel to allow him to look for firmer ground. There is none to be found, cliffs and crevasses in turn defeating every effort.

Ever the pragmatist, always sure that there must be a solution and it is just a matter of working it out, Wilkins thinks he might have found one as rough as it is ready. Why not work *backwards*, extending the 'start' of the runway back over ground they'd previously dismissed as it ran 'half a mile over rough surface, up over a hill, down across ditches, up another rough slope and down into the harbour'.[58]

Look, it might not do for the King, but it must work for them. All it will need is for Ben to push *Los Angeles* as fast and furious as he can over the goat-track of a path they can hack into the slope, so that by the time he gets to the start of the runway proper, he will already have a head-start of speed and they can get airborne before finding themselves skating on thin ice. After a solid day and a half spent trying to smooth out as many bumps as possible, they are ready to give it a burl.

Los Angeles steels itself, as does her pilot, and after Ben opens the throttle they are soon jolting, bolting, bashing, smashing, shuddering and shaking along the runway, shooting down slopes and diving over ditches . . . 20 miles per hour . . . 30 miles per hour . . . with Wilkins wincing as the landing gear smacks against the earth and ice, before they get to the smoother part of the runway and . . . sure enough . . . just before the water looms . . . lift off!

In the air with enough petrol for 1400 miles of flying, now and only now does Wilkins allow himself to enjoy the fact that, 'we were off to explore the Antarctic in the same Lockheed-Vega 1 which had carried us over the top of the Arctic half a year before'.[59]

They are back in business, and what a business it is! Ben slices through the narrow gap between mountains and heads south at 125 miles per hour. Not that Wilkins doesn't have a back-up plan. On the chance they crash somewhere out in the middle of the Antarctic, and have to walk back out, the Vega has on board emergency packs and rations for two months, which should allow them, he has already calculated, about a 500-mile hike.

Adding one last bit of solace as they sail off into the wild white yonder is the fact they also have on board a state-of-the-art radio built by Heintz & Kaufman of San Francisco. It weighs just 50 pounds and is so ingeniously built that, because Wilkins is now keeping a single key pressed down, a signal is being continuously sent to the wireless operator on Deception Island, who is able to track their precise course accordingly. In 1913, he and Stefansson had been lost to the world in an instant. But now, in these stunningly modern times, everything has changed and, whatever else, their location will be broadly known.

Their journey on this day, in stunning contrast to the endless dramas that seemed to have come with Arctic flying, proceeds like a dream. Soaring over Bransfield Strait, warm and comfortable, they watch the pack ice break from icebergs as they pass. Trinity Island is spotted and passed, its 6000-foot peaks magnificently thrusting skywards, but still no match for them as Ben adroitly flies around them and makes for Graham Land. There is no cloud, just a dazzling view never seen before by man:

> The panorama was magnificent – the jagged mountains of black and green rock and glittering snow slopes of Trinity towering beside us, above us the clear sky, below us blue-black water and icebergs – everything frozen and still, black and blue-black and black green and glittering white.[60]

Relishing their speed and ease of passage, Wilkins can't help but remember his bitter journeys with Cope almost a decade earlier: 'It had taken us three months, on foot, to map forty miles; now we were covering forty miles in twenty minutes.'[61]

Pass the hot coffee from the thermos flask, as I sharpen my pencil!

While Ben concentrates on flying, Wilkins continues to sketch the new terrain opening up before their eyes, something which, curiously, has proved to better capture landscapes than mere photographs in this shiny white on white land.

Wilkins' hands now become a blur of movement as he seeks to furiously form each fresh piece of terrain before the next new vista appears. Of particular interest, they are the first to see the top of Graham Plateau, an ice mountain previously estimated as jutting skywards to 6000 feet, but now found to be 9000 feet – which they can mercifully vault over without trouble.

'As we sailed in over the continent,' Wilkins will record with rare pride, 'Ben Eielson, I and our little Lockheed-Vega were making another record; for the first time in history new land was being discovered from the air.'[62]

The plateau of the Palmer Peninsula is crossed, the shelf ice of the Weddell Sea recognised and then: nothing.

'The area before us was blank on the charts since no one had ever visited it before.'[63]

It is an experience a lucky few humans have experienced before: true discovery. Wilkins and Ben will be among the last to ever know the sensation – that wild surmise being confirmed at last; nothingness replaced with knowledge.

With newly discovered lands inevitably comes the prerogative of bestowing new names, and while British explorers have often named things after the royalty and aristocracy that have provided them the largesse to be there in the first place – Antarctica's Victoria Land and George Island being classic cases in point – Wilkins must honour a different class all together: sponsors and benefactors: 'The mountains behind the cape we called Lockheed Mountains'[64], while one of their key sponsors, Mr William Scripps of Detroit, gets an island, Scripps Island, which at least has a touch of poetry to it; which is more than can be said for – ahem – 'Mobiloil Bay'. The Santa Claus of the South is flying high above and casting immortality for certain names as he goes, some men get an inlet, some get a mountain, while you William Randolph Hearst, get no less than a land: Hearstland (it is a huge chunk of ice that in some ways resembles Mr Hearst himself, only a little warmer). For you Stef, a strait – and you are most welcome – while Dr Finley, President of the American Geographical Society of New York, gets the Finley Islands added to his casebook.

At least in Antarctica there is no problem with giving names to things that locals have called a different name for millennia, but still Wilkins can't help but note the absurdity of this exercise:

> To which country and to whom does the pride of these discoveries belong? We might well have claimed the new territory for the United States, England, Australia or Norway, whose people have materially assisted us and every other expedition without exception visiting the Antarctic the last few years. But it would have been presumptuous on our part and meaningless to claim it for anyone. If it is agreed that we can effectively claim country discovered from an aeroplane it would lead to many complications. Could one claim all the country seen together with all the area we might reasonably assume to lie between that part seen and the next known area? If it were agreed that we could claim only country mapped, what degree of accuracy would he needed in the mapping? Would it be necessary to alight on the country and for how long, and then would our claim cover only the country seen from the ground or from what height? And if once having set foot on the ground for a few hours or days, does that mean absolute right of possession?[65]

But no matter. His not to reason why, his but to drop a flag and then speed by.

Dashing discovery is the order of the day and they are, right now, aviation's answer to Captain Cook, with a champagne cork on top.

On days like this, you feel you could fly forever ... but a looming storm ahead persuades them to turn back to Deception Island and – after dropping another Union Jack attached to a ski at 71° 20' south, 64° 15' west, and watching it flutter down – manage to land safely, despite 'all the winds of the storm trying to pull us to pieces'.[66]

A triumph!

While the Antarctic continent was thought to have a coastline extending some 12,000 miles, to this point no more than 500 miles of it had even been seen by the eyes of man, let alone charted. Now that has all changed.

'In one day's flight,' *The Brisbane Courier* will subsequently report, 'Sir George Wilkins has been able to complete more accurate mapping of the Weddell Sea and the Antarctic continent from the South American side than has been achieved by all the previous explorers. Sir George has substantiated the opinion of previous explorers about the mountainous, jagged, inhospitable nature of the country.'[67]

The two men yell and shake hands, Wilkins reeling with the unreality of that day's events.

'We had left at 8.30 that morning, had covered 1300 miles – nearly a thousand miles of it over unknown territory – and had returned in time to cover the plane with its storm hood, go to the *Hektoria*, bathe and dress and sit down to our eight o'clock dinner as usual.'[68]

As dinner conversations go, there can be few in history between companions so filled with achievement in just that day.

'We had discovered altogether eight new islands, three channels and a strait, as well as an unknown land that seemed to be of the Antarctic continent.'[69]

There can be no disputing it.

'Aeroplanes,' Wilkins will triumphantly claim, 'enable more work to be done in a day without exertion in the Antarctic than earlier explorers achieved in years of hardship and danger.'[70]

Mr Hearst will receive his dispatch *after* dinner: 'We are tired and weary; but our successful flight has enabled us to prove that our luck still holds.'[71]

Not that all is joy.

The German philosopher Friedrich Nietzsche had once noted, 'the melancholy of all things completed', and Wilkins feels more than a little of that now.

'What were our feelings after such a journey?' he will ask rhetorically. 'Were we thrilled with the knowledge of success? For my part, I was sadly disappointed. We had passed through the gamut of human experiences.'[72]

A man born to conquer new worlds, the truth of it is, he is running out of new worlds to conquer.

'Yet did that bring happiness? Positively, no . . . Nearly two thousand miles in a bee-line awaits to be discovered.'[73]

But that will have to wait. For, having already accomplished so much, and their supplies dwindling, Wilkins decides further exploration cannot safely be done until next season, and gets on the radio to call in a vessel from the Falklands to come and collect them.

They remain on Deception Island long enough to see in the New Year of 1929 with the whalers and experience one of their legendary celebrations, for that night is, 'the only break in the terrific work of the whaling season and they make the most of it with games, songs, enormous eating and plenty of drink'.[74]

Ah yes, the drinking.

It is somewhere between Herculean and insane, which, as ever, results in men embarking on projects that seem like a good idea at the time . . . Like those two drunken Norwegian sailors you can see there. They have decided they wish to see off 1928 with a bang, and know just the way to do it. That barge, moored down by the beach, the one that lies so low in the water? Well, it does so because it is laden down with no less than 70 tons of explosive put aside for black powder and bombs for the harpoons. They know all about it, for they are the very ones who have been put in charge of its security.

Wouldn't it make for some fabulous fireworks on this night? Don't you think it would . . . go off!

They do!

So, what shall we do with the drunken sailors, what shall we do?

Wilkins watches idly as they attempt to blow the barge sky high – though fortunately the trail of gunpowder they light has its own ideas and heads towards them instead of the explosives. Thus, they manage only to set their own hair on fire in the process, but otherwise escape unscathed, even as 1929 arrives!

Oh, how they all laugh, even hairless sailors.

'This was our farewell to Deception Island and the hearty whalers,'[75] Wilkins will recount. The hungover whalers wave them farewell the next day as their steamer leaves for the Falklands, and from there to South America, through the Panama Canal then on to New York.

As his third Antarctic odyssey comes to an end, Wilkins cannot help but compare it to his first, with that charlatan, John Cope.

> Years ago, toiling at the end of a heavy oar, tossed in an open life-boat, I struggled for 14 days at the feet of the Antarctic's majestic mountains, and in that time mapped some 30 miles of unknown coastline. Throughout every waking hour I was awed by the stupendous majesty of the scenes unfolded. Every foot of the way was fought for. Every point had to be landed upon, and a bitter half-hour spent with freezing fingers manipulating a surveyor's transit.[76]

Then there had been that extraordinary episode on his second trip, after Shackleton had died, when Antarctica had held their ship tightly to her frozen breast.

> It appeared as though we would be for ever held in the arms of Her Majesty the Queen of the Antarctic. I learned to adore her and her whims and fancies as each day she appeared in spangled icy dresses, scintillating colour, and fascinating everyone who could comprehend her presence. I, for one, would have felt pleasure for eternity in her embraces, but a mighty iceberg, no doubt jealous of our attentions, came along, smashed the pack ice, and released our ship.[77]

This year had been different, an entirely different exercise.

> I journeyed south this year, hoping to woo the royal lady of my acquaintance. I have returned conscious of having done nothing more than rudely stumbled upon her bathing, naked, in her private pool.[78]

13 March 1929, New York, Lead Zeppelin

Eyes right and all hail the conquering heroes.

Waiting to greet them the moment they step ashore from the tender of their ship are none other than the New York Mayor, Jimmy Walker, together with various local identities, and most importantly . . . Suzanne. There are speeches, much flag-waving, and of course a ticker-tape parade.

Ben must quickly depart for Washington where he is to be honoured with the Distinguished Flying Cross, while the humble Australian, George Wilkins, chooses to remain in New York, both to spend time with his love and to prophetically proselytise on the importance of Arctic weather to the great and the good (more of the former than the latter), most of them maritime magnates, at the British Empire Chamber of Commerce luncheon. Held on 21 March in the distinguished Whitehall Club, which gazes out on the Hudson River from 30 storeys up, proceedings are presided over by an enormous portrait of one of the most brilliant of the founding fathers of America, Alexander Hamilton, as a stunning new polymath for a stunning new age is handed the floor.

Gentlemen, please put your hands together for one of the great explorers of our time, just returned from his latest adventure . . . Sir Hubert . . . Wilkins!

After making some preliminary remarks, about what an honour it is to address them, Wilkins gets to the nub of it.

> You might be interested to know just why we go back from time to time to these regions, and why there is a direct connection between our work and yours. If we can forecast conditions so that the agriculturalists, the primary producers will know what type of crops to plant, when to plant them, whether it will be an early season or a dry one; if graziers can know just how much stock they can carry on a certain acreage in a certain year, it is going to do a great deal toward the stabilization of prices and it will mean a great deal to every one of us here.[79]

Oh yes, Wilkins has no hesitation in predicting that, for the comparatively minimal sum of just $10 million dollars invested, it will only take a single year before the accuracy of forecasts will improve by a stunning 50 per cent! And a decade from now, gentlemen, there will be accuracy off the scale. It is the polar regions that are the key. To understand them is to understand the planet. For the moment, true, no-one fronts up with such a sum, and the only person who does seem willing to hand over cash is William Randolph Hearst, who wants Wilkins to take another trip on the incredible *Graf Zeppelin*, with a surprising role: correspondent.

Look, in some ways this is rather a demotion from being commander of an expedition – calling all the shots, and firing a few yourself – down to merely commenting on the shots of others on an air flight, but as Mr Hearst is funding Wilkins, just as he is funding *Graf Zeppelin*, he proves to be a difficult man to say no to, especially when he pays so well. Making things odder still, however, is that Wilkins will be restricted to writing about the 'technical viewpoint', while one Karl Von Wiegand will cover the news side of the story even as Lady Drummond-Hay will be all over the society and gossip side of the trip.

For Wilkins, at least, it means a rare chance to ease back and travel the world in luxury, rather than perpetually bouncing from the seat of his pants to the skin of his teeth and back again, as per usual.

Graf Zeppelin represents 775 feet of high society in the air, the passengers flying in a grand manner never seen before or again. The plan is for it to fly around the world in four 'long-hops' with 48 crew and 20 passengers, who will pay no less than $12,000, an astonishing sum that would buy you not one but three very comfortable family homes in the United States at the time. Despite that colossal price, the tickets are sold out within days and one society lady even offers Wilkins $30,000, 'if I would somehow fail to show up in time to start, so that she could get my place'.[80]

The journey begins at New York as the 20 passengers file on board, inevitably accompanied by porters who bring their many suitcases along behind. (Many more suitcases are *not* on board; for on *Graf Zeppelin* one is only allowed 66 pounds of personal luggage. Luckily, your ticket entitles you the privilege of having a further 286 pounds of luggage to be transported via a steamer to your destination.)

The atmosphere is one of rather louche luxury, mixed with nervous anticipation as the hour of takeoff approaches. The most nervous passenger of the lot is female and frequently runs screeching from one end of the airship to another chased by the crew, which is perhaps understandable given that she is Susie the celebrity chimpanzee, but still a little wearing all the same. At least, for half a day at a time she is back in her box before escaping once more.

Frankly, Wilkins would not mind putting some other passengers in a box, particularly those who seem to regard him as some kind of jumped up air steward; even asking what sort of furs they should wear as they cross Siberia?

Yes, well.

Given that their airship is really no more than a floating hotel going *over* Siberia, rather than *in* Siberia, Wilkins points out that appropriate dress should perhaps approximate what they would wear in a New York hotel, but it seems to have little effect. His advice is ignored, and for good reason.

For once, Wilkins is wrong!

'We have a million cubic feet of gas but no heat,' Lady Drummond-Hay chronicles. 'Merciless cold driving through the canvas walls of this flying tent ... We have drawn ourselves lovely pictures of dining elegantly in mid-air with Commodore Eckener at the head of a flower-decked table ... but ... leather coats, woollies and furs will be our evening dress. Hot soup and steaming stew more welcome than cold caviar and chicken salad.'[81]

At least they hear each other's teeth chattering.

'The ship was almost noiseless. We hardly heard the engines except when we were leaning out of the windows ...'[82]

Wilkins is amazed to visit the kitchens on board and find that all the electrical equipment there, 'made no more noise than the whirr of an electric fan'.[83] As to the cabins, all mahogany and brass, at least they are luxurious enough to remind one of the Orient Express.

It is not all smooth sailing for all that, as Wilkins is every bit as amused as the captain of the Zeppelin is annoyed to find that their official Russian translator for the journey, a Mr Karklin, does indeed know Russian perfectly – but does *nyet* speak English. As the Zeppelin approaches Russian airspace, radio messages are sent to the Zeppelin and Mr Karklin clearly understands them precisely, even looking worried on occasion, but for the life of him he cannot make anyone else understand what they mean. Desperate, the ship's officers write out messages for him to send back, in English and German, but Mr Karklin pores over the messages until it becomes clear to all that he is translating them letter by letter, and creating gibberish as he does so. The whole absurd situation really does become dangerous when flying over a Russian river, the crew fire off an 'echolet', an altitude determining device that measures height from the sound echo of a shot.

But the Russians don't know that.

All they hear is a shot.

Are they being fired on?

Wilkins watches as the Russians beneath start running and a considerable number, 'grabbed up their rifles. I expected them to send up a volley towards us, but when we made no further warlike demonstration and, in fact, continued to wave at them from the windows, they let us continue in peace.'[84]

After traversing the Atlantic, Europe and all of Asia, they prepare to land in Tokyo for a replenishing of hydrogen, and where Wilkins will have just enough time to give a lecture at the airport, for what he expects will be a small crowd of enthusiasts. But to his amazement, upon approach, they find that when flying over Tokyo, 'the roofs and every vantage point were occupied by cheering and waving citizens',[85] and when they get to the remote hangar for their berth at Kasumigaura, some 20 miles out of the city, no fewer than a hundred thousand Japanese citizens have gathered and most of them have walked!

And as they circle to land all of them below are cheering – certainly for the Zeppelin, but also because Japan is now clearly connecting to the world as never before after centuries of isolation. Oh, how they roar!

> It is impossible to imagine that people as undemonstrative as the Japanese should show such emotion. Admirals in uniform danced and shouted like schoolboys. High ranking officials tossed their hats into the air. The ground crew clapped and clapped. Thousands of people roared *Banzai!*[86]

As the most famous man among them, Wilkins is asked to speak, and does his best, knowing that few will understand anything he says in any case. Lady Drummond-Hay also delivers some of her most tantalising bits of society gossip, for those Japanese peasantry who have been wondering about such matters. Tea at the Imperial Palace follows, and Wilkins is struck by the level of questions the Japanese ask him about flight: 'I recognised they were as up to date in aviation as the Europeans and Americans.'[87]

When *Graf Zeppelin* flies over California – let's see . . . 79 hours and 22 minutes later – its commander, Dr Hugo Eckener orders that it be flown over San Simeon to salute its owner Mr Hearst. True, it so happens that Mr Hearst is asleep at the time and there is not a light on, in the entire estate, but still. Dr Eckener, realising that they have a chance to break the round-the-world speed record as they approach Los Angeles, announces the airship will stop briefly and depart before the evening. A decision that underwhelms Mr Hearst when he wakes.

No matter . . . !

When you are a billionaire of Mr Hearst's stature, rectifying such matters takes one barked command: *Tell Eckener to remain in Los Angeles, and I expect him for dinner at the Ambassador Hotel.* Of course, Eckener and Wilkins are both in attendance, and both speak at the dinner, Eckener to affirm Mr Hearst's foresight and genius, while Wilkins chooses to talk on his favourite topic: the advantages of accurate meteorological forecast. As the Australian talks of the virtues of establishing weather stations in polar regions, Eckener slips away to prepare *Graf* for takeoff, but there proves to be a problem. Once Wilkins is back on board, *Graf* refuses to rise! Despite meticulous calculations and every person or thing on board being carefully weighed; she stubbornly remains ground-bound.

Wilkins is focused on the high tension wires they will have to clear to leave the airport. He and Eckener both know what danger they represent:

'If we should touch it, the contact would let loose a more terrific flash of lightning than ever came out of a storm. It would blow us all to pieces in a mighty explosion of hydrogen – and not only us, but the crowd on the ground below.'[88]

True, no airship had ever exploded in such fashion, so it is unlikely, but you never know.

Eckener orders the ship to be 'walked around', and it is done – a giant version of children walking around with balloons on a string at the county fair – as the vast craft is lightly pulled with ropes around the ground, searching for a patch of air where the 'inversion layer', the warmed air that is preventing levity, is not present. No such levity presents itself down on the field, and there is certainly no levity on the bridge of the airship, as Dr Eckener watches the time ticking away. Wilkins' own level of confidence is not raised when he hears the crew yell to some sailors watching the non-departure to push the craft up into the air!

'Give her all you can! We need it!'[89]

The sailors are only too happy to oblige and they experience the novelty of personally launching a ship into the air.

Eckener orders full speed ahead on all five engines. The nose of the Zeppelin now rises but the tail drags and all watching can see the rear is headed straight for the high-tension wires. Standing next to Wilkins in the 'gondola' of the craft is Lieutenant Commander Charles Rosendahl, and he is in no doubt as to what will happen next: 'This will be the end of us,' he says. 'Sure as apples we'll hit those wires and the ship will go up in flames.'[90]

Wilkins tries to reassure him, but Rosendahl is yelling now, he has had enough Teutonic cool: 'Doctor, bring her down!' he yells at Eckener. 'You can't get above those wires. BRING HER DOWN!'[91]

Eckener disagrees and orders that all the stern ballast be dropped, all that can be thrown overboard is, including an avalanche of water as *Graf*'s *entire* water supply is dumped on the crowd below. The deluge does the trick and, missing the wires by mere feet, they are away.

True, the loss of the water due to the last-minute dumping is a little problematic when it comes to crossing an entire continent, but Eckener has a solution: free champagne instead. Yes, America is now under strict laws of Prohibition, but that is no problem. Those laws are for down there, not up here! (And in any case, no police can get up here to arrest them.)

But what about when you want to have a bath? No problems, that can be done in pure *eau de cologne*.

And so the journey proceeds, with most passengers pleasantly sloshed on champagne, while smelling very pleasant indeed.

There is one odd happening that calls for another drink. 'The only time we were shot at,' Wilkins will note, 'was over Arizona where some wild cowboy put a couple of bullets through the envelope. He just wondered if he could hit it, probably.'[92]

And yet, while the cowboy is on target, happily, so too is Dr Eckener, who is thrilled upon landing in New York to have established the new world record for global circumnavigation – 21 days. Some will never recover from the nerves of that Los Angeles lift-off, but as for Wilkins: 'The whole trip had been a most enjoyable vacation for me.'[93]

That said, it is clear Wilkins is pleased to be a passenger no more: 'I can start my own work now; I will leave in three weeks' time, and expect to see Commander Byrd in the South Polar Sea.'[94]

CHAPTER SIXTEEN

'HELLO, GOODBYE'

The Lord knows what we may find, dear lass,
And The Deuce knows what we may do –
But we're back once more on the old trail, our own trail, the
out trail,
We're down, hull-down, on the Long Trail – the trail that is
always new!

Rudyard Kipling, 'The Long Trail'

August 1929, New York, loping elopement

It is one thing to conquer both ends of the globe, but quite another to conquer another's heart and have – yes, he is very nearly sure – his own conquered in turn.

The question is always on his mind: what to do about Suzanne? Theirs has not been a romance without ructions.

While on board *Graf Zeppelin*, Wilkins had been stunned to receive a telegram from her, saying that she had been struck down with rheumatic fever; that she would now only be a burden to him and that the marriage should be called off – but whether it was genuine health concerns, or uncertainty about him ever being a husband to her, home and hearth is less certain. Truly, he is less than completely certain that marrying her is a good idea.

Against that, he is 40 years old now, and has already broken off one engagement, with Lorna, which had been awkward. Another broken engagement wouldn't be a scandal, but it would be . . .

No. Enough. He will marry Suzanne. If she will have him; it will not be an easy life but it will be theirs.

Turning up at her door and finding her, if not in rude good health, at least not rude to him, he quickly convinces her to come with him to Cleveland, so they can be married on 30 August 1929. Suzanne will later recall that the city in Ohio was chosen because they had to fit the wedding into Wilkins' current program of attending a trade show there, watching the National Air Races, and buying a tractor.

'He talked all the way to Cleveland about the races, and the tractor, so I wasn't at all sure whether he'd marry a new plane, a Caterpillar or me.'[1]

For his part, Wilkins teases Suzanne that she is lucky he is making it so fast to the altar, a propulsion that is a hangover from the Zeppelin.

'You got me while I was already on the move,'[2] he says.

Upon arrival, Sir Hubert escorts his 28-year-old bride-to-be to the marriage licence counter in Probate Court in downtown Cleveland, where the papers are duly signed. Look, if all else had been equal, it would have been great to have dear Ben as his best man – but that is simply not the way things have turned out. He can explain to Ben later.

The witness to the marriage, John Caine, will tell the papers that he 'had never seen a bridegroom so nervous'.[3]

True!

Wilkins' hands shake so much during the ceremony that he cannot place the ring on his bride's finger.

For some reason, this man who has narrowly escaped death on at least 20 occasions, who has trekked the polar regions, crossed the oceans, faced the Germans, flown over the roof of the world – and barely blinked throughout – is a quivering wreck in the face of a simple marriage ceremony.

The rickety ring placement goes so long that the justice who is presiding over the ceremony finally tells the soon-to-be Lady Wilkins to 'Put it on yourself'.[4]

She does so, they kiss and sign the register.

Despite the out-of-the-way and low-key nature of the nuptials, news of it makes the front page of no less than *The New York Times*:

SIR HUBERT WILKINS MARRIES IN CLEVELAND ...

Cleveland, Ohio, Aug. 30. Captain Sir George Hubert Wilkins, the daring Australian aviator explorer who flew over the 'top of the world' in 1928 and on Thursday completed the round world flight on the *Graf Zeppelin*, stole a march on Cleveland today and was married in the back room of Justice of the Peace Joseph E. Chizek's office. His bride was Miss Suzanne Bennett, an actress, also of Australia.[5]

And for a honeymoon? Well, it can begin with the newlyweds attending the air races that afternoon, and the National Aeronautical Association banquet that evening.

Wilkins' diary for the day contains but one word: 'Married.'[6]

Perhaps he might have written more, but he is busy, don't you see?

In less than a month, Wilkins must set off to the Antarctic once more, to complete the discoveries begun last year. As Lady Wilkins will sum up their relationship: 'And so it went for thirty years, hello, good-bye and hello again.'[7]

For to some extent, Wilkins proves to be like his one-time hero Sir Ernest Shackleton, who in a period of relaxation immediately after the Great War had forlornly written to his wife, the ever-faithful Emily, 'I feel I am no use to anyone unless I am out facing the storm in wild lands.'[8]

Staying at home with slippers and a pipe would never have done for Sir Ernest, and it does not do for Sir Hubert.

This journey to the Antarctic – where he intends to 'wrest from Mother Nature her last secret'[9] – is different this time. For he will be doing it without Ben, who has bowed out of exploration and is now attempting to start his own commercial Alaskan Air Service. In his place, Ben has warmly recommended his friend Parker Kramer, who he knows has a recent history that will appeal to Wilkins. For he is a man after our own heart, and experience, George. He once ditched a plane over Greenland and had to walk 70 miles, alone, before he encountered help – and the following year did much the same after a forced landing on the north coast of Labrador.

William Randolph Hearst asks how much more money Wilkins will be requiring from him *this* time, and is most surprised to be told the answer is none and that: 'I consider that Mr Hearst has paid me for exclusive news of my Antarctic work, no matter whether I spend one season or a dozen in the South.'[10]

One reason for Wilkins' magnanimity is the fact that he is now being supplied with £10,000 by the Colonial Office of the British Empire, together with exclusive use of the boat *William Scoresby* to act as his sea headquarters for this next leg of the expedition – all of it a legacy of his secretly agreeing to drop the Union Jacks on his last famous flight!

As it happens, such financial security for Wilkins will prove to be as precious as it is rare, as, the day he sets out for the South Pole, 24 October 1929, turns out to be 'Black Thursday' and Wall Street is left in smoking ruins as share prices crash by over 10 per cent. It is just the beginning of a precipitous fall as the Monday that follows becomes known as 'Black Monday', with 'Black Tuesday' off the starboard bow and closing fast! The Great Depression has begun. No business is safe, least of all newspapers, and for explorers like Wilkins, who have relied on the public and the papers for funds, there will be no more 'subscription drives' and 'lecture tours'. For all new projects, he will have to rely for cash on the millionaires and magnates that have survived the crash. A Rockefeller, a Hearst, a Getty, these are the new masters for the man of adventure.

In the meantime, Wilkins attaches another member to the crew, Canadian commercial flyer Al Cheesman, who is familiar with flying in extremely low temperatures. However, when they arrange a ride on a whaling vessel and leave Montevideo on the way to the Antarctic, Wilkins is surprised and concerned to find *no* low temperatures of note. Wilkins and Cheesman fly scout missions to look for solid ice on which to establish a base but even as they approach Deception Island they cannot see any ice. It is hard to believe, but their own eyes tell the story. Wilkins thought last year was an anomaly, instead it appears to be the start of an unrecorded pattern.

'The previous year had been unaccountably warm, yet we had encountered pack ice three hundred miles north of Deception. Now the sea was open and the ice in the harbour thinner than in 1928.'[11]

In another ominous sign, the planes that had been left in storage the year before are 'not even covered with snow'.[12] They find themselves in the bizarre position of having to fly *from* Deception Island to look for ice. They find it, but the edge of the pack is over 300 miles away! The ice field had retreated 600 miles in a single year and the weather is still warm!

'The continuing warmth had now melted the floating bits and the great mass of ice at the South Pole was still diminishing in size. Thinking back on it later, I believed that this unusual phenomenon may have had a long-range effect on climactic conditions throughout the world, or may have been somehow related to them. This six hundred mile retreat of the ice in the Antarctic summer of 1929–1930 was followed by the memorable drought in the summer of 1930, during which many parts of the United States, for example, saw the ground turn as dry and hard as any season in memory.'[13]

Well, there is no way round it – Wilkins and his crew will just have to wait at Deception Island until *William Scoresby* appears, and it will become their floating base courtesy of the fact that, with its crane, it is capable of lifting the planes from the deck to the ice and back when the time comes.

With the arrival of *William Scoresby*, the Antarctic expedition continues, as much as it can, given the warmth that makes all plans problematic. Wilkins has arranged for the British steamer to carry a baby Austin car, so that they can experience the experiment of driving in the Antarctic; and the sailors and very curious penguins are treated to the bizarre sight of Wilkins ferrying petrol and supplies across Deception Island by motor vehicle. Wilkins had driven a Ford across Australia and may well have driven the Austin across the only continent more southern than that; but again the ice refuses to play along. Solid ice becomes slush in hours, and 'the car was soon down to axles in watery slush'.[14]

The same thing happens to their planes, which have to be unloaded at speed and then dragged to more solid ice quickly after landing.

Things are so absurd that the men unloading the planes start to break through the ice, which seems to bend like rubber with each step taken. Wilkins despairingly notes: 'By noon the temperature had risen to fifty-one degrees, the highest ever recorded in the Antarctic, to our knowledge. Even the penguins stood around panting.'[15]

Well, there is no way around it. Just as Mohammed had to go to the mountain, so too must the captain of *William Scoresby* follow Wilkins' orders and take his vessel further south than any ship has ever been, in search of ice solid enough to be a runway for taking off and landing.

Wilkins and Cheesman make some exploratory flights, which are hampered by storms, but are still able to add another 300 miles to the existing Antarctic map. More British flags are dropped just on principle, but with the weather warming there is little hope of any grand fresh discovery. The irony is apparent: three years earlier, Wilkins and Eielson proved that a plane could land and take off on polar ice, but now there is no ice for Wilkins to use their discovery to make more discoveries. With the official warm season approaching there is little to do but to turn and head for home. They shall wait just a few days to see if the weather shall turn. But while they wait, Wilkins' world is upended.

•

It is a cold morning in mid-November when the news first comes. A station operator from San Francisco compiles a news brief every morning that is handed on from station to station until the missive reaches Wilkins waiting at Deception Island. Expecting to hear of Wall Street collapses or political fray; the first American headline is simply 'Ben Eielson is lost in the Arctic'.[16] *What? How? When?*

Bit by bit, dribs and drabs of devastating news come in.

Ironically, Ben and his mechanic had disappeared while flying a rescue mission in a lashing Siberian storm while trying to evacuate sailors and fur traders caught in a cargo vessel, *Nanuk*, which had become trapped in the ice.

At Wilkins' urgent request, the light daily briefing now becomes an hourly update, day and night. He is sure Ben will emerge, as he so often has before. He must. After all, Wilkins had *taught* his dear friend how to live on the ice! He is sure that if Ben survived the plane coming down (and Lord knows he has seen Ben make landings that would have killed any other), he could do the rest and walk his way out.

But the news is there is no news – and this is *not* good news. Half a world away, the press come to speak to Ole Eielson, Ben's father. It is an awkward thing, to ask a father if he thinks his son is dead, but Ole Eielson is dwelling in a twilight world, where his son is present and gone at once. Why, he intends to look for the boy himself, north and south, if . . . if he is not found. And if he is found, and found dead, well, he *lived*. Ole Eielson speaks now, his tenses mingling as he grows more tense; a eulogy and an exultation fighting through his words: 'I think young men should get out and see the world. That's what my boys have done, and I'm proud of them. I believe Ben has contributed to aviation with his work in the Arctic and the Antarctic, and if he must be sacrificed I am a proud and a grieved father.'[17]

At last, there is news, but it is tragic.

Ben's plane has been found smashed on the Siberian coast, on the Bering Strait. His battered body is found several days later, alongside that of his mechanic Earl Borland.

Wilkins is profoundly moved by Ben's death. All those miles they had flown together, all those risks they had taken, the narrow escapes. And now this.

> Eielson and I together had seen, in the North and the South, more than half a million square miles of the earth's surface that no human eyes had ever seen . . . None of these would have been known so soon but for Ben Eielson's superb mastery of his job, his cool head, and his gallant spirit . . . He was a sincerely modest man who did his work for the work's sake entirely, but it made him the hero of hundreds of thousands of boys and won him the respect and admiration of older people everywhere in the world . . . His death was a great loss to aviation and to everyone who knew him. It was a great loss to me.[18]

They are two men bound together by their achievement, their friendship and by history. Now Wilkins is a man alone once more, the survivor of one of the most remarkable teams in the history of exploration.

Wilkins leaves immediately for the funeral, half a world away, to pay his profound respects, and console Ben's proud and grieved father – knowing that, if he pushes it, he might get back from Antarctica before they get Ben's corpse back from Siberia.

It will prove to be a close-run thing.

In the final days of March 1930, in Hatton, North Dakota, Ben's body lies under two American flags surrounded by a profusion of flowers. He lies watched over by a Guard of Honour made up by the North Dakota National Guardsmen. On the morn he will be buried next to his mother, and tonight his father mourns, bidding his son farewell before he is swallowed by the crowds and the ceremony that will come with the sun.

Just about 800 souls dwell in Hatton. Today, 10,000 more join them. The *Grand Forks Herald* reports on the sight that no local has ever seen, to farewell a local the like of which none will see again.

> From dawn until 2 p.m., the funeral hour, automobiles and carriages streamed into Hatton over the country roads. Special trains from Fargo and Grand Forks brought additional thousands, and all day passenger airplanes plied between Hatton and Grand Forks, swelling the great total.[19]

The weather is bleak, the snow blows in flurries along the prairie, the wind bites and cuts, but still they come. The crowd is eerily quiet, as the gathering grows to watch a procession that will only last half a mile, from Ole Eielson's beautiful home to the cemetery. But, oh, what a journey to arrive at journey's end: 'to traverse that short span it took him 32 years and a million miles of travel in the far corners of the earth. From the stately house that once heard his boyish shouts to the little graveyard where his mother sleeps, some 10,000 friends and neighbors today watched him ride in glory to his grave, laden with the honors and tributes of a sorrowing state.'[20]

Irony of ironies, there is one who will not be watching him today because his plane has been delayed in Detroit: Sir Hubert Wilkins. It is the sort of twist that would make Ben grin; together they conquered the Arctic and Antarctic, but there *was* always a little trouble when it came to Detroit.

In Hatton, the crowd surrounds the church, a congregation that cannot be contained. The roof of every building in town hosts spectators, young boys climb the trees of the churchyard. They hear the last note of the organ die out, and the dead hero is borne to the family plot, on the south edge of their cemetery, in their little town that now swells with a silent world of mourners. Ashes to ashes, dust to dust . . .

The next day a single figure stands at the grave. It is Wilkins, placing a wreath and saying a quiet prayer. Inevitably, a journalist approaches for comment, and Wilkins' words show the depth of his feeling as he struggles to maintain his composure.

'I cannot say enough about him,' says Wilkins as he chokes into silence. *What did you think of him? prompts the Press.*

'I thought of him . . .' Wilkins' voice shakes as he rallies. 'I thought of him as a friend and a man.' He can give no higher tribute. *But as a pilot, Sir Hubert?* 'He was a skillful flyer and the best pal a man ever had.'[21] *Unfortunate that you could not attend the funeral itself?* Wilkins agrees, but the weather that delayed him brought Ben's memory closer still.

'The storms in Chicago and Detroit reminded me of the flights Ben and I made in the Arctic. It made it just a little harder to think of them burying him out here.'[22]

And here Ben is, in body. In spirit, he is where he should be, in the air, beyond, somewhere. Sir Hubert takes a long loping walk half a mile up the street, to visit Ben's home for the first and last time; to shake the hand of Ole Eielson and to talk of the son and brother they have lost but will never forget.

•

Wilkins announces that he is finished with exploration by flight. But there is one duty left to perform; thanks to Wilkins' instructions, in Seattle the fuselage and engine of a plane are taken out of storage and transported by train to Bismark, North Dakota. For there a museum room is to be dedicated to Carl Benjamin Eielson, pilot and explorer. What remains of *Alaskan*, the Fokker that first carried him and Wilkins into the Arctic, will rest there forever.

•

Married to Sir Hubert Wilkins, Suzanne must become used to the unusual, the exotic and the extraordinary – starting with their delayed honeymoon. Not for them, a couple of weeks in a nice cabin at Niagara Falls, as might be expected for a couple of their ilk. Oh no, they begin by taking a vacation with Hubert's good friend George Bernard Shaw on the ship *Europa*, bound from New York to Southampton. That is until William Randolph Hearst calls to ask if they would mind changing their honeymoon to another luxury trip on *Graf Zeppelin*, soon leaving from New York to go to Friedrichshafen in Germany. Mr Shaw graciously accepts their apologies, Mr Hearst pick up their bills and they are now honeymooning correspondents.

'We will spend a month in Switzerland,' Sir Hubert tells the press. 'And then I will work on a book about my four expeditions.'[23]

Making her way on board behind all the porters carrying her luggage, Lady Wilkins carries a small black cat, with white paws and whiskers, which, she says, will bring her 'good luck'[24] – a curious move when, by most measures, the man sharing her bed is already the luckiest man alive.

Once again stepping on to blessed firm ground in Lenzburg, Switzerland, Suzanne and Wilkins are greeted by a young fellow in the curiously formal wear of a three-piece suit, a pinched Windsor knot and a boater. Wilkins recognises him straight away. It is the enigmatic American multi-millionaire and would-be explorer Lincoln Ellsworth, who had achieved a small measure of fame in 1925 by pouring no less than $100,000 into the failed attempt by Amundsen to fly a plane to the North Pole. He had been stuck on the ice with the Norwegian before they had managed to build themselves a runway to fly out. Ellsworth had also been a passenger on Amundsen's successful attempt to fly to the Pole, on the airship *Norge*, which had so thrilled Wilkins when it passed over him at Point Barrow.

And yes, Mr Ellsworth, what can I do for you?

Well, there is something he would like to discuss. What?

Suzanne doesn't have to even ask. Just one look at Ellsworth caught up in Wilkins' spell, and she recognises the obvious: clearly there is another expedition in the offing. This time however, it is not just your run-of-the-mill staggering adventure that is being mooted.

For Ellsworth proposes, and Wilkins quickly agrees, to attempt to take a submarine beneath the North Pole! It is a story that creates a sensation, a Jules Verne tale brought to life and combined with the enticing mystery of the Poles that still fascinate the world. The notion of a submarine expedition under the ice seems so preposterous that, from the start, Wilkins finds himself under suspicion of staging some sort of grand publicity stunt rather than embarking on a serious exploration.

'Really it is not as dangerous as it appears,' Wilkins assures the public. 'I have been considering the difficulties for fifteen years. It is no wild idea. I have carefully examined all the aspects and am confident I can accomplish there much scientific work only to be achieved by means of a submarine. I will be in London in April 1930 ready to tackle the Arctic.'[25]

Even at the fabled Explorers Club, situated at 544 Cathedral Parkway, New York City – where the good and the great of the day in the field of expanding the frontiers of the known world, men like Roald Amundsen and Teddy Roosevelt in the past, gather to carouse – he is greeted with extreme scepticism.

'Wilkins, you're daft,' he will chronicle the broad thrust of the members' remarks. 'In the first place your men will freeze to death, or you'll run into an iceberg and knock a hole in the vessel. It will be dark beneath the ice and you'll never find your way. The ice is hundreds of feet thick and too deep in the water to let a submarine get by, and you won't be able to come up for air.'[26]

'Not quite so fast,' Wilkins answers. 'You say the men will freeze to death. Remember that the submarine travels in water and so long as the medium through which you travel is water, how can it be colder than water? So how would we freeze to death?'[27]

Wilkins effortlessly parries similar scepticism about negotiating the icebergs and supposedly impenetrable Arctic ice packs.

'Who said it's hundreds of feet thick? All Arctic travelers – and I happen to be one – have reported that away from the coast the ice is rarely more than ten or twelve feet thick.'[28]

Another objection is posited: 'But you will get lost. It will be so dark that you won't be able to navigate.'[29]

Nup. Nup. Nup.

Wilkins has thought his way through that one, too.

'Surely, if I could find Spitsbergen from an aeroplane after travelling 2200 miles over an unmapped ocean at a hundred miles an hour, I will be able to find my way if I am travelling at less than ten miles an hour and coming up often to fix my position.'[30]

'But how are you going to come up to the surface?'

'There are open leads, thousands of them, but where we find none I plan to drill a hole up through the ice and put a pipe up to bring in fresh air, or even allow the men to get to the surface if the pipe is big enough. Men drill thousands of feet down to get oil; surely we can drill fifteen feet up through the ice.'[31]

He makes a strong case that, in theory, he has an answer for all problems presented – which is particularly important when it comes to getting funding from backers who ask similar questions. It is, after all, no small thing to be able to secure their funding while answering their queries and assuring them they are not financing a catastrophe that will see men drowned like rats in a tin can and their names forever associated with failure.

Wonderfully, Lincoln Ellsworth guarantees money, while Randolph Hearst promises support and President Herbert Hoover agrees to help things along by 'selling' the expedition a retired submarine for the sum of precisely $1 a year, a generous lease to say the least.

After fitting the sub out to purpose, Wilkins' plan is to broadly retrace his flight across the roof of the world, doing the 2000 miles from Point Barrow to Spitsbergen, except this time they would go *under* the ice.

The whole thing should take about 30 days, but they will take provisions for a year, on the chance of being caught somewhere, in which case – in the grand Wilkins tradition – the survivors will walk out!

There are, however, many problems to solve before beginning such a trip, starting with the state of the sub itself. With one look, Wilkins does not say it was overpriced at $1, but it is certainly obvious why the US Government was happy to hand it over. Constructed in 1917 by the most famous submarine maker of his day, the New Jersey native Simon Lake, this *Nautilus* – a 175 feet long steel tube, just under 16 feet at its widest point, and 18 feet at its highest, with a flat top – *nominally* has a capability to submerge for just under 125 miles at a time, proceeding at a rate of three knots, for 24 hours. But could this shambles of a sub actually do that in reality? (This sub even has permanent port-holes, which, while useful for observations, just makes the whole thing *feel* vulnerable. Still, any port-hole in a storm.)

To say that Wilkins is unimpressed with the new *Nautilus* is to say that as a body of water the Atlantic is on the large side of things. Although in public he is careful to only purr about the advantages and improvements that repair will bring, privately he is shocked and pessimistic.

'I was appalled at the amount of work which would have to be done to put the boat in shape for the seas . . . We were smothered in grease and oil and I almost decided then and there that the task of refitting the [USS O-12 Class submarine] was beyond anyone's power.'[32]

And he is not the only one. The last chief mechanic to work on it, from the US Navy, had left them a note in the engine room which was nothing if not on point: 'TO AVOID TROUBLE YOU MUST SCRAP THIS ENGINEROOM MACHINERY AND INSTAL A NEW SINGLE UNIT.'[33]

And yet, just as he had done with Cope, the fact that Wilkins has already given his word that the trip will be done with the vessel specified, means he cannot even consider pulling out: 'People thought it would take great courage to go underneath in a submarine but I can assure you that it took more courage and even more than I had to go back on our word. We had put our hand to the plough, it was necessary to go on.'[34]

And if it doesn't work? If it can't cross under the entire Pole?

Well, he can hopefully demonstrate that it is at least possible, and with that demonstration get more funding and a better vessel next time, perhaps even with the backing of the US Navy. And so they set to work.

With the aid of Simon Lake and his partner Sloan Danenhower – who will be the Master and Commander once they go under the polar ice cap – they begin the modifications, starting with installing new engines and batteries and attaching on top, 'a toboggan-like super-structure which will enable it to slide along beneath the underside of the ice cap under its own power'.[35]

They also must mount pneumatic drills, at the head of the conning tower, able to penetrate through as much as 13 feet of solid ice – and if that fails, they also carry explosives to break up the ice – and hopefully not themselves – in an emergency.

Upon consideration, Wilkins changes directions too, deciding – on the reckoning that Europe will provide better facilities to get them underway than Alaska – to start from Spitsbergen and head west, instead of making the Norwegian island their destination.

They will take the sub from New York to London on the surface, before heading up to Norway, where they will submerge before the polar ice cap, and eventually surface in the water off Alaska, finally finishing their journey by travelling down the west coast of the USA and taking berth at San Francisco.

'Of the crew of eighteen men, twelve will operate the boat and six will be engaged in scientific research,' the press report. The course, via the Pole to the Bering Sea, is 2100 miles. The submarine will travel slowly, about 50 miles a day, and at the end of each 50-mile section it will stop for eight hours for scientific study. The entire voyage is estimated to take 42 days. This will allow a few days' halt at the North Pole. At every stop Sir Hubert will broadcast messages to the world via an antenna poked through a hole drilled in the ice.

And this will not be the usual manner of submarine travel, moving through the water freely. Instead . . .

'The *Nautilus* is to make use of the advantages of travel on the ocean's bottom without undergoing the dangers – by travelling, instead, on the underside of the surface ice. Lighter than water, it will press against the ice with long runners.'[36]

While under the cap they intend to take constant readings on such things as the depth of the ocean and content of its floor, the direction and force of the current, the animal and mineral content of water and magnetic variations.

In the meantime, the man that Wilkins has chosen to skipper *Nautilus*, Commander Sloan Danenhower – the son of polar explorer John Danenhower, and an ex-US Navy submarine officer into the bargain – feels no compunction to be upbeat with the press, and says flatly: 'Should we fail over any extended period to get air, even by using our emergency drills and dynamite, our batteries would lie dead and useless, we could not move, we could not breathe, we should die. Not a nice picture!'[37]

Quite.

As negative as Danenhower is, however, he is at least matched by the *Washington Post* which, after sober reflection, denounces Wilkins for taking on such risks: 'The chances of meeting with disaster are incalculable. Everyone admires Sir Hubert's dauntless spirit, but the popular opinion of his judgement just now is far below par. It is no small responsibility to imperil the lives of 22 men on what appears to be a hopeless quest.'[38]

Another who feels the same, when he finds out the details of the plan, is none other than Wilkins' old friend, Frank Hurley, now a highly esteemed documentary maker.

'The general public, I am sure,' Hurley will say, 'does not appreciate the ghastly risks of a voyage like that . . . in hourly danger of striking a rock pinnacle in those uncharted seas, on which his craft would . . . split. In the Arctic Ocean, the birthplace of pack ice, he stood the chance at any time of colliding with the root of an iceberg. The roots of giant 'bergs go down 500, 600, and even 1000 feet in those frozen seas. Never for a moment was Wilkins free from the sickening suspense of collision with the stalagmites or the stalactites of the Arctic seas – to say nothing of the menace of unpredicted (and unpredictable) trouble with the mechanism and the fabric of his submarine. Furthermore, there were the problems of coming to the surface through ice 30 or 40 feet thick, which would require for shattering such heavy detonations as to endanger the comparatively frail shell of the submarine. In short, no one who does not know from experience the Polar regions can begin to envisage the appalling array of perils and difficulties with which Wilkins was confronted on this fantastic journey. Not for any consideration would I have taken on the *Nautilus* cruise and faced the perils, known and unknowable, against which Wilkins pitted his brains and his matchless courage.'[39]

At least *The New York Times* still has faith in his star: 'It seems foolish to talk of luck in the case of such a man . . . He has been taking chances all his life. Fate has let him alone, as if he were invulnerable.'[40]

Yes, but could fate now be sorely tempted to knock off one who has so long denied it, never more vulnerable than now – heading under the Arctic ice cap in an unseaworthy submarine – and that includes his time on the Western Front where his daily survival was a miracle?

In the meantime, Hearst now outlines the detail of *his* funding. Wilkins will receive the colossal sum of $150,000 for his daily reports on the voyage, transmitted over the radio, with the only complication being that the money will only be handed over *after* success is achieved. Very well then. In a keenly innovative step of his own, Wilkins decides to write the book about the expedition before the expedition even starts!

Under the North Pole is duly published, replete with essays by Wilkins and other adventurers on the possibilities and likelihoods of the voyage. The proceeds of the book help give Wilkins the wherewithal to continue hiring his preferred team for the adventure, including the distinguished Norwegian Professor Harald Sverdrup, who will serve as chief scientist for *Nautilus*, a coup that quells sceptical voices that have been muttering that this is not a serious venture. Renowned in his field of meteorology, Professor Sverdrup is extremely serious, and even severe, a man who has spent no less than seven years of exploration and investigation with Amundsen. He is a little like Stefansson in his combination of academic and survival skills, and his hiring is a coup. Most importantly, he agrees with Wilkins' notion that a submarine is the ideal way to solve the mysteries of Arctic drift. (Relics from the American ship *Jeannette*, which had sunk off northern Siberia in 1881, had turned up three years later off the south-west coast of Greenland, which had seen Norwegian scientist Dr Henrik Mohn publish his view that there was an east–west current across the entire Arctic Ocean. This was very difficult to prove atop the ice pack but would be a simple matter to prove under it.)

Typically, Lincoln Ellsworth hands his own funds over to Wilkins, by actually giving him a giant cheque for $125,000, as the cameras click – just one more, Mr Ellsworth, and please smile, Sir Hubert – in a press conference in New York. He also gives him a good luck token, a shoe from Wyatt Earp's horse! Now, given that Earp had been a gun-slinger on both sides of the law, and had died only months before at the age of 80, it really would have to be lucky, wouldn't it?

The gift seems a perfect fit.

'I like Wilkins,' Ellsworth tells the press, 'because he is of that virile pioneer type of the old American West that I've always so greatly admired. He's a man of great faith, courage and simplicity, who says little and does much. That is why I like him and joined forces with him.'

All up, it is a good bit of publicity, though it does not prevent acerbic comments from some quarters which note that, for all his public bravado, Ellsworth will not be travelling on *Nautilus*.

Sir Hubert does not care either way and, typically, demonstrates nothing but confidence.

'The submarine will be tested in March and April,' he tells the press, 'and will leave in May for Spitsbergen by way of England and Norway. The expedition will cost about $250,000 . . .'[41]

A good portion of that, Wilkins soon finds to his distress, is intended by Simon Lake to go towards installing his own expensive and unproven inventions, rather than specifically making the whole sub seaworthy, which seems to the Australian to be the *sine qua non* of the whole exercise. Compounding the frustration is that, as the work begins in Philadelphia Naval Yard, run by the US Government, every non-improvement is performed with exquisite attention to entirely useless detail – proceeding at the pace of bureaucracy, not business. At Wilkins' insistence the craft is moved further down the Delaware River to a non-Government dock in New Jersey, which has the added advantage of being very inconvenient for Simon Lake to get to. The rate of work on things that truly count increases commensurately, with one improvement being the nose of *Nautilus* is filled with concrete so that if there is a heavy collision with the ice, it will be the sub that wins.

Wilkins dubs the re-kitted *Nautilus*, 'a marvel of inconvenience'.[42] It is still a marvel in other ways though: 175 feet long, 560 tons, 'drawing her power from huge Exide batteries big enough to supply the entire electricity load of a small city'.[43] Twin 5000-watt headlights are also installed to light up the underside of the ice, which will then be viewed by another of Lake's inventions, a 'collapsible' periscope. (This periscope has indeed collapsed, just not in a good way. Still, Lake assures Wilkins, it will . . . be all right on the polar night.)

While Lake continues with all the modifications, Wilkins continues to conduct interviews with potential crew members. For the most part they are a rough and ready lot. If you are a mariner by trade, the general rule is you have to be a desperate one to agree to go on an old, experimental submarine beneath the polar ice cap. At least let it be said though, most of them are under no illusions when it comes to understanding the risks they are facing.

One gloomy fellow offers to bring his own enclosed steel cylinder device in which he will live and all he asks is that Wilkins give him two years' worth of food.

And what would be the purpose of this, Wilkins asks, not unreasonably. 'When you are lost with the *Nautilus* I will drift with the currents until I am free from the ice and am picked up by a whaler. In that way I will be able to tell the world what happened.'[44]

'Next!'

What about Mickey Mouse?

Entranced by the whole project, no less than Walt Disney offers the services of Mickey Mouse; sending Wilkins a cartoon of Mickey 'waving hello from the conning tower of a watermelon-shaped midget submarine'.[45]

Genuinely charmed, Wilkins nevertheless decides this is not a Mickey Mouse operation.

Others are signed to the crew as the week goes on, and if it is true that some of the newly signed are distinctly underwhelmed once they see the sub itself, and do a little quiet grumbling about it, that does not hold true for all.

One who will not hear any such downbeat talk is the man Wilkins has selected as his quartermaster, Willard Grimmer, a very upbeat man from Philadelphia, who had been in the US Navy as an electrician and radio expert, and was now trying his luck in this new and fascinating venture. Most excitingly for him, in the time between signing the contract with Wilkins and actually departing, he had met a Pennsylvanian beauty, Miss Mary Fountain, and married her just a week later. Life has never been better.

'She is a lucky ship,' he tells his ship-mates. 'The *Nautilus* has brought me luck in the last month; a chance to meet my wife and a chance to take one of the greatest trips ever planned, something that happens once in two or three generations.'[46]

Onwards!

Wilkins continues to raise funds by lecturing, even as William Randolph Hearst puts *another* $150,000 on the table, if *Nautilus* can possibly meet up with *Graf Zeppelin* at the North Pole and exchange both crew members and mail! (Wilkins is publicly grateful while privately recognising it for what it is, just another Hearst publicity stunt.)

Alas, publicity of an entirely different kind attends the whole project.

Once the modifications are finally finished in Philadelphia, it is time to get *Nautilus* to their starting point of New York where, after conducting some sea trials to confirm she is not only ship-shape, but sub-shape to go beneath the ice, they are towed by a tug up the coast.

Entering New York Harbor, much of the crew are up by the conning tower to take in the sights, when from seemingly out of nowhere a freak wave breaks over the bow and hits quartermaster Willard Grimmer, who is giving his weighty Arctic kit, including his heavy boots, a cautious first trial!

With a cry, Grimmer loses his footing and is swept sideways until – *man overboard!* – he goes over the side. The cries go up, and when those on deck look over the side they are just in time to see poor Grimmer wildly thrashing about in the Hudson, before disappearing from view. A life-belt is thrown in his general direction, but there is no sign that he gets to it.

At 2.47 pm the distress signal is received by the Duty Officer at the Brooklyn Navy Yard: 'We have a man overboard off Swinburne Island. Request police plane immediately.'[47]

Amid much shouting and feverish activity the tug and the submarine immediately detach so that both vessels can look for Grimmer in the immediate area, but there is no sign. Neither the US Navy plane alerted by the police nor the Coast Guard have any more luck, and it is quickly apparent that Grimmer's new bride is now a widow, though she doesn't know it yet.

The episode casts a pall over the entire expedition. If this is what can happen even in a safe harbour, what is it going to be like at the ends of the earth, beneath the polar ice cap?

All they can do for the moment, however, is to soldier on and the following day, even as *Nautilus* is tied up at Coast Guard Base 2, pier 18, Staten Island, a memorial service is conducted for poor Grimmer. All of them, with their hats removed and their heads bowed, stand glumly as Lieutenant Commander Sloan Danenhower takes them through prayers and hymns before providing a brief eulogy.

All of them are affected, but none more than Wilkins, who is inevitably reminded of another young man, also tangentially involved in one of his projects, who also came to an early tragic death just as the whole thing was about to start. For yes, of course the Australian can't help but think of poor Palmer Hutchinson, the spirited reporter who was cut to shreds by the propeller of *Detroiter* before she even took to the air. Now, poor Grimmer, taken by the water before *Nautilus* has even taken to the sea. Palmer Hutchinson, Walter Grimmer – they are two footnotes to history that are always noted in Wilkins' mind, haunting him as the only two men ever lost under his command. So far.

Despite the pall of gloom that settles over the sub and on the spirits of all who are to sail within her, just two days later, on 24 March 1931 it is time for *Nautilus* to be christened beneath the Brooklyn Bridge, and before a large crowd of the public and various dignitaries including Rear Admiral William H. Phelps, Commander G.W. Simpson and Mr Lincoln Ellsworth. No less than the grandson of Jules Verne, Jean Jules Verne, is on hand to exult, exalt and laud them: 'I wish to express my admiration for you in attempting this feat. It almost surpasses anything my grandfather dreamed of. I am sure you will succeed.'[48]

Given that the famed author of such masterpieces as *Around the World in Eighty Days* had dreamed of such things as journeying to the centre of the world, travelling to the moon and how to float cities in the sky, this is no small claim, and Wilkins is uncharacteristically moved. Next, he invites Simon Lake to the podium to say a few words, only to find the inarticulate and overwhelmed inventor cannot manage even that.

It will be up to Lady Wilkins to restore order to the proceedings and a touch of Broadway by observing the ancient tradition of offering a sacrifice to the Gods as the ship is launched. Five millennia ago this had invariably involved the likes of the Babylonians slaughtering oxen. These days it has become breaking a bottle of champagne against the hull, with the added belief that if the bottle doesn't break – as was said to have happened with *Titanic* – it brings bad luck. The problem on this day is that as New York, with the rest of the USA, is under the strict laws of Prohibition, there is no actual champagne in the bottle but . . . never mind.

Lady Wilkins speaks up as she addresses both the submarine and the crowd.

'Ship,' she begins in portentous tones, 'I name you *Nautilus*. Go on your wonderful adventure. In your heart is great treasure. Bring that treasure safely back to me.'[49]

Sir Hubert smiles grimly.

He certainly hopes to come back, but is not 100 per cent positive that his wife is quite as keen on that as she says. (Like all marriages, the Wilkins' one has its moments, but the two participants also have *long* periods when they are apart. Later, when asked by his old friend Stef why at his age he continued to go on such risky ventures, he would say ruefully, 'The one time I *stopped* exploring I married a chorus girl.'[50] Ah.)

In any case, whatever the coming risks, the explorers will remain safe for the next few weeks as more frantic repairs are conducted to correct the *last* frantic repairs to make *Nautilus* seaworthy, let alone Pole-worthy.

But perhaps they should have stuck some real champagne in the bottle after all? For on a bad day, and there are many, it really can feel like the whole project is cursed, and that those who have called them the 'Suicide Club' are right.

On their first deep dive off the coast of Connecticut, just after leaving New York, going down is not the problem. It is getting back up again, as *Nautilus* goes straight to the bottom and rests there, 80 yards beneath the surface. Fortunately, they have taken the precaution of having a specifically designed submarine rescue craft, *The Falcon*, topside and are in instant communication with them. On that vessel the discussion is focused on possible rescue methods, from dispatching a deep-sea diver to place small explosives beneath the keel – which should get them moving in some fashion – to perhaps releasing the pressurised chamber, when from their radio they hear a crew member below yell, 'My God! Look at the depth gauge!'[51]

With a mind of its own, almost like a sick whale knowing it needs some fresh air, *Nautilus* surfaces of its own accord. The culprit for the sudden dive is eventually discovered, some rubbish has blocked a vent valve in a ballast tank, but this hardly inspires confidence in any aboard. Some members of the Suicide Club are clearly already eyeing the exits. At least things are patched up enough that they are able to set off across the Atlantic on 4 June 1931, enabling Wilkins to write the first of his daily 1000-word articles for the Hearst newspapers on the day's events, or at least the events he is happy for the world to know about. But even the surface trip to England is far from smooth, with the seas stormy, the ventilation appalling and most of the crew sick – not from *mal de mer*, but because those who had tried to freshen up the water tanks before departure had used a heavy dose of fresh lead paint!

Ah, but still we haven't got to the bad news: that comes when, right in the middle of the Atlantic, the generators flood with seawater, which seeps into the engine and is fortunately detected before a cylinder can break through the hull, which would have sunk them entirely. As Wilkins records: 'It would doubtless have meant the end of the *Nautilus* and all her crew. There were no lifeboats carried . . .'[52]

Limping its way forward, the submarine continues its hellish progress, as storms continue to lash it. Why not escape to the calm beneath the surface?

Simply because no-one, and least of all the Commander and Wilkins, trusts this sub underwater. It's like having a horse you don't want to make gallop, owning a dog you're afraid might bark. And so they stay, shaking and bobbing on the surface, which means, as Wilkins records, that on one day all 18 men vomit in the enclosed space. Open the vents then?

Not in these storms. Too risky.

But why is everything suddenly so much quieter?

That's because the port engine has stopped, for reasons best known to itself. Hoping that conditions will improve and that they can fix the engine, Commander Danenhower refuses to so humiliate them all by ordering an SOS, which would reveal their pathetic predicament and powerlessness. Nevertheless, after three straight days of no progress at all, either in fixing the engine or in proceeding across the Atlantic, with a sigh he orders Ray Meyers, the radio operator, to send out an SOS.

'Are you a spiritualist?' asks Meyers.

'What's that got to do with it?' replies the Commander.

'That's the only way I can think of to get a message off.'[53]

The radio has given up the ghost, too.

Christ Almighty!

Are they really in the middle of the Atlantic, cooped up in a tin can filled with the stench of fumes and human vomit, with everyone also suffering from food poisoning, and with seemingly no hope of rescue? It is the classic case of the old definition of bad sea-sickness where for the first 12 hours you are afraid you are going to die, and then for the next 12 hours you're afraid you're *not* going to die.

At last however, after a solid 18 hours of tinkering, Meyers thinks he might have got the radio working, and got a message through! There is no certainty, but there is hope, and it is better than nothing.

The best thing? In the wee hours of that night, sitting in their black tin can with no lights – for no-one can bear more fumes from running the generators to power the batteries – an eerie light suddenly plays through their port-holes!

Looking out, they see a blinding light playing upon them, the search-light of a ship coming to their rescue!

It proves to be the American battleship *Wyoming*, on which, in an extraordinary coincidence, Ray Meyers has previously spent five years of service. Greetings are exchanged, lines are thrown and attached, and they are towed across to the nearest port, which is Ireland's Cork.

Wilkins thanks the men for putting up with 'sustained and trying conditions'[54] and asks them all to say nothing to the press, please. Have no fear, repairs will be made, *Nautilus* will be made right. Or righter. And even though their vessel has badly broken down, on the surface, in benign conditions, it will soon be ship-shape to go under the polar ice cap.

Such blithe assurances are met with the kind of silence that wavers between 'gloomy' and 'bristling'.

Though making their entrance by being towed ignominiously into Cork – arriving on 22 June 1931 – the church bells play 'The Star-Spangled Banner' in their honour and, as it happens, some of the crew quickly become so spangled in the port's many bars they are unable to stand up for three days.

Wilkins, dressed for the occasion – putting on a suit, a tie, and a brave face – fronts the press and predicts a quick fortnight of repairs, while waxing both hot and cold on their experiences so far.

'The sustained worry of the voyage,' he says frankly, 'was the worst thing I have known. It was the most thrilling Atlantic crossing of modern times.'[55]

But, not to worry.

'Our experiences have no bearing on our getting to the Arctic. The *Nautilus* is a fine sea boat sure, and we have the best vessel to carry our scientific equipment to the Pole. We expect to reach it before September.'[56]

So, despite the near-disaster, and the obvious inadequacies of their vessel, it is all still going ahead then?

Some drink up even more than before.

For his part Ray Meyers makes a deliberate trip to kiss the Blarney Stone and wish for their luck at sea to change.

After a month of repairs, they sail across the Irish Sea and around to Plymouth for ... more repairs ... (The Admiralty have offered to repair *Nautilus* there rather than at Newcastle, given the odds against an unrepaired *Nautilus* making it to Newcastle. In fact, with one look, the Royal Navy decides the submarine must have new engines.) It is while at Plymouth's Devonport dockyard they receive a surprise visitor. It is no less than the Prince of Wales! The future King of England is of course given an impromptu tour of the vessel, insisting on going into such dark nooks that when he re-emerges the press notes that his once smart-looking attire is now the worse for wear and his hands are greasy. No matter, each crew member is given a blackened Royal handshake of farewell, and wished the best of British luck.

'I expect,' the Prince says, shaking Sir Hubert's hand last of all, 'to hear great things from you.'[57]

Wilkins, a knight of the realm after all, is thrilled with the visit, but more than a little embarrassed at the filthy conditions the future sovereign has seen up close. Things are only middling, and on the eve of departure from Britain, Wilkins gives a candid interview to the *News Chronicle* where for once he is a little less than upbeat, and all he can manage is to sound hopeful.

> We've certainly had a lot of bad luck, but maybe now we've had all that was coming to us and it was certainly better to have it here than in the Polar seas. Due to the lateness of the season we shall be confronted with greater difficulties than we hoped for and perhaps this year's effort will not be the best criterion of the usefulness of submarines in Arctic waters. It is naturally a disappointment that we cannot expect to carry out the voyage completely across the top of the world but from the beginning of our plans it has been the opinion of our scientific associates that a journey confined to Arctic waters between Spitsbergen and the pole is of greater scientific value than anything we could achieve in a hurried dash from side to side.[58]

In an expansive mood Wilkins also displays to the same journal his growing interest in things . . . telepathic.

'Eventually,' he tells the journalist, 'we shan't need to write at all. We shall broadcast straight from the mind. There is one broadcast station – I don't know which, but I think it is German, because they use German signal signs – which I can hear in my head, without being near the receiving apparatus. We shall know some day that there is nothing supernatural, everything is capable of a scientific explanation.'[59]

If only they could come up with a scientific explanation for why the US Government has given them such a hopeless vessel!

For while the redoubtable Randolph Hearst might remain keen on selling the wonder of going from one continent to another under the polar ice cap, Wilkins has come to an obvious conclusion – this sub is sub-par not sub-*par-excellence*, and they must tailor their ambitions to suit their capacities. But if they can't go all the way under the ice cap, Hearst would rather they abandon the whole thing altogether.

Nautilus leaves Plymouth on 27 July 1931 and takes four days to reach Norway. As they leave Bergen bound for Spitsbergen, 1250 miles to the north, Meyers receives a cable from Hearst himself:

NOTIFY WILKINS THAT I'M VIGOROUSLY OPPOSED TO HIS TRYING TO
REACH POLE IN PRESENT NAUTILUS WHICH HAS PROVED UNSEAWORTHY
AND DANGEROUSLY UNRELIABLE EVERY WAY STOP URGE POSTPONEMENT
TRIP MEANWHILE SECURING BEST POSSIBLE SUBMARINE AND MAKING
ADEQUATE TEST SIGNED HEARST. [60]

Wilkins replies equally frankly:

WILL PROCEED NORTH FROM SPITSBERGEN FAR AS CONDITIONS
AND BEHAVIOUR OF THE VESSEL PERMITS IF VESSEL FUNCTIONS
AS REQUIRED MIGHT REACH POLE TWO WEEKS AFTER LEAVING
SPITSBERGEN STOP CONFIDENTIAL HAVE LITTLE CONFIDENCE IN
THIS VESSEL AND ITS OPERATIONS OF SUBMARINE BUT HOPEFULLY
IT WILL PROVE BETTER THAN NOW APPEARS WILKINS. [61]

For make no mistake, Mr Hearst, you are a key sponsor of this venture, but not the one running the show. I, Sir Hubert Wilkins, am running it.

Onwards!

Nautilus continues towards the polar cap, battling stormy seas and its own poor ventilation. The ship's electrician, Arthur Blumberg, regards the fact that they are all still able to breathe as a minor miracle. 'The ventilation system was outrageous. Why we didn't now perish from suffocation will remain a mystery to me forever. Somehow or other no one had thought of an adequate ventilation scheme or at least they had failed to perceive the uselessness of the one that we had.'[62]

The other issue is the danger of suffocating in debt.

When they reach Spitsbergen – the very scene of his previous triumph – Wilkins is expecting to receive a further payment from Hearst, but there is no sign of it. A furious flurry of cables reveals that Mr Hearst's editorial managers have formed the view that Wilkins has breached, and what is worse is continuing to breach, the terms of their contract. For despite his promises, his commercial commitment: he is clearly *not* going to go beneath the entire Arctic and come out the other side – which is the very thing that would drive their readership – instead he is doing no more than going on a sort of 'limited scientific cruise'.[63]

Who cares?

Wilkins does.

And he is sure the Hearst papers will get their money's worth if they continue to back him. He says as much in a final cable before departure, saying that he expects Hearst to keep his word in regard to funds. And yet, in the meantime, the Hearst view does propel Wilkins to go further than he otherwise might have, as he now gives the word to Commander Danenhower and crew: do a final check, batten down the hatches, and get ready – we will go beneath the Pole.

The news is *not* well received.

Wilkins' Log of 23 August 1931 stoically records what happens next: 'All set for diving as soon as conditions improved when Danenhower, looking through clear water over the stern, noticed something missing from behind the propellers.'[64]

Oh.

Christ.

With considerable care, *Nautilus* is put in the lee of an iceberg, to give it as much protection as possible while an inspection is done. No matter that the hour is late and the darkness falling, this cannot wait, and within 20 minutes, a brave volunteer, Frank Crilley – the Second Officer of *Nautilus* – is in his heavy diving suit with brass helmet connected to a hose to the surface that will allow him to breathe, and with a rope around his belly, being lowered over the side with the deep admiration of all.

(Oh, the horror! The *cold!*)

As soon as Crilley disappears from view, Wilkins – always calm when *he* is the one in extreme danger – nervously paces back and forth along the deck, smoking furiously. He happens to be on the outer lap when, minutes later, Crilley re-surfaces and is quickly hauled to the surface, as the crew rush to take his helmet off. His teeth chattering with the shattering cold, he gets the news to Ray Meyers, who – stunned – breaks the news to Professor Sverdrup.

'Doc, we've lost the rudders,'[65] he says, so softly that the Professor has to ask him to clarify.

'Have we lost the *rudders*?' says Sverdrup, not quite believing his ears.

Meyers' face gives him the answer before his mouth does.

'Yes, it's gone, we can't dive – we can't get under the ice.'[66]

Both men pause before Meyers gets to the worst part.

'I have to tell Wilkins.'

And here he is now, notable for the fact that he is the only man on deck with no protection from the cold for his hands.

The Australian looks at both men, his eyebrows raised. What news?

'Captain, we've lost the diving rudders.'

'It isn't possible!'[67] replies Wilkins. But sadly, it not only is possible, it has happened. The rudders are *gone*, Meyers confirms it once more. Both men watch Wilkins to see how he will react to the news that the very things that control the depth of the submarine once it gets underwater are gone.

The answer is, the same way he always reacts to *triumphs and disasters*. He treats *those two impostors just the same*.

As Sverdrup notes with wonder, 'Not one muscle indicated that all of his plans had collapsed.'[68]

If you can make one heap of all your winnings/ And risk it on one turn of pitch-and-toss/ And lose, and start again at your beginnings/ And never breathe a word about your loss . . .

Wilkins closely questions Crilley and establishes that at the junction point of where the rudders used to join the main body of the submarine, it looked like someone had been at it with a hacksaw, so smooth is the surface – and they had correctly counted on the fact that once the base of the diving rudders were weakened then the drag of the water would eventually do the rest.

Very well, then. Wilkins puts his mittens on and goes to pace astern, before retreating to his cabin to calmly note the event in his journal, beginning by commending Crilley for his valour.

'He was down for several minutes then reported that both sides diving rudder had gone. But steering rudder intact and in good order. Is mystery what became of diving rudders and how they could have left ship without damaging steering rudder. Fact remains they are gone.'[69]

To Wilkins, it is absolutely clear that sabotage has occurred. It is obvious that at least one of the crew, or several of them, have nobbled the rudders while doing the final checks in Spitsbergen.

Elsewhere on the sub there is deep satisfaction. Rather than stay loyal to Sir Hubert and become part of a gloriously doomed history, they would prefer to *live*, obscurely, on the surface, and so had nobbled the sub in such a manner that getting under the ice now will be impossible.

So they thought.

But didn't Wilkins just tell you?

That was not disaster. It was an impostor. And he will not be defeated by lowly saboteurs in the crew! (His chief suspects are John Janson, the Second Mate from Chicago, and a carpenter named Harry Zoeller.)

And *Nautilus* will dive anyway.

Those are his orders and he will not even be dissuaded when, later in the day, it is discovered that a cell in the after battery[70] has also cracked. 'Gas fumes filled the ship . . .'[71]

Things are grim and getting grimmer.

Surely, Wilkins doesn't still want to dive?

He does!

To submerge safely without a rudder is impossible, but as impossibilities are Wilkins' personal specialty, he does not give up.

'I can, I believe, safely state,' Scientific Officer Sverdrup will recall, 'that on 22 August everyone on board except Wilkins would have been willing to return acknowledging complete defeat, but he did not for one moment consider the possibility of returning.'[72]

The mood in response turns so very nearly mutinous that Wilkins feels obliged to call a meeting of all hands.

'It was found that, without exception, the others in the vessel wanted to immediately turn back; to make no further attempt to go into the ice this year.'[73]

Their views are dutifully noted.

But Wilkins never has a moment's doubt on two things: firstly, this is *not* a democracy, and secondly, they will be diving anyway.

After all . . .

'To [not dive] would have admitted complete failure. As commander of the expedition I ordered the trials to continue.'[74]

Still they steam north to the polar ice cap.

Still, the mutinous rumblings do not abate.

Is Wilkins stark, staring mad?

Does he really want to drown them all?

> 28 August 1931. The men argue that as we no longer have a 'submarine' [owing to loss of horizontal rudder] we should give up the effort and return. But . . . I am determined the vessel will go under and carry out as many experiments as possible beneath [the ice].[75]

Yes, going all the way under the polar ice cap and coming out the other side is now out of the question, but just going under a short way to prove it could be done will be victory enough. Without their horizontal rudder, going deep will simply not be possible, but at least through a combination of nudging the ice pack to force the nose of their sub under, and adjusting the ballast tanks enough to allow them to descend a little, they should be able to get well under the ice.

That's the plan?

That's the plan.

As if things are not gloomy enough, the weather has turned Shakespearean – steeped in *Macbeth* with a torrent of *The Tempest* thrown in – reflecting the bitter mood of all. It is bad enough to give cause for pause even to the perpetually upbeat Wilkins: 'The sun shines low and coldly upon the horizon. High wind from the east continues. The seas are lumpy and the high wind whips the tops of the waves into spray . . . On the deck it is miserably cold and when the hatch is opened, air rushed in and creates a biting blast in the control room and galley.'[76]

Conditions below deck are so abominable even a snowman would shiver.

> Men sleep in their Worumba, a camel hair sleeping bag and some of them have dragged two extra blankets over them as well. We are colder in the ship than I have been in a snow house when the temperature outside was 50 below zero.[77]

Under such intolerable conditions, as mutinous rumblings grow, it is clear that if they are going to go under the ice cap it must be done soon – and there is more to it than merely proving it could be done.

'In the first place,' Sverdrup will recount, '[I wanted] to test whether the submarine could navigate safely under the ice, whether the lanes and openings in the ice could be seen from underneath, and whether the submarine could reach the surface in such openings. In the second place, we had to learn whether our scientific program could be carried out under the unfavorable conditions onboard a submarine, and especially if the oceanographic work could be undertaken from the diving compartment.'[78]

And so, after consultation with Commander Danenhower, on 22 August, right beside a large ice floe, all hands are on deck and at their stations, all keenly aware of just what is at stake.

'It was a risky thing to do,' one of the crew will recount, 'but we consented because we didn't want to be accused of cowardice when back in port again. Before we took the dive Sir Hubert Wilkins sent a wireless message to Spitsbergen giving our position.'[79]

This is it.

'Flood the main ballast!' orders Danenhower with an authority that nevertheless betrays his tension.

The air in the ballast tanks is vented, and seawater floods into the forward tanks, making them heavier, and tilted downwards, even as the propellers turn – powered now by batteries only, as the diesel engines underwater would consume all their oxygen – they push towards the ice ahead.

There is a jolt of contact, a crunch as the top of *Nautilus* starts sliding on the underside of the ice.

'It was a veritable drum or sound box,' Wilkins will report, 'with the faintest scratch of the ice sounding like the ripping of giant strips of calico. Heavy bumps set up tremors like the continuous shocks of earthquakes.'[80]

And now every eye watches the water line rise, over the port-holes. But will they ever see daylight again? 'And what a water line! It was dotted closely with ice cakes and floes of fantastic shape and size, delineating every conceivable form – minarets, domes, majestic cathedrals, and geometrical blocks. I could see a beautiful colouring in crystal clear blue water.'[81]

The ship settles down – ever further down, down, down – into the water, not yet 20,000 leagues under the sea, as Jules Verne had imagined his own *Nautilus*, but getting there . . .

'Beneath us the bottom lay two thousand fathoms down.'[82]

It is hard to conceive that this is actually happening.

'Ahead!' says Danenhower, and all aboard fall silent. They approach the floe that Wilkins has designated, they can see the ice looming towards them, beckoning.

'Crunch! Bang! Crack! She was under!'[83]

'Stop the motor!'[84] orders Danenhower. And now all is completely silent; they are *beneath* the polar ice, gliding, the first men ever to do such an extraordinary thing.

> What a beautiful sight greeted our gaze through the eye-ports![85]
> As far as one could see through the crystal blue water stretched the roughened, flat undersurface of the floe with occasional small upside-down peaks and valleys. There was a great variety of colouring as the light changes with the passage of clouds above the floe no human eyes had ever before looked on this site.[86]

The vision is a strange combination of beauty and menace, as they see 'steel-like fangs of ice moving stealthily through the water, which changes in color . . . throughout the entire range of blues'.[87]

And look at the sea-life!

All around them are, 'prawn-like and cockroach-shaped creatures, sea fleas, medusa jelly fish'.[88]

It is remarkable, but all have the same thought.

'We were under all right for the first time in history,' Wilkins will speak for them all, 'but could we get out again? The ice was too thick to think of breaking through it.'[89]

Indeed, the real worry is that the ice might break through the vessel, as the previous sound of tearing calico becomes so much more ominous, replaced with a grating noise which penetrates the ship, and the skull of every man within it. 'The noise of the ice scaping along the top of the vessel was terrifying,' Wilkins records. 'It sounded as though the whole superstructure was being demolished.'[90]

Are they to be like *Titanic*, opened up like a tin can by a single shard of cruel ice, sending them straight to the bottom – the first ship sunk by the iceberg *above* them?

Meyers repeats prayers to himself, every and any prayer he can think of, starting with the completely inappropriate 'Now I Lay Me Down To Sleep'. The recent diver, Crilley, announces one other important piece of equipment is now missing: 'a spare set of rosary beads – I just pulled mine apart!'[91]

They continue for 10 minutes – even managing to serve the cause of science as, through compression chambers, they take samples of water, and measure temperatures, current direction and strength – until at last the moment the crew has been waiting for comes.

'Full astern!' Danenhower says sternly.

After the tiniest of pauses, the propellers reverse, as does the pulse of every man on board, and the submarine, slowly and steadily, rises.

Wilkins carefully logs each movement:

> 6½ degrees by the bow and 30 foot depth. Then five degrees and twenty-eight feet, four and twenty-six, three and twenty-four, two and twenty, and suddenly brilliant sunlight flooded in through the dripping ports and we were up clear of the ice.[92]

They have done it!

A roar erupts from the men as not only is the extraordinary feat completed but, even better, it looks like they will live!

Several more brief dives are conducted before Wilkins insists on chronicling a dive by setting up on a nearby ice floe with Emile Dored, the cameraman from the cinema chain Pathé they have with them. Just to be on the safe side – on the chance *Nautilus* won't come back up and they are left on their own – they take with them 'some provisions, a small wireless-set, firearms and ammunition'.[93]

It is an odd thing for the two of them to be on their own in this manner – 'far enough off from the *Nautilus* to get some good photographs of her lying among the ice cakes like a peculiar kind of hump-back whale'[94] – but not altogether uncomfortable.

If it comes to it, Wilkins really does feel they could find a way to survive, and make their way back to civilisation.

Mercifully, there is no such disaster and *Nautilus* soon surfaces and they are able to board it without incident.

Now to tell the world of their triumph! Alas, in going beneath the ice the radio has been damaged and, as far as the world is concerned, *Nautilus* has gone ominously silent. Their last communication had been when they were about to go under the ice and now . . . nothing. After two days, Norway sends ships and planes to the area, fearing the worst.

At least Lady Wilkins, in London, maintains faith.

'I have a hunch,' she tells the press after six days of nothing, 'tonight will bring good news.'[95]

And she is right!

For that night Ray Meyers gets the radio working again, and Wilkins is able to get a message through the Norwegian Meteorological Radio Service that all is well. Before long they are receiving cables of congratulations, together with one unctuous plea, cabled from Mr Hearst:

```
DEAR SIR HUBERT I AM EXCEEDINGLY HAPPY TO HEAR GOOD NEWS
FROM YOU . . . I URGENTLY BEG YOU TO RETURN PROMPTLY TO
SAFETY AND TO DEFER ANY FURTHER ADVENTURE UNTIL ANOTHER
MORE FAVOURABLE TIME AND WITH A BETTER BOAT . . . WE ARE ALL
ALARMED ABOUT THE DANGERS YOU ARE INCURRING WHICH SEEMS TO
US NEEDLESS. WILL YOU NOT PLEASE COME BACK NOW AND DEVOTE
YOUR ENERGIES TO PREPARATIONS FOR ANOTHER EXPEDITION AT A
BETTER TIME AND IN A BETTER VESSEL[96]
```

As that is already Wilkins' intent, he cables back his agreement, and *Nautilus* continues to head south to the Norwegian mainland, an incredible feat achieved and a firm point made:

There was no doubt in my mind that ours was but the first of a great submarine fleet that would one day cruise at will beneath the Arctic ice cap, the shortest distance between the great American and Eurasian land masses.[97]

On 8 September 1931 *Nautilus* is secured alongside the pier at Longyearbyen, Spitsbergen, thus ending the expedition – and the last time that Sir Hubert Wilkins will ever command an expedition. His extraordinary career as an adventurer in the public eye is at an end.

If you can fill the unforgiving minute
With sixty seconds' worth of distance run,
Yours is the Earth and everything that's in it,
And – which is more – you'll be a Man, my son!

EPILOGUE

If down here I chance to die,
Solemnly I beg you take
All that is left of 'I'
To the Hills for old sake's sake,
Pack me very thoroughly
In the ice that used to slake

Rudyard Kipling, 'Ballade of Burial'

All of this brings us back to the question posed by Stefansson to Monash in 1924.

How is it that such a man as this is all but unknown in his own country? Apposite at the time, it is even more extraordinary now. Stefansson posed that question even before Wilkins had pulled off his greatest feats, by flying over the roof of the world, and then being 30 years before his time in taking a submarine beneath the ice.

While his accomplishments were at least the equal of such contemporaries as Sir Charles Kingsford Smith, Sir Douglas Mawson and Sir John Monash – all three of whom were honoured by featuring on Australian banknotes – those men remain iconic figures decades after their deaths, while Wilkins is the most forgotten famous man the country has produced.

The answer is multifaceted, and goes well beyond Wilkins' refusal to blow his own trumpet – in strict contrast to the *modus operandi* of most explorers, many of whom have gone well beyond mere trumpets and employed a brass band. For one thing, as opposed to a dramatic death like many living on the edge – see Kingsford Smith and Burke and Wills – after his own heroic days were over, Wilkins lived long and strong, if sometimes mysteriously, leaving deliberately little trace of what he did.

His final expedition has left him in debt, owing a lot of money to various creditors, as Randolph Hearst definitively turned off the taps, including the money he had promised. Nevertheless, still living out Bean's observation that Wilkins was the embodiment of Kipling's poem *If*, the South Australian's response resonated with the stanza: '*If you can make one heap of all your winnings/And risk it on one turn of pitch-and-toss,/And lose, and start again at your beginnings,/And never breathe a word about your loss.*' That is, he just gets on with it, focusing first on ensuring that all the crew who had been with him on the *Nautilus* had their outstanding wages fully paid. True, such personal penury, together with the remains of the Great Depression, making money very difficult to obtain from other sources, prevented all chance of his mounting another expedition on his own account. But at least by turning himself into one of the busiest and highest-paid lecturers in the USA, and through guiding others in their feats – most notably acting as a well-paid master-mind and hired hand for one of his creditors, Lincoln Ellsworth, as the American millionaire made another four attempts at polar flight glory in the Antarctic, trying to cross the entire continent in one go – he first staves off bankruptcy and then starts to restore his finances.

Given how concerned Ellsworth was, however, about Wilkins taking any of the spotlight from him, he insisted that while Wilkins was welcome to be second-in-command, chief of the base party and special press corre-spondent he was *not* welcome as a participant on the flights.

In 1937 Wilkins was asked by the Soviets to search for the lost Soviet flyer Sigizmund Levanevsky and his five comrades, who had been lost attempting to fly over the North Pole. While the Soviets themselves searched their own side of the North Pole, Wilkins led the attempt on the Alaskan side of the Arctic. Despite the outstanding effort, searching 170,000 square miles of the Arctic Ocean, the most extensive and sustained air searches ever undertaken up to that time, a good deal of it by moonlight, and all of it in winter, which had never been tried before – alas, there was not the slightest sign of them and they were never seen again. Nevertheless, to recognise his prodigious efforts, Wilkins was invited to Moscow to have the Order of Lenin bestowed upon him by the Soviets. On his return, a curious Suzanne asked how the country was compared to the young Soviet Empire he first saw back in 1922–1923.

'Lots of caviar and vodka, but far too many photographs of Stalin,'[1] Wilkins replied. Speaking of which, Stalin sent Lady Wilkins a personal photo, bordered in sumptuous red linen.

'I used it,' she would gleefully recount, 'as a marker in a book on how to invest stocks.'[2]

It was at this time that – perhaps feeling there was little left to explore in the visible world – his focus began to turn to the unseen world, that of the mind and spirit. While searching for the Soviet flyers, Wilkins had also been engaging in what were essentially experiments in mental telepathy, or 'thought transmissions' as he and his collaborator, one Harold Sherman, called them. Every day at a given time, Wilkins would intensely concentrate on what had happened that day, and in New York Sherman would record these thoughts. After all, the advent of radio had accomplished what no-one had ever thought possible, the transmission of a voice from one part of the planet to another. So perhaps the same could be done with thoughts?

To the discomfort of the Australian's friends and his reputation as a scientist, a book jointly authored by the two would actually be published of the result: *Thoughts through Space: A Remarkable Adventure in the Realm of the Mind.*

In his introduction to a later edition of the book, Wilkins would profess to be more than pleased with the results.

> ... after rereading the manuscript, it seems to me that while the tale is told in straightforward simplicity, there lies within it a rousing, staggering indication of the potential powers of the mind, and a signpost toward not necessarily the use of the mind for the purpose of thought transference, but toward the potential of mind development and the wider and better uses to which it may be put.[3]

He would also write: 'Man has found the tools with which to handle the intricacies of nuclear physics; perhaps the basic principles of the mind are more evasive and intricately involved because they are not entirely physical, and the mind of man has not yet come to accept the spiritual as a field offering desired reward for research capital invested. But doubtless a time will come when a verdict will announce, "It has now been proven that thought communication, without mechanical aid, is available to those whose minds are trained and whose will is directed toward the betterment of mankind."'[4]

The book itself, however, was far from convincing, with Wilkins, for example, chronicling his amazement that Sherman could provide 'a complete description of our plane, which he had never seen'.[5] This, despite the fact that photos of the newly celebrated Lockheed 10E were all through the New York newspapers at the time!

Yes, there were critics, but Wilkins was unmoved. 'We may not have proved that telepathy between two people at some distance apart is beyond doubt,' Wilkins wrote, 'but I was personally pleased to have been engaged in the experiment, and feel that we have proved that the subject is entirely worthy of much further attention.'[6]

Meantime, as late as 1938, Wilkins was still actively planning a return submarine expedition to the Arctic, as Lady Wilkins stated in a charming and revealing interview with the *Australian Women's Weekly* conducted in the Monterey Hotel in Sydney on a flying visit, where she quoted her husband affirming that while her desires to accompany him were fine, it altered not a jot the reality, which was his firm reply.

'It would not be right for the leader of an expedition to take his wife. A leader must look after the whole party, not just one person.'[7]

The *Women's Weekly* detailed the couple's plans, which would see them return to Australia in a couple of years to set up their base. From there, Sir Hubert intended to keep travelling for another year or two, whereupon he would 'sit down and write and think about it'.[8]

In the meantime, though, things are busy.

> Sir Hubert is off already. Started appointments at breakfast and will be busy every minute of the time in Sydney, Melbourne and Adelaide. From there we go to South Africa, where he sets off for the Antarctic and I return to New York. Then I shall wait, all excitement, for next summer's submarine trip to the North Pole. Do you know, sometimes I feel as if I am not married at all. How would you like to be married to a man for nine years and only have seen him for four months?[9]

The danger of a return venture was too much for any sponsor or government to risk. The second submarine trip did not eventuate and Wilkins' interests then took a turn even more curious than his embrace of the concept of telepathy.

Once a Methodist, he became a proponent of a new movement whose effective bible would become *The Urantia Book*, a vast collection of 'divine revelations' purportedly transmitted by a sleeping man in Chicago, while under the care of a surgeon and self-taught psychiatrist, Dr William S. Sadler, who, with others started to take down every word. (I know.) The thrust of *Urantia* sought to combine religion, science and philosophy in the one belief structure. 'Until the last minute before rushing to catch a train to New York,' Wilkins would write of it to a friend, 'I was reading and just managed to get through the last chapter. The whole thing is so immense and astounding that one is lost for words to describe it. It is, however, likely to be subject to criticism by the established churches, both in respect to its origin and its revelations, which deprive some of the sects of the basis of their foundation. To me, however, it is the first connected, clear exposition of the faith – the need for faith and the rewards of faith, a combination of considerations which seem to confront everyone no matter what race or colour, and notwithstanding their stature of education.'[10]

Urantia resonated with Wilkins, and there was one quote in particular he kept coming back to: 'Man must learn to feast on uncertainty, fatten on disappointment, enthuse over apparent defeat, to invigorate in the presence of difficulties, to exhibit indomitable courage in the face of immensity, and exercise faith when confronted with the challenge of the inexplicable.'[11]

In the meantime, in that period where it even seemed likely that Wilkins might be home for months at a time, he and Suzanne bought their first and only property, a charming and rambling house on a hill in Pennsylvania that Lady Wilkins christened 'Walhalla', the name of the small town where she was born in Australia.

Here at last Wilkins was momentarily at peace, living the most domestic and calmest period of his life, most particularly when he would take his hearing aid out, which he pointedly did when Suzanne's friends came for long visits. One trip they took together was on the maiden voyage of *Hindenburg* across the Atlantic to New York, attended by much the same fanfare as when *Titanic* had left a quarter of a century earlier. There was relief, of course, when *Hindenburg* arrived safely. And yet, only a year later – '*Oh, the humanity!*' – it famously exploded while attempting to dock in New Jersey, with 36 lives lost. (Others might have marvelled at their narrow escape. Given Wilkins' life, it barely registered, with him only noting, 'when that tragedy did take place in 1937, I could not help thinking back to my carefree hours aboard her . . .'[12])

With the arrival of World War II the Australian again found his services called for, with constant requests seeking his advice on matters to do with flight, the Arctic, photography, topography and so many other matters that one day Suzanne snapped as the phone rang for the twentieth time that day and gave the caller a piece of her mind.

'If you value his opinion so much, why don't you stop picking his brains and give him a job instead?'[13]

'That's not a bad idea,'[14] came the reply.

Only a short time later Wilkins had a job in Washington as a 'consultant' in the mysteriously named 'Military Planning Division', which saw him vanish from their Pennsylvania home for weeks and sometimes months at a time, always returning with precisely *no* news.

'What have you been up to?' Suzanne asked him curiously upon his return from one trip.

'Can you keep a secret?' Wilkins replied conspiratorially.

Suzanne nodded.

'So can I,' her husband replied, returning to his work.[15]

One day in 1940 while reading the paper, Lady Wilkins was interested to learn that not only was her husband in France, but he had been in a plane shot down by the Germans.

Do tell?

Well, Sir Hubert won't.

Wilkins would only concede that it was 'a great lark',[16] while roaring with laughter at the fact that the downed plane had contained French officers fleeing with mountains of loot and silver, only to be stranded in a far-off field with no way to transport the booty. Wilkins, typically, having no concern for treasure, left them to it and set off to make his way out of occupied France all while the Germans were looking for just such people as him – without papers, and unable to speak French without an accent. How many men in their fifties could have managed such a thing? Very few. But for a man with Wilkins' comprehensive skills in living off the land as he had done all over the world, moving at night as he had done for months at a time in the Arctic, blending in as he had done so effortlessly in the early days of the Soviet Union, it meant that mere German check-points had proved no match for him and he had made his way back alone to unoccupied France, and from Nantes had hopped on an RAF plane back to England!

But what were you doing there in the first place, George?

Suzanne, I already told you. I can keep a secret.

Wilkins, in the guise of conducting 'economic surveys', also travelled to Japan and China.

At the Raffles Hotel in Singapore he met an old acquaintance, the Japanese Consul General, who had seen Wilkins speak on the *Graf Zeppelin*'s first landing in Japan. The Consul General has a few drinks with Wilkins, then a few more; then says something extraordinary: 'Eighteen months from now, around the middle of December 1941, the Japanese will attack Pearl Harbor.'[17]

'Why are you telling me this?' asks Wilkins.

Very simple. Because nobody will believe you if you tell them. You are, after all, the same man who thought he could take a submarine under the North Pole, and one and the same who believed he could send thought messages from the Arctic to New York. So yes, of course the Consul General doesn't mind telling him.

'Go back and make a report to your government and they'll laugh you out of the room . . . They'll think you're crazy.'[18]

(The author Simon Nasht has concluded that the Japanese man in question must have been Lieutenant Commander Ryunosuke Kusaka, who met Wilkins in these circumstances, and had written a proposal to bomb Pearl Harbor before *Graf Zeppelin* had even landed in Japan!)

One way or another the war kept Wilkins busier than ever, and he was home for as little as a fortnight a year, leaving him with even less time at home and hearth than when he was exploring the polar regions. Suzanne seemed to mind less and less, and made her own arrangements. Just before the war the two met a handsome singer and actor by the name of Winston Ross in a nightclub. He was 10 years younger than Suzanne but he soon moved into their home, nominally as Wilkins' 'personal assistant' and 'secretary', an odd role given that Wilkins was never home.

(I know. The neighbours talked, for good reason, and at some length . . . given that Ross remained in residence for the next 35 years.)

'See that Suzanne . . . stays out of trouble,' Wilkins told him before heading off on a trip to China in 1939.

'What are you going there for, a spy mission?' Ross asked.

'Well, it is information gathering,' replied Wilkins.

'What happens if some tough Burmese bandits catch you and stick you up against a wall?' joked Ross.

'It will be all right as long as it's a good stud wall,'[19] answered Wilkins drolly.

With which, the man that Suzanne referred to as 'my Hello, Goodbye Man' departed.

At the conclusion of the war, Sir Hubert stayed working for the US Government in various guises, including being the Geographer and Consultant in residence at the Quartermaster Research Command at Natick, west of Boston; an outpost of military research and experimentation. Every year for the next 16 years he journeyed to the Arctic to test newly invented equipment.

One freezing Massachusetts night, a constable on his beat noticed what he presumed was a hobo lying in thick snow. *Sigh*. Down to the lockup for you, old man, till you can sober up. Otherwise, you will freeze to death out here.

Oh. Oh, it is you, Sir Hubert, doing *what*, you say?

Testing your latest invention, a chicken feather sleeping bag, suitable for soldiers in the field and downed airmen?

Very well then. As you were.

Wilkins continued to live eccentrically, insisting on staying in an establishment near the central railway station that was more hovel than hotel and forever forgetting to cash his paychecks. But the old dog could still teach the young 'uns new tricks, and one who trained with him in Alaska would later recount his experience as he and other young bucks tried to keep up with Wilkins, now in his early sixties, on a trek: 'The crew were hand-picked military and civilian men, most of whom were on skis, he was on snow shoes. Not only did he stay well in advance of us, but following his tracks, we saw how he repeatedly went off the trail and returned, making side studies. To top off everything, he beat us back to the bivouac area . . . and had time to film the young soldiers straggling back in his wake.'[20]

On one memorable day, the young US personnel are gathered to attend a seminar on Arctic dress and survival with the grand old man, the only knight involved with the US armed forces, who they often address in a curious manner.

'Sir Wilkins,' the young questioner asks on this day, 'what is the most challenging aspect of cold weather suit design?'

Sir Hubert considers the question carefully then answers quickly: 'How to get a four inch penis through five inches of insulation and direct the flow *out* of the suit!'[21]

The men roar with laughter as Wilkins grins; but he is not joking!

Wilkins' work was significant. Brigadier-General C.G. Galloway would later write to the Australian Minister for External Affairs, Mr R.G. Casey (later the 16th Governor General of Australia, 1965–69), after a memorial ceremony during which a portrait of Wilkins was placed in the Army's climatological laboratory at Natick: 'Let me say that among all the consultants and advisors to the Quartermaster-General of the US Army, I know of no one who has made a more significant contribution to the improvement of clothing and equipment of Military Forces. Also, I would like to say that the friendly association of the United States and Australia has been signally advanced by the contribution Sir Hubert has made to our efforts.'[22]

There was one new military invention christened in 1951 that, not surprisingly, fascinated Wilkins most of all – the nuclear submarine. In terms of sophistication the first sub built was to *Nautilus* what a television set is to a tin can, but the US Navy still named it *Nautilus* in honour of the polar pioneer who had shown them the way – and Wilkins was heavily consulted about everything he had learnt in his submarine foray beneath the ice.

The next one built was *Skate* and, in 1958, both submarines proved that what Wilkins had claimed all along really was possible – they successfully made a crossing beneath the North Pole, moving as a duo.

Emerging triumphantly on the other side, the young Commander of *Skate*, James Calvert, sent Wilkins a message: 'The experiences of this summer followed by conversations with your old associates in Bergen have left us deeply aware of the accuracy of your insight and vision in regard to the use of submarines in the Arctic. The majority of your aims and predictions of nearly thirty years ago were realized this summer. The men of the *Skate* send a sincere salute to a man who has many times shown the way.'[23]

Deeply moved, Wilkins sent his reply: 'Sir Hubert Wilkins sincerely appreciates the message from the men of the *Skate* and extends to them hearty congratulations upon their skillful and efficient accomplishment of measures which no doubt will lead to new and far reaching developments in science, economics and defense.'[24]

After *Skate* returned to America, on 18 October 1958, the crew received a visit from Wilkins. Calvert felt almost embarrassed to show Wilkins the luxuries they had in comparison with the ship and conditions he'd suffered. But Wilkins was as stunned as he was delighted by the tour and by their incredible ship, as Commander Calvert would recount.

'Everything, which for him had proved bulky or useless, had, after twenty-seven years of intensive submarine development, either been replaced by something far better or improved beyond recognition.'[25]

The old knight paused, and Calvert waited for his words.

'Now that you have everything you need to do the job,' he finally said ruminatively, 'you must go in wintertime.'[26]

•

It was around this time that Wilkins appeared on an American game show *What's My Line?* where the panellists asked questions of an unknown guest to try and determine just which famous person he was. And those panellists do their best, plying the playful Wilkins with any number of probing questions, which he answers with no little amusement – as they get precisely nowhere.

When the guest is finally revealed as 'great polar explorer, Sir Hubert Wilkins'[27] the panellists applaud politely though it is not obvious that all can quite recall him – not that Sir Hubert himself looked remotely fussed.

Rather, the footage reveals a seemingly healthy and certainly charming man with warm and humorous eyes, who is, as ever, interested in advancing the cause of science. In recent times he has been helping NASA develop spacesuits for astronauts, advising on the practicalities of keeping a body warm in extreme cold.

He looks, frankly, if not like a man in the prime of his life, at least nowhere near being on his last limping legs.

And yet, only six weeks later, those eyes of his that had seen more fresh horizons than any human who had ever lived – before or since – closed for the last time.

Sir Hubert Wilkins was found dead in his hotel room, lying neatly on the floor, dressed in a suit and tie, his hands clasped over his heart, for all the world as if this world conqueror did not wish to cause anyone any inconvenience and was ready for a simple transfer to his coffin.

A small curiosity of that discovery was that it resembled a scenario for his own suicide that he had put in an odd and mostly fictional account based on his life experiences that he had penned 40 years earlier. In that account he had dressed up, and taken poison. In this case, to be sure, there was no solid evidence of any poison *per se*. But at least to Suzanne Bennett's lover, Winston Ross, things were odd enough that he would later chronicle in a letter to the CIA, 'I accompanied Lady Wilkins to Framingham, Mass., where her husband was found dead in his hotel room, the death certificate less than convincing, the doctor not to be consulted, the funeral home edgy and secretive, and a definite sense of the F.B.I's presence.'[28]

I know. Curiouser and curiouser.

And curiouser still . . .

Lady Wilkins will even go public with her theories, telling the press the US Army had prevented her seeing the body for two days, she was convinced her husband had been poisoned, and had asked the FBI to investigate.

> Lady Wilkins told Australian Associated Press that a friend told her . . . that a man named Emmanuel Lampel had been saying that Sir Hubert was poisoned by Russian agents because they knew 'Hubert had a terrific brain'. She said Lampel, an electronics executive, told her friend he knew of the poisoning because he did a great deal of work for the United States Government.[29]

When the said Emmanuel Lampel is contacted, however, he denies all knowledge and says he is not an electronics executive and, what is more, he had never talked to Lady Wilkins.

To be clear, the strong likelihood is that Wilkins died of natural causes, the Grim Reaper finally bringing him down at a time when he was *not* expecting it – but I do find the circumstance and that letter more than interesting.

Either way, so highly regarded was he by those who knew most intimately of his work that there was even talk that Wilkins would be buried in Westminster Abbey or that a memorial to him would be placed there.[30]

Alas nothing came of that and at his memorial service in St Andrew's Church, Framingham, Massachusetts, a young American flyer to whom the dearly departed had once lent a polar parka recalled that the widow Wilkins had disgraced herself.

> Later, after Sir Hubert died, I went to his memorial service, intending to give the parka to Lady Wilkins. But she turned up drunk with her lover. It's no secret there was no love lost between Sir Hubert and Suzanne. Anyway, I decided not to give her the parka and donated it to the Christchurch Museum.[31]

Shortly thereafter, Lady Wilkins received a condolence letter from Stefansson offering his own reason why his old friend and fellow explorer had never received the fame and kudos that was his natural due.

'Wilkins had but one weakness that militates seriously against the permanence of his fame,' the Icelandic–American wrote. 'He wrote too little for publications, too often turning from one successful job to the next without taking time to explain in books what he had done and why . . .'[32]

Inevitably Wilkins' death saw an increased interest in someone writing a book that would do justice to chronicling his extraordinary life. And yet, there were major difficulties for all who ventured there.

All of his papers, diaries, letters, newspaper cuttings, mementos and artefacts – which filled a good chunk of the home *and* barn in Pennsylvania – were of course inherited by Suzanne and, as they say in the classics, that's where the trouble started . . .

To go through such material was an overwhelming task; and it overwhelmed Suzanne. What it needed was a devoted writer to go through it, and when that opportunity seemed to arrive in the person of an author, aviation administrator and long-distance pilot by the name of John Grierson – a friend of both George Bernard Shaw and Charles Lindbergh – well, Suzanne quickly agreed the Englishman could have access to everything she had in order to write a biography of Wilkins. The problem came a few months later, when the widow Wilkins decided that she did *not* like the impertinent nature of some of Grierson's questions, *especially* when it came to other women who had been in Sir Hubert's life. She certainly did not wish to discuss Lorna Maitland, and as a matter of fact was not happy there was any documentation concerning Maitland in there at all.

Grierson's access was immediately blocked, various papers were wantonly destroyed if they mentioned any old love, and the baton was passed to the famed US journalist Lowell Thomas, who had known Wilkins well and had first begun writing a book on him in the early 1930s. Thomas had even travelled with Wilkins to the North Pole the year before he died, which had quickened his interest in reviving the book. Thomas, too, however, fell out with Suzanne, after he unwisely publicly told one of Wilkins' own jokes – the one about the reason he kept exploring was because the one time he stopped he married a chorus girl – which occasioned some furious correspondence between the two. One highlight of their exchanges was Thomas opining that she should not be mad at him, as he remained a fan of hers.

After all, how could he not be, when, 'I'm not allergic to the beautiful and well-stacked girls who are successful on the stage.'[33]

A what?

A *what?*

Lady Wilkins was, thank you very much, a leading actress, not a mere 'chorus girl'[34], and she will thank Thomas to get that into his head – and here are five more angry letters to help you, starting with this one . . .

> *Dear Lowell,*
> *I was intrigued to hear through the grapevine that Hubert had married a chorus girl at some time he wasn't busy exploring. This is a side of the great man's life I haven't been acquainted with at all . . .*
> *Several members who were at the dinner rang up to ask me if Hubert had been married before, though not one was so uninformed about my career in the theatre as to suppose I was the individual employed to produce the laugh the remark was intended to achieve . . .*[35]

Lady Bracknell could take lessons from Lady Wilkins about how to bite a man's head off with half a sentence.

The doors were snapped shut on the Wilkins' estate and papers until Suzanne's own death in December 1974, whereupon all the boxes of documentation were inherited by . . . Winston Ross . . . who was also overwhelmed.

There is just so much of it! Ross's thoughts are clearly expressed in a note he wrote on the back of one of hundreds of photographs of Suzanne, now also placed in boxes: 'I am deluged with duty – should I bother? All rooms up hill and down dale are filled. Does posterity care? Junk sells at yards and we have historical junk.'[36]

The people who most cared at the time, bizarrely enough, were stamp collectors. For Wilkins had, as part of his fundraising, taken thousands of 'covers' (a stamped envelope) to Antarctica, where they were then 'cancelled' at postal stations at the literal end of the earth – thus creating that rarest of things, an envelope bearing the postmark of Antarctica. Hundreds of stamp collectors had paid up to two dollars for these envelopes in the 1920s but there were still hundreds and hundreds of the envelopes left. Ever practical, Wilkins used the envelopes to store his mass of correspondence from the twentieth century. In 1980, in a chance conversation with a philatelist, Winston Ross came to realise the value of these peculiar envelopes with their odd stamps and postmarks and, in short order, the correspondence of Wilkins was being sold and dispersed around the world to people who had no interest in the precious contents and only cared for the odd packaging they came in!

In 1985 the University of Ohio, realising just what a treasure trove was being dispersed, inquired about purchasing some of Wilkins' papers. For the sum of $125,000 Ross was happy to part with most of what he still had on hand. (The remainder was split between quarrelling heirs – mostly descended from Ross himself – who joined snippy stamp collectors in the endless and enduring argument over who exactly owned what of Sir Hubert Wilkins' possessions.)

It got worse still. In 1998, after Winston Ross's own death, hundreds of boxes of Wilkins' life were sold by Ross's son as curios and trinkets, for a few dollars a pop. Those boxes that weren't sold were taken – I weep – to the dump.

And yet, despite the scattering to the winds of so much of the detail of Sir Hubert's life, the last three decades or so have seen things starting to fall into place for him to regain his proper place in Australian history – right at the forefront.

In 1993, the Australian businessman and historical aviation enthusiast to beat them all, Dick Smith, visited Winston Ross at the old Wilkins' home in Pennsylvania and was able to get some insights into the marriage of Wilkins and Suzanne Bennett. Smith's work in the field gave him ever more understanding both of what Wilkins had done and how staggering it was that his achievements were not better known.

In 2001, Wilkins' old farmhouse home near Hallett in South Australia was painstakingly and lovingly restored.

The growing interest in Wilkins in recent years has been in no small part powered by dedicated researchers finding ever more pieces of his extraordinary jigsaw puzzle and starting to put it all back together. My own favourite of the genre concerns the writer Jeffrey Maynard who, a decade ago, discovered among Wilkins' collection of books, *Volume*

XII Photographic Record of the War of Bean's *Official History of Australia in the War of 1914–1918*, the one that Wilkins had contributed so many photographs to from the front lines of the Western Front.

In 1923, Bean had written to Wilkins telling him that the book had been published. There was no copy to go with the letter, and Wilkins had bought his own. And what does Maynard discover, carefully going through it? That Wilkins had gone through it himself, lightly marking with pencil the initials GHW on those photos he had taken. It had only been for his own satisfaction, and he had never protested that he had not been sufficiently credited – and he had no way of knowing that one day a dedicated researcher and writer like Maynard would shine a light on such fascinating proof of Wilkins' extraordinary humility, and ability, for the photos are stunning. And yes, there is something wonderfully moving about one of Wilkins' own countrymen, 60 years after his death, journeying to the other side of the world to learn more about him, to pull that book down off the shelf in the dusty barn, and put one more piece of the scattered jigsaw puzzle in place.

Kudos, too, to Simon Nasht, not just for his wonderful biography of Wilkins, but for his outstanding ABC film, *The Voyage of the Nautilus*.

One of the other forces that has played towards Wilkins' re-emergence has been the rising crisis of climate change, which has seen huge swathes of the scientific world now focused on the thing that was concerning him more than a century ago: the impact of such things as devastating droughts, the interconnectedness of weather and the importance of measuring it across the globe – and in his case in the polar regions – in order to understand it.

Of those weather stations that he had long been advocating in Antarctica, no fewer than 20 were established in his lifetime, and there are now double that, all of them minutely recording the changing weather. For the Arctic it is more problematic, as there is no land to build on near the North Pole, but there are many stations within the Arctic Circle regions of Russia, Canada and Scandinavia. The data collected by such stations does indeed help predict future weather patterns such as *El Niño* and *La Niña*, though weather stations in other parts of the world also help with the predictions.

All up, could there be any better site for Wilkins' final resting place than in those polar extremes?

Gather close, for here now is the final part of his story, his final polar adventure, this one posthumous.

For on 17 March 1959, just as the long Arctic night is starting to ebb, right at the top of the world, that twilight world of white is suddenly shattered as – *rrrrammming* speed – a massive steel tube splits it all asunder, as shards of ice are thrown hither and thither. It is, of course, the nuclear submarine USS *Skate*, with Commander James Calvert at the helm, and soon emerging from the conning tower. It is not just that in every direction he sees a storm that is significant. It is that every direction from here is south.

They have emerged precisely at the North Pole – the first submarine to do so – and are here on a mission.

Sir Hubert Wilkins is on board, at least in a manner, his ashes brought out to the surface in a small bronze urn by Commander Calvert. As the crew of *Skate* assemble on the ice and bow their heads, Calvert reads these whispering words to the wind:

'On this day we pay humble tribute to one of the great men of our century. His indomitable will, his adventurous spirit, his simplicity and his courage have all set high marks for those of us who follow him. He spent his life in the noblest of callings, the attempt to broaden the horizons of the minds of men.'[37]

Ashes to ashes, Pole to Pole, Wilkins' remains are thrown to the air as an ensign fires his rifle three times in a naval salute.

The ashes vanish into the snow, enveloped into the endless white of the Arctic, his final home.

Farewell, George.

What a man he was, an extraordinary, unique man who through hard work, intellectual curiosity, an innate confidence and desire to learn new skills went from humble beginnings out the back o' Bourke, to the world itself being not just his oyster but his to explore, by land, air, sea and under-the-sea. Could there be any better place for his ashes to be scattered than the top of the world?

He filled every unforgiving minute with sixty seconds' worth of distance run, squeezed every day for every hour he could get out of it and seemed to most enjoy leap years for the fact they gave him an extra day to go on his adventures. His survival against all odds, time and time again, marks him as one of the luckiest, bravest and most extraordinarily active men in history.

And he was one of ours – a great Australian who made his mark in half-a-dozen fields, saw it all, did it all, met them all, chronicled the never-before seen, pushed back the frontiers of science and made the world he was born to a much smaller place.

He was, the Incredible Wilkins.

Vale.

ENDNOTES

1 Wilkins, Papers, Ohio State University, Box 1, Folder 14, p. 31 [reported speech].

Introduction

1 *Time* magazine, Vol. 186, No. 5, 2015, p. 11.

Prologue

1 *The Herald*, 12 June 1924, p. 5.
2 *Adelaide News*, 24 June 1924, p. 4.
3 *Adelaide News*, 24 June 1924, p. 4.
4 Maynard, *The Unseen Anzac*, Scribe, Victoria, 2017, p. 196.
5 Maynard, *The Unseen Anzac*, p. 196.
6 *The Advertiser*, 13 July 1932, p. 8.
7 *The Herald*, 12 June 1924, p. 5.
8 *Adelaide News*, 24 June 1924, p. 4.
9 Maynard, *The Unseen Anzac*, p. 191.
10 Maynard, *The Unseen Anzac*, p. 191.
11 *The Argus*, 12 July 1924, p. 6.
12 Maynard, *The Unseen Anzac*, p. 196.
13 *Adelaide News*, 24 June 1924, p. 4.
14 *Adelaide News*, 24 June 1924, p. 4.

Chapter 1

1 Wilkins, *Under the North Pole: The Wilkins-Ellsworth Submarine Expedition*, Brewer, Warren & Putnam, New York, 1931, p. 76.
2 Maynard, *Wings of Ice*, Thistle, London, 2015, p. 49.
3 Maynard, *Wings of Ice*, p. 49.
4 Maynard, *Wings of Ice*, p. 49.
5 Wilkins, Papers, Ohio State University, unpublished manuscript.
6 Wilkins, Papers, Ohio State University, Box 1.1, Folder 22, Family, ancestors and childhood, no page number.
7 Wilkins, unpublished manuscript.
8 Wilkins, unpublished manuscript.
9 Wilkins, unpublished manuscript.
10 Wilkins, unpublished manuscript.
11 Wilkins, unpublished manuscript.
12 Wilkins, Papers, Ohio State University, Box 1.1, Folder 22, p. 8.
13 Wilkins, Papers, Ohio State University, Box 1.1, Folder 22, p. 8.
14 *Adelaide Express and Telegraph*, 31 December 1895.
15 *Adelaide Express and Telegraph*, 31 December 1895.

16 *The Advertiser*, 29 December 1913, p. 9.
17 *The Advertiser*, 29 December 1913, p. 9.
18 *The Advertiser*, 29 December 1913, p. 9.
19 *The Express and Telegraph*, 31 December 1895, p. 3.
20 *The Express and Telegraph*, 31 December 1895, p. 3.
21 Nasht, *The Last Explorer,* Hachette, Sydney, 2007, p. 11.
22 Nasht, *The Last Explorer*, p. 11.
23 Nasht, *The Last Explorer*, p. 12.
24 Wilkins, Papers, Ohio State University, Box 1.1, Folder 22, p. 8.
25 Nasht, *The Last Explorer*, p. 12.
26 Nasht, *The Last Explorer*, p. 12.
27 Nasht, *The Last Explorer*, p. 12.
28 Lines from the famous Rudyard Kipling poem 'If'.
29 Wilkins, Papers, Ohio State University, Box 1.1, Folder 22, no page number.
30 *The Telegraph*, 8 October 1931, p. 6.
31 Maynard, *The Unseen Anzac*, p. 7.
32 Maynard, *The Unseen Anzac*, p. 7.
33 Maynard, *The Unseen Anzac*, p. 7.
34 Maynard, *The Unseen Anzac*, p. 7.
35 Maynard, *The Unseen Anzac*, p. 7.
36 Maynard, *The Unseen Anzac*, p. 7.
37 Wilkins, unpublished manuscript.
38 Maynard, *The Unseen Anzac*, p. 7.
39 Wilkins, Papers, Ohio State University, Box 1.1, Folder 22, no page number.
40 Wilkins, unpublished manuscript.
41 Wilkins, unpublished manuscript.
42 Wilkins, unpublished manuscript.
43 Wilkins, unpublished manuscript.
44 Wilkins, unpublished manuscript.
45 'In the Valley Where the Bluebirds Sing', lyrics by Monroe H. Rosenfeld, 1902.
46 Nasht, *The Last Explorer*, p. 17.
47 Nasht, *The Last Explorer*, p. 17.
48 Nasht, *The Last Explorer*, p. 17.
49 Maynard, *The Unseen Anzac*, p. 10.
50 Maynard, *The Unseen Anzac*, p. 10.
51 Maynard, *The Unseen Anzac*, p. 10.
52 *The Sun* (Sydney), 27 April 1930, p. 17.
53 Nasht, *The Last Explorer*, p. 17.
54 Maynard, *The Unseen Anzac*, p. 11.
55 *The Sun* (Sydney), 27 April 1930, p. 17.
56 Kalush and Sloman, *The Secret Life of Houdini: The Making of America's First Superhero*, Simon & Schuster, UK, 2007, p. 252.
57 Thomas, *Sir Hubert Wilkins – His World of Adventure*, Colorgravure, Melbourne, 1963, p. 19.
58 Thomas, *Sir Hubert Wilkins – His World of Adventure*, p. 19.
59 Maynard, *The Unseen Anzac*, p. 13.
60 Maynard, *The Unseen Anzac*, p. 13.
61 Maynard, *The Unseen Anzac*, p. 13.
62 Wilkins, 'Early Flying Experiences', unpublished manuscript.
63 Wilkins, 'Early Flying Experiences'.
64 Wilkins, 'Early Flying Experiences'.
65 Wilkins, 'Early Flying Experiences'.
66 Wilkins, 'Early Flying Experiences'.
67 Nasht, *The Last Explorer*, p. 23.
68 Wilkins, 'Early Flying Experiences'.
69 Nasht, *The Last Explorer*, p. 23.
70 Wilkins, 'Early Flying Experiences', [reported speech].
71 Wilkins, 'Early Flying Experiences'.

72 Thomas, *Sir Hubert Wilkins – His World of Adventure*, p. 26.
73 Thomas, *Sir Hubert Wilkins – His World of Adventure*, p. 26.
74 Nasht, *The Last Explorer*, p. 24.
75 Nasht, *The Last Explorer*, p. 24.
76 Thomas, *Sir Hubert Wilkins – His World of Adventure*, p. 27.
77 Author's note: Unbeknownst to Wilkins, Wilbur Wright had already done the same thing, in Rome in 1909.
78 *Canberra Times*, 26 May 1928, p. 1.

Chapter 2

1 Nasht, *The Last Explorer*, p. 25.
2 Maynard, *The Unseen Anzac*, p. 17.
3 *Adelaide Observer*, 9 November 1929, p. 11.
4 *Adelaide Observer*, 9 November 1929, p. 11.
5 Nasht, *The Last Explorer*, p. 26.
6 Nasht, *The Last Explorer*, p. 27.
7 Author's note: Although Wilkins refers to this pilot as Fezil Bey, his name was 'Fazil' Bey.
8 Nasht, *The Last Explorer*, p. 30.
9 Nasht, *The Last Explorer*, p. 30.
10 Wilkins, 'Early Flying Experiences'.
11 Grant, *To the Four Corners, The Memoirs of a News Photographer*, Hutchison & Co. Ltd, London, 1933, p. 140.
12 Grant, *To the Four Corners*, p. 140.
13 Maynard, *The Unseen Anzac*, p. 17.
14 Grant, *To the Four Corners*, p. 156.
15 Maynard, *The Unseen Anzac*, p. 17.
16 Maynard, *The Unseen Anzac*, p. 18.
17 Maynard, *The Unseen Anzac*, p. 18.
18 Wilkins, unpublished lecture draft.
19 *The Advertiser*, 12 February 1913, p. 17.
20 Grant, *To the Four Corners*, p. 141.
21 Gibbs and Grant, *The Balkan War: Adventures of War with Cross and Crescent*, Small, Maynard & Co., Boston, 1913, p. 172.
22 Gibbs and Grant, *The Balkan War*, pp. 172–173.
23 Nasht, *The Last Explorer*, p. 28.
24 Maynard, *The Unseen Anzac*, p. 19.
25 Grant, *To the Four Corners*, pp. 141–142.
26 Grant, *To the Four Corners*, pp. 141–142.
27 Nasht, *The Last Explorer*, p. 29.
28 *The Advertiser*, 12 February 1913, p. 17.
29 *The Advertiser*, 12 February 1913, p. 17.
30 Maynard, *The Unseen Anzac*, p. 21.
31 Maynard, *The Unseen Anzac*, p. 21.
32 *The Advertiser*, 12 February 1913, p. 17.
33 Wilkins, Diary, 31 October 1913.
34 Thomas, *Sir Hubert Wilkins – His World of Adventure*, pp. 33–34.
35 Thomas, *Sir Hubert Wilkins – His World of Adventure*, pp. 33–34.
36 Thomas, *Sir Hubert Wilkins – His World of Adventure*, pp. 33–34.
37 Gibbs and Grant, *The Balkan War*, p. 231 [reported speech].
38 Gibbs and Grant, *The Balkan War*, p. 231 [reported speech].
39 Gibbs and Grant, *The Balkan War*, p. 231 [reported speech].
40 Thomas, *Sir Hubert Wilkins – His World of Adventure*, pp. 33–34.
41 Wilkins, Papers, Ohio State University, Box 13, Folder 8, Balkan War, p. 32.
42 Thomas, *Sir Hubert Wilkins – His World of Adventure*, pp. 33–34.
43 Thomas, *Sir Hubert Wilkins – His World of Adventure*, pp. 33–34.
44 Gibbs and Grant, *The Balkan War*, p. 231 [reported speech].
45 Gibbs and Grant, *The Balkan War*, p. 231.

46 Wilkins, Papers, Ohio State University, Box 13, Folder 8, no page number.
47 Thomas, *Sir Hubert Wilkins – His World of Adventure*, pp. 33–34 [reported speech].
48 Thomas, *Sir Hubert Wilkins – His World of Adventure*, pp. 33–34.
49 Wilkins, Papers, Ohio State University, Box 13, Folder 8, no page number.
50 Wilkins, Papers, Ohio State University, Box 13, Folder 8, no page number.
51 Thomas, *Sir Hubert Wilkins – His World of Adventure*, pp. 33–34.
52 Grant, *To the Four Corners*, pp. 141–142.
53 Thomas, *Sir Hubert Wilkins – His World of Adventure*, p. 39.
54 Nasht, *The Last Explorer*, p. 31.
55 Nasht, *The Last Explorer*, p. 31.
56 Nasht, *The Last Explorer*, p. 31.
57 Author's note: Yes, I know, an extraordinary story. But far from unprecedented. Among many others, the Russians did this to Dostoyevsky in his rebellious student days – fake executions were an established method of breaking prisoners down.
58 Maynard, *The Unseen Anzac*, p. 21.
59 Maynard, *The Unseen Anzac*, p. 21.
60 Maynard, *The Unseen Anzac*, pp. 21–22.
61 Maynard, *The Unseen Anzac*, p. 22.
62 Maynard, *The Unseen Anzac*, p. 22.
63 *The Herald* (Melbourne), 17 February 1926, p. 12.
64 Thomas, *Sir Hubert Wilkins – His World of Adventure*, p. 43.
65 Thomas, *Sir Hubert Wilkins – His World of Adventure*, p. 43.
66 Thomas, *Sir Hubert Wilkins – His World of Adventure*, p. 44.
67 Thomas, *Sir Hubert Wilkins – His World of Adventure*, p. 44.
68 Thomas, *Sir Hubert Wilkins – His World of Adventure*, p. 44.
69 Thomas, *Sir Hubert Wilkins – His World of Adventure*, p. 44.
70 Thomas, *Sir Hubert Wilkins – His World of Adventure*, p. 44.
71 Wilkins, letter written on back of Sandow Chocolate-Cocoa flyer; the same flyers Penfold was dropping from the balloon. Transcribed from Ohio State University archives by Simon Nasht.
72 Thomas, *Sir Hubert Wilkins – His World of Adventure*, p. 44.
73 Thomas, *Sir Hubert Wilkins – His World of Adventure*, p. 45.
74 Thomas, *Sir Hubert Wilkins – His World of Adventure*, p. 45.
75 Thomas, *Sir Hubert Wilkins – His World of Adventure*, p. 45.
76 Thomas, *Sir Hubert Wilkins – His World of Adventure*, p. 45.
77 Thomas, *Sir Hubert Wilkins – His World of Adventure*, p. 45.
78 *East Oregonian*, 25 February 1913, p. 10.
79 *East Oregonian*, 25 February 1913, p. 10.
80 *East Oregonian*, 25 February 1913, p. 10.
81 *Daily Herald*, 24 December 1912.
82 Wilkins, 'Early Flying Experiences', p. 6.
83 Nasht, *The Last Explorer*, p. 33.
84 Wilkins, 'From a hundred above to fifty below', unpublished manuscript, pp. 50–53.
85 Wilkins, 'From a hundred above to fifty below', pp. 50–53.
86 Wilkins, 'From a hundred above to fifty below', pp. 50–53.
87 Wilkins, 'From a hundred above to fifty below', pp. 50–53.
88 Wilkins, 'From a hundred above to fifty below', pp. 50–53.
89 Wilkins, 'From a hundred above to fifty below', pp. 50–53.
90 Wilkins, 'From a hundred above to fifty below', pp. 50–53.
91 Wilkins, 'From a hundred above to fifty below', p. 1.
92 Wilkins, 'From a hundred above to fifty below', pp. 50–53.
93 Wilkins, 'From a hundred above to fifty below', pp. 50–53.
94 Wilkins, 'From a hundred above to fifty below', pp. 50–53.
95 Wilkins, 'From a hundred above to fifty below', pp. 50–53.
96 Wilkins, 'From a hundred above to fifty below', pp. 50–53.
97 Wilkins, 'From a hundred above to fifty below', pp. 50–53.
98 Nasht, *The Last Explorer*, p. 32.
99 Nasht, *The Last Explorer*, p. 32.

100 Nasht, *The Last Explorer*, p. 32.
101 Nasht, *The Last Explorer*, p. 32.
102 Nasht, *The Last Explorer*, p. 33.
103 Nasht, *The Last Explorer*, p. 32.
104 Wilkins, 'From a hundred above to fifty below', pp. 1–4.
105 Wilkins, Papers, Ohio State University, Box 13, Folder 13, Canadian Arctic Expedition.
106 Wilkins, 'From a hundred above to fifty below', p. 5.
107 Wilkins, Papers, Ohio State University, Box 13, Folder 13.
108 Wilkins, Report to the Department of the Naval Service, Ottawa, in *Canadian Arctic Expedition Reports*, Pravana Books, Delhi, 2020, p. 56.
109 Noice, *With Stefansson in the Arctic*, Harrap, London, 1925, p. 24.
110 Wilkins, 'From a hundred above to fifty below', p. 5.
111 Wilkins, Papers, Ohio State University, Box 13, Folder 13, p. 58.
112 Wilkins, Papers, Ohio State University, Box 13, Folder 13, p. 58.

Chapter 3

1   Wilkins, *Flying the Arctic*, Grosset & Dunlap, New York, 1928, p. v.
2   Rearden, *Alaska's First Bush Pilots, 1923–30*, Pictorial Histories Publishing, Missoula, 2009, p. 87.
3   Thomas, *Sir Hubert Wilkins – His World of Adventure*, p. 53.
4   Thomas, *Sir Hubert Wilkins – His World of Adventure*, p. 53.
5   Thomas, *Sir Hubert Wilkins – His World of Adventure*, p. 53.
6   Thomas, *Sir Hubert Wilkins – His World of Adventure*, p. 53.
7   Thomas, *Sir Hubert Wilkins – His World of Adventure*, p. 53.
8   Thomas, *Sir Hubert Wilkins – His World of Adventure*, p. 53.
9   Thomas, *Sir Hubert Wilkins – His World of Adventure*, p. 53.
10 Thomas, *Sir Hubert Wilkins – His World of Adventure*, p. 53.
11 Thomas, *Sir Hubert Wilkins – His World of Adventure*, p. 54.
12 Wilkins, Papers, Ohio State University, Box 13, Folder 13, p. 8.
13 Anderson, Rudolph, *Report of the Canadian Arctic Expedition 1913–18*, F.A. Acland, Ottawa, 1923, Ohio State University, p. 3.
14 Wilkins, Papers, Ohio State University, Box 13, Folder 13, p. 7.
15 Wilkins, Papers, Ohio State University, Box 13, Folder 13, p. 31.
16 Wilkins, Papers, Ohio State University, Box 13, Folder 13, p. 31.
17 Maynard, *The Unseen Anzac*, p. 25.
18 Thomas, *Sir Hubert Wilkins – His World of Adventure*, p. 54.
19 Thomas, *Sir Hubert Wilkins – His World of Adventure*, p. 54.
20 Chipman, Diary, 3 August, 1913.
21 Thomas, *Sir Hubert Wilkins – His World of Adventure*, p. 55.
22 Thomas, *Sir Hubert Wilkins – His World of Adventure*, p. 55.
23 Thomas, *Sir Hubert Wilkins – His World of Adventure*, p. 55.
24 Bartlett, *The Last Voyage of the Karluk: Flagship of Vilhjalmar Stefansson's Canadian Arctic Expedition of 1913–16*, McClelland, Toronto, 1916, p. 12.
25 Bartlett, *The Last Voyage of the Karluk*, p. 14.
26 Bartlett, *The Last Voyage of the Karluk*, p. 15.
27 Diubaldo, *Stefansson and the Canadian Arctic*, McGill-Queen's University Press, Montreal, 1999, p. 78.
28 Nasht, *The Last Explorer*, p. 36.
29 Wilkins, Diary, 19 September 1913.
30 Thomas, *Sir Hubert Wilkins – His World of Adventure*, p. 56.
31 Wilkins, Diary, 19 September 1913.
32 Wilkins, Diary, 19 September 1913.
33 Wilkins, Diary, 19 September 1913.
34 Wilkins, Diary, 19 September 1913.
35 *The Advertiser*, 6 March 1914, p. 17.
36 Bartlett, *The Last Voyage of the Karluk*, p. 11.
37 Wilkins, Papers, Ohio State University, Box 13, Folder 13, p. 18.

38 Wilkins, Diary, 19 September 1913.
39 Wilkins, Diary, 19 September 1913.
40 Wilkins, Diary, 19 September 1913.
41 Wilkins, Diary, 19 September 1913.
42 Wilkins, Diary, 22 September 1913.
43 Wilkins, Diary, 23 September 1913.
44 *The Advertiser*, 6 March 1914, p. 17.
45 *The Advertiser*, 6 March 1914, p. 17.
46 *The Advertiser*, 6 March 1914, p. 17.
47 Wilkins, Diary, 24 September 1913.
48 Wilkins, Diary, 24 September 1913.
49 Wilkins, Diary, 26 September 1913.
50 Wilkins, Diary, 26 September 1913.
51 Wilkins, Diary, 26 September 1913.
52 Wilkins, Diary, 24 September 1913.
53 Wilkins, Diary, 27 September 1913.
54 Wilkins, Diary, 27 September 1913.
55 Wilkins, Diary, 27 September 1913.
56 Wilkins, Diary, 27 September 1913.
57 Wilkins, Diary, 29 September 1913.
58 Wilkins, Papers, Ohio State University, Box 13, Folder 13, p. 30.
59 Wilkins, Papers, Ohio State University, Box 13, Folder 13, p. 31.
60 Wilkins, Papers, Ohio State University, Box 13, Folder 13, p. 31.
61 Wilkins, Diary, 1 October 1913.
62 Wilkins, Diary, 3 October 1913.
63 Thomas, *Sir Hubert Wilkins – His World of Adventure*, p. 58.
64 Thomas, *Sir Hubert Wilkins – His World of Adventure*, p. 58.
65 Thomas, *Sir Hubert Wilkins – His World of Adventure*, p. 58.
66 Thomas, *Sir Hubert Wilkins – His World of Adventure*, p. 58.
67 Thomas, *Sir Hubert Wilkins – His World of Adventure*, p. 59.
68 Thomas, *Sir Hubert Wilkins – His World of Adventure*, p. 59.
69 Thomas, *Sir Hubert Wilkins – His World of Adventure*, p. 59.
70 Thomas, *Sir Hubert Wilkins – His World of Adventure*, p. 58.
71 *The Advertiser*, 6 March 1914, p. 17.
72 Wilkins, Papers, Ohio State University, Box 13, Folder 13, p. 14.
73 Wilkins, Diary, 3 October 1913.
74 Author's note: This is Wilkins' own phonetic spelling of what we more commonly call 'Igloo', but I have retained it partly for its charm and for the fact it illustrates the point that so much of the 'Eskimo' language was recorded through Western ears; not via the hand of the Inuit.
75 Wilkins, Diary, 5 October 1913.
76 Wilkins, Diary, 5 October 1913.
77 Wilkins, Diary, 5 October 1913.
78 Wilkins, Diary, 5 October 1913.
79 Wilkins, Diary, 5 October 1913.
80 Wilkins, unpublished manuscript.
81 Wilkins, Diary, 5 October 1913.
82 Wilkins, Report to Department of the Naval Service, Ottawa, 1917, p. 55.
83 Wilkins, Report to Department of the Naval Service, Ottawa, 1917, p. 55.
84 Wilkins, Diary, 7 October 1913.
85 Wilkins, Diary, 8 October 1913.
86 Wilkins, Diary, 9 October 1913.
87 Wilkins, Diary, 10 October 1913.
88 Wilkins, Diary, 11 October 1913.
89 Wilkins, Diary, 11 October 1913.
90 Wilkins, Diary, 11 October 1913.
91 Thomas, *Sir Hubert Wilkins – His World of Adventure*, p. 60.

Chapter 4

1 Thomas, *Sir Hubert Wilkins – His World of Adventure*, p. 60.
2 Wilkins, Diary, 26 October 1913.
3 Wilkins, Diary, 26 October 1913.
4 Thomas, *Sir Hubert Wilkins – His World of Adventure*, p. 61.
5 Thomas, *Sir Hubert Wilkins – His World of Adventure*, p. 61.
6 Wilkins, Diary, 31 October 1913.
7 Wilkins, Diary, 31 October 1913.
8 Wilkins, Diary, 31 October 1913.
9 Wilkins, Diary, 31 October 1913 [reported speech].
10 Wilkins, Diary, 31 October 1913 [reported speech].
11 Thomas, *Sir Hubert Wilkins – His World of Adventure*, pp. 61–62.
12 Wilkins, Diary, 31 October 1913.
13 Wilkins, Diary, 31 October 1913.
14 Wilkins, Diary, 31 October 1913.
15 Wilkins, Diary, 31 October 1913.
16 Thomas, *Sir Hubert Wilkins – His World of Adventure*, p. 62.
17 Wilkins, Diary, 31 October 1913.
18 Wilkins, 'True Adventure Thrills', manuscript, unpublished, undated radio script, episode 20.
19 Nasht, *The Last Explorer*, p. 41.
20 Wilkins, Diary, 31 October 1913.
21 *The Advertiser*, 6 March 1914, p. 17.
22 *The Advertiser*, 6 March 1914, p. 17.
23 Wilkins, Diary, 20 November 1913.
24 Wilkins, Diary, 20 November 1913.
25 Wilkins, Diary, 21 November 1913.
26 Wilkins, Diary, 21 November 1913 [reported speech].
27 Wilkins, Diary, 15 December 1913.
28 Wilkins, Diary, 15 December 1913.
29 Wilkins, Diary, 15 December 1913.
30 Wilkins, Diary, 24 December 1913.
31 Thomas, *Sir Hubert Wilkins – His World of Adventure*, p. 63.
32 *Morning Bulletin*, 27 March 1926, p. 10.
33 Wilkins, Diary, 25 December 1913.
34 Wilkins, Diary, 25 December 1913.
35 Wilkins, Diary, 25 December 1913.
36 Wilkins, Diary, 26 December 1913.
37 Wilkins, Diary, 26 December 1913.
38 Wilkins, Diary, 26 December 1913.
39 Wilkins, Diary, 26 December 1913.
40 Wilkins, Diary, 26 December 1913.
41 Wilkins, Diary, 26 December 1913.
42 Wilkins, Diary, 26 December 1913.
43 Wilkins, Diary, 22 January 1914.
44 Wilkins, Diary, 22 January 1914.
45 Wilkins, Diary, 22 January 1914.
46 Wilkins, Diary, 22 January 1914.
47 Wilkins, Diary, 9 March 1914.
48 Wilkins, Diary, 9 March 1914.
49 Wilkins, Diary, 9 March 1914.
50 Thomas, *Sir Hubert Wilkins – His World of Adventure*, p. 64.
51 Stefansson, *My Life With The Eskimo*, The Macmillan Company, New York, 1913, p. 339.
52 Arctic Institute of North America, *Arctic*, Vol. 45, No. 1, p. 92
53 Wilkins, Diary, 19 March 1914.
54 Wilkins, Diary, 19 March 1914.
55 Thomas, *Sir Hubert Wilkins – His World of Adventure*, p. 65.
56 Wilkins, Manuscript, 'True Adventure Thrills', p. 64.

57 Wilkins, Manuscript, 'True Adventure Thrills', p. 64.
58 Wilkins, Manuscript, 'True Adventure Thrills', p. 64.
59 Wilkins, Manuscript, 'True Adventure Thrills', p. 64.
60 Wilkins, Manuscript, 'True Adventure Thrills', p. 64.
61 Wilkins, Manuscript, 'True Adventure Thrills', p. 64.
62 Wilkins, Manuscript, 'True Adventure Thrills', p. 64.
63 Wilkins, Manuscript, 'True Adventure Thrills', p. 67.
64 Wilkins, Manuscript, 'True Adventure Thrills', p. 67.
65 Wilkins, Manuscript, 'True Adventure Thrills', p. 64.
66 Nasht, *The Last Explorer*, p. 43.

Chapter 5

1 Wilkins, Papers, Ohio State University, Box 13, Folder 13, p. 13.
2 Thomas, *Sir Hubert Wilkins – His World of Adventure*, p. 66.
3 Thomas, *Sir Hubert Wilkins – His World of Adventure*, p. 66.
4 Thomas, *Sir Hubert Wilkins – His World of Adventure*, p. 66.
5 Thomas, *Sir Hubert Wilkins – His World of Adventure*, p. 66.
6 Wilkins, Diary, 23 April 1914.
7 Wilkins, Diary, 24 April 1914.
8 Wilkins, Diary, 26 April 1914.
9 Thomas, *Sir Hubert Wilkins – His World of Adventure*, p. 67.
10 Thomas, *Sir Hubert Wilkins – His World of Adventure*, p. 67.
11 Thomas, *Sir Hubert Wilkins – His World of Adventure*, p. 68.
12 Thomas, *Sir Hubert Wilkins – His World of Adventure*, p. 68.
13 Thomas, *Sir Hubert Wilkins – His World of Adventure*, p. 68.
14 Thomas, *Sir Hubert Wilkins – His World of Adventure*, p. 68.
15 Thomas, *Sir Hubert Wilkins – His World of Adventure*, p. 68.
16 Thomas, *Sir Hubert Wilkins – His World of Adventure*, p. 69.
17 Thomas, *Sir Hubert Wilkins – His World of Adventure*, p. 69.
18 Thomas, *Sir Hubert Wilkins – His World of Adventure*, pp. 69–70.
19 Wilkins, Diary, 6 June 1914.
20 Wilkins, Diary, 6 June 1914.
21 Wilkins, Diary, 6 June 1914.
22 Wilkins, Diary, 6 June 1914.
23 Wilkins, Diary, 6 June 1914.
24 Wilkins, Diary, 6 June 1914.
25 Wilkins, Diary, 6 June 1914.
26 Thomas, *Sir Hubert Wilkins – His World of Adventure*, p. 71.
27 Thomas, *Sir Hubert Wilkins – His World of Adventure*, p. 71.
28 Thomas, *Sir Hubert Wilkins – His World of Adventure*, p. 71.
29 Wilkins, Diary, 5 August 1914.
30 Wilkins, Diary, 10 August 1914.
31 Wilkins, Diary, 10 August 1914.
32 Thomas, *Sir Hubert Wilkins – His World of Adventure*, p. 71.
33 Thomas, *Sir Hubert Wilkins – His World of Adventure*, p. 72.
34 Wilkins, Diary, 11 August 1914.
35 Thomas, *Sir Hubert Wilkins – His World of Adventure*, p. 72.
36 Thomas, *Sir Hubert Wilkins – His World of Adventure*, p. 72.
37 Thomas, *Sir Hubert Wilkins – His World of Adventure*, p. 72.
38 Wilkins, Diary, 17 August 1914.
39 Wilkins, Diary, 15 August 1914.
40 Thomas, *Sir Hubert Wilkins – His World of Adventure*, p. 73.
41 Thomas, *Sir Hubert Wilkins – His World of Adventure*, p. 73.
42 *Burra Record*, 2 July 1924, p. 4.
43 Thomas, *Sir Hubert Wilkins – His World of Adventure*, p. 73.
44 Thomas, *Sir Hubert Wilkins – His World of Adventure*, p. 73.
45 Thomas, *Sir Hubert Wilkins – His World of Adventure*, p. 73.

46 Thomas, *Sir Hubert Wilkins – His World of Adventure*, pp. 73–74.
47 Wilkins, Diary, 21 August 1914.
48 Wilkins, Diary, 22 August 1914.
49 Thomas, *Sir Hubert Wilkins – His World of Adventure*, p. 74.
50 Thomas, *Sir Hubert Wilkins – His World of Adventure*, p. 74.
51 Wilkins, Diary, 22 August 1914.
52 Thomas, *Sir Hubert Wilkins – His World of Adventure*, p. 75.
53 Thomas, *Sir Hubert Wilkins – His World of Adventure*, p. 75.
54 Wilkins, Diary, 11 September 1914.
55 Wilkins, *Report to Mr V. Stefansson, Commander, Canadian Arctic Expedition*.
56 Thomas, *Sir Hubert Wilkins – His World of Adventure*, p. 75.
57 Thomas, *Sir Hubert Wilkins – His World of Adventure*, pp. 75–76.
58 Thomas, *Sir Hubert Wilkins – His World of Adventure*, p. 76.
59 Wilkins, Diary, 11 September 1914.
60 Thomas, *Sir Hubert Wilkins – His World of Adventure*, p. 76.
61 Wilkins, Diary, 11 September 1914.
62 Thomas, *Sir Hubert Wilkins – His World of Adventure*, p. 76.
63 Thomas, *Sir Hubert Wilkins – His World of Adventure*, p. 76.
64 Wilkins, Diary, 13 September 1914.
65 Wilkins, Diary, 13 September 1914.
66 Nasht, *The Last Explorer*, p. 46.
67 Wilkins, Diary, 27 September 1914.
68 Thomas, *Sir Hubert Wilkins – His World of Adventure*, p. 77.
69 Wilkins, Diary, 23 October 1914.
70 Wilkins, Diary, 14 November 1914.
71 Thomas, *Sir Hubert Wilkins – His World of Adventure*, p. 77.
72 Nasht, *The Last Explorer*, p. 47.
73 Thomas, *Sir Hubert Wilkins – His World of Adventure*, p. 77.
74 Thomas, *Sir Hubert Wilkins – His World of Adventure*, p. 77.
75 Thomas, *Sir Hubert Wilkins – His World of Adventure*, p. 77.
76 Wilkins, Papers, Ohio State University, Box 13, Folder 13, p. 27.
77 Wilkins, Papers, Ohio State University, Box 13, Folder 13, p. 27.
78 Wilkins, Papers, Ohio State University, Box 13, Folder 13, p. 18.
79 Wilkins, Papers, Ohio State University, Box 13, Folder 13, p. 27.

Chapter 6

1 Thomas, *Sir Hubert Wilkins – His World of Adventure*, p. 78.
2 Thomas, *Sir Hubert Wilkins – His World of Adventure*, p. 79.
3 Thomas, *Sir Hubert Wilkins – His World of Adventure*, p. 79.
4 Grierson, *Sir Hubert Wilkins: Enigma of Exploration*, Robert Hale, London, 1960, p. 54.
5 Thomas, *Sir Hubert Wilkins – His World of Adventure*, p. 79.
6 Thomas, *Sir Hubert Wilkins – His World of Adventure*, p. 79.
7 Thomas, *Sir Hubert Wilkins – His World of Adventure*, p. 79.
8 Thomas, *Sir Hubert Wilkins – His World of Adventure*, p. 79.
9 Thomas, *Sir Hubert Wilkins – His World of Adventure*, p. 79.
10 Thomas, *Sir Hubert Wilkins – His World of Adventure*, p. 80.
11 Thomas, *Sir Hubert Wilkins – His World of Adventure*, p. 80.
12 Thomas, *Sir Hubert Wilkins – His World of Adventure*, p. 80.
13 Thomas, *Sir Hubert Wilkins – His World of Adventure*, p. 80.
14 Thomas, *Sir Hubert Wilkins – His World of Adventure*, p. 80.
15 Thomas, *Sir Hubert Wilkins – His World of Adventure*, p. 81.
16 Wilkins, Papers, Ohio State University, Box 13, Folder 13, p. 25.
17 Wilkins, Papers, Ohio State University, Box 13, Folder 13, p. 25.
18 Wilkins, Papers, Ohio State University, Box 13, Folder 13, p. 26.
19 Wilkins, Papers, Ohio State University, Box 13, Folder 13, p. 26.
20 Wilkins, Papers, Ohio State University, Box 13, Folder 13, pp. 27–28.
21 Wilkins, Papers, Ohio State University, Box 13, Folder 13, p. 28.

22 Wilkins, Papers, Ohio State University, Box 13, Folder 13, p. 28.
23 Wilkins, Papers, Ohio State University, Box 13, Folder 13, p. 28.
24 Wilkins, Papers, Ohio State University, Box 13, Folder 13, p. 28.
25 Wilkins, Papers, Ohio State University, Box 13, Folder 13, p. 28.
26 Wilkins, Lecture, 'A Visit to the "Blonde" Eskimo'.
27 Wilkins, Lecture, 'A Visit to the "Blonde" Eskimo', p. 1.
28 Wilkins, Lecture, 'A Visit to the "Blonde" Eskimo', p. 2.
29 Wilkins, Lecture, 'A Visit to the "Blonde" Eskimo', p. 1.
30 Wilkins, Lecture, 'A Visit to the "Blonde" Eskimo', p. 2.
31 Wilkins, Lecture, 'A Visit to the "Blonde" Eskimo', p. 3.
32 *Colac Reformer*, 30 December 1916, p. 5 [reported speech to the journalist].
33 Wilkins, Lecture, 'A Visit to the "Blonde" Eskimo', pp. 4–5.
34 Wilkins, Lecture, 'A Visit to the "Blonde" Eskimo', p. 5.
35 Wilkins, Lecture, 'A Visit to the "Blonde" Eskimo', p. 5.
36 Wilkins, Lecture, 'A Visit to the "Blonde" Eskimo', p. 5.
37 Wilkins, Lecture, 'A Visit to the "Blonde" Eskimo', p. 5.
38 Wilkins, Lecture, 'A Visit to the "Blonde" Eskimo', p. 5.
39 *The Register*, 12 March 1917, p. 4.
40 Thomas, *Sir Hubert Wilkins – His World of Adventure*, p. 82.
41 Bartlett, *The Last Voyage of the Karluk*, p. 91.
42 Bartlett, *The Last Voyage of the Karluk*, p. 91.
43 Bartlett, *The Last Voyage of the Karluk*, p. 91.
44 Thomas, *Sir Hubert Wilkins – His World of Adventure*, p. 82.
45 Thomas, *Sir Hubert Wilkins – His World of Adventure*, p. 82.
46 Macfarlane, 'The Scot Who Was Left Out in the Cold', *Scottish Field*, 6 March 2020.
47 Nasht, *The Last Explorer*, p. 48.
48 Nasht, *The Last Explorer*, p. 48.
49 Thomas, *Sir Hubert Wilkins – His World of Adventure*, p. 83.
50 Anderson, *Report of the Canadian Arctic Expedition 1913–18*, F.A. Acland, Ottawa, 1923, Ohio State University, p. 46.
51 Bean, Diary, 12/13 August 1916, AWM38 DRL606/54/1, p. 195. Author's note: Reading back on his diary entry later, Bean would note, 'This was an exaggeration . . . They were mostly better men than I; but I have deliberately let this stand, unfair though it is, so that this record should be unaltered. I was pretty hard worked at the time when I wrote this, and rather sore.'
52 Bean, Diary, 12/13 August 1916, AWM38 DRL606/54/1, p. 195.
53 Maynard, *The Unseen Anzac*, p. 51.
54 Scott, *Scott's Last Expedition, Vol. I: Being the journals of Captain R.F. Scott, R.N., C.V.O.*, Smith, Elder & Co., London, 1914, p. 439.
55 Young, *A Great Task of Happiness: The life of Kathleen Scott*, Macmillan, London, 1995, p. 157.
56 *The Telegraph*, 8 October 1931, p. 6.
57 Thomas, *Sir Hubert Wilkins – His World of Adventure*, p. 83.
58 *The Advertiser*, 12 February 1913, p. 17.
59 *The Advertiser*, 6 March 1914, p. 4.
60 Pearson, *The Sealed Train*, Macmillan, London, 1975, pp. 77–78.
61 Pearson, *The Sealed Train*, pp. 77–78.
62 Pearson, *The Sealed Train*, p. 128.
63 Churchill, *The World Crisis*, Vol. V, T. Butterworth, London, 1923, p. 73.
64 Thomas, *Sir Hubert Wilkins – His World of Adventure*, p. 83.
65 Thomas, *Sir Hubert Wilkins – His World of Adventure*, p. 84.
66 Thomas, *Sir Hubert Wilkins – His World of Adventure*, p. 84.
67 Thomas, *Sir Hubert Wilkins – His World of Adventure*, p. 84.
68 Thomas, *Sir Hubert Wilkins – His World of Adventure*, p. 84.
69 National Archives of Australia (NAA) B2455, Wilkins George Hubert.
70 *The Advertiser*, 29 March 1917, p. 8.
71 Thomas, *Sir Hubert Wilkins – His World of Adventure*, p. 84.
72 Thomas, *Sir Hubert Wilkins – His World of Adventure*, pp. 84–85.
73 Thomas, *Sir Hubert Wilkins – His World of Adventure*, p. 85.

74 Thomas, *Sir Hubert Wilkins – His World of Adventure*, p. 85.
75 Thomas, *Sir Hubert Wilkins – His World of Adventure*, p. 85.
76 Wilkins, Papers, Ohio State University, Box 1, Folder 14, WWI Service, pp. 1–2.
77 Thomas, *Sir Hubert Wilkins – His World of Adventure*, p. 85.
78 Thomas, *Sir Hubert Wilkins – His World of Adventure*, p. 85.
79 Thomas, *Sir Hubert Wilkins – His World of Adventure*, p. 85.
80 Nasht, *The Last Explorer*, p. 46.
81 Thomas, *Sir Hubert Wilkins – His World of Adventure*, p. 85.
82 Military Records of George Hubert Wilkins, 1914–1920, National Archives of Australia, Series No. B2455, Item ID 838937, p. 41.
83 Thomas, *Sir Hubert Wilkins – His World of Adventure*, p. 85.
84 Thomas, *Sir Hubert Wilkins – His World of Adventure*, p. 85.
85 Wilkins, Papers, Ohio State University, Box 1, Folder 14.
86 Wilkins, Papers, Ohio State University, Box 1, Folder 14.
87 Thomas, *Sir Hubert Wilkins – His World of Adventure*, p. 85.
88 Thomas, *Sir Hubert Wilkins – His World of Adventure*, p. 86.
89 Wilkins, Papers, Ohio State University, Box 1, Folder 14.
90 Thomas, *Sir Hubert Wilkins – His World of Adventure*, p. 85.
91 *The Telegraph*, 8 October 1931, p. 6.
92 Hurley, Papers of Frank Hurley, My Diary, Official War Photographer Commonwealth Military Forces, 21 August 1917 – 31 August 1918, NLA, MS 883, Series 1, Item 5, p. 1.

Chapter 7

1 *The Telegraph*, 6 October 1931, p. 6.
2 Maynard, *The Unseen Anzac*, p. xiii.
3 Hurley, My Diary, pp. 1–2.
4 Author's note: In October 1917 Charles Bean sought Fred Cutlack's services as assistant official war correspondent. By January 1918, Cutlack had taken over the tasks Bean had previously performed, and in March 1918 when Bean returned to France, Cutlack remained his assistant.
5 Maynard, *The Unseen Anzac*, p. 48.
6 Wilkins, Papers, Ohio State University, Box 1, Folder 14, p. 2.
7 Bean, AWM 38 Official History 1914–18 War, Diaries and Notebooks 606/87/1, August 1917, p. 8.
8 Wilkins, Papers, Ohio State University, Box 1, Folder 14, pp. 2–3.
9 Wilkins, Papers, Ohio State University, Box 1, Folder 14, pp. 2–3.
10 Hurley, My Diary, p. 2.
11 Maynard, *The Unseen Anzac*, p. 47.
12 Maynard, *The Unseen Anzac*, p. 48.
13 Wilkins, Papers, Ohio State University, Box 1, Folder 14, p. 3 [reported speech].
14 Wilkins, Papers, Ohio State University, Box 1, Folder 14, p. 3 [reported speech].
15 Wilkins, Papers, Ohio State University, Box 1, Folder 14, p. 2.
16 Maynard, *The Unseen Anzac*, p. 75.
17 Thomas, *Sir Hubert Wilkins – His World of Adventure*, p. 87.
18 Wilkins, Papers, Ohio State University, Box 1, Folder 14, p. 2.
19 Hurley, My Diary, p. 5.
20 Hurley, My Diary, p. 26.
21 Wilkins, Papers, Ohio State University, Box 1, Folder 14, p. 9.
22 Wilkins, Papers, Ohio State University, Box 1, Folder 14, p. 9.
23 Bean, AWM 38 Official History 1914–18 War, Diaries and Notebooks 606/88/1, September 1917, pp. 48–49.
24 Bean, AWM 38 Official History 1914–18 War, Diaries and Notebooks 606/88/1, September 1917, pp. 50–51.
25 Bean, AWM 38 Official History 1914–18 War, Diaries and Notebooks 606/116/1, June–September 1918, p. 71.
26 Hurley, My Diary, p. 50.
27 Hurley, My Diary, pp. 50–51.
28 Hurley, My Diary, p. 51.

29 Hurley, My Diary, pp. 68–69.
30 Thomas, Sir Hubert Wilkins – His World of Adventure, p. 88.
31 Thomas, Sir Hubert Wilkins – His World of Adventure, p. 88.
32 Wilkins, Papers, Ohio State University, Box 1, Folder 14, pp. 9–10.
33 Wilkins, Papers, Ohio State University, Box 1, Folder 14, p. 21.
34 Thomas, Sir Hubert Wilkins – His World of Adventure, p. 88.
35 Hurley, My Diary, p. 42.
36 Hurley, My Diary, p. 44.
37 Thomas, unpublished Wilkins biography manuscript draft, p. 190. Author's note: This now famous macabre anecdote was cut from Lowell Thomas's manuscript presumably on the grounds of taste. Its stark horror and literally hysterical humour were far too close to the bone for those so close to the Great War. Clearly the image and the moment stuck in Wilkins' mind; how could they ever escape it?
38 Hurley, My Diary, p. 42.
39 Maynard, The Unseen Anzac, p. 88.
40 Maynard, The Unseen Anzac, p. 88.
41 Maynard, The Unseen Anzac, p. 88.
42 Bean, AWM, Official History, 1914–18 War, Diaries and Notebooks, 38 3DRL 606 88–1, pp. 84–85 [reported speech].
43 Thomas, Sir Hubert Wilkins – His World of Adventure, p. 89.
44 Andrews, Hubert Who?, HarperCollins, Sydney, 2011, pp. 67–68.
45 Wilkins, Papers, Ohio State University, Box 1, Folder 14, p. 12.
46 Thomas, Sir Hubert Wilkins – His World of Adventure, p. 89.
47 Maynard, The Unseen Anzac, p. 94.
48 Wilkins, Papers, Ohio State University, Box 1, Folder 14, p. 21.
49 Nasht, The Last Explorer, p. 60.
50 Adelaide Observer, 9 November 1929, p. 11.
51 Maynard, The Unseen Anzac, p. 93.
52 Maynard, The Unseen Anzac, p. 94.
53 Maynard, The Unseen Anzac, p. 94.
54 Maynard, The Unseen Anzac, p. 93.
55 Hurley, My Diary, p. 80.
56 Hurley, My Diary, p. 80.
57 Hurley, My Diary, p. 80.
58 Wilkins, Papers, Ohio State University, Box 1, Folder 14, p. 25.
59 Wilkins, Papers, Ohio State University, Box 1, Folder 14, p. 25.
60 Wilkins, Papers, Ohio State University, Box 1, Folder 14, p. 25.
61 Wilkins, Papers, Ohio State University, Box 1, Folder 14, p. 25.
62 Taylor, The First World War: An Illustrated History, Hamish Hamilton, London, 1963, p. 179.
63 Wilkins, Papers, Ohio State University, Box 1, Folder 14, p. 25.
64 Monash, Australian Victories in France in 1918, Naval and Military Press, Uckfield, UK, 2009, p. 96.
65 Hurley, My Diary, p. 59.
66 Maynard, The Unseen Anzac, p. 140.
67 McCarthy, Gallipoli to the Somme, Cooper, London, 1983, p. 191.
68 Hurley, My Diary, 4 June 1918.
69 Maynard, The Unseen Anzac, p. 141; Maynard's footnote is Bean, War Records of C.E.W. Bean, Diaries and Notebooks, AWM 38 3DRL 606/114/1, p. 1.
70 Burness, ed., The Western Front Diary of Charles Bean, Australian War Memorial, NewSouth, 2018, p. 505.
71 Hurley, My Diary, 6–8 June 1918.
72 Hurley, My Diary, p. 59.
73 Hurley, My Diary, pp. 90–91.
74 Bean, AWM 38 Official History 1914–18 War, Diaries and Notebooks 3DRL 606/107/1, April 1918, p. 54.
75 Nasht, The Last Explorer, p. 65.

Chapter 8

1 Wilkins, Papers, Ohio State University, Box 1, Folder 14, pp. 9–10.
2 Bean, *Anzac to Amiens*, Australian War Memorial, Canberra, 1946, p. 415.
3 Bean, *Anzac to Amiens*, p. 415.
4 Bean, *The Official History of Australia in the War of 1914–1918*, Vol. V, Angus & Robertson, Sydney, 1937, p. 177 [extemporised].
5 Bean, *The Official History of Australia in the War of 1914–1918*, Vol. V, p. 177.
6 Bean, Diary, AWM38 3DRL, 606/103/1, March 1918, p. 4.
7 Kennedy, *From Anzac Cove to Villers-Bretonneux*, unpublished memoir, AWM PR02032, p. 63.
8 Bean, *The Official History of Australia in the War of 1914–1918*, Vol. V, p. 122.
9 Bean, *The Official History of Australia in the War of 1914–1918*, Vol. V, p. 123 [reported speech].
10 Bean, Diary, AWM38 3DRL 606/105/1, p. 20.
11 Bean, Diary, AWM38 3DRL 606/105/1, p. 22.
12 Bean, Diary, AWM38 3DRL 606/105/1, pp. 23–24 [reported speech].
13 Bean, Diary, AWM38 3DRL 606/105/1, p. 24.
14 Bean, Diary, AWM38 3DRL 606/105/1, p. 25.
15 Bean, Diary, AWM38 3DRL 606/105/1 – April 1918, p. 25.
16 Bean, Diary, AWM38 3DRL 606/105/1 – April 1918, p. 26.
17 Bean, Diary, AWM38 3DRL 606/105/1 – April 1918, p. 26.
18 Bean, Diary, AWM38 3DRL 606/105/1 – April 1918, p. 33.
19 Holton, typed narrative, AWM PR05317, p. 33.
20 Wilkins, Papers, Ohio State University, Box 1, Folder 14, pp. 22–23.
21 Day, 'Unsung No. 1 with a Bullet', *The Australian*, 11 April 2007.
22 Younger (Director), *Unsolved History: Death of the Red Baron*, Termite Art Productions, USA, 2002.
23 Joy, *The Aviators*, Shakespeare Head Press, Sydney, Australia, 1971, p. 99.
24 Scott, *A Soldier on the Somme, Diary of Private Edwin Need*, Eastern Press, Mulgrave, 2014, p. 132.
25 Turnbull, Diary, 24 April 1918, AWM PR91/015, Digital RCDIG0001110, p. 49 [reported speech].
26 Thomas, *Sir Hubert Wilkins – His World of Adventure*, p. 92.
27 Bean, AWM 38 Official History 1914–18 War, Diaries and Notebooks 606/96/1, January 1918, p. 36.
28 Thomas, *Sir Hubert Wilkins – His World of Adventure*, p. 92.
29 Bean, *The Official History of Australia in the War of 1914–1918*, Vol. V, p. 582.
30 Downing, *To the Last Ridge*, Grub Street, London, 2013, p. 117.
31 Downing, *To the Last Ridge*, p. 118.
32 Bean, *The Official History of Australia in the War of 1914–1918*, Vol. V, p. 604.
33 McKenna papers, in Schneider, *Pompey Elliott's Left Hand Man: Lieutenant Colonel Charles Denehy*, self-published, Camberwell, 2015, p. 247.
34 Bean, *The Official History of Australia in the War of 1914–1918*, Vol. V, p. 602.
35 Scott, *A Soldier on the Somme, Diary of Private Edwin Need*, pp. 136–137.
36 Bean, *The Official History of Australia in the War of 1914–1918*, Vol. V, p. 602.
37 Downing, *To the Last Ridge*, p. 118.
38 J.J. McKenna papers, in Schneider, *Pompey Elliott's Left Hand Man*, p. 247.
39 Downing, *To the Last Ridge*, p. 118.
40 McKenna papers, in Schneider, *Pompey Elliott's Left Hand Man*, p. 247.
41 Browning, *Fix Bayonets: The Unit History of the 51st Battalion*, self-published, Perth, 2000, p. 157.
42 Bean, Diary, AWM38 3DRL 606/108/1, April–May 1918, p. 51.
43 Thomas, *Sir Hubert Wilkins – His World of Adventure*, p. 92.
44 Thomas, *Sir Hubert Wilkins – His World of Adventure*, p. 92.
45 Thomas, *Sir Hubert Wilkins – His World of Adventure*, p. 92.
46 Thomas, *Sir Hubert Wilkins – His World of Adventure*, p. 93.
47 Thomas, *Sir Hubert Wilkins – His World of Adventure*, p. 93.
48 Maynard, *The Unseen Anzac*, p. 153.
49 Thomas, *Sir Hubert Wilkins – His World of Adventure*, p. 93.

50   Thomas, *Sir Hubert Wilkins – His World of Adventure*, p. 93.
51   Thomas, *Sir Hubert Wilkins – His World of Adventure*, p. 94.
52   Nasht, *The Last Explorer*, p. 72.
53   Nasht, *The Last Explorer*, p. 73.
54   Nasht, *The Last Explorer*, p. 73.
55   Thomas, *Sir Hubert Wilkins – His World of Adventure*, p. 94.
56   Author's note: In 1915 Monash was made Companion of the Order of the Bath, the lowest level of knighthood of that order. At Bertangles, in August 1918, he was raised to the highest level, Knight Commander of the Order of the Bath.
57   *The Bulletin*, 1918, Vol. 101, p. 117.
58   Bean, AWM 38 Official History 1914–18 War, Diaries and Notebooks, 3DRL 606/116/1, p. 77.
59   Bean, AWM 38 Official History 1914–18 War, Diaries and Notebooks, 3DRL 606/116/1, p. 78.
60   Bean, AWM 38 Official History 1914–18 War, Diaries and Notebooks, 3DRL 606/116/1, p. 78.
61   Serle, *John Monash: A Biography*, Melbourne University Press in association with Monash University, Melbourne, 1982, p. 351.
62   Serle, *John Monash*, p. 351.
63   Kieza, *Monash – The Soldier who shaped Australia*, ABC Books HarperCollins, Sydney, 2015, p. 484.
64   Thomas, *Sir Hubert Wilkins – His World of Adventure*, p. 94 [reported speech].
65   Wilkins, Papers, Ohio State University, Box 1, Folder 14, p. 20.
66   Thomas, *Sir Hubert Wilkins – His World of Adventure*, p. 94.
67   Thomas, *Sir Hubert Wilkins – His World of Adventure*, p. 95.
68   Thomas, *Sir Hubert Wilkins – His World of Adventure*, p. 95.
69   Thomas, *Sir Hubert Wilkins – His World of Adventure*, p. 95.
70   Monash, *Australian Victories in France in 1918*, p. 290.
71   Maynard, *The Unseen Anzac*, p. 163.
72   Wilkins, Papers, Ohio State University, Box 12, Folder 6, correspondence 1908–1919, p. 31.
73   Thomas, *Sir Hubert Wilkins – His World of Adventure*, p. 95 [reported speech].
74   Thomas, *Sir Hubert Wilkins – His World of Adventure*, p. 95.
75   Thomas, *Sir Hubert Wilkins – His World of Adventure*, p. 95.
76   Thomas, *Sir Hubert Wilkins – His World of Adventure*, p. 95.
77   Thomas, *Sir Hubert Wilkins – His World of Adventure*, p. 95.
78   Thomas, *Sir Hubert Wilkins – His World of Adventure*, pp. 95–96.
79   *The Bendigo Advertiser*, 23 September 1916, p. 7.
80   *The Bendigo Advertiser*, 23 September 1916, p. 7.
81   Bean, AWM, Official History, 1914–18 War, 38 3DRL 606 116-1, p. 114 [reported speech].
82   *The Bendigonian*, 12 September 1918, p. 10.
83   Bean, AWM, Official History, 1914–18 War, 38 3DRL 606 116-1, p. 119 [reported speech].
84   Maynard, *The Unseen Anzac*, pp. 168–69.
85   Thomas, *Sir Hubert Wilkins – His World of Adventure*, p. 96.
86   Thomas, *Sir Hubert Wilkins – His World of Adventure*, p. 96.
87   Wilkins, Papers, Ohio State University, Box 1, Folder 14.
88   Wilkins, Papers, Ohio State University, Box 1, Folder 14.
89   Wilkins, Papers, Ohio State University, Box 1, Folder 14.
90   Wilkins, Papers, Ohio State University, Box 1, Folder 14.
91   Bean, AWM, Official History, 1914–18 War, 38 3DRL 606 117-1, p. 9 [reported speech].
92   Bean, AWM, Official History, 1914–18 War, 38 3DRL 606 117-1, p. 9 [reported speech].
93   Bean, AWM, Official History, 1914–18 War, 38 3DRL 606 117-1, p. 9 [reported speech].
94   Bean, *Official History of Australia in the War of 1914–1918*, Vol. VI, Angus & Robertson, Sydney, 1942, p. 962.
95   Thomas, *Sir Hubert Wilkins – His World of Adventure*, p. 96.
96   Thomas, *Sir Hubert Wilkins – His World of Adventure*, p. 96.
97   Thomas, *Sir Hubert Wilkins – His World of Adventure*, p. 97.
98   Andrews, *Hubert Who?*, p. 73. Author's note: The puzzle of precisely for which action Wilkins gained his bar persists to this day, with a few contenders for the recommendation that confirmed his second MC. Wilkins himself was unsure, leaving differing accounts as to the action that earned the honour.

99 *The Sun* (NSW), 1 January 1929, p. 12.
100 Wilkins, Papers, Ohio State University, Box 1, Folder 14, no page number.
101 *The Times of London*, 12 November 1918, p. 1.
102 *The Times of London*, 12 November 1918, p. 1.

Chapter 9

1 Wilkins, *Flying the Arctic*, p. 60.
2 Hughes, *Billy Hughes: Prime Minister and Controversial Founding Father of the Australian Labor Party*, John Wiley & Sons, Brisbane, 2005, p. 77.
3 Telegram, 18 February 1919, quoted in Nasht, *The Last Explorer*, p. 76.
4 Hamilton, *Goodbye Cobber, God Bless You*, Pan Macmillan, Sydney, 2005, p. 243.
5 Bean, *The Official History of Australia in the War of 1914–1918*, Vol. II, Angus & Robertson, Sydney, 1941, p. 612.
6 Simpson, *Maygar's Boys: A biographical history of the 8th Light Horse Regiment A.I.F. 1914–19*, Just Soldiers, Military Research & Publications, Moorooduc, 1998, p. 279.
7 McCarthy, *Gallipoli to the Somme*, p. 174.
8 *The West Australian*, 28 September 1915, p. 5.
9 Bean, *Gallipoli Mission*, AWM, 1948, p. 109.
10 Bean, *Gallipoli Mission*, p. 109.
11 Bean, *Gallipoli Mission*, p. 50.
12 Bean, *Gallipoli Mission*, p. 16.
13 Maynard, *The Unseen Anzac*, p. 184.
14 Bean, *Gallipoli Mission*, p. 280.
15 Bean, *Gallipoli Mission*, p. 280.
16 Maynard, *The Unseen Anzac*, p. 182 [reported speech].
17 Thomas, *Sir Hubert Wilkins – His World of Adventure*, p. 97.
18 Thomas, *Sir Hubert Wilkins – His World of Adventure*, p. 97.
19 Andrews, *Hubert Who?*, pp. 78–79.
20 Bean, *Gallipoli Mission*, p. 104.
21 Bean, *Gallipoli Mission*, pp. 233–234.
22 *The Capricornian*, 13 December 1919, p. 31.
23 *The Capricornian*, 13 December 1919, p. 31.
24 *The Capricornian*, 13 December 1919, p. 31.
25 *Townsville Daily Bulletin*, 23 September 1938, p. 12.
26 Joy, *The Aviators*, p. 41.
27 Kingsford Smith, letter to his parents in Sydney, sent from Hitchin, UK, 17 April 1919, Norman Ellison Collection, NLA, Canberra.
28 *The Queenslander*, 5 April 1928, p. 61.
29 *Queensland Times*, 25 November 1919, p. 5.
30 *The Sydney Morning Herald*, in *The Capricornian*, 13 December 1919, p. 31.
31 *The Herald*, 24 November 1919, p. 1.
32 *The Herald*, 24 November 1919, p. 1.
33 *The Herald*, 24 November 1919, p. 1.
34 *The Herald*, 24 November 1919, p. 1.
35 *The Herald*, 24 November 1919, p. 1.
36 *The Herald*, 24 November 1919, p. 1 [reported speech].
37 *The Herald*, 24 November 1919, p. 1.
38 Wilkins, Papers, Ohio State University, Box 13, Folder 14, England to Australia Race, p. 43.
39 Wilkins Foundation, 'Where in Crete is the Blackburn Kangaroo'.
40 Wilkins, Papers, Ohio State University, Box 13, Folder 14, p. 43.
41 Wilkins, Papers, Ohio State University, Box 13, Folder 14, p. 43.
42 Wilkins, Papers, Ohio State University, Box 13, Folder 14, p. 43.
43 Wilkins, Papers, Ohio State University, Box 13, Folder 14, pp. 43–44.
44 Wilkins to Bean, Cable, 27 January 1920, Wilkins, Papers, Ohio State University.
45 *The New York Times*, 24 January 1920, p. 6.
46 *The New York Times*, 24 April 1919, p. 3.
47 Thomas, *Sir Hubert Wilkins – His World of Adventure*, p. 107.

48 *The Sun* (Kalgoorlie), 17 October 1920, p. 5.
49 *The Sydney Morning Herald*, 8 October 1920, p. 8.
50 Wilkins, Papers, Ohio State University, Box 13, Folder 16, British Imperial Antarctic Expedition 1919–1920, p. 40.
51 Wilkins, Papers, Ohio State University, Box 13, Folder 16, p. 40.
52 Thomas, *Sir Hubert Wilkins – His World of Adventure*, p. 107 [reported speech].
53 Thomas, *Sir Hubert Wilkins – His World of Adventure*, p. 107 [reported speech].
54 *Nature*, Vol. 104, no. 93, 1919, p. 93.
55 *Border Morning Mail and Riverina Times*, 4 November 1919, p. 2.
56 Maynard, *The Unseen Anzac*, p. 197.
57 Maynard, *The Unseen Anzac*, p. 197.
58 Wilkins, Papers, Ohio State University, Box 13, Folder 16, p. 49.
59 Wilkins, Papers, Ohio State University, Box 13, Folder 16, p. 47.
60 Wilkins, Papers, Ohio State University, Box 13, Folder 16, p. 47.
61 Wilkins, Papers, Ohio State University, Box 13, Folder 16, no page number [reported speech].
62 Thomas, *Sir Hubert Wilkins – His World of Adventure*, p. 110.
63 Thomas, *Sir Hubert Wilkins – His World of Adventure*, p. 110.
64 Thomas, *Sir Hubert Wilkins – His World of Adventure*, p. 110.
65 Thomas, *Sir Hubert Wilkins – His World of Adventure*, p. 110.
66 Thomas, *Sir Hubert Wilkins – His World of Adventure*, p. 111.
67 Thomas, *Sir Hubert Wilkins – His World of Adventure*, p. 111.
68 Thomas, *Sir Hubert Wilkins – His World of Adventure*, p. 111.
69 Thomas, *Sir Hubert Wilkins – His World of Adventure*, pp. 111–112.
70 Thomas, *Sir Hubert Wilkins – His World of Adventure*, p. 112.
71 Thomas, *Sir Hubert Wilkins – His World of Adventure*, p. 112.
72 Thomas, *Sir Hubert Wilkins – His World of Adventure*, p. 112.
73 Thomas, *Sir Hubert Wilkins – His World of Adventure*, p. 113.
74 Thomas, *Sir Hubert Wilkins – His World of Adventure*, p. 113.
75 Bagshawe, *Two Men in the Antarctic: An Expedition to Graham Land, 1920–1922*, The Macmillan Company, New York, 1939, p. 57.
76 Thomas, *Sir Hubert Wilkins – His World of Adventure*, p. 114.
77 Thomas, *Sir Hubert Wilkins – His World of Adventure*, p. 114.
78 Thomas, *Sir Hubert Wilkins – His World of Adventure*, p. 114.
79 Thomas, *Sir Hubert Wilkins – His World of Adventure*, p. 114.
80 Thomas, *Sir Hubert Wilkins – His World of Adventure*, p. 114.
81 Thomas, *Sir Hubert Wilkins – His World of Adventure*, p. 116.
82 Thomas, *Sir Hubert Wilkins – His World of Adventure*, p. 116.
83 Wilkins, Papers, Ohio State University, Box 13, Folder 16, pp. 47–48.
84 Thomas, *Sir Hubert Wilkins – His World of Adventure*, p. 116.
85 Author's note: Unbeknownst to Wilkins, another Norwegian whaler did in fact call on Bagshawe and Lester and offered them a trip home, but they dutifully refused. Such was Cope's spell over these men that his subterfuge and threats were entirely unnecessary; both of them enjoyed their year in the Antarctic and felt they were doing valuable work, despite the non-return of their dubious leader.
86 Thomas, *Sir Hubert Wilkins – His World of Adventure*, p. 117.
87 Thomas, *Sir Hubert Wilkins – His World of Adventure*, p. 117.
88 Thomas, *Sir Hubert Wilkins – His World of Adventure*, p. 117.
89 Thomas, *Sir Hubert Wilkins – His World of Adventure*, p. 117.
90 Thomas, *Sir Hubert Wilkins – His World of Adventure*, p. 117.
91 Bagshawe, *Two Men in the Antarctic*, Introduction.
92 Thomas, *Sir Hubert Wilkins – His World of Adventure*, p. 118 [reported speech].
93 Thomas, *Sir Hubert Wilkins – His World of Adventure*, p. 119.
94 J.M.W., 'Obituary: Sir Ernest Henry Shackleton, C. V. O., O. B. E.', *The Geographical Journal*, Vol. 59, No. 3, March 1922, p. 230.
95 R.E.P., 'Obituary: Frank Wild', *The Geographical Journal*, Vol. 95, No. 3, March 1940, p. 239.
96 Thomas, *Sir Hubert Wilkins – His World of Adventure*, p. 122.
97 Thomas, *Sir Hubert Wilkins – His World of Adventure*, p. 123.

98 Thomas, *Sir Hubert Wilkins – His World of Adventure*, p. 123.

99 Thomas, *Sir Hubert Wilkins – His World of Adventure*, p. 123.

100 Thomas, *Sir Hubert Wilkins – His World of Adventure*, p. 117.

101 Thomas, *Sir Hubert Wilkins – His World of Adventure*, p. 124.

102 Nasht, *The Last Explorer*, p. 106.

103 Nasht, *The Last Explorer*, p. 106.

104 Wild, *Shackleton's Last Voyage. The story of the Quest. By Commander Frank Wild, C.B.E*, Cassell, London, 1923, p. 64.

105 Thomas, *Sir Hubert Wilkins – His World of Adventure*, p. 125.

106 *The West Australian*, 7 June 1923, p. 8.

107 *The New York Times*, 22 April 1928, p. 4.

108 *The New York Times*, 22 April 1928, p. 4.

109 *The New York Times*, 22 April 1928, p. 4.

Chapter 10

1 Nasht, *The Last Explorer*, p. 112.

2 *The Herald*, 2 January 1926, p. 5.

3 Thomas, *Sir Hubert Wilkins – His World of Adventure*, p. 126.

4 Thomas, *Sir Hubert Wilkins – His World of Adventure*, p. 127.

5 Thomas, *Sir Hubert Wilkins – His World of Adventure*, p. 127.

6 Thomas, *Sir Hubert Wilkins – His World of Adventure*, p. 128.

7 Thomas, *Sir Hubert Wilkins – His World of Adventure*, p. 129.

8 Wilkins, Manuscript, 'Some Notes on Conditions Existing in Soviet Russia'.

9 Wilkins, Manuscript, 'Some Notes on Conditions Existing in Soviet Russia'.

10 Wilkins, Manuscript, 'Some Notes on Conditions Existing in Soviet Russia'.

11 Wilkins, Manuscript, 'Some Notes on Conditions Existing in Soviet Russia'.

12 *The Sun* (NSW), 1 April 1923, p. 13.

13 Wilkins, Manuscript, 'Some Notes on Conditions Existing in Soviet Russia'.

14 Wilkins, Manuscript, 'Some Notes on Conditions Existing in Soviet Russia'.

15 Wilkins, Manuscript, 'Some Notes on Conditions Existing in Soviet Russia'.

16 Wilkins, Manuscript, 'Some Notes on Conditions Existing in Soviet Russia'.

17 Wilkins, Manuscript, 'Some Notes on Conditions Existing in Soviet Russia'.

18 Wilkins, *True Adventure Thrills*, episode 43.

19 Nasht, *The Last Explorer*, p. 118.

20 Nasht, *The Last Explorer*, p. 118.

21 *Tribune*, 19 October 1988, p. 8

22 Nasht, *The Last Explorer*, p. 119.

23 Albig, *Public Opinion*, McGraw Hill, New York, 1939, p. 322.

24 Nasht, *The Last Explorer*, p. 119.

25 Nasht, *The Last Explorer*, p. 119.

26 Wilkins, *Undiscovered Australia*, Ernest Benn, London, 1928, p. 11.

27 Wilkins, *Undiscovered Australia*, pp. 13–14.

28 Wilkins, *Undiscovered Australia*, p. 14.

29 Wilkins, *Undiscovered Australia*, p. 13.

30 Wilkins, *Undiscovered Australia*, p. 15.

31 Wilkins, *Undiscovered Australia*, p. 15.

32 *The Queenslander*, 28 April 1923, p. 9

33 *The Sun* (NSW), 31 March 1923, p. 4.

34 *The Queenslander*, 14 April 1923, p. 15.

35 *The Queenslander*, 14 April 1923, p. 15.

36 Wilkins, *Undiscovered Australia*, p. 21.

37 Wilkins, *Undiscovered Australia*, p. 24.

38 *The Daily Telegraph*, 1 June 1923, p. 7.

39 Wilkins, *Undiscovered Australia*, p. 24.

40 Wilkins, *Undiscovered Australia*, p. 26.

41 Wilkins, *Undiscovered Australia*, p. 26.

42 *West Australian*, 7 June 1923, p. 8.

43 Wilkins, *Undiscovered Australia*, p. 269.
44 Wilkins, *Undiscovered Australia*, p. 27.
45 Wilkins, *Undiscovered Australia*, p. 28.
46 Wilkins, *Undiscovered Australia*, p. 28.
47 Wilkins, *Undiscovered Australia*, p. 28.
48 Wilkins, *Undiscovered Australia*, p. 28.
49 Wilkins, *Undiscovered Australia*, p. 28.
50 Wilkins, *Undiscovered Australia*, p. 28.
51 Wilkins, *Undiscovered Australia*, p. 20.
52 Wilkins, *Undiscovered Australia*, p. 29.
53 *The Advertiser*, 2 July 1923, p. 11.
54 *The Daily Telegraph*, 1 June 1923, p. 7.
55 *The Daily Mail*, 11 June 1923, p. 7.
56 *The Daily Mail*, 11 June 1923, p. 7.
57 *The Daily Telegraph*, 12 September 1923, p. 4.
58 *The Advertiser*, 2 July 1923, p. 11.
59 *The Advertiser*, 2 July 1923, p. 11.
60 *The Examiner*, 2 July 1923, p. 7.
61 *The Advertiser*, 2 July 1923, p. 11.
62 *The Advertiser*, 2 July 1923, p. 11.
63 Wilkins, *Undiscovered Australia*, p. 30.
64 Wilkins, *Undiscovered Australia*, p. 33.
65 Wilkins, *Undiscovered Australia*, p. 33.
66 Wilkins, *Undiscovered Australia*, p. 34.
67 Wilkins, *Undiscovered Australia*, p. 34.
68 Wilkins, *Undiscovered Australia*, p. 37.
69 Author's note: Given the publishing edicts and censorship of the time, Wilkins liberally used '-------' and '-------' to indicate the language used by those he encountered. Fortunately, I am no longer under such ------ing stupid restrictions and have made my best supposition at the language used. (It was in fact quite daring of Wilkins to use so much '-----', usually 'bloody', 'bastard' and 'bugger' were the only words dared to be hinted at, but Wilkins rejoiced in the profuse profanity used and brilliantly conveys the variety and depths of terms casually employed by the bushmen of the time.)
70 Wilkins, *Undiscovered Australia*, p. 37 ['dashes' of swearing filled in].
71 Wilkins, *Undiscovered Australia*, p. 37.
72 Wilkins, *Undiscovered Australia*, p. 37.
73 Wilkins, *Undiscovered Australia*, p. 37.
74 Wilkins, *Undiscovered Australia*, p. 37.
75 Wilkins, *Undiscovered Australia*, p. 38.
76 Wilkins, *Undiscovered Australia*, p. 39.
77 Wilkins, *Undiscovered Australia*, p. 39.
78 Wilkins, *Undiscovered Australia*, p. 39.
79 Wilkins, *Undiscovered Australia*, p. 40.
80 Wilkins, *Undiscovered Australia*, p. 40.
81 Wilkins, *Undiscovered Australia*, p. 40.
82 Wilkins, *Undiscovered Australia*, p. 40.
83 Wilkins, *Undiscovered Australia*, p. 40.

Chapter 11

1 Wilkins, *Undiscovered Australia*, p. 9.
2 *Daily Mercury*, 31 October 1923, p. 7.
3 Wilkins, *Undiscovered Australia*, p. 51.
4 Wilkins, *Undiscovered Australia*, p. 51.
5 Wilkins, *Undiscovered Australia*, p. 51 [reported speech].
6 Wilkins, *Undiscovered Australia*, p. 51.
7 Wilkins, *Undiscovered Australia*, p. 51.
8 Wilkins, *Undiscovered Australia*, p. 51.

9   Wilkins, *Undiscovered Australia*, p. 52.
10 Wilkins, *Undiscovered Australia*, p. 52.
11 Wilkins, *Undiscovered Australia*, p. 53.
12 Wilkins, *Undiscovered Australia*, p. 53.
13 Wilkins, *Undiscovered Australia*, p. 63.
14 Wilkins, *Undiscovered Australia*, p. 65.
15 Wilkins, *Undiscovered Australia*, p. 65.
16 Wilkins, *Undiscovered Australia*, p. 65.
17 Wilkins, *Undiscovered Australia*, p. 65.
18 Wilkins, *Undiscovered Australia*, p. 72.
19 Wilkins, *Undiscovered Australia*, p. 72.
20 Wilkins, Papers, Ohio State University, Box 13.1, Folder 8, Wilkins Australia and Island Expedition, p. 108.
21 Ellwood, 'The Aboriginal Miners and Prospectors of Cape York Peninsula, 1870 to ca. 1950s', *Journal of Australasian Mining History*, Vol. 16, October 2018.
22 Wilkins, *Undiscovered Australia*, p. 73.
23 Wilkins, *Undiscovered Australia*, p. 74.
24 Wilkins, *Undiscovered Australia*, p. 73.
25 Wilkins, *Undiscovered Australia*, p. 74.
26 Wilkins, *Undiscovered Australia*, p. 85.
27 Wilkins, *Undiscovered Australia*, p. 99.
28 Wilkins, *Undiscovered Australia*, p. 112.
29 Wilkins, *Undiscovered Australia*, p. 134.
30 *The Leader*, 6 November 1925, p. 6.
31 *The Leader*, 6 November 1925, p. 6.
32 *The Leader*, 6 November 1925, p. 6.
33 *Morning Bulletin*, 3 February 1925, p. 12.
34 Wilkins, *Undiscovered Australia*, p. 136.
35 Wilkins, *Undiscovered Australia*, p. 138.
36 Wilkins, *Undiscovered Australia*, p. 138.
37 Wilkins, *Undiscovered Australia*, p. 138.
38 Wilkins, *Undiscovered Australia*, p. 138.
39 Wilkins, *Undiscovered Australia*, p. 138.
40 Wilkins, *Undiscovered Australia*, p. 138.
41 Wilkins, *Undiscovered Australia*, p. 138.
42 Wilkins, *Undiscovered Australia*, p. 138.
43 Wilkins, *Undiscovered Australia*, p. 138.
44 Wilkins, *Undiscovered Australia*, p. 139.
45 *Northern Territory Times and Gazette*, 9 September 1924, p. 3.
46 Wilkins, *Undiscovered Australia*, p. 144.
47 Wilkins, *Undiscovered Australia*, p. 144.
48 Wilkins, *Undiscovered Australia*, p. 145.
49 Wilkins, *Undiscovered Australia*, p. 145.
50 Wilkins, *Undiscovered Australia*, p. 145.
51 Wilkins, *Undiscovered Australia*, p. 145.
52 Wilkins, *Undiscovered Australia*, p. 146.
53 Wilkins, *Undiscovered Australia*, p. 146.
54 Wilkins, *Undiscovered Australia*, p. 146.
55 Wilkins, *Undiscovered Australia*, p. 146.
56 Wilkins, *Undiscovered Australia*, p. 146.
57 Wilkins, *Undiscovered Australia*, p. 146.
58 Wilkins, *Undiscovered Australia*, p. 146.
59 Wilkins, *Undiscovered Australia*, p. 147.
60 Wilkins, *Undiscovered Australia*, p. 147.
61 Wilkins, *Undiscovered Australia*, p. 147.
62 Wilkins, *Undiscovered Australia*, p. 147.
63 Wilkins, *Undiscovered Australia*, p. 147.

64 Wilkins, *Undiscovered Australia*, p. 147.
65 Wilkins, *Undiscovered Australia*, p. 147.
66 *Northern Standard*, 31 August 1923, p. 2.
67 Wilkins, *Undiscovered Australia*, p. 147.
68 Wilkins, *Undiscovered Australia*, p. 148.
69 Wilkins, *Undiscovered Australia*, p. 148.
70 Wilkins, *Undiscovered Australia*, p. 148.
71 Wilkins, *Undiscovered Australia*, p. 149.
72 Wilkins, *Undiscovered Australia*, p. 150.
73 Wilkins, *Undiscovered Australia*, p. 150.
74 Wilkins, *Undiscovered Australia*, p. 156.
75 Wilkins, *Undiscovered Australia*, p. 188.
76 Wilkins, *Undiscovered Australia*, p. 200.
77 Wilkins, *Undiscovered Australia*, p. 200.
78 Wilkins, *Undiscovered Australia*, p. 201.
79 Wilkins, *Undiscovered Australia*, p. 202.
80 Wilkins, *Undiscovered Australia*, p. 197.
81 Wilkins, *Undiscovered Australia*, p. 204.
82 Wilkins, *Undiscovered Australia*, p. 204.
83 Wilkins, *Undiscovered Australia*, p. 205.
84 Wilkins, *Undiscovered Australia*, p. 206.
85 Wilkins, *Undiscovered Australia*, p. 257.
86 Wilkins, *Undiscovered Australia*, pp. 258–259.
87 Wilkins, *Undiscovered Australia*, p. 261.
88 Wilkins, *Undiscovered Australia*, p. 261.
89 Wilkins, *Undiscovered Australia*, p. 261.
90 Wilkins, *Undiscovered Australia*, p. 261.
91 Wilkins, *Undiscovered Australia*, pp. 261–262.
92 *Truth*, 28 June 1925, p. 8.
93 *Truth*, 28 June 1925, p. 8.
94 Wilkins, Papers, Ohio State University, Box 13.1, Folder 8.
95 Sidney F. Harmer, British Museum Director, to Wilkins 28 Oct 1925, from Wilkins, Papers, Ohio State University.
96 Wilkins, *Undiscovered Australia*, p. 41.
97 *News* (SA), 7 August 1925, p. 10.
98 Maynard, *The Unseen Anzac*, p. 203.
99 Maynard, *The Unseen Anzac*, p. 203.

Chapter 12

1 *Daily Advertiser*, 19 February 1926, p. 2
2 *The Queenslander*, 5 April 1928, p. 61.
3 *Morning Bulletin*, 4 January 1926, p. 7.
4 *Morning Bulletin*, 4 January 1926, page 7 [reported speech, with generic introduction].
5 *Morning Bulletin*, 4 January 1926, p. 7 [reported speech].
6 *Morning Bulletin*, 4 January 1926, p. 7.
7 *Morning Bulletin*, 4 January 1926, p. 7.
8 *Morning Bulletin*, 4 January 1926, p. 7.
9 *Morning Bulletin*, 4 January 1926, p. 7.
10 *The Sun* (NSW), 17 February 1926, p. 14.
11 *Daily Advertiser*, 19 February 1926, p. 2.
12 Thomas, *Sir Hubert Wilkins: His World of Adventure*, McGraw Hill, New York, USA, 1961, p. 184.
13 Wilkins, *Flying the Arctic*, p. 10.
14 *The Sun* (NSW), 17 February 1926, p. 14.
15 *The Telegraph*, 8 October 1931, p. 6.
16 Wilkins, *Flying the Arctic*, p. 11.
17 Wilkins, Papers, Ohio State University, Box 13.1, Folder 10, Detroit Arctic Expeditions.

18 Wilkins, *Flying the Arctic*, p. 40.
19 *The Daily Mail*, 26 March 1926, p. 3.
20 Nasht, *The Last Explorer*, p. 141.
21 Nasht, *The Last Explorer*, p. 141.
22 *Weekly Times*, 30 January 1926, p. 5.
23 Wilkins, *Flying the Arctic*, p. 13.
24 Wilkins, *Flying the Arctic*, p. 13.
25 Wilkins, *Flying the Arctic*, pp. 14–15.
26 Wilkins, *Flying the Arctic*, pp. 14–15.
27 Wilkins, *Flying the Arctic*, p. 16.
28 Wilkins, *Flying the Arctic*, p. 16.
29 Wilkins, *Flying the Arctic*, p. 217.
30 Wilkins, *Flying the Arctic*, p. 18.
31 *The Telegraph*, 8 October 1931, p. 6.
32 *Chronicle*, 1 May 1926, p. 73.
33 *Leader*, 15 February 1926, p. 2.
34 *Weekly Times*, 30 January 1926, p. 5.
35 *Weekly Times*, 30 January 1926, p. 5.
36 *Weekly Times*, 30 January 1926, p. 5.
37 Wilkins, *Flying the Arctic*, p. 20.
38 Wilkins, *Flying the Arctic*, p. 19.
39 Wilkins, *Flying the Arctic*, p. 24.
40 Wilkins, *Flying the Arctic*, p. 23.
41 *The Evening News*, 15 February 1926, p. 5.
42 *The Sun* (NSW), 14 February 1926, p. 1
43 Wilkins, *Flying the Arctic*, p. 26.
44 *The Daily Mail*, 26 March 1926, p. 3.
45 *The Advertiser*, 8 March 1926, p. 12.
46 Wilkins, *Flying the Arctic*, p. 30.
47 Wilkins, *Flying the Arctic*, p. 30.
48 Wilkins, *Flying the Arctic*, p. 31.
49 Wilkins, *Flying the Arctic*, p. 32.
50 FitzSimons, *Charles Kingsford Smith and Those Magnificent Men in Their Flying Machines*, HarperCollins, Sydney, 2010, p. 134.
51 Wilkins, *Flying the Arctic*, p. 32.
52 Wilkins, *Flying the Arctic*, p. 33.
53 Wilkins, *Flying the Arctic*, p. 35.
54 Wilkins, *Flying the Arctic*, p. 35.
55 Wilkins, *Flying the Arctic*, p. 36.
56 Wilkins, *Flying the Arctic*, p. 37.
57 Wilkins, *Flying the Arctic*, p. 38.
58 Wilkins, *Flying the Arctic*, p. 38.
59 Wilkins, *Flying the Arctic*, p. 38.
60 Wilkins, *Flying the Arctic*, p. 38.
61 Wilkins, *Flying the Arctic*, p. 42.
62 Wilkins, *Flying the Arctic*, p. 42.
63 Wilkins, *Flying the Arctic*, p. 42.
64 Wilkins, *Flying the Arctic*, p. 43.
65 Wilkins, *Flying the Arctic*, p. 43.
66 Wilkins, *Flying the Arctic*, p. 44.
67 Wilkins, *Flying the Arctic*, p. 44.
68 Wilkins, *Flying the Arctic*, p. 45.
69 Wilkins, *Flying the Arctic*, p. 46.
70 Wilkins, *Flying the Arctic*, p. 46.
71 Wilkins, *Flying the Arctic*, p. 47.

Chapter 13

1 Wilkins, *Flying the Arctic*, p. ix.
2 Nasht, *The Last Explorer*, p. 314.
3 Wilkins, *Flying the Arctic*, p. 47.
4 *News*, 19 May 1926, p. 9.
5 Wilkins, *Flying the Arctic*, p. 53.
6 Wilkins, *Flying the Arctic*, p. 54.
7 Wilkins, *Flying the Arctic*, p. 54.
8 Wilkins, *Flying the Arctic*, p. 54.
9 Wilkins, *Flying the Arctic*, p. 54.
10 Wilkins, *Flying the Arctic*, p. 55.
11 Wilkins, *Flying the Arctic*, p. 55.
12 Wilkins, *Flying the Arctic*, p. 55.
13 Wilkins, *Flying the Arctic*, p. 56.
14 Wilkins, *Flying the Arctic*, pp. 56–57.
15 Wilkins, *Flying the Arctic*, p. 57.
16 Wilkins, *Flying the Arctic*, p. 57.
17 Wilkins, *Flying the Arctic*, p. 58.
18 Wilkins, *Flying the Arctic*, p. 58.
19 Wilkins, *Flying the Arctic*, p. 59.
20 Wilkins, *Flying the Arctic*, p. 60.
21 Wilkins, *Flying the Arctic*, p. 61.
22 Wilkins, *Flying the Arctic*, p. 62.
23 Wilkins, *Flying the Arctic*, p. 63.
24 Thomas, *Sir Hubert Wilkins: His World of Adventure*, McGraw Hill, p. 150.
25 Wilkins, *Flying the Arctic*, p. 64.
26 Wilkins, *Flying the Arctic*, p. 64.
27 *The Herald*, 1 May 1926, p. 9.
28 *The Daily Mail*, 2 May 1926, p. 1.
29 Wilkins, *Flying the Arctic*, p. 67.
30 Wilkins, *Flying the Arctic*, p. 67 [reported speech].
31 Wilkins, *Flying the Arctic*, p. 255.
32 Wilkins, *Flying the Arctic*, p. 75.
33 Wilkins, *Flying the Arctic*, p. 75.
34 *Daily Standard*, 8 April 1926, p. 1.
35 *Daily Standard*, 8 April 1926, p. 1.
36 *Chronicle*, 29 May 1926, p. 17.
37 Wilkins, *Flying the Arctic*, p. 83.
38 Author's note: Although not named by Wilkins in his account in *Flying the Arctic*, an examination of his route reveals Allakaket to be the existing village of that time that meets the geographical markers laid out.
39 *Newcastle Sun*, 12 April 1926, p. 1.
40 Wilkins, *Flying the Arctic*, p. 85.
41 Wilkins, *Flying the Arctic*, p. 84.
42 Wilkins, *Flying the Arctic*, p. 85.
43 Wilkins, *Flying the Arctic*, pp. 85–86.
44 Wilkins, *Flying the Arctic*, pp. 85–86.
45 Wilkins, *Flying the Arctic*, p. 86.
46 *Weekly Times*, 8 May 1926, p. 6.
47 *Weekly Times*, 8 May 1926, p. 6.
48 Wilkins, *Flying the Arctic*, p. 96.
49 Wilkins, *Flying the Arctic*, p. 97.
50 Wilkins, *Flying the Arctic*, p. 97.
51 Wilkins, *Flying the Arctic*, p. 93.
52 Wilkins, *Flying the Arctic*, pp. 101–102.
53 Wilkins, *Flying the Arctic*, p. 102.
54 *The Advertiser*, 5 February 1927, p. 5.

55 Wilkins, *Flying the Arctic*, p. 114.
56 Wilkins, *Flying the Arctic*, p. 133.
57 Wilkins, *Flying the Arctic*, pp. 133–134.
58 Wilkins, *Flying the Arctic*, p. 132.
59 Wilkins, *Flying the Arctic*, p. 129.
60 Wilkins, Papers, Ohio State University, Box 13.1, Folder 10, Detroit Arctic Expeditions, An Arctic Adventure by Howard Mason, p. 3.
61 Wilkins, *Flying the Arctic*, p. 141.
62 Wilkins, Papers, Ohio State University, Box 13.1, Folder 10, p. 3.
63 Wilkins, *Flying the Arctic*, p. 143.
64 Wilkins, *Flying the Arctic*, p. 144. Author's note: I note that accounts of this incident, including Wilkins' own book *Flying the Arctic* have the above conversation recorded as a thought that Eielson later told Wilkins. However, Wilkins' contemporary account in his archives records this as being *said* by Ben *at the time*. My speculation is that Eielson may have been embarrassed by these words as it made him look as though he was complaining or being defeatist, and so Wilkins changed it to a thought Ben kept to himself until success had been achieved. I regard it as a wonderful example of gallows humour and am pleased to restore Eielson's dark wit to its rightful place in this incredible story.
65 Wilkins, *Flying the Arctic*, p. 147.
66 Wilkins, *Flying the Arctic*, p. 149.
67 Wilkins, *Flying the Arctic*, p. 149.
68 *The Telegraph*, 11 July 1927, p. 5.
69 Wilkins, *Flying the Arctic*, p. 149.
70 *The West Australian*, 28 December 1928, p. 8.
71 Wilkins, *Flying the Arctic*, p. 150.
72 *The Telegraph*, 11 July 1927, p. 5.
73 Wilkins, *Flying the Arctic*, p. 151.
74 Wilkins, Papers, Ohio State University, Box 13.1, Folder 10, p. 3.
75 Wilkins, *Flying the Arctic*, p. 152.
76 Wilkins, *Flying the Arctic*, p. 152.
77 Wilkins, *Flying the Arctic*, p. 154.
78 *The Telegraph*, 11 July 1927, p. 5.
79 Wilkins, *Flying the Arctic*, p. 154.
80 *Geraldton Express*, 25 May 1927, p. 1.
81 *Geraldton Express*, 25 May 1927, p. 1.
82 Wilkins, *Flying the Arctic*, p. 157.
83 Wilkins, *Flying the Arctic*, p. 157.
84 Wilkins, *Flying the Arctic*, p. 159.
85 Wilkins, *Flying the Arctic*, p. 159.
86 Wilkins, *Flying the Arctic*, p. 159.
87 Wilkins, *Flying the Arctic*, p. 159.
88 Wilkins, *Flying the Arctic*, pp. 159–160.
89 Wilkins, *Flying the Arctic*, p. 161.
90 Wilkins, *Flying the Arctic*, p. 161.
91 Wilkins, *Flying the Arctic*, p. 163.
92 Thomas, *Sir Hubert Wilkins – His World of Adventure*, p. 165.
93 Wilkins, *Flying the Arctic*, p. 164.
94 Wilkins, *Flying the Arctic*, p. 166.
95 Wilkins, *Flying the Arctic*, p. 166.
96 Thomas, *Sir Hubert Wilkins – His World of Adventure*, p. 165.
97 Wilkins, *Flying the Arctic*, p. 168.
98 Wilkins, *Flying the Arctic*, p. 169.
99 Wilkins, Papers, Ohio State University, Box 13.1, Folder 10, p. 4.
100 Wilkins, Papers, Ohio State University, Box 13.1, Folder 10, p. 4 [reported speech].
101 Wilkins, *Flying the Arctic*, p. 171.
102 Wilkins, *Flying the Arctic*, p. 171.
103 Wilkins, *Flying the Arctic*, p. 180.

104 Wilkins, *Flying the Arctic*, p. 180.
105 Wilkins, *Flying the Arctic*, p. 185.
106 Wilkins, *Flying the Arctic*, p. 185.
107 Wilkins, *Flying the Arctic*, p. 186.
108 Wilkins, *Flying the Arctic*, p. 193.

Chapter 14

1 Wilkins, *Flying the Arctic*, p. 198.
2 Wilkins, *Flying the Arctic*, p. 198.
3 Wilkins, *Flying the Arctic*, p. 200.
4 'Lockheed Vega', FiddlersGreen.net, http://www.fiddlersgreen.net/models/aircraft/Lockheed-Vega.html
5 Kingsford Smith, *The Old Bus*, Herald Press, Melbourne, Australia, 1932, p. 9.
6 Stannage, *Smithy*, Oxford University Press, London, UK, 1950, p. 23.
7 Stannage, *Smithy*, p. 23.
8 Wilkins, letter to a Mr Byrne, sent from New York City, 29 October 1936.
9 Wilkins, *Flying the Arctic*, p. 211.
10 Wilkins, *Flying the Arctic*, p. 213.
11 Wilkins, *Flying the Arctic*, p. 214.
12 Wilkins, Papers, Ohio State University, Box 14, Folder 1, Wilkins-Hearst Antarctic Expedition Correspondence, 1928–1930.
13 Wilkins, *Flying the Arctic*, p. 215.
14 Wilkins, *Flying the Arctic*, p. 217.
15 Wilkins, *Flying the Arctic*, p. 217.
16 Wilkins, *Flying the Arctic*, p. 221.
17 Wilkins, *Flying the Arctic*, pp. 233–234.
18 *The Daily Express*, 6 April 1926, p. 2.
19 Wilkins, *Flying the Arctic*, p. 235.
20 Wilkins, *Flying the Arctic*, pp. 238–239.
21 Wilkins, *Flying the Arctic*, p. 241.
22 Wilkins, *Flying the Arctic*, p. 243.
23 Wilkins, *Flying the Arctic*, p. 245.
24 Wilkins, *Flying the Arctic*, p. 246.
25 *Tweed Daily*, 9 June 1928, p. 5.
26 Wilkins, *Flying the Arctic*, p. 255.
27 Wilkins, *Flying the Arctic*, p. 255.
28 Wilkins, *Flying the Arctic*, p. 257.
29 *Kalgoorlie Miner*, 25 April 1928, p. 6.
30 Wilkins, *Flying the Arctic*, p. 268.
31 Wilkins, *Flying the Arctic*, p. 268.
32 Wilkins, *Flying the Arctic*, p. 269.
33 *The Argus*, 26 April 1928, p. 6.
34 Wilkins, *Flying the Arctic*, p. 269.
35 Wilkins, *Flying the Arctic*, pp. 269–270.
36 Wilkins, *Flying the Arctic*, p. 270.
37 Wilkins, *Flying the Arctic*, p. 270.
38 Wilkins, *Flying the Arctic*, p. 271.
39 Wilkins, *Flying the Arctic*, p. 271.
40 *Kalgoorlie Miner*, 25 April 1928, p. 6.
41 *Kalgoorlie Miner*, 25 April 1928, p. 6.
42 Wilkins, *Flying the Arctic*, p. 272.
43 Wilkins, *Flying the Arctic*, p. 274.
44 *The Sun* (NSW), 15 April 1926, p. 1.
45 Wilkins, *Flying the Arctic*, p. 279.
46 Wilkins, *Flying the Arctic*, p. 282.
47 Wilkins, *Flying the Arctic*, p. 282.
48 *Geelong Advertiser*, 24 April 1928, p. 5.

49 Wilkins, *Flying the Arctic*, p. 283.
50 Wilkins, *Flying the Arctic*, p. 283.
51 Wilkins, *Flying the Arctic*, p. 285.
52 Wilkins, *Flying the Arctic*, p. 284.
53 Wilkins, *Flying the Arctic*, p. 284.
54 *The Telegraph*, 25 April 1928, p. 6.
55 Wilkins, *Flying the Arctic*, p. 285.
56 Wilkins, *Flying the Arctic*, p. 286.
57 *Geelong Advertiser*, 24 April 1928, p. 5.
58 Wilkins, *Flying the Arctic*, p. 287.
59 *Geelong Advertiser*, 24 April 1928, p. 5.
60 *The Brisbane Courier*, 26 April 1928, p. 20.
61 *Geelong Advertiser*, 24 April 1928, p. 5.
62 *Geelong Advertiser*, 24 April 1928, p. 5.
63 *Geelong Advertiser*, 24 April 1928, p. 5.
64 Wilkins, *Flying the Arctic*, p. 289.
65 Wilkins, *Flying the Arctic*, p. 296.
66 *The Brisbane Courier*, 26 April 1928, p. 20.
67 *Geelong Advertiser*, 24 April 1928, p. 5.
68 Wilkins, *Flying the Arctic*, p. 296.
69 Wilkins, *Flying the Arctic*, p. 297.
70 Wilkins, *Flying the Arctic*, p. 297.
71 Wilkins, *Flying the Arctic*, p. 297.
72 Wilkins, *Flying the Arctic*, p. 298.
73 Wilkins, *Flying the Arctic*, p. 298.
74 *Geelong Advertiser*, 24 April 1928, p. 5.
75 Wilkins, *Flying the Arctic*, p. 299.
76 Wilkins, *Flying the Arctic*, p. 300.
77 Wilkins, *Flying the Arctic*, p. 300.
78 Wilkins, *Flying the Arctic*, p. 301.
79 *Geelong Advertiser*, 24 April 1928, p. 5.
80 Wilkins, *Flying the Arctic*, p. 303.
81 Wilkins, *Flying the Arctic*, p. 304.
82 Wilkins, *Flying the Arctic*, p. 304.
83 Wilkins, *Flying the Arctic*, p. 305.
84 Wilkins, *Flying the Arctic*, p. 305.
85 Wilkins, *Flying the Arctic*, p. 306.
86 Wilkins, *Flying the Arctic*, p. 309.
87 Wilkins, *Flying the Arctic*, p. 310.
88 Wilkins, *Flying the Arctic*, p. 310.
89 Wilkins, *Flying the Arctic*, p. 311.
90 Wilkins, *Flying the Arctic*, p. 311.
91 Wilkins, *Flying the Arctic*, p. 315.
92 Wilkins, *Flying the Arctic*, p. 315.
93 Wilkins, *Flying the Arctic*, p. 315.
94 Wilkins, *Flying the Arctic*, p. 316.
95 Wilkins, *Flying the Arctic*, p. 318.
96 *The Register*, 8 September 1928, p. 5.
97 *The Brisbane Courier*, 27 April 1928, p. 15.
98 Wilkins, *Flying the Arctic*, p. 319.
99 Wilkins, *Flying the Arctic*, p. 321.
100 *Geelong Advertiser*, 24 April 1928, p. 5.
101 Wilkins, *Flying the Arctic*, p. 321.
102 Wilkins, *Flying the Arctic*, p. 325.
103 Wilkins, *Flying the Arctic*, p. 325.
104 Wilkins, *Flying the Arctic*, p. 329.
105 Wilkins, *Flying the Arctic*, p. 330.

106 Wilkins, Papers, Ohio State University, Box 13.1, Folder 10, Detroit Arctic Expedition, Correspondence.
107 *Geelong Advertiser*, 24 April 1928, p. 5.
108 *The Brisbane Courier*, Monday 23 April 1928, p. 13.
109 *The Brisbane Courier*, Monday 23 April 1928, p. 13.
110 *Tweed Daily*, 9 June 1928, p. 5 [second sentence is reported speech].
111 *The Argus*, 26 April 1928, p. 6.
112 *The Australasian*, 28 April 1928, p. 11.
113 *Geelong Advertiser*, 24 April 1928, p. 5.
114 *The New York Times*, 22 April 1928, p. 1.
115 *The New York Times*, 22 April 1928, p. 1.
116 *The Herald*, 23 April 1928, p. 1.

Chapter 15

1 *Kalgoorlie Miner*, 16 May 1928, p. 5.
2 *Nidaros – Trondhjems morgenavis* [*Nidaros* – Trondhjem's morning newspaper], 25 April 1928, p. 1.
3 *Nidaros – Trondhjems morgenavis*, 25 April 1928, p. 1.
4 *Nidaros – Trondhjems morgenavis*, 25 April 1928, p. 1.
5 *Nidaros – Trondhjems morgenavis*, 25 April 1928, p. 1.
6 *Nidaros – Trondhjems morgenavis*, 25 April 1928, p. 1.
7 Wilkins, *Flying the Arctic*, p. 331.
8 *Daily Telegraph*, 25 May 1928, p. 4.
9 Wilkins, Papers, Ohio State University, Box 13.1, Folder 10, Detroit Arctic Expedition, Correspondence.
10 Thomas, *Sir Hubert Wilkins – His World of Adventure*, p. 181.
11 *The Advertiser*, 26 May 1928, p. 13.
12 *Warwick Daily News*, 26 May 1928, p. 1.
13 *The Forbes Advocate*, 23 April 1928, p. 2.
14 *Warwick Daily News*, 26 May 1928, p. 1 [reported speech].
15 Thomas, *Sir Hubert Wilkins – His World of Adventure*, p. 182.
16 *Tweed Daily*, 9 June 1928, p. 5.
17 Thomas, *Sir Hubert Wilkins – His World of Adventure*, p. 182.
18 Thomas, *Sir Hubert Wilkins – His World of Adventure*, p. 182.
19 Nasht, *The Last Explorer*, p. 173.
20 Nichols, 'Sir Hubert Wilkins of the Nautilus', *Explorers Journal*, December 1964, p. 209.
21 *The News*, 4 June 1928, p. 5.
22 *The Telegraph*, 8 October 1931, p. 6 [reported speech].
23 *The Telegraph*, 8 October 1931, p. 6.
24 *The Telegraph*, 11 June 1928, p. 9.
25 Huntford, *Scott and Amundsen: The Last Place on Earth*, Little, Brown Book Group, London, 1979, p. 544.
26 Huntford, *Scott and Amundsen*, p. 541.
27 Huntford, *Scott and Amundsen*, p. 541.
28 *The New York Times*, 6 June 1928, p. 23.
29 *The New York Times*, 31 August 1929, p. 1.
30 *The New York Times*, 31 August 1929, p. 1.
31 Thomas, *Sir Hubert Wilkins – His World of Adventure*, p. 240.
32 Thomas, *Sir Hubert Wilkins – His World of Adventure*, p. 240.
33 Thomas, *Sir Hubert Wilkins – His World of Adventure*, p. 240.
34 Thomas, *Sir Hubert Wilkins – His World of Adventure*, p. 240.
35 Thomas, *Sir Hubert Wilkins – His World of Adventure*, p. 240.
36 Thomas, *Sir Hubert Wilkins – His World of Adventure*, p. 240.
37 *The Sydney Morning Herald*, 4 July 1928, p. 15.
38 Thomas, *Sir Hubert Wilkins – His World of Adventure*, p. 240.
39 Thomas, *Sir Hubert Wilkins – His World of Adventure*, p. 240.
40 Thomas, *Sir Hubert Wilkins – His World of Adventure*, p. 241.

41 Thomas, *Sir Hubert Wilkins – His World of Adventure*, p. 241.
42 Thomas, *Sir Hubert Wilkins – His World of Adventure*, p. 241.
43 Thomas, *Sir Hubert Wilkins – His World of Adventure*, p. 241.
44 *The Mercury*, 24 September 1928, p. 10.
45 Thomas, *Sir Hubert Wilkins – His World of Adventure*, p. 241.
46 Thomas, *Sir Hubert Wilkins – His World of Adventure*, p. 241.
47 *Border Watch*, 30 May 1929, p. 6.
48 Thomas, *Sir Hubert Wilkins – His World of Adventure*, p. 183.
49 *The New York Times*, 21 September 1928.
50 *The New York Times*, 21 September 1928.
51 Wilkins, Papers, Ohio State University, Box 14, Folder 1, Antarctic Expeditions.
52 Nasht, *The Last Explorer*, p. 180.
53 Wilkins, Papers, Ohio State University, Box 13.1, Folder 36, Wilkins Hearst Antarctic Expedition, Correspondence.
54 *Toowoomba Chronicle and Darling Downs Gazette*, 14 November 1928, p. 7.
55 Thomas, *Sir Hubert Wilkins – His World of Adventure*, p. 185.
56 Thomas, *Sir Hubert Wilkins – His World of Adventure*, p. 187.
57 Thomas, *Sir Hubert Wilkins – His World of Adventure*, p. 188.
58 Thomas, *Sir Hubert Wilkins – His World of Adventure*, p. 188.
59 Thomas, *Sir Hubert Wilkins – His World of Adventure*, p. 189.
60 Thomas, *Sir Hubert Wilkins – His World of Adventure*, p. 190.
61 Thomas, *Sir Hubert Wilkins – His World of Adventure*, p. 190.
62 Thomas, *Sir Hubert Wilkins – His World of Adventure*, p. 190.
63 Thomas, *Sir Hubert Wilkins – His World of Adventure*, pp. 190–191.
64 *The Week*, 17 May 1929, p. 40.
65 *The Week*, 17 May 1929, p. 40.
66 Thomas, *Sir Hubert Wilkins – His World of Adventure*, p. 191.
67 *The Brisbane Courier*, 26 December 1928, p. 10.
68 Thomas, *Sir Hubert Wilkins – His World of Adventure*, p. 192.
69 Thomas, *Sir Hubert Wilkins – His World of Adventure*, p. 191.
70 *Sunday Times*, 12 May 1929, p. 1.
71 Nasht, *The Last Explorer*, p. 184.
72 *The Week*, 17 May 1929, p. 40.
73 Nasht, *The Last Explorer*, pp. 184–185.
74 Thomas, *Sir Hubert Wilkins – His World of Adventure*, p. 192.
75 Thomas, *Sir Hubert Wilkins – His World of Adventure*, p. 193.
76 *The Queenslander*, 16 May 1929, p. 7.
77 *The Queenslander*, 16 May 1929, p. 7.
78 *The Queenslander*, 16 May 1929, p. 7.
79 Nasht, *The Last Explorer*, p. 191.
80 Thomas, *Sir Hubert Wilkins – His World of Adventure*, p. 200.
81 'Graf Zeppelin's Interior: The Gondola', Airships.net.
82 Thomas, *Sir Hubert Wilkins – His World of Adventure*, p. 200.
83 Thomas, *Sir Hubert Wilkins – His World of Adventure*, p. 200.
84 Thomas, *Sir Hubert Wilkins – His World of Adventure*, p. 201.
85 *The Week*, 30 August 1929, p. 23.
86 Nasht, *The Last Explorer*, p. 199.
87 Thomas, *Sir Hubert Wilkins – His World of Adventure*, p. 204.
88 Thomas, *Sir Hubert Wilkins – His World of Adventure*, p. 207.
89 Thomas, *Sir Hubert Wilkins – His World of Adventure*, p. 207.
90 Wilkins, 'True Adventure Thrills', episode 56.
91 Nasht, *The Last Explorer*, p. 200.
92 Wilkins, Papers, Ohio State University, Box 14.1, Folder 13, Graf Zeppelin, p. 31.
93 Thomas, *Sir Hubert Wilkins – His World of Adventure*, p. 208.
94 *The Examiner*, 31 August 1929, p. 10.

Chapter 16

1 Nasht, *The Last Explorer*, p. 203.
2 Wilkins, Papers, Ohio State University, Box 14.1, Folder 13, Graf Zeppelin, p. 30 [reported speech].
3 *The New York Times*, 31 August 1929, p. 1.
4 Nasht, *The Last Explorer*, p. 203.
5 *The New York Times*, 31 August 1929, p. 1.
6 Nasht, *The Last Explorer*, p. 203.
7 Thomas, *Sir Hubert Wilkins – His World of Adventure*, p. 242.
8 Fisher, *Shackleton and the Antarctic*, Houghton Mifflin, Boston, 1958, p. 435.
9 *Maryborough Chronicle*, 16 November 1928, p. 6.
10 Thomas, *Sir Hubert Wilkins – His World of Adventure*, p. 209.
11 Thomas, *Sir Hubert Wilkins – His World of Adventure*, p. 210.
12 Thomas, *Sir Hubert Wilkins – His World of Adventure*, p. 210.
13 Thomas, *Sir Hubert Wilkins – His World of Adventure*, p. 211.
14 Thomas, *Sir Hubert Wilkins – His World of Adventure*, p. 213.
15 Thomas, *Sir Hubert Wilkins – His World of Adventure*, p. 213.
16 Thomas. *Sir Hubert Wilkins – His World of Adventure*, p. 211.
17 Thomas, *Sir Hubert Wilkins – His World of Adventure*, pp. 211–212.
18 Thomas, *Sir Hubert Wilkins – His World of Adventure*, pp. 211–212.
19 Bonham, 'Hatton's Hero: Today Marks 80th Anniversary of Carl Ben Eielson Funeral', *Grand Forks Herald*, 25 March 2010.
20 Bonham, 'Hatton's Hero', *Grand Forks Herald*, 25 March 2010.
21 *The Argus*, 31 March 1930, p. 6.
22 *The Argus*, 31 March 1930, p. 6.
23 *The Maitland Daily Mercury*, 4 June 1930, p. 4.
24 *The Maitland Daily Mercury*, 4 June 1930, p. 4.
25 *Sunday Times*, 12 May 1929, p. 1.
26 Thomas, *Sir Hubert Wilkins – His World of Adventure*, p. 218.
27 Thomas, *Sir Hubert Wilkins – His World of Adventure*, p. 218.
28 Thomas, *Sir Hubert Wilkins – His World of Adventure*, p. 218.
29 Thomas, *Sir Hubert Wilkins – His World of Adventure*, p. 218.
30 Thomas, *Sir Hubert Wilkins – His World of Adventure*, p. 218.
31 Thomas, *Sir Hubert Wilkins – His World of Adventure*, p. 218.
32 Nasht, *The Last Explorer*, p. 215.
33 Nasht, *The Last Explorer*, p. 216.
34 Nasht, *The Last Explorer*, p. 216.
35 *Sunday Mail*, 5 October 1930, p. 19.
36 *Daily Mercury*, 21 May 1931, p. 10.
37 Nasht, *The Last Explorer*, p. 216.
38 Nasht, *The Last Explorer*, p. 217.
39 *The Telegraph*, 8 October 1931, p. 6.
40 Nasht, *The Last Explorer*, p. 217.
41 *The Argus*, 2 February 1931, p. 10.
42 Nasht, *The Last Explorer*, p. 220.
43 Thomas, *Sir Hubert Wilkins – His World of Adventure*, p. 220.
44 Thomas, *Sir Hubert Wilkins – His World of Adventure*, p. 222.
45 Thomas, *Sir Hubert Wilkins – His World of Adventure*, p. 222.
46 *The New York Times*, 23 March 1931, p. 1 and p. 7.
47 *The New York Times*, 23 March 1931, p. 1 and p. 7.
48 Williams, *Submarines Under Ice: The U.S. Navy's Polar Operations*, Naval Institute Press, Annapolis Maryland, 1998, p. 8.
49 Nasht, *The Last Explorer*, p. 226.
50 Lady Wilkins, letter to Lowell Thomas, 3 June 1968.
51 Williams, *Submarines Under Ice*, p. 30.
52 Nasht, *The Last Explorer*, p. 230.
53 Williams, *Submarines Under Ice*, p. 15.

54 Nasht, *The Last Explorer*, p. 233.
55 *The Examiner*, 25 June 1931, p. 3.
56 *The Examiner*, 25 June 1931, p. 3.
57 *The Herald*, 16 July 1931, p. 7.
58 *News Chronicle* interview, reported in *The Mail*, 5 September 1931, p. 7.
59 *News Chronicle*, 29 July 1931.
60 Nasht, *The Last Explorer*, p. 237.
61 Nasht, *The Last Explorer*, p. 237.
62 Nasht, *The Last Explorer*, p. 239.
63 Nasht, *The Last Explorer*, p. 241.
64 Thomas, *Sir Hubert Wilkins – His World of Adventure*, p. 225.
65 Sverdrup (translation Eeliassen), *How and Why with the Nautilus*, Gylendal Norsk Forlag, Oslo, 1931, p. 106.
66 Sverdrup, *How and Why with the Nautilus*, p. 106.
67 Sverdrup, *How and Why with the Nautilus*, p. 106.
68 Sverdrup, *How and Why with the Nautilus*, p. 106.
69 Thomas, *Sir Hubert Wilkins – His World of Adventure*, p. 225.
70 Wilkins, Papers, Ohio State University, Box 14.1, Folder 36, Wilkins-Ellsworth Trans Arctic Submarine Expedition Report Nautilus Expedition 1931, cp. 52.
71 Wilkins, Papers, Ohio State University, Box 14.1, Folder 36, cp. 54.
72 Sverdrup, 'Scientific Results of the Nautilus Expedition, 1931', Massachusetts Institute of Technology, *Papers in Physical Oceanography and Meteorology*, Vol. II, No. 1, Cambridge Massachusetts, 1933, p. 11.
73 Wilkins, Papers, Ohio State University, Box 14.1, Folder 36, cp. 54.
74 Wilkins, Papers, Ohio State University, Box 14.1, Folder 36, cp. 54.
75 Wilkins, Papers, Ohio State University, Box 14.1, Folder 36, p. 66.
76 Wilkins, Papers, Ohio State University, Box 14.1, Folder 36, Wilkins Journal, 25 August 1931, p. 60.
77 Wilkins, Papers, Ohio State University, Box 14.1, Folder 36, Wilkins Journal, 25 August 1931, p. 60.
78 Sverdrup, 'Scientific Results of the Nautilus Expedition, 1931', p. 7.
79 *Geraldton Guardian and Express*, 12 July 1932, p. 3.
80 *Time Magazine*, 14 September 1931.
81 Thomas, *Sir Hubert Wilkins – His World of Adventure*, p. 227.
82 Thomas, *Sir Hubert Wilkins – His World of Adventure*, p. 227.
83 Thomas, *Sir Hubert Wilkins – His World of Adventure*, p. 227.
84 Thomas, *Sir Hubert Wilkins – His World of Adventure*, p. 227.
85 Wilkins, Papers, Ohio State University, Box 14.1, Folder 36, p. 70.
86 Thomas, *Sir Hubert Wilkins – His World of Adventure*, p. 227.
87 *Time* magazine, 14 September 1931.
88 *Time* magazine, 14 September 1931.
89 Thomas, *Sir Hubert Wilkins – His World of Adventure*, p. 227.
90 Ingersoll, *Explorers Journal*, Vol. 60, 1982, p. 34.
91 Nasht, *The Last Explorer*, p. 249.
92 Thomas, *Sir Hubert Wilkins – His World of Adventure*, p. 228.
93 *The Labor Daily*, 12 July 1932, p. 1.
94 Thomas, *Sir Hubert Wilkins – His World of Adventure*, p. 225.
95 *Time* magazine, 14 September 1931.
96 Nasht, *The Last Explorer*, pp. 253–254.
97 Nasht, *The Last Explorer*, p. 229.

Epilogue

1 Thomas, *Sir Hubert Wilkins – His World of Adventure*, p. 244.
2 Thomas, *Sir Hubert Wilkins – His World of Adventure*, p. 244.
3 Wilkins, Sir Hubert and Harold M. Sherman, *Thoughts through Space: A Remarkable Adventure in the Realm of the Mind*, Hampton Roads Publishing, Charlottesville, VA, 2004, p. xvi.
4 Wilkins and Sherman, *Thoughts through Space*, p. xvii.
5 Wilkins and Sherman, *Thoughts through Space: A Remarkable Adventure in the Realm of the Mind*, C & R Anthony, New York, 1951, p. 95.

6   Wilkins and Sherman, *Thoughts through Space*, 2004, p. xvii.
7   *The Australian Women's Weekly*, 17 September 1938, p. 3.
8   *The Australian Women's Weekly*, 17 September 1938, p. 3.
9   *The Australian Women's Weekly*, 17 September 1938, p. 3.
10 Letter from Sir Hubert Wilkins to Harold Sherman, 17 May 1942, in Praamsma, Saskia and Matthew Block, *The New Urantia Notebook of Sir Hubert Wilkins*, Square Circles Publishing, Pahrump, NV, 2015, p. 81.
11 *The Urantia Book*, p. 29.
12 Thomas, *Sir Hubert Wilkins – His World of Adventure*, p. 230.
13 Thomas, *Sir Hubert Wilkins – His World of Adventure*, p. 245.
14 Thomas, *Sir Hubert Wilkins – His World of Adventure*, p. 245.
15 Thomas, *Sir Hubert Wilkins – His World of Adventure*, p. 245 [reported speech].
16 Nasht, *The Last Explorer*, p. 281.
17 Nasht, *The Last Explorer*, p. 285.
18 Nasht, *The Last Explorer*, p. 285.
19 Dick Smith Interview with Winston and Marley Ross, 24 April 1993, supplied to author. Mike Ross, the son of Winston Ross, later told Dick Smith that it was his presumption that his father and Suzanne were lovers. He maintained the same in an interview with Stephen Carthew.
20 Nasht, *The Last Explorer*, p. 289.
21 Recollection of Wilkins' friend, George Schneider, to Stephen Carthew.
22 Letter from Brigadier-General C.G. Galloway to Australian Minister for External Affairs, Mr R.G. Casey, n.d., circa 1952–57, in Grierson, John, *Sir Hubert Wilkins: Enigma of Exploration*, Robert Hale Ltd, London, 1960, p. 205.
23 Calvert, *Surface at the Pole: The Extraordinary Voyages of the U.S.S. Skate*, Scholastic Book Services, New York, 1966, p. 174.
24 Calvert, *Surface at the Pole*, p. 174.
25 Maynard, *Wings of Ice*, p. 265.
26 Maynard, *Wings of Ice*, p. 265.
27 *What's My Line?*, episode 16 March 1958. https://www.youtube.com/watch?v=X2n_0qEY8h4.
28 Winston Ross, letter to 'CIA Chief N.Y. Office', 21 April 1965, Wilkins, Papers, Ohio State University.
29 *The Advertiser*, 18 August 1959.
30 *Canberra Times*, 6 December 1958, p. 3.
31 Maynard, *Wings of Ice*, p. 263.
32 Wilkins, Papers, Ohio State University, Box 20, Folder 36.
33 Thomas to Lady Wilkins, letter, 16 May 1968.
34 Lady Wilkins to Thomas, letter, 13 April 1965.
35 Lady Wilkins to Thomas, letter, 13 April 1965.
36 Maynard, *The Unseen Anzac*, p. 226.
37 Calvert, *Surface at the Pole*, pp. 243–244.

BIBLIOGRAPHY

Books

Albig, William, *Public Opinion*, McGraw Hill, New York, 1939

Andrews, Malcolm, *Hubert Who?*, HarperCollins, Sydney, 2011

Bagshawe, Thomas Wyatt, *Two Men in the Antarctic: An Expedition to Graham Land, 1920–1922*, The Macmillan Company, New York, 1939

Bartlett, Robert Abram, *The Last Voyage of the Karluk: Flagship of Vilhjalmar Stefansson's Canadian Arctic Expedition of 1913–16*, McClelland, Toronto, 1916

Bean, Charles, *Anzac to Amiens*, Australian War Memorial, Canberra, 1946

Bean, Charles, *Gallipoli Mission*, Australian War Memorial, Canberra, 1948

Bean, Charles, *The Official History of Australia in the War of 1914–1918*, Vol. II, Angus & Robertson, Sydney, 1941

Bean, Charles, *The Official History of Australia in the War of 1914–1918*, Vol. V, Angus & Robertson, Sydney, 1937

Bean, Charles, *Official History of Australia in the War of 1914–1918*, Vol. VI, Angus & Robertson, Sydney, 1942

Browning, Neville, *Fix Bayonets: The Unit History of the 51st Battalion*, self-published, Perth, 2000

Burness, Peter, Ed., *The Western Front Diary of Charles Bean*, Australian War Memorial, NewSouth, 2018

Calvert, James, *Surface at the Pole: The Extraordinary Voyages of the U.S.S. Skate*, Scholastic Book Services, New York, 1966

Churchill, Winston, *The World Crisis*, Vol. V, T. Butterworth, London, 1923

Diubaldo, Richard J., *Stefansson and the Canadian Arctic*, McGill-Queen's University Press, Montreal, 1999

Downing, Walter, *To the Last Ridge*, Grub Street, London, 2013

Fisher, Margery, *Shackleton and the Antarctic*, Houghton Mifflin, Boston, 1958

FitzSimons, Peter, *Charles Kingsford Smith and Those Magnificent Men in Their Flying Machines*, HarperCollins, Sydney, 2010

Gibbs, Philip and Grant, Bernard, *The Balkan War: Adventures of War with Cross and Crescent*, Small, Maynard & Co., Boston, 1913

Grant, Bernard, *To the Four Corners, The Memoirs of a News Photographer*, Hutchison & Co. Ltd, London, 1933

Grierson, John, *Sir Hubert Wilkins: Enigma of Exploration*, Robert Hale, London, 1960

Hamilton, John, *Goodbye Cobber, God Bless You*, Pan Macmillan, Sydney, 2005

Hughes, Aneurin, *Billy Hughes: Prime Minister and Controversial Founding Father of the Australian Labor Party*, John Wiley & Sons, Brisbane, 2005

Huntford, Roland, *Scott and Amundsen: The Last Place on Earth*, Little, Brown Book Group, London, 1979

Joy, William, *The Aviators*, Shakespeare Head Press, Sydney, 1971

Kalush, William and Sloman, Larry, *The Secret Life of Houdini: The Making of America's First Superhero*, Simon & Schuster, UK, 2007

Kieza, Grantlee, *Monash – The Soldier who Shaped Australia*, ABC Books HarperCollins, Sydney, 2015

Kingsford Smith, Sir Charles, *The Old Bus*, Herald Press, Melbourne, 1932

McCarthy, Dudley, *Gallipoli to the Somme*, Cooper, London, 1983

Maynard, Jeff, *The Unseen Anzac*, Scribe, Victoria, (2015), 2017

Maynard, Jeff, *Wings of Ice*, Thistle, London, 2015

Monash, John, *Australian Victories in France in 1918*, Naval and Military Press, Uckfield, 2009

Nasht, Simon, *The Last Explorer*, Hachette, Sydney, 2007

Noice, Harold, *With Stefansson in the Arctic*, Harrap, London, 1925

Pearson, Michael, *The Sealed Train*, Macmillan, London, 1975

Praamsma, Saskia and Block, Matthew, *The New Urantia Notebook of Sir Hubert Wilkins: Fact Finder and Truth Seeker*, Square Circles Publishing, Pahrump, NV, 2015

Rearden, Jim, *Alaska's First Bush Pilots, 1923–30*, Pictorial Histories Publishing, Missoula, 2009

Schneider, Kristen, J.J. McKenna papers in, *Pompey Elliott's Left Hand Man: Lieutenant Colonel Charles Denehy*, self-published, Camberwell, 2015

Scott, Joan, *A Soldier on the Somme, Diary of Private Edwin Need*, Eastern Press, Mulgrave, 2014

Scott, Robert Falcon, *Scott's Last Expedition, Vol. I: Being the journals of Captain R.F. Scott, R.N., C.V.O.*, Smith, Elder & Co., London, 1914

Serle, Geoffrey, *John Monash: A Biography*, Melbourne University Press in association with Monash University, Melbourne, 1982

Simpson, Cameron Victor, *Maygar's Boys: A biographical history of the 8th Light Horse Regiment A.I.F. 1914–19*, Just Soldiers, Military Research & Publications, Moorooduc, 1998

Stannage, John, *Smithy*, Oxford University Press, London, 1950

Stefansson, Vilhjálmur, *My Life With The Eskimo*, The Macmillan Company, New York, 1913

Sverdrup, H.U. (translation Trond Eeliassen), *How and Why with the Nautilus*, Gylendal Norsk Forlag, Oslo, 1931

Taylor, A.J.P., *The First World War: An Illustrated History*, Hamish Hamilton, London, 1963

Thomas, Lowell, *Sir Hubert Wilkins: His World of Adventure*, McGraw-Hill, New York, USA, 1961

Thomas, Lowell, *Sir Hubert Wilkins – His World of Adventure*, Colorgravure, Melbourne, 1963

Wild, Frank, *Shackleton's last voyage. The story of the Quest. By Commander Frank Wild, C.B.E.*, Cassell, London, 1923

Wilkins, Capt. G.H., *Flying the Arctic*, Grosset & Dunlap, New York, 1928

Wilkins, Capt. Sir G.H, *Undiscovered Australia*, Ernest Benn, London, 1928

Wilkins, George, 'Report to the Department of the Naval Service, Ottawa', in *Canadian Arctic Expedition Reports*, Pravana Books, Delhi, 2020

Wilkins, George, *Under the North Pole: The Wilkins-Ellsworth Submarine Expedition*, Brewer, Warren & Putnam, New York, 1931

Wilkins, G. H. and Sherman, H.M, *Thoughts through Space: A Remarkable Adventure in the Realm of the Mind*, C & R Anthony, New York, 1951

Wilkins, Sir Hubert and Sherman, Harold M., *Thoughts through Space: A Remarkable Adventure in the Realm of the Mind*, Hampton Roads Publishing, Charlottesville, VA, 2004

Williams, Marion, D., *Submarines Under Ice: The U.S. Navy's Polar Operations*, Naval Institute Press, Annapolis Maryland, 1998

Young, Louisa, *A Great Task of Happiness: The life of Kathleen Scott*, Macmillan, London, 1995

Journals

Condon, Richard, 'Natkusiak', *Arctic*, Vol. 45, No. 1, http://pubs.aina.ucalgary.ca/arctic/Arctic45-1-90.pdf

Ellwood, Galiina, 'The Aboriginal Miners and Prospectors of Cape York Peninsula, 1870 to ca. 1950s', *Journal of Australasian Mining History*, Vol. 16, October 2018, https://www.mininghistory.asn.au/wp-content/uploads/5-EllwoodV16-compressed.pdf

Ingersoll, Ernst, *Explorers Journal*, Vol. 60, 1982

J.M.W., 'Obituary: Sir Ernest Henry Shackleton, C. V. O., O. B. E', *The Geographical Journal*, Vol. 59, No. 3, March 1922

Nichols, Herbert B., 'Sir Hubert Wilkins of the Nautilus', *Explorers Journal*, Vol. 42(4), December 1964

R.E.P., 'Obituary: Frank Wild', *The Geographical Journal*, Vol. 95, No. 3, March 1940

Sverdrup, Harald Ulrik, 'Scientific Results of the Nautilus Expedition, 1931', Massachusetts Institute of Technology, *Papers in Physical Oceanography and Meteorology*, Vol. II, No. 1, Cambridge, Massachusetts, 1933

Magazines

Arctic, Arctic Institute of North America, Vol. 45, No. 1.
Nature, Vol. 104, No. 93, 1919
The Australian Women's Weekly, 17 September 1938
Time magazine, Vol. 186, No. 5, 2015

Newspapers

Adelaide Express and Telegraph (SA)
Adelaide Observer (SA)
Border Morning Mail and Riverina Times (NSW)
Border Watch (SA)
Burra Record (SA)
Canberra Times
Chronicle (SA)
Colac Reformer (Vic)
Daily Advertiser (NSW)
Daily Herald (London, UK)
Daily Mercury (Qld)
Daily Standard (Qld)
East Oregonian (Oregon, US)
Geelong Advertiser (Vic)
Geraldton Express (WA)
Geraldton Guardian and Express (WA)
Grand Forks Herald (US)
Kalgoorlie Miner (WA)
Leader (NSW)
Maryborough Chronicle (Vic)
Morning Bulletin (Qld)
Newcastle Sun (NSW)
News (SA)
News Chronicle (UK)
Nidaros – Trondhjems morgenavis (Norway)
Northern Standard (NT)
Northern Territory Times and Gazette (NT)
Queensland Times (Qld)
Sunday Mail (Qld)
Sunday Times (WA)
The Advertiser (SA)
The Argus (Vic)
The Australasian (Vic)
The Australian
The Bendigo Advertiser (Vic)
The Bendigonian (Vic)
The Brisbane Courier (Qld)
The Bulletin
The Canberra Times (ACT)
The Capricornian (Qld)
The Daily Express (NSW)
The Daily Mail (Qld)
The Daily Telegraph (NSW)
The Evening News (Qld)
The Examiner (Tas)
The Forbes Advocate (NSW)
The Herald (Vic)
The Labor Daily (NSW)
The Leader (WA)

The Mail (SA)
The Maitland Daily Mercury (NSW)
The Mercury (Tas)
The New York Times (US)
The News (Adelaide, SA)
The Queenslander (Qld)
The Register (SA)
The Sun (NSW)
The Sun (WA)
The Sydney Morning Herald (NSW)
The Telegraph (Qld)
The Times of London (UK)
The Week (Qld)
The West Australian (WA)
Toowoomba Chronicle and Darling Downs Gazette (Qld)
Townsville Daily Bulletin (Qld)
Tribune (NSW)
Truth (NSW)
Tweed Daily (NSW)
Warwick Daily News (Qld)
Weekly Times (Vic)

Archives/Collections

Anderson, Rudolph, *Report of the Canadian Arctic Expedition 1913–18*, F.A. Acland, Ottawa, 1923, Ohio State University
Bean, Charles, AWM 38 Official History 1914–18 War: Records of C.E.W. Bean, Official Historian, Diaries and Notebooks, Australian War Memorial, Canberra
Chipman, K. Diary, 3 August, 1913
Holton, Alfred, typed narrative, Australian War Memorial, Canberra
Hurley, Frank, Papers of Frank Hurley 1912–1962, My Diary, Official War Photographer Commonwealth Military Forces, 21 August 1917 – 31 August 1918, National Library of Australia, MS 883, Series 1, Item 5
Kennedy, Walter, *From Anzac Cove to Villers-Bretonneux*, unpublished memoir, Australian War Memorial, Canberra
Kingsford Smith, Charles, Letter to his parents in Sydney, sent from Hitchin, UK, 17 April 1919, Norman Ellison Collection, National Library of Australia, Canberra
Thomas, Lowell, unpublished Wilkins biography manuscript draft, Wilkins, Papers, Ohio State University, Box 1.1, Folder 11
Turnbull, John, Diary, Australian War Memorial, Canberra,
Wilkins, George Hubert, Military Records of George Hubert Wilkins, 1914–1920, National Archives of Australia, Series No. B2455, Item ID 838937
Wilkins, Sir George Hubert, Papers, including correspondence, diaries, reports, lecture notes and unpublished manuscripts, Byrd Polar and Climate Research Center Archival Program, Ohio State University
    Wilkins, 'Early Flying Experiences', unpublished manuscript
    Wilkins, 'From a hundred above to fifty below', unpublished manuscript
    Wilkins, manuscript, 'True Adventure Thrills', Box 1, Folder 16
    Wilkins, lecture, 'A Visit to the "Blonde" Eskimo'
    Wilkins, manuscript, 'Some Notes on Conditions Existing in Soviet Russia'
    Wilkins, letter to a Mr Byrne, sent from New York City, 29 October 1936
    Lady Wilkins, letter to Lowell Thomas, 3 June 1968
    Winston Ross, letter to 'CIA Chief N.Y. Office', 21 April 1965
    Thomas to Lady Wilkins, letter, 16 May 1968
    Lady Wilkins to Thomas, letter, 13 April 1965

Online Sources

Airships.Net, Graf Zeppelin's Interior: The Gondola, https://www.airships.net/lz127-graf-zeppelin/interiors/

Fiddlers Green, 'Lockheed Vega', http://www.fiddlersgreen.net/models/aircraft/Lockheed-Vega.html

Flinders Ranges Research, *Sir Hubert Wilkins* http://www.southaustralianhistory.com.au/wilkins.htm

MacFarlane, B. 'The Scot Who Was Left Out in the Cold', *Scottish Field*, 6 March 2020, https://www.scottishfield.co.uk/culture/the-scot-who-was-left-out-in-the-cold/

Wilkins Foundation, *The Wilkins Chronicle* https://www.wilkinsfoundation.org.au/domains/the-wilkins-chronicle/

Wilkins Foundation, Where in Crete is the Blackburn Kangaroo? https://wilkinsfoundation.org.au/research/Where-in-Crete-is-the-Blackburn-Kangaroo.pdf

Interviews

Dick Smith interview with Winston and Marley Ross, 24 April 1993, supplied to author

Songs

'In the Valley Where the Bluebirds Sing', lyrics by Monroe H. Rosenfeld, 1902

TV and Film

What's My Line?, episode 16 March 1958 https://www.youtube.com/watch?v=X2n_0qEY8h4

Younger, James (director), *Unsolved History: Death of the Red Baron*, Termite Art Productions, USA, 2002

INDEX